Lecture Notes in Computer Science   14838

Founding Editors

Gerhard Goos
Juris Hartmanis

The series Lecture Notes in Computer Science (LNCS), including its subseries Lecture Notes in Artificial Intelligence (LNAI) and Lecture Notes in Bioinformatics (LNBI), has established itself as a medium for the publication of new developments in computer science and information technology research, teaching, and education.

LNCS enjoys close cooperation with the computer science R & D community, the series counts many renowned academics among its volume editors and paper authors, and collaborates with prestigious societies. Its mission is to serve this international community by providing an invaluable service, mainly focused on the publication of conference and workshop proceedings and postproceedings. LNCS commenced publication in 1973.

Leonardo Franco · Clélia de Mulatier ·
Maciej Paszynski · Valeria V. Krzhizhanovskaya ·
Jack J. Dongarra · Peter M. A. Sloot
Editors

Computational Science – ICCS 2024

24th International Conference
Malaga, Spain, July 2–4, 2024
Proceedings, Part VII

 Springer

Editors
Leonardo Franco (iD)
University of Malaga
Malaga, Spain

Clélia de Mulatier (iD)
University of Amsterdam
Amsterdam, The Netherlands

Maciej Paszynski (iD)
AGH University of Science and Technology
Krakow, Poland

Valeria V. Krzhizhanovskaya (iD)
University of Amsterdam
Amsterdam, The Netherlands

Jack J. Dongarra (iD)
University of Tennessee
Knoxville, TN, USA

Peter M. A. Sloot (iD)
University of Amsterdam
Amsterdam, The Netherlands

ISSN 0302-9743   ISSN 1611-3349 (electronic)
Lecture Notes in Computer Science
ISBN 978-3-031-63785-8   ISBN 978-3-031-63783-4 (eBook)
https://doi.org/10.1007/978-3-031-63783-4

This Springer imprint is published by the registered company Springer Nature Switzerland AG
The registered company address is: Gewerbestrasse 11, 6330 Cham, Switzerland

If disposing of this product, please recycle the paper.

Preface

Welcome to the proceedings of the 24th International Conference on Computational Science (https://www.iccs-meeting.org/iccs2024/), held on July 2–4, 2024 at the University of Málaga, Spain.

In keeping with the new normal of our times, ICCS featured both in-person and online sessions. Although the challenges of such a hybrid format are manifold, we have always tried our best to keep the ICCS community as dynamic, creative, and productive as possible. We are proud to present the proceedings you are reading as a result.

ICCS 2024 was jointly organized by the University of Málaga, the University of Amsterdam, and the University of Tennessee.

Facing the Mediterranean in Spain's Costa del Sol, Málaga is the country's sixth-largest city, and a major hub for finance, tourism, and technology in the region.

The University of Málaga (Universidad de Málaga, UMA) is a modern, public university, offering 63 degrees and 120 postgraduate degrees. Close to 40,000 students study at UMA, taught by 2500 lecturers, distributed over 81 departments and 19 centers. The UMA has 278 research groups, which are involved in 80 national projects and 30 European and international projects. ICCS took place at the Teatinos Campus, home to the School of Computer Science and Engineering (ETSI Informática), which is a pioneer in its field and offers the widest range of IT-related subjects in the region of Andalusia.

The International Conference on Computational Science is an annual conference that brings together researchers and scientists from mathematics and computer science as basic computing disciplines, as well as researchers from various application areas who are pioneering computational methods in sciences such as physics, chemistry, life sciences, engineering, arts, and the humanities, to discuss problems and solutions in the area, identify new issues, and shape future directions for research.

The ICCS proceedings series have become a primary intellectual resource for computational science researchers, defining and advancing the state of the art in this field.

We are proud to note that this 24th edition, with 17 tracks (16 thematic tracks and one main track) and close to 300 participants, has kept to the tradition and high standards of previous editions.

The theme for 2024, "Computational Science: Guiding the Way Towards a Sustainable Society", highlights the role of Computational Science in assisting multidisciplinary research on sustainable solutions. This conference was a unique event focusing on recent developments in scalable scientific algorithms; advanced software tools; computational grids; advanced numerical methods; and novel application areas. These innovative novel models, algorithms, and tools drive new science through efficient application in physical systems, computational and systems biology, environmental systems, finance, and others.

ICCS is well known for its excellent lineup of keynote speakers. The keynotes for 2024 were:

- David Abramson, University of Queensland, Australia
- Manuel Castro Díaz, University of Málaga, Spain
- Jiří Mikyška, Czech Technical University in Prague, Czechia
- Takemasa Miyoshi, RIKEN, Japan
- Coral Calero Muñoz, University of Castilla-La Mancha, Spain
- Petra Ritter, Berlin Institute of Health & Charité University Hospital Berlin, Germany

This year we had 430 submissions (152 to the main track and 278 to the thematic tracks). In the main track, 51 full papers were accepted (33.5%); in the thematic tracks, 104 full papers (37.4%). The higher acceptance rate in the thematic tracks is explained by their particular nature, whereby track organizers personally invite many experts in the field to participate. Each submission received at least 2 single-blind reviews (2.6 reviews per paper on average).

ICCS relies strongly on our thematic track organizers' vital contributions to attract high-quality papers in many subject areas. We would like to thank all committee members from the main and thematic tracks for their contribution to ensuring a high standard for the accepted papers. We would also like to thank Springer, Elsevier, and Intellegibilis for their support. Finally, we appreciate all the local organizing committee members for their hard work in preparing this conference.

We hope the attendees enjoyed the conference, whether virtually or in person.

July 2024

Leonardo Franco
Clélia de Mulatier
Maciej Paszynski
Valeria V. Krzhizhanovskaya
Jack J. Dongarra
Peter M. A. Sloot

Organization

Conference Chairs

General Chair

Valeria Krzhizhanovskaya   University of Amsterdam, The Netherlands

Main Track Chair

Clélia de Mulatier   University of Amsterdam, The Netherlands

Thematic Tracks Chair

Maciej Paszynski   AGH University of Krakow, Poland

Thematic Tracks Vice Chair

Michael Harold Lees   University of Amsterdam, The Netherlands

Scientific Chairs

Peter M. A. Sloot   University of Amsterdam, The Netherlands
Jack Dongarra   University of Tennessee, USA

Local Organizing Committee

Leonardo Franco (Chair)   University of Malaga, Spain
Francisco Ortega-Zamorano   University of Malaga, Spain
Francisco J. Moreno-Barea   University of Malaga, Spain
José L. Subirats-Contreras   University of Malaga, Spain

Thematic Tracks and Organizers

Advances in High-Performance Computational Earth Sciences: Numerical Methods, Frameworks & Applications (IHPCES)

Takashi Shimokawabe   University of Tokyo, Japan
Kohei Fujita   University of Tokyo, Japan
Dominik Bartuschat   FAU Erlangen-Nürnberg, Germany

Artificial Intelligence and High-Performance Computing for Advanced Simulations (AIHPC4AS)

Maciej Paszynski   AGH University of Krakow, Poland

Biomedical and Bioinformatics Challenges for Computer Science (BBC)

Mario Cannataro   University Magna Graecia of Catanzaro, Italy
Giuseppe Agapito   University Magna Graecia of Catanzaro, Italy
Mauro Castelli   Universidade Nova de Lisboa, Portugal
Riccardo Dondi   University of Bergamo, Italy
Rodrigo Weber dos Santos Federal University of Juiz de Fora, Brazil
Italo Zoppis   University of Milano-Bicocca, Italy

Computational Diplomacy and Policy (CoDiP)

Roland Bouffanais   University of Geneva, Switzerland
Michael Lees   University of Amsterdam, The Netherlands
Brian Castellani   Durham University, UK

Computational Health (CompHealth)

Sergey Kovalchuk   Huawei, Russia
Georgiy Bobashev   RTI International, USA
Anastasia Angelopoulou University of Westminster, UK
Jude Hemanth   Karunya University, India

Computational Optimization, Modelling, and Simulation (COMS)

Xin-She Yang   Middlesex University London, UK
Slawomir Koziel   Reykjavik University, Iceland
Leifur Leifsson   Purdue University, USA

Generative AI and Large Language Models (LLMs) in Advancing Computational Medicine (CMGAI)

Ahmed Abdeen Hamed   State University of New York at Binghamton,
                            USA
Qiao Jin   National Institutes of Health, USA
Xindong Wu   Hefei University of Technology, China
Byung Lee   University of Vermont, USA
Zhiyong Lu   National Institutes of Health, USA
Karin Verspoor   RMIT University, Australia
Christopher Savoie   Zapata AI, USA

Machine Learning and Data Assimilation for Dynamical Systems (MLDADS)

Rossella Arcucci   Imperial College London, UK
Cesar Quilodran-Casas   Imperial College London, UK

Multiscale Modelling and Simulation (MMS)

Derek Groen   Brunel University London, UK
Diana Suleimenova   Brunel University London, UK

Network Models and Analysis: From Foundations to Artificial Intelligence (NMAI)

Marianna Milano   Università Magna Graecia of Catanzaro, Italy
Giuseppe Agapito   University Magna Graecia of Catanzaro, Italy
Pietro Cinaglia   University Magna Graecia of Catanzaro, Italy
Chiara Zucco   University Magna Graecia of Catanzaro, Italy

Numerical Algorithms and Computer Arithmetic for Computational Science (NACA)

Pawel Gepner	Warsaw Technical University, Poland
Ewa Deelman	University of Southern California, Marina del Rey, USA
Hatem Ltaief	KAUST, Saudi Arabia

Quantum Computing (QCW)

Katarzyna Rycerz	AGH University of Krakow, Poland
Marian Bubak	Sano and AGH University of Krakow, Poland

Simulations of Flow and Transport: Modeling, Algorithms, and Computation (SOFTMAC)

Shuyu Sun	King Abdullah University of Science and Technology, Saudi Arabia
Jingfa Li	Beijing Institute of Petrochemical Technology, China
James Liu	Colorado State University, USA

Smart Systems: Bringing Together Computer Vision, Sensor Networks and Artificial Intelligence (SmartSys)

Pedro Cardoso	University of Algarve, Portugal
João Rodrigues	University of Algarve, Portugal
Jânio Monteiro	University of Algarve, Portugal
Roberto Lam	University of Algarve, Portugal

Solving Problems with Uncertainties (SPU)

Vassil Alexandrov	Hartree Centre – STFC, UK
Aneta Karaivanova	IICT – Bulgarian Academy of Science, Bulgaria

Teaching Computational Science (WTCS)

Evguenia Alexandrova   Hartree Centre – STFC, UK
Tseden Taddese   UK Research and Innovation, UK

Reviewers

Ahmed Abdelgawad   Central Michigan University, USA
Samaneh Abolpour Mofrad   Imperial College London, UK
Tesfamariam Mulugeta Abuhay   Queen's University, Canada
Giuseppe Agapito   University of Catanzaro, Italy
Elisabete Alberdi   University of the Basque Country, Spain
Luis Alexandre   UBI and NOVA LINCS, Portugal
Vassil Alexandrov   Hartree Centre – STFC, UK
Evguenia Alexandrova   Hartree Centre – STFC, UK
Julen Alvarez-Aramberri   Basque Center for Applied Mathematics, Spain
Domingos Alves   Ribeirão Preto Medical School, University of São
                                    Paulo, Brazil
Sergey Alyaev   NORCE, Norway
Anastasia Anagnostou   Brunel University London, UK
Anastasia Angelopoulou   University of Westminster, UK
Rossella Arcucci   Imperial College London, UK
Emanouil Atanasov   IICT – Bulgarian Academy of Sciences, Bulgaria
Krzysztof Banaś   AGH University of Krakow, Poland
Luca Barillaro   Magna Graecia University of Catanzaro, Italy
Dominik Bartuschat   FAU Erlangen-Nürnberg, Germany
Pouria Behnodfaur   Curtin University, Australia
Jörn Behrens   University of Hamburg, Germany
Adrian Bekasiewicz   Gdansk University of Technology, Poland
Gebrail Bekdas   Istanbul University, Turkey
Mehmet Ali Belen   Iskenderun Technical University, Turkey
Stefano Beretta   San Raffaele Telethon Institute for Gene Therapy,
                                    Italy
Anabela Moreira Bernardino   Polytechnic Institute of Leiria, Portugal
Eugénia Bernardino   Polytechnic Institute of Leiria, Portugal
Daniel Berrar   Tokyo Institute of Technology, Japan
Piotr Biskupski   IBM, Poland
Georgiy Bobashev   RTI International, USA
Carlos Bordons   University of Seville, Spain
Bartosz Bosak   PSNC, Poland
Lorella Bottino   University Magna Graecia of Catanzaro, Italy

Roland Bouffanais University of Geneva, Switzerland
Marian Bubak Sano and AGH University of Krakow, Poland
Aleksander Byrski AGH University of Krakow, Poland
Cristiano Cabrita Universidade do Algarve, Portugal
Xing Cai Simula Research Laboratory, Norway
Carlos Calafate Universitat Politècnica de València, Spain
Victor Calo Curtin University, Australia
Mario Cannataro University Magna Graecia of Catanzaro, Italy
Karol Capała AGH University of Krakow, Poland
Pedro J. S. Cardoso Universidade do Algarve, Portugal
Eddy Caron ENS-Lyon/Inria/LIP, France
Stefano Casarin Houston Methodist Hospital, USA
Brian Castellani Durham University, UK
Mauro Castelli Universidade Nova de Lisboa, Portugal
Nicholas Chancellor Durham University, UK
Thierry Chaussalet University of Westminster, UK
Sibo Cheng Imperial College London, UK
Lock-Yue Chew Nanyang Technological University, Singapore
Pastrello Chiara Krembil Research Institute, Canada
Su-Fong Chien MIMOS Berhad, Malaysia
Marta Chinnici enea, Italy
Bastien Chopard University of Geneva, Switzerland
Maciej Ciesielski University of Massachusetts, USA
Pietro Cinaglia University of Catanzaro, Italy
Noelia Correia Universidade do Algarve, Portugal
Adriano Cortes University of Rio de Janeiro, Brazil
Ana Cortes Universitat Autònoma de Barcelona, Spain
Enrique Costa-Montenegro Universidad de Vigo, Spain
David Coster Max Planck Institute for Plasma Physics, Germany
Carlos Cotta University of Málaga, Spain
Peter Coveney University College London, UK
Alex Crimi AGH University of Krakow, Poland
Daan Crommelin CWI Amsterdam, The Netherlands
Attila Csikasz-Nagy King's College London, UK/Pázmány Péter Catholic University, Hungary
Javier Cuenca University of Murcia, Spain
António Cunha UTAD, Portugal
Pawel Czarnul Gdansk University of Technology, Poland
Pasqua D'Ambra IAC-CNR, Italy
Alberto D'Onofrio University of Trieste, Italy
Lisandro Dalcin KAUST, Saudi Arabia

Bhaskar Dasgupta	University of Illinois at Chicago, USA
Clélia de Mulatier	University of Amsterdam, The Netherlands
Ewa Deelman	University of Southern California, Marina del Rey, USA
Quanling Deng	Australian National University, Australia
Eric Dignum	University of Amsterdam, The Netherlands
Riccardo Dondi	University of Bergamo, Italy
Rafal Drezewski	AGH University of Krakow, Poland
Simon Driscoll	University of Reading, UK
Hans du Buf	University of the Algarve, Portugal
Vitor Duarte	Universidade NOVA de Lisboa, Portugal
Jacek Długopolski	AGH University of Krakow, Poland
Wouter Edeling	Vrije Universiteit Amsterdam, The Netherlands
Nahid Emad	University of Paris Saclay, France
Christian Engelmann	ORNL, USA
August Ernstsson	Linköping University, Sweden
Aniello Esposito	Hewlett Packard Enterprise, Switzerland
Roberto R. Expósito	Universidade da Coruna, Spain
Hongwei Fan	Imperial College London, UK
Tamer Fandy	University of Charleston, USA
Giuseppe Fedele	University of Calabria, Italy
Christos Filelis-Papadopoulos	Democritus University of Thrace, Greece
Alberto Freitas	University of Porto, Portugal
Ruy Freitas Reis	Universidade Federal de Juiz de Fora, Brazil
Kohei Fujita	University of Tokyo, Japan
Takeshi Fukaya	Hokkaido University, Japan
Wlodzimierz Funika	AGH University of Krakow, Poland
Takashi Furumura	University of Tokyo, Japan
Teresa Galvão	University of Porto, Portugal
Luis Garcia-Castillo	Carlos III University of Madrid, Spain
Bartłomiej Gardas	Institute of Theoretical and Applied Informatics, Polish Academy of Sciences, Poland
Victoria Garibay	University of Amsterdam, The Netherlands
Frédéric Gava	Paris-East Créteil University, France
Piotr Gawron	Nicolaus Copernicus Astronomical Centre, Polish Academy of Sciences, Poland
Bernhard Geiger	Know-Center GmbH, Austria
Pawel Gepner	Warsaw Technical University, Poland
Alex Gerbessiotis	NJIT, USA
Maziar Ghorbani	Brunel University London, UK
Konstantinos Giannoutakis	University of Macedonia, Greece
Alfonso Gijón	University of Granada, Spain

Jorge González-Domínguez	Universidade da Coruña, Spain
Alexandrino Gonçalves	CIIC – ESTG – Polytechnic University of Leiria, Portugal
Yuriy Gorbachev	Soft-Impact LLC, Russia
Pawel Gorecki	University of Warsaw, Poland
Michael Gowanlock	Northern Arizona University, USA
George Gravvanis	Democritus University of Thrace, Greece
Derek Groen	Brunel University London, UK
Loïc Guégan	UiT the Arctic University of Norway, Norway
Tobias Guggemos	University of Vienna, Austria
Serge Guillas	University College London, UK
Manish Gupta	Harish-Chandra Research Institute, India
Piotr Gurgul	SnapChat, Switzerland
Oscar Gustafsson	Linköping University, Sweden
Ahmed Abdeen Hamed	State University of New York at Binghamton, USA
Laura Harbach	Brunel University London, UK
Agus Hartoyo	TU Kaiserslautern, Germany
Ali Hashemian	Basque Center for Applied Mathematics, Spain
Mohamed Hassan	Virginia Tech, USA
Alexander Heinecke	Intel Parallel Computing Lab, USA
Jude Hemanth	Karunya University, India
Aochi Hideo	BRGM, France
Alfons Hoekstra	University of Amsterdam, The Netherlands
George Holt	UK Research and Innovation, UK
Maximilian Höb	Leibniz-Rechenzentrum der Bayerischen Akademie der Wissenschaften, Germany
Huda Ibeid	Intel Corporation, USA
Alireza Jahani	Brunel University London, UK
Jiří Jaroš	Brno University of Technology, Czechia
Qiao Jin	National Institutes of Health, USA
Zhong Jin	Computer Network Information Center, Chinese Academy of Sciences, China
David Johnson	Uppsala University, Sweden
Eleda Johnson	Imperial College London, UK
Piotr Kalita	Jagiellonian University, Poland
Drona Kandhai	University of Amsterdam, The Netherlands
Aneta Karaivanova	IICT-Bulgarian Academy of Science, Bulgaria
Sven Karbach	University of Amsterdam, The Netherlands
Takahiro Katagiri	Nagoya University, Japan
Haruo Kobayashi	Gunma University, Japan
Marcel Koch	KIT, Germany

Harald Koestler	University of Erlangen-Nuremberg, Germany
Georgy Kopanitsa	Tomsk Polytechnic University, Russia
Sotiris Kotsiantis	University of Patras, Greece
Remous-Aris Koutsiamanis	IMT Atlantique/DAPI, STACK (LS2N/Inria), France
Sergey Kovalchuk	Huawei, Russia
Slawomir Koziel	Reykjavik University, Iceland
Ronald Kriemann	MPI MIS Leipzig, Germany
Valeria Krzhizhanovskaya	University of Amsterdam, The Netherlands
Sebastian Kuckuk	Friedrich-Alexander-Universität Erlangen-Nürnberg, Germany
Michael Kuhn	Otto von Guericke University Magdeburg, Germany
Ryszard Kukulski	Institute of Theoretical and Applied Informatics, Polish Academy of Sciences, Poland
Krzysztof Kurowski	PSNC, Poland
Marcin Kuta	AGH University of Krakow, Poland
Marcin Łoś	AGH University of Krakow, Poland
Roberto Lam	Universidade do Algarve, Portugal
Tomasz Lamża	ACK Cyfronet, Poland
Ilaria Lazzaro	Università degli studi Magna Graecia di Catanzaro, Italy
Paola Lecca	Free University of Bozen-Bolzano, Italy
Byung Lee	University of Vermont, USA
Mike Lees	University of Amsterdam, The Netherlands
Leifur Leifsson	Purdue University, USA
Kenneth Leiter	U.S. Army Research Laboratory, USA
Paulina Lewandowska	IT4Innovations National Supercomputing Center, Czechia
Jingfa Li	Beijing Institute of Petrochemical Technology, China
Siyi Li	Imperial College London, UK
Che Liu	Imperial College London, UK
James Liu	Colorado State University, USA
Zhao Liu	National Supercomputing Center in Wuxi, China
Marcelo Lobosco	UFJF, Brazil
Jay F. Lofstead	Sandia National Laboratories, USA
Chu Kiong Loo	University of Malaya, Malaysia
Stephane Louise	CEA, LIST, France
Frédéric Loulergue	University of Orléans, INSA CVL, LIFO EA 4022, France
Hatem Ltaief	KAUST, Saudi Arabia
Zhiyong Lu	National Institutes of Health, USA

Fernando Nobrega Santos   University of Amsterdam, The Netherlands
Joseph O'Connor   University of Edinburgh, UK
Frederike Oetker   University of Amsterdam, The Netherlands
Arianna Olivelli   Imperial College London, UK
Ángel Omella   Basque Center for Applied Mathematics, Spain
Kenji Ono   Kyushu University, Japan
Hiroyuki Ootomo   Tokyo Institute of Technology, Japan
Eneko Osaba   TECNALIA Research & Innovation, Spain
George Papadimitriou   University of Southern California, USA
Nikela Papadopoulou   University of Glasgow, UK
Marcin Paprzycki   IBS PAN and WSM, Poland
David Pardo   Basque Center for Applied Mathematics, Spain
Anna Paszynska   Jagiellonian University, Poland
Maciej Paszynski   AGH University of Krakow, Poland
Łukasz Pawela   Institute of Theoretical and Applied Informatics,
                                   Polish Academy of Sciences, Poland
Giulia Pederzani   Universiteit van Amsterdam, The Netherlands
Alberto Perez de Alba Ortiz   University of Amsterdam, The Netherlands
Dana Petcu   West University of Timisoara, Romania
Beáta Petrovski   University of Oslo, Norway
Frank Phillipson   TNO, The Netherlands
Eugenio Piasini   International School for Advanced Studies
                                   (SISSA), Italy
Juan C. Pichel   Universidade de Santiago de Compostela, Spain
Anna Pietrenko-Dabrowska   Gdansk University of Technology, Poland
Armando Pinho   University of Aveiro, Portugal
Pietro Pinoli   Politecnico di Milano, Italy
Yuri Pirola   Università degli Studi di Milano-Bicocca, Italy
Ollie Pitts   Imperial College London, UK
Robert Platt   Imperial College London, UK
Dirk Pleiter   KTH/Forschungszentrum Jülich, Germany
Paweł Poczekajło   Koszalin University of Technology, Poland
Cristina Portalés Ricart   Universidad de Valencia, Spain
Simon Portegies Zwart   Leiden University, The Netherlands
Anna Procopio   Università Magna Graecia di Catanzaro, Italy
Ela Pustulka-Hunt   FHNW Olten, Switzerland
Marcin Płodzień   ICFO, Spain
Ubaid Qadri   Hartree Centre – STFC, UK
Rick Quax   University of Amsterdam, The Netherlands
Cesar Quilodran Casas   Imperial College London, UK
Andrianirina Rakotoharisoa   Imperial College London, UK
Celia Ramos   University of the Algarve, Portugal

Robin Richardson	Netherlands eScience Center, The Netherlands
Sophie Robert	University of Orléans, France
João Rodrigues	Universidade do Algarve, Portugal
Daniel Rodriguez	University of Alcalá, Spain
Marcin Rogowski	Saudi Aramco, Saudi Arabia
Sergio Rojas	Pontifical Catholic University of Valparaiso, Chile
Diego Romano	ICAR-CNR, Italy
Albert Romkes	South Dakota School of Mines and Technology, USA
Juan Ruiz	University of Buenos Aires, Argentina
Tomasz Rybotycki	IBS PAN, CAMK PAN, AGH, Poland
Katarzyna Rycerz	AGH University of Krakow, Poland
Grażyna Ślusarczyk	Jagiellonian University, Poland
Emre Sahin	Science and Technology Facilities Council, UK
Ozlem Salehi	Özyeğin University, Turkey
Ayşin Sancı	Altinay, Turkey
Christopher Savoie	Zapata Computing, USA
Ileana Scarpino	University "Magna Graecia" of Catanzaro, Italy
Robert Schaefer	AGH University of Krakow, Poland
Ulf D. Schiller	University of Delaware, USA
Bertil Schmidt	University of Mainz, Germany
Karen Scholz	Fraunhofer MEVIS, Germany
Martin Schreiber	Université Grenoble Alpes, France
Paulina Sepúlveda-Salas	Pontifical Catholic University of Valparaiso, Chile
Marzia Settino	Università Magna Graecia di Catanzaro, Italy
Mostafa Shahriari	Basque Center for Applied Mathematics, Spain
Takashi Shimokawabe	University of Tokyo, Japan
Alexander Shukhman	Orenburg State University, Russia
Marcin Sieniek	Google, USA
Joaquim Silva	Nova School of Science and Technology – NOVA LINCS, Portugal
Mateusz Sitko	AGH University of Krakow, Poland
Haozhen Situ	South China Agricultural University, China
Leszek Siwik	AGH University of Krakow, Poland
Peter Sloot	University of Amsterdam, The Netherlands
Oskar Slowik	Center for Theoretical Physics PAS, Poland
Sucha Smanchat	King Mongkut's University of Technology North Bangkok, Thailand
Alexander Smirnovsky	SPbPU, Russia
Maciej Smołka	AGH University of Krakow, Poland
Isabel Sofia	Instituto Politécnico de Beja, Portugal
Robert Staszewski	University College Dublin, Ireland

Magdalena Stobińska	University of Warsaw, Poland
Tomasz Stopa	IBM, Poland
Achim Streit	KIT, Germany
Barbara Strug	Jagiellonian University, Poland
Diana Suleimenova	Brunel University London, UK
Shuyu Sun	King Abdullah University of Science and Technology, Saudi Arabia
Martin Swain	Aberystwyth University, UK
Renata G. Słota	AGH University of Krakow, Poland
Tseden Taddese	UK Research and Innovation, UK
Ryszard Tadeusiewicz	AGH University of Krakow, Poland
Claude Tadonki	Mines ParisTech/CRI – Centre de Recherche en Informatique, France
Daisuke Takahashi	University of Tsukuba, Japan
Osamu Tatebe	University of Tsukuba, Japan
Michela Taufer	University of Tennessee, USA
Andrei Tchernykh	CICESE, Mexico
Kasim Terzic	University of St Andrews, UK
Jannis Teunissen	KU Leuven, Belgium
Sue Thorne	Hartree Centre – STFC, UK
Ed Threlfall	United Kingdom Atomic Energy Authority, UK
Vinod Tipparaju	AMD, USA
Pawel Topa	AGH University of Krakow, Poland
Paolo Trunfio	University of Calabria, Italy
Ola Tørudbakken	Meta, Norway
Carlos Uriarte	University of the Basque Country, BCAM – Basque Center for Applied Mathematics, Spain
Eirik Valseth	University of Life Sciences & Simula, Norway
Rein van den Boomgaard	University of Amsterdam, The Netherlands
Vítor V. Vasconcelos	University of Amsterdam, The Netherlands
Aleksandra Vatian	ITMO University, Russia
Francesc Verdugo	Vrije Universiteit Amsterdam, The Netherlands
Karin Verspoor	RMIT University, Australia
Salvatore Vitabile	University of Palermo, Italy
Milana Vuckovic	European Centre for Medium-Range Weather Forecasts, UK
Kun Wang	Imperial College London, UK
Peng Wang	NVIDIA, China
Rodrigo Weber dos Santos	Federal University of Juiz de Fora, Brazil
Markus Wenzel	Fraunhofer Institute for Digital Medicine MEVIS, Germany

Lars Wienbrandt	Kiel University, Germany
Wendy Winnard	UKRI STFC, UK
Maciej Woźniak	AGH University of Krakow, Poland
Xindong Wu	Hefei University of Technology, China
Dunhui Xiao	Tongji University, China
Huilin Xing	University of Queensland, Australia
Yani Xue	Brunel University, UK
Abuzer Yakaryilmaz	University of Latvia, Latvia
Xin-She Yang	Middlesex University London, UK
Dongwei Ye	University of Amsterdam, The Netherlands
Karol Życzkowski	Jagiellonian University, Poland
Gabor Závodszky	University of Amsterdam, Hungary
Sebastian Zając	SGH Warsaw School of Economics, Poland
Małgorzata Zajęcka	AGH University of Krakow, Poland
Justyna Zawalska	ACC Cyfronet AGH, Poland
Wei Zhang	Huazhong University of Science and Technology, China
Yao Zhang	Google, USA
Jinghui Zhong	South China University of Technology, China
Sotirios Ziavras	New Jersey Institute of Technology, USA
Zoltan Zimboras	Wigner Research Center, Hungary
Italo Zoppis	University of Milano-Bicocca, Italy
Chiara Zucco	University Magna Graecia of Catanzaro, Italy
Pavel Zun	ITMO University, Russia

Contents – Part VII

Smart Systems: Bringing Together Computer Vision, Sensor Networks, and Artificial Intelligence

Solving Problems with Uncertainties

Teaching Computational Science

Simulations of Flow and Transport: Modeling, Algorithms and Computation

Capillary Behaviors of Miscible Fluids in Porous Media: A Pore-Scale Simulation Study

Ronghao Cui⬤ and Shuyu Sun(✉)⬤

Computational Transport Phenomena Laboratory (CTPL), Physical Science and Engineering (PSE) Division, King Abdullah University of Science and Technology (KAUST), Thuwal 23955-6900, Saudi Arabia
shuyu.sun@kaust.edu.sa

Abstract. Capillary behaviors of two phases in porous media are widely investigated by pore-scale simulations. Flow properties of miscible two-phase fluids pose challenges because compositional species in miscible fluids exist in both of two phases. In this work, pore-network modeling is developed to analyze capillary behaviors of miscible fluids. First, the pore-network structure is generated by introducing the Halton sequence and the Delaunay triangulation. Then, we perform the primary drainage to simulate the variation of capillary pressure as a function of saturation under equilibrium conditions. The miscibility details of two-phase fluids are determined by two-phase flash calculations. The effect of representative network sizes on capillary behaviors is analyzed to balance computational demands and interpretability. Our models demonstrate the efficiency and the robustness of the network generation procedure. Finally, our simulation results indicate that the miscibility of fluids significantly influences capillary behaviors, highlighting the necessity of species-specific compositional modeling for miscible two-phase fluids in porous media.

Keywords: Pore-scale simulation · Capillary pressure · Miscible fluids

1 Introduction

Two-phase flow of fluids in subsurface porous media has been playing an important role in energy resources exploitation and carbon geological storage. Compared with gas-water or oil-water two phases that have been traditionally modeled as immiscible phases, gas-oil two phases can be miscible and then simulated by species-specified compositional flow [1,2]. Species in miscible fluids can exist in both of two phases, leading to complex phase and interfacial behaviors. These behaviors are heavily affected by temperature, pressure, and composition conditions. Understanding the flow properties of miscible phases is one of the key issues to enhance sustainable recovery of hydrocarbon resources and stable storage of greenhouse gases.

Capillary pressure p_c plays an essential role in residual trapping of fluids during the two-phase flow in porous media. The variation of p_c as a function of

L. Franco et al. (Eds.): ICCS 2024, LNCS 14838, pp. 3–10, 2024.
https://doi.org/10.1007/978-3-031-63783-4_1

the wetting-phase saturation S_w is known as the p_c-S_w curve. In fact, the p_c-S_w curve is one of the most common constitutive properties of two-phase flow in porous media. The p_c-S_w curve can be obtained from laboratory experiments and numerical simulations. Compared with laboratory experiments, numerical simulations overcome the limitations at spatial and temporal scales to investigate the p_c-S_w curve in porous media under equilibrium conditions [3]. The prevalent numerical method for the p_c-S_w curve is pore-network modeling with the low computational cost. A pore-network model is composed of pore bodies and pore throats with simple geometries. Pore bodies mainly contribute to network saturations and serve as computational nodes for flow simulations [4]. The miscible two-phase fluids involve the equation of state (EoS) to characterize their physical properties. As one of the most prevalent EoSs, the Peng-Robinson EoS has shown excellent estimation for hydrocarbon phase behaviors. Two-phase flash calculations are the important part of phase equilibrium calculations to obtain details of compositional species in each phase. Successive substitution iteration and Newton's method are popular methods to solve the formulations of two-phase flash problems [5].

The workflow of this work is summarized as follows. We at first introduce the Halton sequence and Delaunay triangulation in the pore-network generator to create the pore-network structure. Then, we perform the primary drainage simulations to compute p_c-S_w curves. We evaluate the effect of representative network sizes based on the outcomes of p_c-S_w curves in order to optimize the computational cost and interpretability. At last, we consider binary miscible fluids whose two-phase information is obtained from flash calculations to investigate the capillary behaviors in porous media.

2   Methodology

2.1   The Network Structure

In this work, we have developed random generator-based pore-network models for flow simulations. The generation procedure is summarized as follows. At first, the network topology was determined by creating a given number of pore bodies distributed in a two-dimensional (2D) square domain. We generated quasirandom pore bodies in this 2D space by use of the Halton sequence. Compared with the source of pseudorandom numbers, pore bodies created by the Halton sequence is of lower discrepancy in distribution but not uniformly cover the domain [6]. Second, we connected pore bodies by pore throats using the Delaunay triangulation method to create triangles in the 2D domain [7]. In practice, we predefined a specific number of pore bodies based on the density of pore bodies within the 2D domain, positioning them at the four domain boundaries to avoid lengthy pore throats. For approximating the statistical properties of the porous media, pore throats between pore bodies were randomly deleted according to the averaged coordination number. The pore-throat deletion was followed by the elimination process which was conducted to delete isolated pore

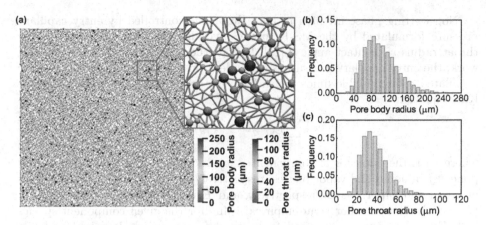

Fig. 1. The pore-network model consisting of 10050 pore bodies and 25094 pore throats with the average coordination number of 5.0: (a) the structure illustration, (b) the inscribed radius distribution of pore bodies, and (c) the inscribed radius distribution of pore throats.

bodies, isolated clusters, and dead-end pore bodies by use of the method proposed by Raoof and Hassanizadeh [8]. Third, we determined pore body sizes by the lognormal distribution function, while pore throat sizes were assigned according to their neighboring pore bodies at both ends [9,10]. We used cube as the geometry of pore bodies and square as the cross-sectional shape of pore throats. These angular geometries can accommodate two phases simultaneously. An example of the pore-network model we generated is shown in Fig. 1a. In this case, we generated 10050 pore bodies and 25094 pore throats with the average coordination number of 5.0. Figure 1b and 1c show the generated inscribed radius distributions of pore bodies and pore throats, respectively.

2.2   Numerical Experiments

Miscible fluids we used in this work can be partitioned into two phases, vapor and liquid. We simulated the primary vapor drainage in a strongly liquid-wet porous medium to obtain p_c-S_w curves, where contact angle was assumed as zero. Therefore, we did not consider the effect of contact angles on capillary pressure. The primary drainage process can be explained as follows. The liquid phase is initially saturated in the pore network at the given temperature and pressure. The vapor phase starts to contact the pore network at the same temperature. With the increase of the vapor phase, there is a gradual increment of capillary pressure between two phases in the pore network. At each step of capillary pressure, equilibrium positions of all gas-liquid interfaces within the network are determined. Then, computations are carried out for wetting-phase saturation S_w within the entire pore network. As a results, a p_c-S_w curve for the network can be obtained.

Non-wetting phase invasion in pore throats is controlled by entry capillary pressure formulated by the MS-P theory involving interfacial tension σ, pore throat radius r_t, contact angle θ [11]. Due to the assumption of $\theta = 0°$ in this work, the entry capillary pressure is dependent on σ with the given r_t. We have calculated σ for miscible fluids which have N_c components using the Weinaug-Katz equation:

$$\sigma = \left[\sum_{j=1}^{N_c} (\chi_j x_j c_w - \chi_j y_j c_n) \right]^E \tag{1}$$

where χ_j is the parachor of component j, x_j and y_j are mole fractions of component j in liquid and vapor phases, respectively, c_w and c_n are molar densities of liquid and vapor phases, respectively, and E is the critical scaling exponent ($E = 4.0$). The parachor is an empirical value for the given component species [12]. For calculating σ, we need to know the compositional variables of two phases, including x_j, y_j, c_w, c_n. Two-phase flash calculations can be used to compute these compositional variables. With the known overall composition z_j for miscible fluids, the fugacities of each species in liquid and vapor phases, f_j^L and f_j^V, are supposed to be equal to each other, i.e., $f_j^L = f_j^V$. Knowing liquid and vapor pressures, p^L and p^V, we have introduced corresponding fugacity coefficients, φ_j^L and φ_j^V, where $f_j^L = \varphi_j^L x_j p^L$ and $f_j^V = \varphi_j^V y_j p^V$. The vapor-liquid equilibrium ratio for component j, K_j, is defined as the ratio of y_j to x_j ($K_j = y_j/x_j$). Based on the fugacity equivalence relation, K_j is computed as given by:

$$K_j = \frac{\varphi_j^L p^L}{\varphi_j^V p^V} \tag{2}$$

EoSs play an important role in two-phase flash calculations, since φ_j^L and φ_j^V are calculated from EoSs [5]. In this work, we have employed the Peng-Robinson EoS for miscible fluids. The vapor-phase mole fraction in miscible fluids is defined as $\beta^V = N^V/N$, where N^V and N are mole numbers of the vapor phase and overall miscible fluids, respectively. With β^V, K_j, and z_j, the Rachford-Rice equation ($F_{RR} = 0$) is given by:

$$F_{RR} = \sum_{j=1}^{N_c} \frac{(K_j - 1)z_j}{1 + \beta^V(K_j - 1)} = 0 \tag{3}$$

Two-phase flash formulations are solved by successive substitution iteration in this work. Without any information about vapor-liquid two-phase equilibrium to calculate K_j from Eq. 2 at the beginning of two-phase flash calculations, Wilson's correlation with thermodynamic properties of fluids is employed to obtain the initial guess of K_j, as given by:

$$K_j^0 = \frac{p_{c,j}}{p^V} \exp(5.37(1 + \omega_j)(1 - \frac{T_{c,j}}{T})) \tag{4}$$

where ω_j is the acentric factor of component j, $T_{c,j}$ and $p_{c,j}$ are critical temperature and critical pressure of component j, respectively, and T is the temperature.

With known K_j ($K_j = K_j^0$ for the first step), the Rachford-Rice equation (Eq. 3) is solved to obtain β^V. Next, the Peng-Robinson EoS is employed to get φ_j^L and φ_j^V for updating K_j according to Eq. 2. Related EoS parameters used in this work can be found in Ref. [12]. Successive substitution iteration repeats updating K_j and β^V until the convergence is reached.

3 Results and Discussion

3.1 The Representative Network Size

The representative elementary volume (REV) size is an important issue when investigating flow properties in porous media. The REV effect is analyzed using our pore-network model to obtain the representative network size. As shown in Fig. 2, we selected 400, 2000, and 4000 pore bodies generated within the square domain. The interfacial tension is set as $72.8\,\mathrm{mN/m}$. We performed the primary vapor drainage and calculated p_c-S_w curves in pore-network models for 20 runs using the same input parameters. It can be seen that the network sizes heavily influence p_c-S_w curves during the drainage. With 400 pore bodies, there are large variations of p_c-S_w curves for different runs. As the number of pore bodies increases, p_c-S_w curves from each run gradually approaches the average curve. Figure 2 shows that p_c-S_w curves are not significantly influenced by the network size when containing 4000 pore bodies. Therefore, we used 4000 pore bodies to simulate the drainage process for miscible fluids. It should be noted that this is only valid for our 2D models. If there is a three-dimensional (3D) pore-network model, the best representative network size will change. This is because 3D pore-network models can have higher coordination number for pore bodies than the 2D version.

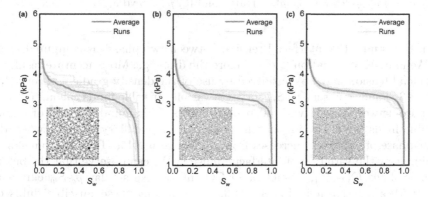

Fig. 2. The effect of representative network sizes on p_c-S_w curves: (a) 400 pore bodies, (b) 2000 pore bodies, and (c) 4000 pore bodies.

3.2   Capillary Behaviors of Miscible Fluids

To investigate capillary behaviors of different miscible fluids in porous media, we consider binary mixtures consisting of one component selected from nitrogen (N_2), methane (C_1), or carbon dioxide (CO_2), and the other one chosen from n-butane (nC_4), n-pentane (nC_5), or n-decane (nC_{10}). It should be noted that we have performed the primary vapor drainage only when there exist two phases. As a matter of fact, one single phase for miscible fluids may occur under appropriate temperature and pressure conditions. Due to the monotonically decreasing function of β^V in the Rachford-Rice equation (F_{RR} in Eq. 3), we roughly used the values of $F_{RR}|_{\beta=0}$ and $F_{RR}|_{\beta=1}$ to determine whether there were two phases. In the case of two phases, it is required that $F_{RR}|_{\beta=0} > 0$ and $F_{RR}|_{\beta=1} < 0$. However, a more rigorous examination of phase states should utilize the procedure of phase stability analysis [5].

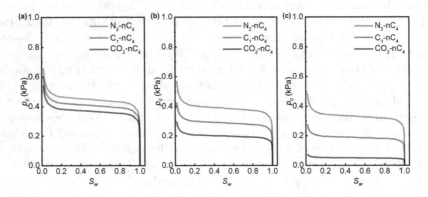

Fig. 3. The capillary behaviors of N_2-nC_4, C_1-nC_4, and CO_2-nC_4 miscible fluids at T = 323 K: (a) p^V = 2 MPa, (b) p^V = 4 MPa, and (c) p^V = 6 MPa.

In this study, the interfacial tension between two phases is computed using the Weinaug-Katz equation (Eq. 1) before the drainage. More accurate results of interfacial tension may be considered by use of the density gradient theory [13]. Figures 3, 4 and 5 present p_c-S_w curves of various miscible fluids calculated from our pore-network model with 4000 pore bodies. It is found that all of binary miscible fluids present the monotonous decrease in capillary pressure when the vapor-phase pressure p^V increases from 2 MPa to 6 MPa. This phenomenon is dominant by the variation of interfacial tension. We can observe that the binary miscible fluids (N_2-nC_4, C_1-nC_4, and CO_2-nC_4) have similar p_c-S_w curves at p^V = 2 MPa, as shown in Fig. 3a. Differences among these miscible fluids on capillary behaviors expand significantly as p^V increases (see Fig. 3b and 3c). In addition, at the same T and p^V conditions, N_2, C_1, and CO_2 exhibit a higher capillary pressure when mixed with alkane with the longer carbon chain, as shown in Figs. 3a, 4a and 5a.

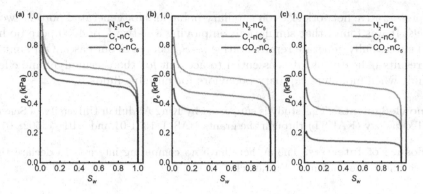

Fig. 4. The capillary behaviors of N_2-nC_5, C_1-nC_5, and CO_2-nC_5 miscible fluids at $T = 323$ K: (a) $p^V = 2$ MPa, (b) $p^V = 4$ MPa, and (c) $p^V = 6$ MPa.

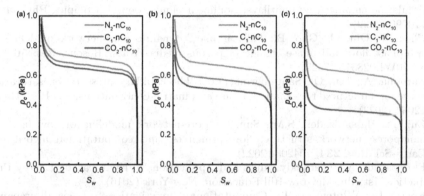

Fig. 5. The capillary behaviors of N_2-nC_{10}, C_1-nC_{10}, and CO_2-nC_{10} miscible fluids at $T = 323$ K: (a) $p^V = 2$ MPa, (b) $p^V = 4$ MPa, and (c) $p^V = 6$ MPa.

Interfacial tension-induced capillary behaviors are very common in miscible fluids. Therefore, the estimation of these complex capillary behaviors using traditional pore-network modeling with the fixed interfacial tension may lead to large errors. Preliminary results in Figs. 3, 4 and 5 have indicated that the thermodynamic analysis on pore-network modeling is needed in the future for the explanation of interactions between flow properties and phase miscibility.

4   Conclusions

We have simulated capillary behaviors of miscible fluids in porous media during the primary drainage process through the pore-network modeling method. Random generator-based pore-network models were developed by mainly introducing the Halton sequence and the Delaunay triangulation. Multiple cases in this work have shown that our network generation is efficient and robust, approaching the probability distribution properties of porous media. We have examined the effect of representative network sizes on capillary behaviors for generated pore-network models. It was found that simulations employing a network of 4000 pore bodies could successfully generate representative p_c-S_w curves in our cases. Our simulation results indicate that it is essential to account for the thermodynamic effect of fluids when modeling capillary behaviors for miscible fluids.

Acknowledgments. This study is supported by King Abdullah University of Science and Technology (KAUST) through the grants BAS/1/1351-01 and URF/1/5028-01.

Disclosure of Interests. The authors have no competing interests to declare that are relevant to the content of this article.

References

1. Li, Y., Yang, H., Sun, S.: Fully implicit two-phase VT-flash compositional flow simulation enhanced by multilayer nonlinear elimination. J. Comput. Phys. **449**, 110790 (2022)
2. Chen, S., Qin, C., Guo, B.: Fully implicit dynamic pore-network modeling of two-phase flow and phase change in porous media. Water Resour. Res. **56**(11), e2020WR028510 (2020)
3. Cardona, A., Liu, Q., Santamarina, J.C.: The capillary pressure vs. saturation curve for a fractured rock mass: fracture and matrix contributions. Sci. Rep. **13**(1), 12044 (2023)
4. Cui, R., Hassanizadeh, S.M., Sun, S.: Pore-network modeling of flow in shale nanopores: network structure, flow principles, and computational algorithms. Earth-Sci. Rev. **234**, 104203 (2022)
5. Firoozabadi, A.: Thermodynamics and Applications in Hydrocarbon Energy Production, 1st edn. McGraw-Hill Education, New York (2016)
6. Kocis, L., Whiten, W.J.: Computational investigations of low-discrepancy sequences. ACM Trans. Math. Softw. **23**(2), 266–294 (1997)
7. Qin, C., Hassanizadeh, S.M., Ebigbo, A.: Pore-scale network modeling of microbially induced calcium carbonate precipitation: insight into scale dependence of biogeochemical reaction rates. Water Resour. Res. **52**(11), 8794–8810 (2016)
8. Raoof, A., Hassanizadeh, S.M.: A new method for generating pore-network models of porous media. Transp. Porous Med. **81**, 391–407 (2010)
9. Joekar-Niasar, V., Hassanizadeh, S., Leijnse, A.: Insights into the relationships among capillary pressure, saturation, interfacial area and relative permeability using pore-network modeling. Transp. Porous Med. **74**, 201–219 (2008)
10. Cui, R., Feng, Q., Chen, H., Zhang, W., Wang, S.: Multiscale random pore network modeling of oil-water two-phase slip flow in shale matrix. J. Petrol. Sci. Eng. **175**, 46–59 (2019)
11. Joekar-Niasar, V., Hassanizadeh, S.M., Dahle, H.K.: Non-equilibrium effects in capillarity and interfacial area in two-phase flow: dynamic pore-network modelling. J. Fluid Mech. **655**, 38–71 (2010)
12. Li, Y., Kou, J., Sun, S.: Thermodynamically stable two-phase equilibrium calculation of hydrocarbon mixtures with capillary pressure. Ind. Eng. Chem. Res. **57**(50), 17276–17288 (2018)
13. Cui, R., Narayanan Nair, A.K., Che Ruslan, M.F.A., Yang, Y., Sun, S.: Interfacial properties of the hexane+ carbon dioxide+ brine system in the presence of hydrophilic silica. Ind. Eng. Chem. Res. **62**(34), 13470–13478 (2023)

Numerical Results and Convergence of Some Inf-Sup Stable Elements for the Stokes Problem with Pressure Dirichlet Boundary Condition

Huoyuan Duan$^{(\boxtimes)}$ and Xianliang Wu

School of Mathematics and Statistics, Wuhan University, Wuhan 430072, China
hyduan.math@whu.edu.cn

Abstract. For the Stokes problem with pressure Dirichlet boundary conditions, we propose an Enriched Mini element. For both the Mini element and the Enriched Mini element, we show that they are inf-sup stable. Unexpectedly, they yield wrong convergent finite element solutions for the singular velocity solution. On the contrary, the Taylor-Hood element, which is still inf-sup stable, gives correct convergence. However, how to analyze the convergence becomes open. We provide extensive numerical studies on the wrong convergence of the inf-sup stable Mini-type elements and the correct convergence of the inf-sup stable Taylor-Hood element and on the inf-sup stability constants.

Keywords: Stokes problem · Pressure Dirichlet boundary condition · Non H^1 velocity · Inf-sup stable element · Wrong convergence

1 Introduction

The Mini element [1,3,7] is inf-sup stable for the Stokes problem. This element uses the nodal linear element enriched with one element bubble for each component of the velocity variable and the nodal linear element for the pressure variable. It gives an optimal convergent approximation of the velocity variable. Unexpectedly, we found that if the pressure variable has its own boundary condition on the part or the whole of the domain boundary and if the velocity variable is singular and does not belong to the Hilbert space H^1, the Mini element gives a wrong solution. The non H^1 singularity does not sound peculiar. In general, the Stokes problem lives with the Dirichlet integral, and the H^1 space is a natural solution space for the velocity variable. This is the case when the velocity variable is imposed with the Dirichlet boundary condition on the whole domain boundary. However, in practical situations (cf. [6]), the velocity variable may only have partial Dirichlet boundary condition while its tangential components have other partial Dirichlet boundary condition. As a supplement, the pressure variable accordingly has partial Dirichlet boundary condition. Under these boundary conditions, the velocity variable would not belong to the H^1 space (cf. [3,5]). Following the theory in [8], we enrich the Mini element adding one degree of freedom in the interior of each elemental side locating on the domain

© The Author(s), under exclusive license to Springer Nature Switzerland AG 2024
L. Franco et al. (Eds.): ICCS 2024, LNCS 14838, pp. 11–19, 2024.
https://doi.org/10.1007/978-3-031-63783-4_2

boundary(just considering the two-dimensional problem). However, the Enriched Mini element is inf-sup stable, but it still wrongly converges. We also develop a new general approach for proving the inf-sup stability of the Mini element for the Stokes equations with pressure Dirichlet boundary conditions. On the contrary, for the Taylor-Hood element [2,3,7], it generates correctly convergent approximations for the pressure Dirichlet boundary condition. The convergence is not optimal for smooth solutions, but a theoretical converge rate for the singular solution can be reached in the numerical results. This element is inf-sup stable, as is shown in [8]. Unexpectedly, it seems extremely difficult to give a convergence analysis, which then becomes an open problem. From the kernel coercivity and the inf-sup stability of the Taylor-Hood element, the convergence cannot be theoretically justified whenever the velocity solution does not belong to the H^1 space. For the Stokes problem with the velocity Dirichlet boundary condition on the whole domain boundary, as is well-known [3,7], the convergence and the optimality of the Taylor-Hood element follow from the kernel coercivity and the inf-sup stability. The Taylor-Hood element has extensive applications, e.g., its application in the multiphysics problems [10]. The inf-sup stability plays a key role in the multigrid method and the preconditioning method [11,12] for the saddle-point systems of the Stokes equations and the vector Laplacian. Note that the Stokes equations (1) becomes the vector Laplacian when $\Gamma_2 = \partial\Omega$.

2 Stokes Problem, Mini Element and Taylor-Hood Element

Let $\Omega \subset \mathbb{R}^2$ be a simply-connected domain, with boundary $\partial\Omega = \bar{\Gamma}_1 \cup \bar{\Gamma}_2, \Gamma_1 \cap \Gamma_2 = \emptyset$. The Stokes problem reads as follows:

$$-\Delta \mathbf{u} + \nabla p = \mathbf{f}, \quad \mathrm{div}\, \mathbf{u} = 0 \quad \text{in } \Omega, \tag{1}$$

$$\mathbf{u} = \mathbf{0} \quad \text{on } \Gamma_1, \quad \mathbf{n} \times \mathbf{u} = \mathbf{0}, \quad p = 0 \quad \text{on } \Gamma_2. \tag{2}$$

When $\Gamma_1 = \partial\Omega$, we require that $\int_\Omega p = 0$. Define $\mathbf{U} = \{\mathbf{v} \in H(\mathbf{curl};\Omega) \cap H(\mathrm{div};\Omega) : \mathbf{v}|_{\Gamma_1} = \mathbf{0}, \mathbf{n} \times \mathbf{v}|_{\Gamma_2} = \mathbf{0}\}$, $Q = \{q \in H^1(\Omega) : q|_{\Gamma_2} = 0\}$. Let $\mathscr{T}_h$ denote the shape-regular triangulation of Ω into triangles. A generic element $T \in \mathscr{T}_h$ has its diameter h_T; $h := \max_{T \in \mathscr{T}_h} h_T$. Let $\mathscr{P}_\ell$ be the space of polynomials of total degree less than or equal to the integer $\ell \geq 0$. Let $\lambda_1, \lambda_2, \lambda_3$ denote the three shape functions of $\mathscr{P}_1(T)$ on a generic element T. Define the element bubble $b_T := \lambda_1\lambda_2\lambda_3$, and $\mathscr{P}_1^+(T) := \mathrm{span}\{\lambda_1, \lambda_2, \lambda_3, b_T\}$. Then, define $\mathbf{U}_h^{\mathrm{Mini}} - Q_h$ the Mini element and $\mathbf{U}_h^{\mathrm{T-H}} - Q_h$ the Taylor-Hood element, respectively,

$$\mathbf{U}_h^{\mathrm{Mini}} = \{\mathbf{v}_h \in \mathbf{U} \cap (H^1(\Omega))^2 : \mathbf{v}_h|_T \in (\mathscr{P}_1^+(T))^2, \forall T \in \mathscr{T}_h\}, \tag{3}$$

$$\mathbf{U}_h^{\mathrm{T-H}} = \{\mathbf{v}_h \in \mathbf{U} \cap (H^1(\Omega))^2 : \mathbf{v}_h|_T \in (\mathscr{P}_2(T))^2, \forall T \in \mathscr{T}_h\}, \tag{4}$$

$$Q_h = \{q_h \in Q : q_h|_T \in \mathscr{P}_1(T), \forall T \in \mathscr{T}_h\}. \tag{5}$$

The finite element problem reads as follows: Find $\mathbf{u}_h \in \mathbf{V}_h$ and $p_h \in Q_h$ such that

$$\begin{cases} a_h(\mathbf{u}_h, \mathbf{v}_h) + b(\mathbf{v}_h, p_h) = (\mathbf{f}, \mathbf{v}_h) & \forall \mathbf{v}_h \in \mathbf{V}_h, \\ b(\mathbf{u}_h, q_h) = 0 & \forall q_h \in Q_h, \end{cases} \qquad (6)$$

where $\mathbf{V}_h$ stands for either $\mathbf{U}_h^{\text{Mini}}$ or $\mathbf{U}_h^{\text{T-H}}$, and $a_h(\mathbf{u}, \mathbf{v}) = (\mathbf{curl}\,\mathbf{u}, \mathbf{curl}\,\mathbf{v}) + \sum_{T \in \mathscr{T}_h} h_T^2 (\text{div}\,\mathbf{u}, \text{div}\,\mathbf{v})_{0,T}$ and $b(\mathbf{v}, q) = (\mathbf{v}, \nabla q)$. The L^2 inner product is denoted by $(\cdot, \cdot)_{0,D}$ and when $D = \Omega$, the subscripts $0, D$ are dropped. It is necessary to adopt the bilinear form $a_h(\cdot, \cdot)$ instead of the classical Dirichlet integral bilinear form $a_1(\mathbf{u}, \mathbf{v}) = (\nabla \mathbf{u}, \nabla \mathbf{v})$ and the curl-div bilinear form $a_2(\mathbf{u}, \mathbf{v}) = (\mathbf{curl}\,\mathbf{u}, \mathbf{curl}\,\mathbf{v}) + (\text{div}\,\mathbf{u}, \text{div}\,\mathbf{v})$. Both $a_1(\cdot, \cdot)$ and $a_2(\cdot, \cdot)$ always lead to wrong convergent approximations for the singular velocity solution. The velocity solution $\mathbf{u} \in \mathbf{U}$, but $\mathbf{U} \subset (H^r(\Omega))^2$ for some $0 < r < 1$, i.e., $\mathbf{U} \not\subset (H^1(\Omega))^2$ unless Ω is smooth enough. The H^1-conforming element cannot be dense in $\mathbf{U}$ with respect to $a_1(\cdot, \cdot)$ or $a_2(\cdot, \cdot)$. But, the density holds for $a_h(\cdot, \cdot)$, cf. [4].

3   Kernel Coercivity, Inf-Sup Stability, Convergence and Open Problem

The bilinear form $a_h(\cdot, \cdot)$ already induces a norm on $\mathbf{V}_h$, and in particular, from $a_h(\mathbf{v}_h, \mathbf{v}_h) = 0$ we have $\mathbf{curl}\,\mathbf{v}_h = \mathbf{0}, \text{div}\,\mathbf{v}_h = 0, \mathbf{v}_h \times \mathbf{n} = \mathbf{0}$, and as a result, $\mathbf{v}_h \equiv \mathbf{0}$. From [8], we have the kernel coercivity with respect to the L^2-norm:

$$a_h(\mathbf{v}_h, \mathbf{v}_h) \geq C \|\mathbf{v}_h\|_0^2 \quad \forall \mathbf{v}_h \in \mathbf{V}_h. \qquad (7)$$

On the other hand, only for the Taylor-Hood element $\mathbf{V}_h := \mathbf{U}_h^{\text{T-H}}$, the inf-sup stability was proven in [8], for some constant $\mu > 0$ independent of h,

$$\sup_{\mathbf{0} \neq \mathbf{v}_h \in \mathbf{V}_h} \frac{b(\mathbf{v}_h, q_h)}{\|\mathbf{v}_h\|_1} \geq \mu \|q_h\|_0 \quad \forall q_h \in Q_h. \qquad (8)$$

Here $\| \cdot \|_1$ is the norm of the Hilbert space $H^1(\Omega)$, i.e., $\|v\|_1^2 = \|v\|_0^2 + \|\nabla v\|_0^2$, and $\|\cdot\|_0$ denotes the L^2 norm. Note that the inf-sup stability holds with respect to $\| \cdot \|_1$, and of course it holds with respect to $\|\| \cdot \|\|_h^2 := \| \cdot \|_0^2 + \| \cdot \|_{a_h}^2$, with $\| \cdot \|_{a_h}^2 := a_h(\cdot, \cdot)$. Regarding the Mini element, following the local argument in [3,7], it is not difficult to show the weaker inf-sup stability: where $|q|_{1,h}^2 := \sum_{T \in \mathscr{T}_h} h_T^2 \|\nabla q\|_{0,T}^2$,

$$\sup_{\mathbf{0} \neq \mathbf{v}_h \in \mathbf{U}_h^{\text{Mini}}} \frac{b(\mathbf{v}_h, q_h)}{\|\mathbf{v}_h\|_1} \geq C |q_h|_{1,h} \quad \forall q_h \in Q_h. \qquad (9)$$

In order to obtain (8) following the theory in [8], we enrich the Mini element in the following way: for every elemental side $F \subset \partial T$ locating on Γ_2, we add one degree of freedom in its midpoint. The base function can be chosen as the midpoint shape function of the quadratic element $\mathscr{P}_2(T)$. Denote this modified Mini element by $\mathbf{U}_h^{\text{Mini},\Gamma_2}$. Below, we call $\mathbf{U}_h^{\text{Mini},\Gamma_2}$ the Enriched Mini element. Then, from the nontrivial argument in [8], we conclude that (8) holds for $\mathbf{V}_h := \mathbf{U}_h^{\text{Mini},\Gamma_2}$.

Theorem 1. *The Enriched Mini element satisfies the inf-sup stability* (8).

However, the theory in [8] does not cover the Mini element, wherein whether the inf-sup stability (8) holds or not for the Mini element is stated as an open problem. Below, we give an approach to complete this problem. This approach uses the weaker inf-sup stability (9) and the regular-singular decomposition [9]. It is new, different from the local stability argument (cf., [1,3,7]) and also different from the argument in [8].

Theorem 2. *The Mini element satisfies the inf-sup stability* (8).

Proof.   From [9], on a Lipschitz polygon Ω, for $\mathbf{w} \in H_0(\mathrm{div}\,;\Omega) \cap H(\mathbf{curl}\,;\Omega)$, we have the regular-singular decomposition $\mathbf{w} = \mathbf{w}^{\mathrm{reg}} + \nabla p^{\mathrm{sing}}$, where $\mathbf{w}^{\mathrm{reg}} \in H_0(\mathrm{div}\,;\Omega) \cap (H^1(\Omega))^2$, $p^{\mathrm{sing}} \in H^1(\Omega)/\mathbb{R}$. Moreover, $\|\mathbf{w}^{\mathrm{reg}}\|_1 \le C(\|\mathbf{curl}\,\mathbf{w}\|_0 + \|\mathrm{div}\,\mathbf{w}\|_0)$. Put $\mathbf{v} := (w_2, -w_1)$ and $\mathbf{v}^{\mathrm{reg}} := (w_2^{\mathrm{reg}}, -w_1^{\mathrm{reg}})$. Then $\mathbf{v}, \mathbf{v}^{\mathrm{reg}} \in H_0(\mathbf{curl}\,;\Omega) \cap H(\mathrm{div}\,;\Omega)$, $\mathbf{v} = \mathbf{v}^{\mathrm{reg}} + \mathbf{curl}\,p^{\mathrm{sing}}$ and $\|\mathbf{v}^{\mathrm{reg}}\|_1 \le C(\|\mathbf{curl}\,\mathbf{v}\|_0 + \|\mathrm{div}\,\mathbf{v}\|_0)$. Here consider the case $\Gamma_2 := \partial\Omega$ only. For any given $q_h \in Q_h$, introduce the problem: Find $\theta \in H_0^1(\Omega)$ such that $-\Delta\theta = q_h$ in Ω and $\theta = 0$ on $\partial\Omega$. Then, $\mathbf{v} := \nabla\theta$. We have $\mathbf{v} \in H_0(\mathbf{curl}\,;\Omega) \cap H(\mathrm{div}\,;\Omega)$, $-\mathrm{div}\,\mathbf{v} = q_h$, and we have the regular-singular decomposition $\mathbf{v} = \mathbf{v}^{\mathrm{reg}} + \mathbf{curl}\,p^{\mathrm{sing}}$. Since $\mathbf{v}^{\mathrm{reg}} \in (H^1(\Omega))^2$, from [3], it is not difficult to find $\mathbf{v}_h^{\mathrm{reg}} \in \mathbf{V}_h := \mathbf{U}_h^{\mathrm{Mini}}$ denoting the finite element interpolation such that

$$\left(\sum_{T \in \mathscr{T}_h} h_T^{-2} \|\mathbf{v}_h^{\mathrm{reg}} - \mathbf{v}^{\mathrm{reg}}\|_{0,T}^2 \right)^{\frac{1}{2}} + \|\mathbf{v}_h^{\mathrm{reg}}\|_1 \le C\|\mathbf{v}^{\mathrm{reg}}\|_1, \tag{10}$$

where $\|\mathbf{v}^{\mathrm{reg}}\|_1 \le C(\|\mathbf{curl}\,\mathbf{v}\|_0 + \|\mathrm{div}\,\mathbf{v}\|_0) = C\|\mathrm{div}\,\mathbf{v}\|_0 = C\|q_h\|_0$. Now,

$$\sup_{\mathbf{0} \ne \mathbf{v}_h \in \mathbf{V}_h} \frac{b(\mathbf{v}_h, q_h)}{\|\mathbf{v}_h\|_1} \ge \frac{b(\mathbf{v}_h^{\mathrm{reg}}, q_h)}{\|\mathbf{v}_h^{\mathrm{reg}}\|_1} = \frac{b(\mathbf{v}^{\mathrm{reg}}, q_h)}{\|\mathbf{v}_h^{\mathrm{reg}}\|_1} + \frac{b(\mathbf{v}_h^{\mathrm{reg}} - \mathbf{v}^{\mathrm{reg}}, q_h)}{\|\mathbf{v}_h^{\mathrm{reg}}\|_1},$$

where, since $(\mathbf{curl}\,p^{\mathrm{sing}}, \nabla q_h) = 0$, there exists a constant $C_1 > 0$ such that

$$\frac{b(\mathbf{v}^{\mathrm{reg}}, q_h)}{\|\mathbf{v}_h^{\mathrm{reg}}\|_1} = \frac{b(\mathbf{v}, q_h)}{\|\mathbf{v}_h^{\mathrm{reg}}\|_1} = \frac{\|q_h\|_0^2}{\|\mathbf{v}_h^{\mathrm{reg}}\|_1} \ge C\frac{\|q_h\|_0^2}{\|\mathbf{v}^{\mathrm{reg}}\|_1} \ge C_1\|q_h\|_0,$$

and from (10), there exists $C_2 > 0$ such that $\dfrac{b(\mathbf{v}_h^{\mathrm{reg}} - \mathbf{v}^{\mathrm{reg}}, q_h)}{\|\mathbf{v}_h^{\mathrm{reg}}\|_1} \ge -C_2|q_h|_{1,h}$.

Hence, it follows that $\sup_{\mathbf{0} \ne \mathbf{v}_h \in \mathbf{V}_h} \dfrac{b(\mathbf{v}_h, q_h)}{\|\mathbf{v}_h\|_1} \ge C_1\|q_h\|_0 - C_2|q_h|_{1,h}$. Combing the weaker inf-sup stability (9), we obtain the conclusion.

The above approach is general, covering any pair $\mathbf{V}_h$ and Q_h, so long as the weaker inf-sup stability (9) holds. However, this approach seems not be applicable to three-dimensional problems, because it relies on the relation between the ∇ operator and the $\mathbf{curl}$ operator in two dimensions while such relation does not hold any longer in three dimensions.

Open Problem. The kernel coercivity (7) and the inf-sup stability (8) are not sufficient to guarantee the correct convergence if the exact velocity solution is singular. The numerical results show that both the Mini element and the Enriched Mini element generate wrong approximations. The Taylor-Hood element yields a correctly convergent solution for the singular velocity solution. The challenging issue is how to prove the convergence.

4 Numerical Results

We report the numerical results. The main purpose is to investigate the stability, the convergence and the error bound for the smooth and singular velocity solution, we always set the exact pressure $p := 0$; meanwhile, we only consider $\Gamma_2 = \partial\Omega$, i.e., $\mathbf{n} \times \mathbf{u} = \mathbf{0}$, $p = 0$ on $\partial\Omega$. In engineering applications, the pressure is very often continuous. Whenever Ω is nonsmooth with reentrant corners and edges, under the above boundary conditions, the velocity solution is usually singular, i.e., lying outside H^1. Choosing the L-shaped domain $\Omega = (-1,1)^2 \setminus ([0,1] \times [-1,0])$ and using the uniform mesh of triangles, we numerically study the following issues:

- The Mini element and the Enriched Mini element. (1) Correct and optimal convergence for smooth velocity solution. (2) Wrong convergence for singular velocity solution. (3) Inf-sup stability holds.
- The Taylor-Hood element. (1) Correct but suboptimal convergence for smooth velocity solution. (2) Correct convergence with the optimal rate the same as the regularity of the singular velocity solution.

Example 1: Smooth Velocity Solution
Let the exact smooth velocity solution $\mathbf{u}$ be $u_1(x,y) = \sin(\pi y)\cos(\pi x)$, $u_2(x,y) = -\sin(\pi x)\cos(\pi y)$. The numerical results are reported in Table 1 and Table 2. The Mini element and the Enriched Mini element give the optimal convergence $O(h^2)$ for both the velocity and the pressure. The Taylor-Hood element gives the suboptimal convergence $O(h^2)$ for the velocity; the pressure shows super-convergence $O(h^3)$.

Table 1. Errors of velocity $\|\mathbf{u} - \mathbf{u}_h\|_0$

h	Mini element		Enriched Mini element		Taylor-Hood element	
	Error	Order	Error	Order	Error	Order
1/4	$9.37e-02$	–	$9.23e-02$	–	$1.01e-02$	–
1/8	$2.03e-02$	2.21	$2.03e-02$	2.18	$1.40e-03$	2.85
1/16	$4.86e-03$	2.06	$4.88e-03$	2.06	$2.03e-04$	2.79
1/32	$1.20e-03$	2.02	$1.20e-03$	2.02	$3.51e-05$	2.53
1/64	$3.00e-04$	2.00	$3.00e-04$	2.01	$7.46e-06$	2.23

Table 2. Errors of pressure $||p - p_h||_0$

h	Mini element		Enriched Mini element		Taylor-Hood element	
	Error	Order	Error	Order	Error	Order
1/4	4.97e − 02	−	5.23e − 02	−	2.07e − 04	−
1/8	2.40e − 02	1.05	2.47e − 02	1.08	6.42e − 05	1.69
1/16	7.09e − 03	1.76	7.18e − 03	1.78	8.87e − 06	2.86
1/32	1.86e − 03	1.93	1.87e − 03	1.94	9.91e − 07	3.16
1/64	4.71e − 04	1.98	4.73e − 04	1.98	9.97e − 08	3.31

Example 2: Singular Velocity

Let the exact singular velocity $\mathbf{u}$ be $u_1(x,y) = -\frac{2}{3}\rho^{-1/3} \cdot \sin\left(\frac{\theta}{3}\right)$, $u_2(x,y) = \frac{2}{3}\rho^{-1/3} \cdot \cos\left(\frac{\theta}{3}\right)$, where (ρ, θ) stands for the polar coordinates originating at the origin, with $\rho = \sqrt{x^2 + y^2}$, $\tan\theta = y/x$. The regularity of $\mathbf{u}$ is $2/3 - \epsilon$ for any small $\epsilon > 0$. The numerical results are reported in Table 3 and Table 4. Both the Mini element and the Enriched Mini element give the wrong approximations for both the velocity and the pressure. The Taylor-Hood element gives a correctly convergent approximation, with a correct convergence rate of about $O(2/3)$. The pressure also converges with super-convergence $O(h)$.

Table 3. Errors of velocity $||\mathbf{u} - \mathbf{u}_h||_0$

h	Mini element		Enriched Mini element		Taylor-Hood element	
	Error	Order	Error	Order	Error	Order
1/4	3.29e − 01	−	5.63e − 01	−	2.33e − 01	−
1/8	3.31e − 01	−0.01	5.91e − 01	−0.07	1.50e − 01	0.64
1/16	3.87e − 01	−0.23	6.44e − 01	−0.12	9.12e − 02	0.71
1/32	4.71e − 01	−0.28	7.01e − 01	−0.12	5.31e − 02	0.78
1/64	5.61e − 01	−0.25	7.48e − 01	−0.09	3.07e − 02	0.79

Table 4. Errors of pressure $||p - p_h||_0$

h	Mini element		Enriched Mini element		Taylor-Hood element	
	Error	Order	Error	Order	Error	Order
1/4	3.87e − 01	−	4.97e − 01	−	1.28e − 01	−
1/8	4.31e − 01	−0.15	5.41e − 01	−0.12	9.09e − 02	0.49
1/16	4.81e − 01	−0.16	5.73e − 01	−0.08	5.65e − 02	0.68
1/32	5.23e − 01	−0.12	5.90e − 01	−0.04	3.09e − 02	0.87
1/64	5.53e − 01	−0.08	5.96e − 01	−0.02	1.54e − 02	1.01

Example 3: More Singular Velocity and Singular Data

Let a more singular velocity $\mathbf{u}$ be given with $u_1(x,y) = -\frac{1}{2}\rho^{-1/2} \cdot \sin\left(\frac{\theta}{2}\right)(x + 1)(y + 1)$ and $u_2(x,y) = \frac{1}{2}\rho^{-1/2} \cdot \cos\left(\frac{\theta}{2}\right)(x + 1)(y + 1)$. The regularity of $\mathbf{u}$ is $1/2 - \epsilon$ for any small $\epsilon > 0$. The right-hand sides $\mathbf{f}$, $g := \operatorname{div}\mathbf{u}$ and the boundary data $\chi := \mathbf{t} \cdot \mathbf{u}$ are all much more singular: they are not L^2 functions; they belong to some negative fractional order Sobolev spaces. The numerical results are reported in Table 5 and Table 6. Likewise, both the Mini element and the Enriched Mini element do not converge, while the Taylor-Hood still gives a convergence with a rate slightly higher than the theoretical rate of about $1/2$.

Table 5. Errors of velocity $\|\mathbf{u} - \mathbf{u}_h\|_0$

h	Mini element		Enriched Mini element		Taylor-Hood element	
	Error	Order	Error	Order	Error	Order
1/4	4.68e − 01	−	8.43e − 01	−	3.61e − 01	−
1/8	4.96e − 01	−0.08	9.59e − 01	−0.19	2.62e − 01	0.47
1/16	6.29e − 01	−0.34	1.15e + 00	−0.26	1.82e − 01	0.53
1/32	8.50e − 01	−0.44	1.39e + 00	−0.28	1.21e − 01	0.58
1/64	1.13e + 00	−0.41	1.67e + 00	−0.26	8.09e − 02	0.58

Table 6. Errors of pressure $\|p - p_h\|_0$

h	Mini element		Enriched Mini element		Taylor-Hood element	
	Error	Order	Error	Order	Error	Order
1/4	4.91e − 01	−	6.53e − 01	−	1.85e − 01	−
1/8	6.30e − 01	−0.36	8.28e − 01	−0.34	1.48e − 01	0.32
1/16	7.97e − 01	−0.34	1.01e + 00	−0.28	1.04e − 01	0.51
1/32	9.81e − 01	−0.30	1.19e + 00	−0.23	6.39e − 02	0.70
1/64	1.18e + 00	−0.26	1.37e + 00	−0.21	3.57e − 02	0.84

Computation on Inf-Sup Constant: (8)

We report the inf-sup constants in (8) for the Mini element, the Enriched Mini element, and the Taylor-Hood element. From Table 7 and Table 8, the inf-sup constants are bounded from below as h tends to zero.

Table 7. Inf-sup constant with norm $|| \cdot ||_1$

h	Mini element	Enriched Mini element	Taylor-Hood element
	μ	μ	μ
1/4	0.5107264	0.5151383	0.9276922
1/8	0.4334440	0.4349372	0.9132119
1/16	0.4011601	0.4014494	0.9039361
1/32	0.3910563	0.3911005	0.8980655
1/64	0.3882729	0.3882791	0.8943652

Table 8. Inf-sup constant with norm $||| \cdot |||_h$

h	Mini element	Enriched Mini element	Taylor-Hood element
	μ	μ	μ
1/4	0.8319936	0.8364393	2.0150314
1/8	0.7974487	0.7987991	2.6165266
1/16	0.7383949	0.7391324	3.1131941
1/32	0.6924892	0.6927582	3.3252386
1/64	0.6637092	0.6637771	3.3637365

Conclusion. The Stokes inf-sup stable elements such as Mini elements and Taylor-Hood elements under the no-slip velocity boundary condition can still be inf-sup stable under the pressure Dirichlet boundary condition, as proven by a new theory developed in this paper and confirmed by the numerical results provided. However, numerical examples of singular and non-H^1 velocity studied have shown the wrong convergence of the Mini-type elements. They have also shown the correct convergence (albeit suboptimal) of the Taylor-Hood elements; but how to prove the convergence of the Taylor-Hood elements for singular velocity is open.

Acknowledgements. The authors were partially supported by the National Natural Science Foundation of China (No. 12371371,12261160361,11971366).

References

1. Arnold, D.N., Brezzi, F., Fortin, M.: A stable finite element for the Stokes equations. Calcolo **21**, 337–344 (1984)
2. Taylor, C., Hood, P.: A numerical solution of the Navier-Stokes equations using the finite element technique. Comput. Fluids **1**, 73–100 (1973)
3. Girault, V., Raviart, P.-A.: Finite Element Methods for Navier-Stokes Equations, Theory and Algorithms. Springer, Heidelberg (1986). https://doi.org/10.1007/978-3-642-61623-5

4. Ern, A., Guermond, J.L.: Mollification in strongly Lipschitz domains with application to continuous and discrete De Rham complex. Comput. Methods Appl. Math. **16**, 51–75 (2015)
5. Amrouche, C., Bernardi, C., Dauge, M., Girault, V.: Vector potentials in three-dimensional non-smooth domains. Math. Methods Appl. Sci. **21**, 823–864 (1998)
6. Amrouche, C., Seloula, N.H.: L^p-theory for vector potentials and Sobolev's inequalities for vector fields: application to the Stokes equations with pressure boundary conditions. Math. Models Methods Appl. Sci. **23**, 37–92 (2013)
7. Brezzi, F., Fortin, M.: Mixed and Hybrid Finite Element Methods. Springer, New York (1991). https://doi.org/10.1007/978-1-4612-3172-1
8. Du, Z.J., Duan, H.Y., Liu, W.: Staggered Taylor-Hood and Fortin elements for Stokes equations of pressure boundary conditions in Lipschitz domain. Numer. Methods PDE **36**, 185–208 (2020)
9. Bonnet-Ben Dhia, A.S., Hazard, C., Lohrengel, S.: A singular field method for the solution of Maxwell's equations in polyhedral domains. SIAM J. Appl. Math. **59**, 2028–2044 (1999)
10. Feng, X.B., Ge, Z.H., Li, Y.K.: Analysis of a multiphysics finite element method for a poroelasticity model. IMA J. Numer. Anal. **38**, 330–359 (2018)
11. Chen, L., Wu, Y., Zhong, L., Zhou, J.: MultiGrid preconditioners for mixed finite element methods of the vector Laplacian. J. Sci. Comput. **77**, 101–128 (2018)
12. Mardal, K.A., Winther, R.: Preconditioning discretizations of systems of partial differential equations. Numer. Linear Algebra Appl. **18**, 1–40 (2011)

Unstructured Flux-Limiter Convective Schemes for Simulation of Transport Phenomena in Two-Phase Flows

Néstor Balcázar-Arciniega[✉][iD], Joaquim Rigola[iD], and Assensi Oliva[iD]

Heat and Mass Transfer Technological Center (CTTC), Universitat Politècnica de Catalunya-BarcelonaTech (UPC), Colom 11, 08222 Terrassa (Barcelona), Spain
nestor.balcazar@upc.edu, nestorbalcazar@yahoo.es

Abstract. Unstructured flux-limiters convective schemes designed in the framework of the unstructured conservative level-set (UCLS) method, are assessed for transport phenomena in two-phase flows. Transport equations are discretized by the finite volume method on 3D collocated unstructured meshes. The central difference scheme discretizes the diffusive term. Gradients are evaluated by the weighted least-squares method. The fractional-step projection method solves the pressure-velocity coupling in momentum equations. Numerical findings about the effect of flux limiters on the control of numerical diffusion and improvement of numerical stability in DNS of two-phase flows are reported.

Keywords: Unstructured Flux-Limiters · Finite-Volume Method · Unstructured Meshes · Unstructured Conservative Level-Set Method · Mass transfer · Direct Numerical Simulation · High-Performance Computing

1 Introduction

Transport phenomena in two-phase flows are prevalent in natural and industrial processes, playing crucial roles in various engineering devices, from steam generators to cooling towers in thermal power plants and the so-called unit operations of chemical engineering. Bubbles and droplets are essential in separation processes such as distillation or promoting chemical reactions in industrial devices. Beyond its scientific significance, a profound comprehension of the intricate interplay between fluid mechanics and transport phenomena in multiphase flows is indispensable for designing and optimising engineering systems involving multiple phases.

The principal author, N. Balcázar-Arciniega, as Serra-Hunter Lecturer (UPC-LE8027), acknowledges the financial support provided by the Catalan Government through this program. Simulations were conducted using computing resources allocated by the Spanish Supercomputing Network (RES) under the grants IM-2023-2-0009, IM-2023-1-0003 on the supercomputer MareNostrum IV in Barcelona, Spain. The financial support of the MINECO, Spain (grant PID2020-115837RB-100), is acknowledged.

L. Franco et al. (Eds.): ICCS 2024, LNCS 14838, pp. 20–32, 2024.
https://doi.org/10.1007/978-3-031-63783-4_3

Experimental exploration of two-phase flows, particularly bubbly flows, is constrained by optical access. Analytical methods involve significant simplifications of physical models. In contrast, computational methods, such as Direct Numerical Simulation (DNS) of two-phase flows [39,49], have been empowered with the advent of supercomputing capabilities. Several approaches, including front-tracking (FT) [48,50], level-set (LS) [27,37,44], conservative level-set (CLS) [7,12,36], Volume of Fluid (VoF) [29], and coupled VoF-LS [8,42,43], have been proposed for DNS of two-phase flow. In further developments these methods have been applied to mass transfer and heat transfer in single bubbles [17,21,23], bubble swarms [1,12–14,33,41], variable surface tension [10,14], and liquid-vapor phase change [16].

Numerical challenges for the DNS of two-phase flows include the so-called numerical diffusion [38], numerical oscillations around discontinuities, the computational cost of solving the Poisson equation in two-phase flow with high-density ratio, numerical coalescence in bubble swarms, and accurate computation of surface tension forces. These issues have been effectively tackled within the framework of the Unstructured Conservative Level-Set (UCLS) method proposed by Balcazar et al. [2,4,7,9,10,13–16]. The UCLS method employs a carefully tuned level-set function to set the interface thickness. Additionally, the least-squares method accurately computes normal vectors perpendicular to the interface, leading to a precise calculation of surface tension forces. Moreover, physical properties are smoothed across the interface using the conservative level-set function to prevent numerical oscillations. The so-called numerical coalescence of bubbles is avoided by the multiple marker approach [9,13].

This research focuses on assessing unstructured flux-limiter convection schemes proposed by Balcazar et al. [7,13] in the framework of the UCLS method. An appropriate selection of flux-limiters in DNS of interfacial transport phenomena leads to minimising the so-called numerical diffusion and preventing numerical oscillations across the fluid interface. This assessment is particularly relevant in bubbly flows with high physical properties ratios and predicting mass transfer coefficients (Sherwood number) for interfacial transport processes. It is worth mentioning that flux-limiter schemes have been designed initially for regular and Cartesian meshes [34,35,46], whereas some further efforts have been reported on unstructured meshes [7,12,22,25,30,52]. Nevertheless, the impact of unstructured flux limiters on the simulation of transport phenomena in two-phase flows, e.g., interfacial heat transfer and mass transfer, needs additional research. This work contributes to filling this gap, as well as to the development of numerical models for transport phenomena in two-phase flows.

This paper is organised as follows: The mathematical formulation and numerical methods are introduced in Sect. 2. Section 3 reports numerical experiments on the impact of flux-limiter schemes. Finally, conclusions and future work are outlined in Sect. 4.

2   Mathematical Formulation and Numerical Methods

2.1   Transport Equations

The Navier-Stokes equations for the dispersed phase (Ω_d) and continuous phase (Ω_c) are solved in the framework of the one-fluid formulation [39,49] and the multi-marker UCLS approach [12,14], as follows:

$$\frac{\partial}{\partial t}(\rho\mathbf{v}) + \nabla\cdot(\rho\mathbf{vv}) = -\nabla p + \nabla\cdot\mu(\nabla\mathbf{v}) + \nabla\cdot\mu(\nabla\mathbf{v})^T + (\rho-\rho_0)\mathbf{g} + \mathbf{f}_\sigma, \quad (1)$$

$$\nabla\cdot\mathbf{v} = 0, \tag{2}$$

where $\mathbf{v}$ is the fluid velocity, p is the pressure, ρ is the fluid density, μ is the dynamic viscosity, $\mathbf{g}$ refers to the gravitational acceleration, $\mathbf{f}_\sigma$ is the surface tension force per unit volume concentrated on the interface. Subscripts d and c refer to the dispersed and continuous phases, respectively. Density and viscosity are constant at each fluid phase, with a jump discontinuity across the interface: $\rho = \rho_c H_c + \rho_d H_d$, $\mu = \mu_c H_c + \mu_d H_d$. Here, H_c is the Heaviside step function, which is one in Ω_c and zero in Ω_d. Furthermore, $H_d = 1 - H_c$. The force $-\rho_0\mathbf{g}$ (Eq. (1)) [3,4,12], is activated if periodic boundary conditions are applied along the y-axis (parallel to $\mathbf{g}$), with $\rho_0 = V_\Omega^{-1}\int_\Omega(\rho_c H_c + \rho_d H_d)\,dV$. Otherwise, $\rho_0 = 0$.

Interface capturing is performed by the Unstructured Conservative Level-Set (UCLS) method proposed by Balcazar et al. [7,12], in the framework of 3D unstructured meshes and the finite-volume method. On the other hand, the numerical coalescence of fluid particles is avoided by the multi-marker UCLS approach [2–4,10,12,14]. In this context, a regularized signed distance function [7,10,12] is used for each marker, $\phi_i = \frac{1}{2}\left(\tanh\left(\frac{d_i}{2\varepsilon}\right)+1\right)$, where d_i is a signed distance function [37,45], and ε sets the thickness of the interface profile [4,7,9,10,12]. The UCLS advection equation is solved in conservative form, for each marker:

$$\frac{\partial\phi_i}{\partial t} + \nabla\cdot\phi_i\mathbf{v} = 0, \quad i = \{1, 2, ..., N_m - 1, N_m\}. \tag{3}$$

Here, N_m is the number of UCLS markers, which equals the number of fluid particles. To maintain the level-set profile, the following unstructured re-initialization equation [7,12] is solved:

$$\frac{\partial\phi_i}{\partial\tau} + \nabla\cdot\phi_i(1-\phi_i)\mathbf{n}_i^0 = \nabla\cdot\varepsilon\nabla\phi_i, \quad i = \{1, 2, ..., N_m - 1, N_m\}, \tag{4}$$

which is computed for the pseudo-time τ up to the steady state. Here, $\mathbf{n}_i^0$ denotes the interface normal unit vector evaluated at $\tau = 0$. At Ω_P, $\varepsilon_P = 0.5(h_P)^\alpha$, $\alpha = [0.9, 1]$ unless otherwise stated, h_P is the characteristic local grid size [7,10,12]. Interface normal vectors ($\mathbf{n}_i$) and curvatures (κ_i) are calculated as follows [4,7,12]: $\mathbf{n}_i = \nabla\phi_i\,\|\nabla\phi_i\|^{-1}$, $\kappa_i = -\nabla\cdot\mathbf{n}_i$.

Computation of the surface tension force ($\mathbf{f}_\sigma$, Eq. (1)) is performed in the framework of the Continuous Surface Force (CSF) model [18], as extended to the multi-marker UCLS approach [3, 4, 10, 12, 14]:

$$\mathbf{f}_\sigma = \sum_{i=1}^{N_m}(\mathbf{f}_{\sigma,i}^{(n)} + \mathbf{f}_{\sigma,i}^{(t)}). \tag{5}$$

Here, the interface tangential component $\mathbf{f}_{\sigma,i}^{(t)}$, is the so-called Marangoni force [24], defined as $\mathbf{f}_{\sigma,i}^{(t)} = \nabla_{\Gamma_i}\sigma\delta_{\Gamma,i}^s$ [3, 7, 12, 14], σ is the surface tension coefficient. If σ is constant, as in the present research, then $\mathbf{f}_{\sigma,i}^{(t)} = \mathbf{0}$. On the other hand, the normal component of the surface tension force, perpendicular to the interface (Γ_i), is calculated as $\mathbf{f}_{\sigma,i}^{(n)} = \sigma\kappa_i\nabla\phi_i$ [7, 9, 12].

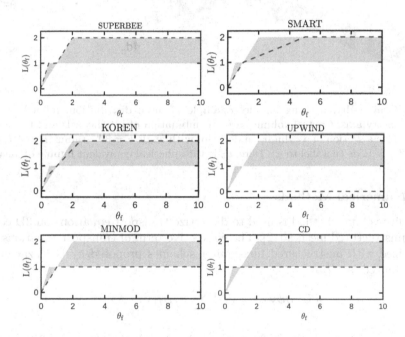

Fig. 1. Flux-limiters (dashed lines) $\psi(\theta_f)$ versus monitor variable θ_f, on the Sweby's diagram (shaded region) [46] for second-order TVD flux-limiter region.

For external mass transfer [20], the concentration of chemical species (C) evolves according to the following transport equation:

$$\frac{\partial C}{\partial t} + \nabla \cdot (\mathbf{v}C) = \nabla \cdot (\mathcal{D}\nabla C), \tag{6}$$

which is solved in Ω_c. Here, $\mathcal{D} = \mathcal{D}_c$ is the diffusivity. The concentration of chemical species inside the bubbles is kept constant, whereas the concentration at the interface cells is calculated according to the unstructured interpolation method proposed by Balcazar-Arciniega et al. [12].

Table 1. Convective term of transport equations. Here, v_i denotes the Cartesian components of $\mathbf{v}$. Correspondence of transport equations and parameters β and ψ in Eq. (7).

Transport equation	Convective term	β	ψ
Eq. (1)	$\nabla \cdot \rho v_i \mathbf{v}$	ρ	v_i
Eq. (3)	$\nabla \cdot \phi \mathbf{v}$	1	ϕ
Eq. (6)	$\nabla \cdot (C \mathbf{v})$	1	C

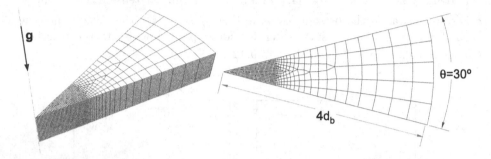

Fig. 2. Gravity-driven rising bubble. Example of mesh distribution, refined around the symmetry axis. Control volumes are a combination of hexahedrals and triangular prisms. Ω is a section of a cylindrical domain, with angle $\theta = 30°$, radius $R = 4\,d_b$ and height $L_y = 12\,d_b$ (parallel to $\mathbf{g}$). Here, d_b is the spherical equivalent bubble diameter.

2.2   Numerical Methods

The finite-volume method is used to discretize transport equations on 3D collocated unstructured meshes [12]. The convective term of transport equations are discretized with unstructured flux-limiter schemes proposed by Balcazar et al. [7,12]. Indeed, at Ω_P:

$$(\nabla \cdot \beta\psi\mathbf{v})_P = V_P^{-1} \sum_f \beta_f \psi_f (\mathbf{v}_f \cdot \mathbf{A}_f), \tag{7}$$

where $\mathbf{A}_f = A_f \mathbf{e}_f$ is the area vector, subindex f denotes the cell-faces, V_P denotes the volume of Ω_P, and $\mathbf{e}_f$ is a unit-vector pointing outside Ω_P. Moreover,

$$\psi_f = \psi_{C_p} + \frac{1}{2}\mathrm{L}(\theta_f)(\psi_{D_p} - \psi_{C_p}), \tag{8}$$

where $\theta_f = (\psi_{C_p} - \psi_{U_p})/(\psi_{D_p} - \psi_{C_p})$, and $L(\theta_f)$ is the flux limiter function. Furthermore, subindex D_p is the downwind point, subindex C_p is the upwind point, subindex U_p is the far-upwind point, according to the stencil proposed for the single marker and multi-marker UCLS method [7,12]. The correspondence of convective term in transport equations and Eq. (7) is outlined in Table 1. Flux-limiter functions used in this research are summarized as follows: [26,28, 32,35,47]:

$$L(\theta_f) \equiv \begin{cases} \max\{0, \min\{2\theta_f, 1\}, \min\{2, \theta_f\}\} & \text{SUPERBEE,} \\ \max\{0, \min\{2\theta_f, (2/3)\,\theta_f + (1/3), 2\}\} & \text{KOREN,} \\ \max\{0, \min\{4\theta_f, 0.75 + 0.25\theta_f, 2\}\} & \text{SMART,} \\ (\theta_f + |\theta_f|)/(1+|\theta_f|) & \text{VANLEER,} \\ \mathrm{minmod}\{1, \theta_f\} & \text{MINMOD,} \\ 0 & \text{UPWIND,} \\ 1 & \text{CD.} \end{cases} \tag{9}$$

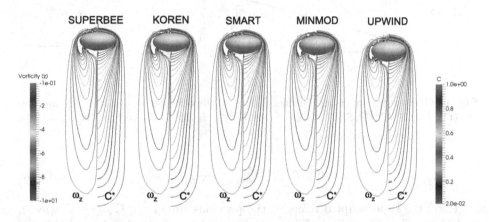

Fig. 3. Gravity-driven rising bubble. Effect of flux-limiter convective schemes on the vorticity $\omega_z = \mathbf{e_z} \cdot (\nabla \times \mathbf{v})$, and normalized concentration $\mathrm{C}^* = C_c/C_{\Gamma,c}$ fields, at $t^* = 5$. Here, $C_{\Gamma,c}$ is the concentration of chemical species at the interface. Eo = 3.125, Mo = 10^{-6}, Sc = 1, $\eta_\rho = \eta_\mu = 100$.

Figure 1 summarizes the so-called Sweby's diagram for second-order Total Variation Diminishing (TVD) flux-limiters [47] and some of the functions outlined in Eq. (9). First-order accurate schemes, e.g., UPWIND, suffer from numerical diffusion, whereas high-resolution schemes (SUPERBEE, KOREN, SMART, VANLEER, MINMOD) tend to minimize this numerical artefact.

Concerning the diffusive term of transport equations, a central-difference scheme [13] is used. Gradients are evaluated by the weighted least-squares method [7,12], whereas values at the cell-faces are linearly interpolated. The pressure-velocity coupling is solved with the fractional-step projection method [19,39,49]:

$$\frac{\rho_P \mathbf{v}_P^* - \rho_P^0 \mathbf{v}_P^0}{\Delta t} = \mathbf{C}_{\mathbf{v},P}^0 + \mathbf{D}_{\mathbf{v},P}^0 + (\rho_P - \rho_0)\mathbf{g} + \mathbf{f}_{\sigma,P}, \tag{10}$$

$$\left(\nabla \cdot \left(\frac{\Delta t}{\rho} \nabla p \right) \right)_P = (\nabla \cdot \mathbf{v}^*)_P, \quad \mathbf{e}_{\partial\Omega} \cdot \nabla p|_{\partial\Omega} = 0. \tag{11}$$

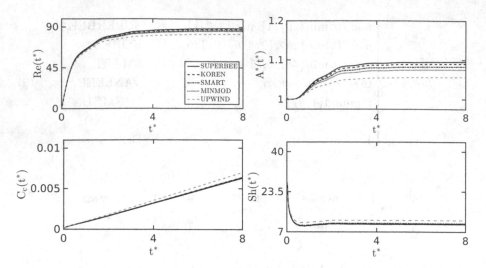

Fig. 4. Gravity-driven rising bubble. Effect of flux-limiter convective schemes on $\{\mathrm{Re}(t^*), \mathrm{C_c}(t^*), \mathrm{A}^*(t^*), \mathrm{Sh}(t^*)\}$. Eo $= 3.125$, Mo $= 10^{-6}$, Sc $= 1$, $\eta_\rho = \eta_\mu = 100$.

$$\frac{\rho_P \mathbf{v}_P - \rho_P \mathbf{v}_P^*}{\Delta t} = -(\nabla p)_P. \tag{12}$$

Here, the superscript 0 refers to the previous time-step, $\mathbf{C_v} = -\nabla \cdot (\rho \mathbf{vv})$, and $\mathbf{D_v} = \nabla \cdot \mu \nabla \mathbf{v} + \nabla \cdot \mu (\nabla \mathbf{v})^T$. The predictor velocity is $\mathbf{v}_P^*$, and the corrected velocity is represented by $\mathbf{v}_P$. A linear system arises from the finite volume approximation of Eq. (11), and it is solved using a preconditioned conjugate gradient method with a Jacobi pre-conditioner [31,51]. The boundary $\partial\Omega$ excludes regions with periodic conditions, where information from the corresponding periodic nodes is employed [4,12]. Finally, a convective velocity ($\mathbf{v}_f$ in Eq. (7)) is interpolated at cell faces to avoid pressure-velocity decoupling on collocated meshes [7,12,40]. This convective velocity is used for the computation of the volume flux ($\mathbf{v}_f \cdot \mathbf{A}_f$) in Eq. (7). For further technical details about the finite-volume methods and computational algorithms used in this work, the reader is referred to Balcazar-Arciniega et al. [12].

3   Numerical Experiments

Systematic validations, verifications, and extensions of the Unstructured Conservative Level-Set (UCLS) method have been extensively documented in our prior works. These include studies on gravity-driven rising bubbles [3,4,7], falling droplets [6], gravity-driven bubbly flows [3,9,12–14], binary droplet collision [9], collision of a droplet against an interface [9], deformation of droplets under shear stresses [8], mass transfer in bubbly flows [2,11–14], and liquid-vapour phase change [5,16]. Consequently, this research represents a further step in assessing unstructured flux-limiter convective schemes for simulating transport phenomena in two-phase flows within the UCLS method framework, as proposed by Balcazar et al. [7,9,12,14,16].

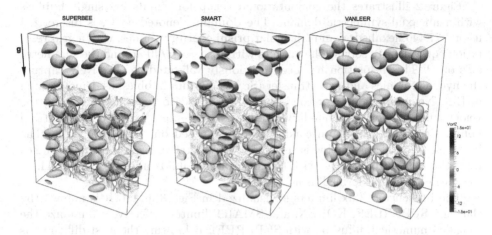

Fig. 5. Gravity-driven bubble swarm. Effect of flux-limiter convective schemes on the vorticity $\omega_z = \mathbf{e}_z \cdot (\nabla \times \mathbf{v})$. Eo = 3, Mo = 10^{-8}, $\alpha = 0.0654$, $\eta_\rho = \eta_\mu = 100$. Fully periodic domain (6 periodic cubes), $L_x = 4\,d_b$, $L_y = 4\,d_b$, $L_z = 4\,d_b$. 150^3 uniform hexahedral control volumes, equivalent to the grid size $h = d_b/37.5$. 192 CPU-cores.

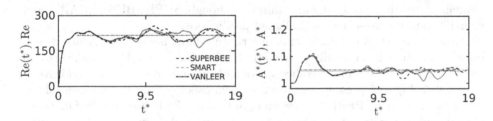

Fig. 6. Gravity-driven bubble swarm. Effect of flux-limiter convective schemes on $\{\mathrm{Re}(\mathrm{t}^*), \mathrm{A}^*(\mathrm{t}^*)\}$. Eo = 3.0, Mo = 10^{-8}, $\alpha = 0.0654$, $\eta_\rho = \eta_\mu = 100$.

Characterization of gravity-driven bubbly flow is performed by the Eötvös number Eo = $g d_b^2 (\rho_c - \rho_d)\sigma^{-1}$, Morton number Mo = $g\mu_c^4(\rho_c - \rho_d)\rho_c^{-2}\sigma^{-3}$, viscosity ratio $\eta_\mu = \mu_c/\mu_d$, density ratio $\eta_\rho = \rho_c/\rho_d$, bubble volume fraction $\alpha = V_{\Omega_d}/V_\Omega$, and Reynolds number $\mathrm{Re}_i(\mathrm{t}^*) = \rho_c \mathrm{U}_{r,i} d_b/\mu_c$, $\mathrm{Re}(\mathrm{t}^*) = N_b^{-1}\sum_{i=1}^{N_b}\mathrm{Re}_i(\mathrm{t}^*)$, $\mathrm{Re}_i = T^{-1}\int_{t_0}^{t_0+T}\mathrm{Re}_i(t)dt$, Re = $N_b^{-1}\sum_{i=1}^{N_b}\mathrm{Re}_i$. Here d_b refers to the spherical equivalent bubble diameter, $V_{\Omega_{d,d}}$ is the volume of bubbles, $\mathrm{U}_{r,i}$ is the relative velocity of the bubble respect to the velocity of Ω_c, subindex

i refers to the ith bubble, and $t^* = t\,g^{1/2}\,d_b^{-1/2}$. Additionally, for the characterization of mass transfer, the following dimensionless numbers are considered: Schmidt number (Sc $= \mu_c/(\rho_c \mathcal{D}_c)$), and Sherwood number (Sh $= k_c d_b/\mathcal{D}_c$), where k_c is the mass transfer coefficient at Ω_c [12].

Figure 2 illustrates the computational setup for simulating single bubbles within an axial-symmetric domain. The domain, denoted as Ω, is discretized using 277 440 hexahedral and triangular prism control volumes, with a grid resolution equivalent to solving the bubbles with $h_{\min} = d_b/40$ around the symmetry axis of Ω. This resolution has been previously validated to effectively capture the hydrodynamics and mass transfer (Sc $= 1$) within bubbles, as demonstrated in [12]. Neumann boundary conditions are applied to $\{\phi, C\}$. Concerning the velocity field $\mathbf{v}$, non-slip boundary conditions are applied to the top and bottom boundaries, while Neumann conditions are imposed on the lateral walls. The initial position of the bubble is set on the symmetry axis, at a distance of $2\,d_b$ from the bottom boundary, with the surrounding fluids initially at rest. These simulations were performed using 14 CPUs.

The influence of flux limiters is visualized in Figs. 3 and 4. It is noteworthy that the SUPERBEE, KOREN, and SMART limiters effectively minimize the so-called numerical diffusion, with SUPERBEE exhibiting the least diffusion as expected. In contrast, MINMOD and UPWIND schemes introduce more numerical diffusion, with UPWIND being the most diffusive. The impact of these limiters is evident in the computation of Sh(t^*), where the UPWIND scheme leads to a higher prediction of the Sherwood number due to this numerical artifact. Similarly, regarding the prediction of Re(t^*), it is observed that the UPWIND scheme underestimates the Reynolds number compared to other limiters. The computation of the normalized area A$^* = A(t)/(\pi\,d_b^2)^{-1}$ is also significantly influenced by the choice of flux limiter, although SUPERBEE, SMART and KOREN limiters present closely aligned predictions.

Now, we investigate the impact of unstructured flux limiters on gravity-driven bubble swarms within a fully periodic cube denoted as Ω. The domain is discretized using 150^3 uniform hexahedral control volumes, resulting in a grid size of $h = d/40.7$. Figure 5 presents instantaneous snapshots of the z-component of the vorticity (ω_z) for SUPERBEE, SMART, and VANLEER flux limiter functions. Dimensionless parameters are set to Eo $= 3$, Mo $= 10^{-8}$, $\alpha = 0.0654$, $\eta_\rho = 100$, and $\eta_\mu = 100$. In the initial conditions, 8 bubbles are randomly distributed in the cubic domain, while the fluids remain at rest.

Figure 6 illustrates $\{\text{Re}(t^*), A^*(t^*)\}$ (black lines) and their time-averaged values $\{\text{Re}, A^*\}$ for the bubble swarm (depicted by red lines). It is noteworthy that the predicted $\{\text{Re}, A^*\}$ values exhibit remarkable consistency across SUPERBEE, SMART and VANLEER flux limiters, emphasizing their reliability and effectiveness in capturing the hydrodynamics and interfacial area of the bubble swarm.

4   Conclusions

The application of unstructured flux limiters [7,12] has proven effective in predicting hydrodynamics and mass transfer within two-phase flows. In the context of single bubble simulations, these flux limiters were specifically applied to the convective term of the transport equations, encompassing momentum, interface advection, and concentration of chemical species, as detailed in Table 1. While all evaluated flux limiters successfully prevent numerical oscillations within the framework of the UCLS method, the numerical predictions of Reynolds number and Sherwood number underscore the superior performance of the SUPER-BEE limiter in minimizing numerical diffusion. Conversely, the UPWIND limiter exhibits the maximum numerical diffusion.

In relation to the hydrodynamics of bubble swarms, we observed consistent predictions for Re and Sh across the SUPERBEE, SMART, and VANLEER limiters. These results recommend their application for DNS of bubbly flows within the UCLS method framework. Our future efforts will focus on applying flux limiter schemes to intricate interface phenomena, such as the transport of surfactants on fluid interfaces and liquid-vapor phase change. These forthcoming studies aim to deepen our understanding and broaden the applicability of flux limiters in capturing complex physical interactions within multiphase systems.

References

1. Aboulhasanzadeh, B., Thomas, S., Taeibi-Rahni, M., Tryggvason, G.: Multiscale computations of mass transfer from buoyant bubbles. Chem. Eng. Sci. **75**, 456–467 (2012). https://doi.org/10.1016/j.ces.2012.04.005, https://linkinghub.elsevier.com/retrieve/pii/S000925091200231X
2. Balcázar, N., Antepara, O., Rigola, J., Oliva, A.: DNS of drag-force and reactive mass transfer in gravity-driven bubbly flows. In: García-Villalba, M., Kuerten, H., Salvetti, M.V. (eds.) DLES 2019. ES, vol. 27, pp. 119–125. Springer, Cham (2020). https://doi.org/10.1007/978-3-030-42822-8_16
3. Balcazar, N., Castro, J., Rigola, J., Oliva, A.: DNS of the wall effect on the motion of bubble swarms. Procedia Comput. Sci. **108**, 2008–2017 (2017). https://doi.org/10.1016/j.procs.2017.05.076, https://linkinghub.elsevier.com/retrieve/pii/S1877050917306142
4. Balcazar, N., Lehmkuhl, O., Jofre, L., Oliva, A.: Level-set simulations of buoyancy-driven motion of single and multiple bubbles. Int. J. Heat and Fluid Flow **56** (2015). https://doi.org/10.1016/j.ijheatfluidflow.2015.07.004
5. Balcazar, N., Rigola, J., Oliva, A.: Unstructured level-set method for saturated liquid-vapor phase change. In: WCCM-ECCOMAS 2020. Volume 600 - Fluid Dynamics and Transport Phenomena, pp. 1–12 (2021). https://doi.org/10.23967/wccm-eccomas.2020.352, https://www.scipedia.com/public/Balcazar_et_al_2021a
6. Balcazar, N., Castro, J., Chiva, J., Oliva, A.: DNS of falling droplets in a vertical channel. Int. J. Comput. Methods Exper. Measur. **6**(2), 398–410 (2017). https://doi.org/10.2495/CMEM-V6-N2-398-410, http://www.witpress.com/doi/journals/CMEM-V6-N2-398-410

7. Balcázar, N., Jofre, L., Lehmkuhl, O., Castro, J., Rigola, J.: A finite-volume/level-set method for simulating two-phase flows on unstructured grids. Int. J. Multiphase Flow **64**, 55–72 (2014). https://doi.org/10.1016/j.ijmultiphaseflow.2014.04.008, https://linkinghub.elsevier.com/retrieve/pii/S030193221400072X
8. Balcazar, N., Lehmkuhl, O., Jofre, L., Rigola, J., Oliva, A.: A coupled volume-of-fluid/level-set method for simulation of two-phase flows on unstructured meshes. Comput. Fluids **124**, 12–29 (2016). https://doi.org/10.1016/j.compfluid.2015.10.005
9. Balcazar, N., Lehmkuhl, O., Rigola, J., Oliva, A.: A multiple marker level-set method for simulation of deformable fluid particles. Int. J. Multiph. Flow **74**, 125–142 (2015). https://doi.org/10.1016/j.ijmultiphaseflow.2015.04.009
10. Balcázar, N., Rigola, J., Castro, J., Oliva, A.: A level-set model for thermocapillary motion of deformable fluid particles. Int. J. Heat and Fluid Flow **62**, 324–343 (2016). https://doi.org/10.1016/j.ijheatfluidflow.2016.09.015, https://linkinghub.elsevier.com/retrieve/pii/S0142727X16301266
11. Balcazar-Arciniega, N., Rigola, J., Oliva, A.: DNS of mass transfer in Bi-dispersed bubbly flows in a vertical pipe. In: ERCOFTAC Series, vol. 31, pp. 55–61 (2024). https://doi.org/10.1007/978-3-031-47028-8_9
12. Balcazar-Arciniega, N., Antepara, O., Rigola, J., Oliva, A.: A level-set model for mass transfer in bubbly flows. Int. J. Heat Mass Transf. **138**, 335–356 (2019). https://doi.org/10.1016/j.ijheatmasstransfer.2019.04.008
13. Balcazar-Arciniega, N., Rigola, J., Oliva, A.: DNS of mass transfer from bubbles rising in a vertical channel. In: Lecture Notes in Computer Science (including subseries Lecture Notes in Artificial Intelligence and Lecture Notes in Bioinformatics) 11539 LNCS, pp. 596–610 (2019). https://doi.org/10.1007/978-3-030-22747-0_45
14. Balcazar-Arciniega, N., Rigola, J., Oliva, A.: DNS of mass transfer in bi-dispersed bubble swarms. In: Computational Science – ICCS 2022. ICCS 2022. Lecture Notes in Computer Science, vol. 13353, pp. 284–296. Springer, Cham (2022). https://doi.org/10.1007/978-3-031-08760-8_24
15. Balcázar-Arciniega, N., Rigola, J., Oliva, A.: DNS of Thermocapillary Migration of a Bi-dispersed Suspension of Droplets. In: Computational Science – ICCS 2023. ICCS 2023. Lecture Notes in Computer Science (2016153612), pp. 303–317 (2023). https://doi.org/10.1007/978-3-031-36030-5_25
16. Balcazar-Arciniega, N., Rigola, J., Oliva, A.: Unstructured Conservative Level-Set (UCLS) simulations of film boiling heat transfer. In: Computational Science – ICCS 2023. ICCS 2023. Lecture Notes in Computer Science, vol. 10477, pp. 318–331 (2023). https://doi.org/10.1007/978-3-031-36030-5
17. Bothe, D., Fleckenstein, S.: A volume-of-fluid-based method for mass transfer processes at fluid particles. Chem. Eng. Sci. (2013). https://doi.org/10.1016/j.ces.2013.05.029
18. Brackbill, J.U., Kothe, D.B., Zemach, C.: A continuum method for modeling surface tension. J. Comput. Phys. **100**(2), 335–354 (1992). https://doi.org/10.1016/0021-9991(92)90240-Y
19. Chorin, A.J.: Numerical solution of the Navier-stokes equations. Math. Comput. **22**(104), 745 (1968). https://doi.org/10.2307/2004575
20. Clift, R., Grace, J., Weber, M.: Bubbles, Drops, and Particles. DOVER Publications, Inc., New York (1978)
21. Darmana, D., Deen, N.G., Kuipers, J.A.M.: Detailed 3D modeling of mass transfer processes in two-phase flows with dynamic interfaces. Chem. Eng. Technol. **29**(9), 1027–1033 (2006). https://doi.org/10.1002/ceat.200600156

22. Darwish, M.S., Moukalled, F.: TVD schemes for unstructured grids. Int. J. Heat Mass Transf. (2003). https://doi.org/10.1016/S0017-9310(02)00330-7
23. Davidson, M.R., Rudman, M.: Volume-of-fluid calculation of heat or mass transfer across deforming interfaces in two-fluid flow. Numer. Heat Trans. Part B: Fund. **41**(3-4), 291–308 (2002). https://doi.org/10.1080/104077902753541023
24. Deen, W.: Analysis of Transport Phenomena. Oxford University Press, Oxford (2011)
25. Denner, F., van Wachem, B.G.: TVD differencing on three-dimensional unstructured meshes with monotonicity-preserving correction of mesh skewness. J. Comput. Phys. **298**, 466–479 (2015). https://doi.org/10.1016/j.jcp.2015.06.008
26. Gaskell, P.H., Lau, A.K.C.: Curvature-compensated convective transport: SMART, a new boundedness- preserving transport algorithm. Int. J. Numer. Meth. Fluids **8**(6), 617–641 (1988). https://doi.org/10.1002/fld.1650080602
27. Gibou, F., Fedkiw, R., Osher, S.: A review of level-set methods and some recent applications. J. Comput. Phys. **353**, 82–109 (2018). https://doi.org/10.1016/j.jcp.2017.10.006, https://linkinghub.elsevier.com/retrieve/pii/S0021999117307441
28. Guenther, C., Syamlal, M.: The effect of numerical diffusion on simulation of isolated bubbles in a gas-solid fluidized bed. Powder Technol. (2001). https://doi.org/10.1016/S0032-5910(00)00386-7
29. Hirt, C., Nichols, B.: Volume of fluid (VOF) method for the dynamics of free boundaries. J. Comput. Phys. **39**(1), 201–225 (1981). https://doi.org/10.1016/0021-9991(81)90145-5, https://linkinghub.elsevier.com/retrieve/pii/0021999181901455
30. Hou, J., Simons, F., Hinkelmann, R.: Improved total variation diminishing schemes for advection simulation on arbitrary grids. Int. J. Numer. Methods Fluids **70**(3), 359–382 (2012). https://doi.org/10.1002/fld.2700
31. Karniadakis, G.E., Kirby II, R.M.: Parallel scientific computing in C++ and MPI. Cambridge University Press (2003). https://doi.org/10.1017/CBO9780511812583, https://www.cambridge.org/core/product/identifier/9780511812583/type/book
32. Koren, B.: A robust upwind discretization method for advection, diffusion and source terms. In: Vreugdenhil, C., Koren, B., eds. Numerical Methods for Advection-Diffusion Problems, vol. 45 of Notes on Numerical Fluid Mechanics, Vieweg, Braunschweig, pp. 117–138 (1993)
33. Koynov, A., Khinast, J.G., Tryggvason, G.: Mass transfer and chemical reactions in bubble swarms with dynamic interfaces. AIChE J. **51**(10), 2786–2800 (2005). https://doi.org/10.1002/aic.10529, https://doi.org/10.1002/aic.10529
34. LeVeque, R.J.: High-resolution conservative algorithms for advection in incompressible flow. SIAM J. Numer. Anal. **33**(2), 627–665 (1996). https://doi.org/10.1137/0733033
35. LeVeque, R.J.: Finite Volume Methods for Hyperbolic Problems (2002). https://doi.org/10.1017/cbo9780511791253
36. Olsson, E., Kreiss, G.: A conservative level set method for two phase flow. J. Comput. Phys. **210**(1), 225–246 (2005). https://doi.org/10.1016/j.jcp.2005.04.007, https://linkinghub.elsevier.com/retrieve/pii/S0021999105002184
37. Osher, S., Sethian, J.A.: Fronts propagating with curvature-dependent speed: algorithms based on Hamilton-Jacobi formulations. J. Comput. Phys. **79**(1), 12–49 (1988). https://doi.org/10.1016/0021-9991(88)90002-2, https://linkinghub.elsevier.com/retrieve/pii/0021999188900022
38. Patankar, S.V.: Numerical Heat Transfer and Fluid Flow. Taylor & Francis (1980)
39. Prosperetti, A., Tryggvason, G.: Computational Methods for Multiphase Flow. Cambridge University Press, Cambridge (2007). https://doi.org/10.1017/CBO9780511607486

40. Rhie, C.M., Chow, W.L.: Numerical study of the turbulent flow past an airfoil with trailing edge separation. AIAA J. **21**(11), 1525–1532 (1983). https://doi.org/10. 2514/3.8284

41. Roghair, I., Van Sint Annaland, M., Kuipers, J.: An improved Front-Tracking technique for the simulation of mass transfer in dense bubbly flows. Chem. Eng. Sci. **152**, 351–369 (2016). https://doi.org/10.1016/j.ces.2016.06.026, https:// linkinghub.elsevier.com/retrieve/pii/S00092509163032G8

42. Sun, D., Tao, W.: A coupled volume-of-fluid and level set (VOSET) method for computing incompressible two-phase flows. Int. J. Heat and Mass Transfer **53**(4), 645–655 (2010). https://doi.org/10.1016/j.ijheatmasstransfer.2009.10.030, https://linkinghub.elsevier.com/retrieve/pii/S0017931009005717

43. Sussman, M., Puckett, E.G.: A coupled level set and volume-of-fluid method for computing 3D and axisymmetric incompressible two-phase flows. J. Comput. Phys. **162**(2), 301–337 (2000). https://doi.org/10.1006/jcph.2000.6537, https:// linkinghub.elsevier.com/retrieve/pii/S0021999100965379

44. Sussman, M., Smereka, P., Osher, S.: A level set approach for computing solutions to incompressible two-phase flow. J. Comput. Phys. **114**(1), 146–159 (1994). https://doi.org/10.1006/jcph.1994.1155, https://linkinghub.elsevier.com/retrieve/ pii/S0021999184711557

45. Sussman, M., Smereka, P., Osher, S.: A level set approach for computing solutions to incompressible two-phase flow. J. Comput. Phys. **114**(1), 146–159 (1994). https://doi.org/10.1006/jcph.1994.1155, https://linkinghub.elsevier.com/retrieve/ pii/S0021999184711557

46. Sweby, P.K.: High resolution schemes using flux limiters for hyperbolic conservation laws. SIAM J. Numer. Anal. (1984). https://doi.org/10.1137/0721062

47. Sweby, P.K.: High resolution schemes using flux limiters for hyperbolic conservation laws. SIAM J. Numer. Anal. **21**(5), 995–1011 (1984). https://doi.org/10.1137/ 0721062, https://doi.org/10.1137/0721062

48. Tryggvason, G., et al.: A front-tracking method for the computations of multiphase flow. J. Comput. Phys. **169**(2), 708–759 (2001). https://doi.org/10.1006/jcph.2001. 6726, https://linkinghub.elsevier.com/retrieve/pii/S0021999101967269

49. Tryggvason, G., Scardovelli, R., Zaleski, S.: Direct numerical simulations of gas-liquid multiphase flows (2001). https://doi.org/10.1017/CBO9780511975264

50. Unverdi, S.O., Tryggvason, G.: A front-tracking method for viscous, incompressible, multi-fluid flows. J. Comput. Phys. **100**(1), 25–37 (1992). https://doi. org/10.1016/0021-9991(92)90307-K, https://linkinghub.elsevier.com/retrieve/pii/ 002199919290307K

51. Van der Vorst, H.A., Dekker, K.: Conjugate gradient type methods and preconditioning. J. Comput. Appl. Math. **24**(1-2), 73–87 (1988). https://doi. org/10.1016/0377-0427(88)90344-5, https://linkinghub.elsevier.com/retrieve/pii/ 0377042788903445

52. Zhang, D., Jiang, C., Cheng, L., Liang, D.: A refined r-factor algorithm for TVD schemes on arbitrary unstructured meshes. Int. J. Numer. Methods in Fluids **80**(2), 105–139 (2016). https://doi.org/10.1002/fld.4073

A Backward-Characteristics Monotonicity Preserving Method for Stiff Transport Problems

Ilham Asmouh[1]([⊠])(iD) and Abdelouahed Ouardghi[2](iD)

[1] Institut für Mathematik, Universität Innsbruck,
Technikerstraße 13, 6020 Innsbruck, Austria
ilham.asmouh@uibk.ac.at
[2] Jülich Supercomputing Centre, Forschungszentrum Jülich GmbH,
Jülich, Germany
a.ouardghi@fz-juelich.de

Abstract. Convection-diffusion problems in highly convective flows can exhibit complicated features such as sharp shocks and shear layers which involve steep gradients in their solutions. As a consequence, developing an efficient computational solver to capture these flow features requires the adjustment of the local scale difference between convection and diffusion terms in the governing equations. In this study, we propose a monotonicity preserving backward characteristics scheme combined with a second-order BDF2-Petrov-Galerkin finite volume method to deal with the multiphysics nature of the problem. Unlike the conventional Eulerian techniques, the two-step backward differentiation procedure is applied along the characteristic curves to obtain a second-order accuracy. Numerical results are presented for several benchmark problems including sediment transport in coastal areas. The obtained results demonstrate the ability of the new algorithm to accurately maintain the shape of the computed solutions in the presence of sharp gradients and shocks.

Keywords: Flow transport · Backward characteristics method · Petrov-Galerkin finite volume method · Backward differentiation formula (BDF2)

1 Introduction

Convection-diffusion problems appear in various fields of science and technology such as heat and mass transfer, environmental protection, fluid dynamics and hydrology. These problems in highly convective flows are essentially characterized by some complicated features such as sharp shocks and shear layers which involve steep gradients in their numerical solutions. As a consequence, construction of efficient computational solvers to highly capture these flow features requires the adjustment of the local scale difference between convection and diffusion terms in the governing equations. Eulerian-based approaches have been widely used for the numerical solution of these problems. However, for convection-dominated cases, those methods exhibit spurious oscillations and numerical instabilities and stringent stability conditions are consequently unavoidable.

L. Franco et al. (Eds.): ICCS 2024, LNCS 14838, pp. 33–47, 2024.
https://doi.org/10.1007/978-3-031-63783-4_4

Backward characteristics methods are among excellent integration schemes by the virtue of their stability properties and good accuracy. Indeed, By discretizing the Lagrangian derivative of the solution in time instead of the Eulerian derivative, we can exceed the maximum allowable time step while maintaining the efficiency of symmetric solvers. However, these methods can experience some difficulties preserving the shape and mass conservation of advected quantities. Several remedies have been proposed including addition of filters to impose monotonicity or positivity in the field of plasma physics [2]. However, this technique is challenging especially in higher order reconstruction due to the necessity of some conditions on the polynomials to be monotone or positive [3]. Weighted essentially non-oscillatory (WENO) reconstructions have been very popular [12,14] even though these techniques are not convenient for long term simulation since they are known to be too dissipative. In this work, we propose combining low and high order interpolation schemes such that the interpolated value remains within the largest and the smallest values of the solution in a set of points surrounding the feet of the characteristics.

To deal with the diffusive terms, we propose a finite volume scheme. It is known that finite volume methods are very popular for their ability in capturing shocks, producing simple stencils and effectively treating Neumann boundary conditions and nonuniform grids, which make them an attractive choice for fluid flow simulation. However, finite volume schemes confront some challenges, in general, related to the accuracy of fluxes approximation which has an immediate impact on truncation errors if the fluxes are not approximated carefully. In contrast, the fluxes in finite volume element (FVE) methods are approximated by replacing the unknowns with a finite element solution. Thus, the best choice of finite element space makes the discretization design process bringing the focus to the local character of the solution rather than the equation. Furthermore, very effective discretization processes are provided for multilevel adaptive methods. Moreover, finite volume element methods are known as one of the preferred approaches as the test space essentially maintains the local conservation of the method without serious restrictions in terms of implementation. The method consists on a volume integral formulation of the problem using a finite partitioning set of volumes for the equation discretization and restriction of admissible functions to a finite element space for the solution discretization.

Several efforts have been investigated in this area. The authors in [5] analyzed some error estimates for finite volume finite element method for nonlinear convection-diffusion problems where nonlinear convective terms are approximated using a monotone vertex-centered finite volume scheme. In [4], the authors analyzed the stability of finite volume element scheme for parabolic problems where the diffusion terms are discretized using Crouzeix-Raviart piecewise linear finite elements on a triangular grid and an upwind barycentric finite volumes for the convective terms. The time integration is carried out using an implicit Euler approach. In the area of solute transport problems, a proper orthogonal decomposition (POD) is combined with classical finite volume element method where an error estimate between the reduced-order POD and conventional FVE solutions is discussed [10]. In [8] the authors presented a discontinuous finite volume element discretization for a coupled Navier–Stokes Cahn–Hilliard phase field model. An analogous approach with mesh adaptivity is presented in [7] to solve the nonlinear Allen-Cahn equation in two-space dimensions, where backward Euler scheme is used for time integration and the nonlinearity is solved using Newton iterative method.

The aim of this work is to construct a new backward-characteristics finite volume element (BC-FVE) method able to accurately approximate stiff transport problems on unstructured grid. This is achieved by merging the advantages of a monotone backward characteristics method and a Petrov-Galerkin finite volume method. The novel method also avoids linearisation process of the convective terms. The accuracy of the new method is tested for stiff transport problems with analytical solution, including pollutant transport in the Loukkos river in northern Morocco. The results presented in this paper show high resolution of the proposed BC-FVE method in simulating transport and dispersion of pollutants on large sea-surface regions.

2 Monotone Backward-Characteristics Scheme for the Convective Term

In this section, we construct an essentially non-oscillatory backward-characteristics scheme to solve highly convective problems. To this end, given a two-dimensional bounded domain $\Omega \subset \mathbb{R}^2$ with Lipschitz boundary Γ and a time interval $[0, T]$, we are interested in solving the following problem:

$$\frac{\partial c}{\partial t} + \mathbf{u}(t, \mathbf{x}, c) \cdot \nabla c = \nabla \cdot (\mathbf{K} \nabla c), \qquad (t, \mathbf{x}) \in (0, T) \times \Omega,$$

$$c(0, \mathbf{x}) = c_0(\mathbf{x}), \qquad \mathbf{x} \in \Omega. \tag{1}$$

Here, $c(t, \mathbf{x})$ denotes the concentration of a species, $\mathbf{u}(t, \mathbf{x}, c)$ is the velocity field which may depend on the solution c, and $c_0(\mathbf{x})$ is a fixed initial condition. We assume that equation (1) is equipped with appropriate boundary conditions. The quantity $\mathbf{K}$ is the diffusivity tensor which is uniformly positive definite on Ω with components in $L^\infty(\Omega)$.

Consider the time interval $[t^n, t^{n+1}]$ and let $\Delta t = t^{n+1} - t^n$ denote the time step size. The spatial discretization of Ω consists of a quasi-uniform regular triangulation $\mathcal{T}_h$ consisting of triangular elements $\mathcal{K}_k$ such that $\Omega = \bigcup_{k=1}^{N} \mathcal{K}_k$ where N denotes the total number of triangles. We introduce a dual mesh $\mathcal{T}_h^*$ associated to $\mathcal{T}_h$ such that a given triangle $\mathcal{K}$ with vertices e_1, e_2 and e_3 is divided into six pieces by satisfying the following relations

$$\frac{|e_1 e_{1,2}^s|}{|e_1 e_2|} = \frac{|e_{2,1}^s e_2|}{|e_1 e_2|} = s, \quad \frac{|e_1 e_{1,2}^r|}{|e_1 \mathcal{E}_2|} = r, \quad \frac{|e_2 e_{2,3}^s|}{|e_2 e_3|} = \frac{|e_{3,2}^s e_3|}{|e_2 e_3|} = s, \quad \frac{|e_2 e_{2,3}^r|}{|e_2 \mathcal{E}_3|} = r,$$

$$\frac{|e_3 e_{3,1}^s|}{|e_3 e_1|} = \frac{|e_{1,3}^s e_1|}{|e_3 e_1|} = s, \quad \frac{|e_3 e_{3,1}^r|}{|e_3 \mathcal{E}_1|} = r,$$

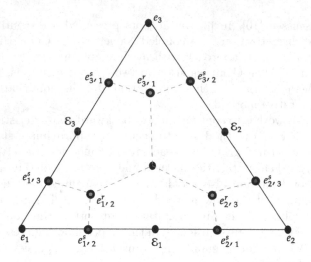

Fig. 1. A schematic diagram showing the partition of a triangle into three quadrilaterals and three pentagons.

where $s \in (0, \frac{1}{2})$ and $r \in (0, \frac{2}{3})$ and $\mathcal{E}_j$ denotes the midpoint of the segment $e_i e_j$. By doing so, we obtain a partition of the triangle $\mathcal{K}$, which consists of three pentagons and three quadrilaterals as illustrated in Fig. 1. Our control volumes $\mathcal{D}_m$ are then polygons surrounding a vertex e_j or a midpoint $\mathcal{E}_j$. Therefore we obtain a family of control volumes covering the domain Ω with a total number n_v of control polygons $\mathcal{D}_m$. Solving the advected part of (1) using the method of characteristics requires the solution of the following problem

$$\frac{d\boldsymbol{X}_m(t)}{dt} = \boldsymbol{u}_m\left(t, \boldsymbol{X}_m(t), c_m\right), \qquad t \in [t_n, t_{n+1}],$$
$$\boldsymbol{X}_m(t_{n+1}) = \boldsymbol{x}_m, \tag{2}$$

where $\boldsymbol{X}_m(t) = (X_m(t), Y_m(t))^T$ denotes the departure point at time t of a particle that will arrive at the centroid $\boldsymbol{x}_m = (x_m, y_m)^T$ of the control volume $\mathcal{D}_m$ at time t_{n+1}. To solve problem (2) we use the well-known third-order Runge-Kutta scheme [6, Table II.1.1]. Next, integrating the transport equation $\frac{\partial c}{\partial t} + \mathbf{u} \cdot \nabla c = 0$ along the characteristic curves yields

$$c(t_{n+1}, \boldsymbol{x}_m) = c\left(t_n, \boldsymbol{X}_m(t_n)\right). \tag{3}$$

In general, the departure points $\boldsymbol{X}_m(t_n)$ do not lie on a mesh point. Therefore, an interpolation is required. This yields

$$c\left(t_n, \boldsymbol{X}_m(t_n)\right) = \sum_{k \in \mathcal{N}} C_k^n \Psi_k(\boldsymbol{X}_m(t_n)), \tag{4}$$

where $\mathcal{N}$ refers to the number of contributed points during local interpolation, C_m^n denotes the known solution at the vertex x_m at time t^n and Ψ_m, $m = 1, \ldots, n_v$ is the basis function evaluated at the departure point $\boldsymbol{X}_m(t_n)$. It should be stressed that high-order interpolation schemes might exhibit nonphysical oscillations, where the solution undergoes strong variations in general [1]. To deal with those undesirable behaviors, we propose a combination of low and high order interpolation schemes such that the interpolated value remain within the largest and the smallest values of the solution in a set of points surrounding the feet of the characteristics. Therefore, the low- and high-order solutions are computed, respectively using the well-known inverse distance weighted (IDW) [9] and the inverse multiquadric (IMQ) interpolations [13] where the corresponding basis functions ψ_k and Θ_l are given by the formulas

$$\psi_k(\boldsymbol{X}_m(t_n)) = \frac{(1/d_{mk})^2}{\sum_{k=1}^{n_{vL}}(1/d_{mk})^2}, \quad \Theta_l(\boldsymbol{X}_m(t_n)) = \frac{1}{\sqrt{1+(\epsilon d_{ml})^2}},$$

where ϵ denotes the shape parameter [11] and d_{mk} is the Euclidean distance between the departure point $\boldsymbol{X}_m(t_n)$ to the point x_k calculated as

$$d_{mk} = \|\boldsymbol{X}_m(t_n) - \boldsymbol{x}_k\| = \sqrt{\left(X_m(t_n) - x_k\right)^2 + \left(Y_m(t_n) - y_k\right)^2}.$$

See Fig. 2 for a graphical depiction. Thus, given the initial solution, the main steps used in the proposed BC-FVE to solve the advected part of problem (1) along a time step $[t^n, t^{n+1}]$ are summarized in Algorithm 1.

Algorithm 1. One time step of the BC-FVE approach for solving the advected part of problem (1).

1: Calculate the departure point $\boldsymbol{X}_m(t_n)$.
2: Compute the high- and low-order solutions

$$c_{mH}^{n+1} = \sum_{l=1}^{n_{vH}} C_l^n \Theta_l(\boldsymbol{X}_m^n) \text{ and } c_{mL}^{n+1} = \sum_{k=1}^{n_{vL}} C_k^n \psi_k(\boldsymbol{X}_m^n).$$

3: Given the host control volume $\mathcal{D}_m^*$, compute

$$c^{\max} = \max(c_1^n, c_2^n, \ldots, c_{n_{vH}}^n) \text{ and } c^{\min} = \min(c_1^n, c_2^n, \ldots, c_{n_{vH}}^n).$$

4: Set

$$\alpha_m = \begin{cases} \min\left(1, \frac{c^{\max}-c_{mL}^{n+1}}{c_{mH}^{n+1}-c_{mL}^{n+1}}\right), & \text{if } c_{mH}^{n+1} - c_{mL}^{n+1} > 0, \\ \min\left(1, \frac{c^{\min}-c_{mL}^{n+1}}{c_{mH}^{n+1}-c_{mL}^{n+1}}\right), & \text{if } c_{mH}^{n+1} - c_{mL}^{n+1} < 0, \\ 0, & \text{if } c_{mH}^{n+1} - c_{mL}^{n+1} = 0. \end{cases}$$

5: Update the solution as

$$c_m^{n+1} = c_{mL}^{n+1} + \alpha_m \left(c_{mH}^{n+1} - c_{mL}^{n+1}\right).$$

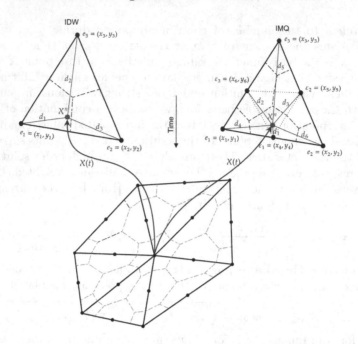

Fig. 2. A schematic diagram showing the main quantities used to interpolate the solution from the computed departure points. For low-order IDW interpolation $n_{vL} = 3$ whereas for high-order IMQ interpolation $n_{vH} = 6$.

3   Petrov-Galerkin Finite Volume Scheme for the Diffusive Term

After dealing with the convective term, we are left with the following parabolic equation expressed in terms of the material derivative

$$\frac{Dc}{Dt} = \nabla \cdot (\mathbf{K} \nabla c). \tag{5}$$

The aim of this section is to discretize the diffusive term using the well-known Petrov-Galerkin finite volume method [5,15]. Assuming the computational domain is discretized as described in Sect. 2. In order to derive the weak form of problem (5), we consider the following finite element space for the primal mesh

$$\mathbb{V}(\mathcal{T}_h) := \left\{ w_h \in H_0^1(\Omega) \;\middle|\; w_h|_{\mathcal{K}_k} \in \mathcal{P}_2, \; \forall \mathcal{K}_k \in \mathcal{T}_h \right\},$$

where $\mathcal{P}_2$ denotes the piecewise quadratic polynomial space. We are given also a piecewise constant function space for the dual mesh

$$\mathbb{W}(\mathcal{T}_h^*) := \left\{ v_h \in L^2(\Omega) \;\middle|\; v_h|_{\mathcal{D}_m} \in \mathcal{P}_0, \; \forall \mathcal{D}_m \in \mathcal{T}_h^* \right\},$$

where $\mathcal{P}_0$ denotes the constant polynomial space. Note that the spaces $\mathbb{V}(\mathcal{T}_h)$ and $\mathbb{W}(\mathcal{T}_h^*)$ are given by

$$\mathbb{V}(\mathcal{T}_h) = \mathrm{span}\{\phi_l(\boldsymbol{x}), \ l = 1, \ldots, n_v\}, \ \mathbb{W}(\mathcal{T}_h^*) = \mathrm{span}\{\mathbb{1}_{\mathcal{D}_m}(\boldsymbol{x}), \ m = 1, \ldots, n_v\},$$

where ϕ_l denotes the standard quadratic Lagrange basis function and the characteristic function $\mathbb{1}_{\mathcal{D}_m}$ is defined over each control volume $\mathcal{D}_m$ as

$$\mathbb{1}_{\mathcal{D}_m}(\boldsymbol{x}) = \begin{cases} 1, & \text{if } \boldsymbol{x} \in \mathcal{D}_m, \\ 0, & \text{elsewhere.} \end{cases}$$

The discrete balance equation associated to problem (5) is given by integrating the whole equation over each control volume $\mathcal{D}_m$ and using the divergence theorem to obtain

$$\int_{\mathcal{D}_m} \frac{Dc_h}{Dt} dV = -\int_{\partial \mathcal{D}_m} (\mathbf{K}\nabla c_h) \cdot \mathbf{n}_e d\sigma, \tag{6}$$

where $\mathbf{n}_e$ denotes the outward normal vector to the edge e of $\mathcal{D}_m$. Applying the BDF2 scheme yields

$$\int_{\mathcal{D}_m} \left(\frac{3}{2\Delta t} c_h^{n+1}(\boldsymbol{x}) - \frac{2}{\Delta t} c_h^n(\boldsymbol{\mathcal{X}}^{\mathbf{n}}(\boldsymbol{x})) - \frac{1}{2\Delta t} c_h^{n-1}(\boldsymbol{\mathcal{X}}^{\mathbf{n-1}}(\boldsymbol{x})) \right) dV = \tag{7}$$
$$-\int_{\partial \mathcal{D}_m} (\mathbf{K}\nabla c_h^{n+1}(\boldsymbol{x})) \cdot \mathbf{n}_e d\sigma,$$

where $\mathcal{X}^n(\boldsymbol{x})$ and $\mathcal{X}^{n-1}(\boldsymbol{x})$ denote the departure points at times t_n and t_{n-1} respectively, of the particle that will reach the point $\boldsymbol{x}$ at time t_{n+1}. Therefore, c_h^n is the solution evaluated using the backward-characteristics method at t^n, and c_h^{n-1} is the solution evaluated two time steps back along the characteristics. To construct a finite volume element scheme, we are seeking for a solution $c_h \in \mathbb{V}(\mathcal{T}_h)$ as in the framework of finite element analysis. Taking into consideration that the semi-discrete solution c_h can be rewritten in terms of the chosen basis functions as

$$c_h^n(\boldsymbol{x}) = \sum_{l=1}^{n_v} C_l^n \phi_l(\boldsymbol{x}), \ m = 1, \ldots, n_v.$$

Inserting the above expression into (7) and rearranging all terms yields the following compact form

$$\frac{3}{2\Delta t} \mathbf{M} C^{n+1} + \mathbf{S} C^{n+1} = \frac{2}{\Delta t} \tilde{\mathbf{H}}^n - \frac{1}{2\Delta t} \hat{\mathbf{H}}^{n-1}, \tag{8}$$

where $\mathbf{M}$ and $\mathbf{S}$ are the mass and stiffness matrices which entries are given by

$$M_{ml} = \int_{\Omega} \phi_l(\boldsymbol{x}) \mathbb{1}_{\mathcal{D}_m}(\boldsymbol{x}) dV, \qquad S_{ml} = -\int_{\partial \mathcal{D}_m} (\mathbf{K}\nabla \phi_l(\boldsymbol{x})) \cdot \mathbf{n}_e(\boldsymbol{x}) d\sigma.$$

The right-hand sides $\tilde{\mathbf{H}}^n$ and $\hat{\mathbf{H}}^{n-1}$ are given by the following formulas, where the corresponding integral are approximated using the Gauss-Legendre quadrature rule

$$\tilde{\mathbf{H}}^n := \left(\int_\Omega \mathbf{c}_h^n(\mathcal{X}^n(\boldsymbol{x}))\phi_l(\boldsymbol{x})\mathbb{1}_{\mathcal{D}_m}\,dV \right), \quad \hat{\mathbf{H}}^{n-1} := \left(\int_\Omega \mathbf{c}_h^{n-1}(\mathcal{X}^{n-1}(\boldsymbol{x}))\phi_l(\boldsymbol{x})\mathbb{1}_{\mathcal{D}_m}\,dV \right).$$

4   Numerical Results

In this section, we examine the accuracy of the novel BC-FVE on several benchmark of transport problems including a system of nonlinear Burgers equations and transport of a pollutant in coastal areas.

4.1   Slotted Cylinder

Our first test is the slotted cylinder. The rotation is driven by $\mathbf{u}(x,y) = (-4y, 4x)^T$ and the diffusive term is taken as zero. For a given (x_0, y_0), let $\rho(x,y) = \sqrt{(x-x_0)^2 + (y-y_0)^2}$. The slotted cylinder is defined as

$$c(0, x, y) = \begin{cases} 4, & \text{if } \rho(x,y) \leq 1 \quad \text{and} \quad (\, |x - x_0| \geq 0.03 \quad \text{or} \quad y \geq 0.22), \\ 0, & \text{elsewhere.} \end{cases}$$

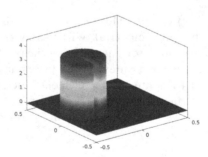

Fig. 3. 3D representation of the initial solution of slotted cylinder. The minimum of c is 0 and its maximum is 4.

This problem has served as a prototype to examine the performance of several algorithms for flow transport. The snap-shot of analytical solution in Fig. 3 is used to quantify the accuracy of the novel method. The computational domain is partitioned into 6274 control volumes. In Fig. 4, we display the obtained numerical solutions, for low-order interpolation, high-order interpolation and limiting approach, for CFL = 10.5. The numerical solutions after one revolution (first column) are then repeated but after four revolutions, as illustrated in the second column. The clear indication from Fig. 4 that the IDW solution is dissipative and substantially greater distortion appears after four revolutions. On the other side, the IMQ solution exhibits steep gradients which become more steep after four revolutions. However, all the non desirable effects are suppressed when merging the two interpolations and the BC-FVE solution results in a more accurate solution where the slotted cylinder shape is well reproduced during time evolution. To further quantify the accuracy of the proposed approach and analyze the effect of CFl on the stability of the method, we list in Fig. 5 the cross-sections along $y = 0$ after one and four revolutions for different values of CFL number. The corresponding cross-sections of the analytical solution are added for comparison reasons. It is known that backward characteristics methods work with high precision for large CFL numbers. Indeed, for CFL$= 0.87$ both IDW and IMQ solutions exhibit nonphysical oscillations. Unfortunately, the BC-FVE solution is also less accurate for same value of CFL. However, the more the CFL number is high the more accurate results we obtain. Indeed, an intriguing finding is that the inaccuracy of backward characteristics schemes decreases as the time step increases in a certain range of parameters.

4.2   Viscous Burgers Equations

In the second test we are concerned with a system of nonlinear Burgers' equations given as

$$\frac{\partial \boldsymbol{u}}{\partial t} + \boldsymbol{u} \cdot \nabla \boldsymbol{u} - \frac{1}{Re} \Delta \boldsymbol{u} = \boldsymbol{0}, \tag{9}$$

where $\boldsymbol{u} = (u, v)^\top$ is the velocity field, u the velocity in x-direction, v the velocity in y-direction, and Re is the Reynolds number. The computational domain is the square domain $\Omega = [-2, 2] \times [-2, 2]$. Initial and boundary conditions are chosen to satisfy the following analytical solution

$$u(t, x, y) = v(t, x, y) = \frac{1}{2}\left(1 - \tanh\left(\frac{Re}{4}(x + y - t)\right)\right),$$

In Table 1 we summarize the minimum (Min), the maximum (Max) of the velocity component u, the L^2-error, the rate of convergence and the computational cost (CPU) obtained for the IDW scheme, the IMQ scheme and the novel BC-FVE for two values of Reynold number Re. It should be noted that the Min and the Max results are taken in the range of large gradients. Roughly speaking, better accuracy is achieved for $Re = 10^4$. In addition, the IMQ solution produces more accurate convergence rates even though the solution is not monotone. on the other side, the IDW solution is very dissipative. The results sound more accurate when merging both interpolation and the BC-FVE solution is monotone where the convergence rates are clearly improved. In term of computational costs, the CPU time required for the IDW solution is about 3/4 times the CPU time required for the IMQ solution. The BC-FVE solution in tern, is about 5/3 times the IMQ solution.

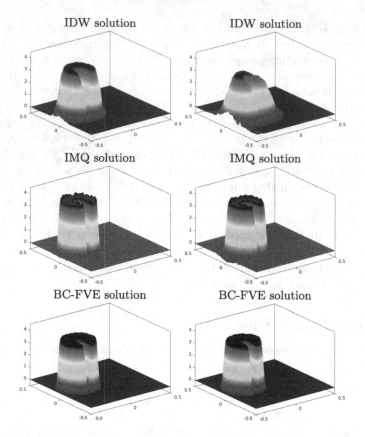

Fig. 4. Snapshots for Example 4.1 obtained using different methods after one revolution (first column) and four revolutions (second colomn).

4.3   Transport in the Loukkos River in Northern Morocco

In the last example we turn our attention to a real application over the physical domain given in Fig. 6, where we consider the transport of some sediments concentration c in the Loukkos river in the northern Morocco. The Loukkos river is one of the largest streams in Morocco with an average flow of $50\,\text{m}^3/\text{s}$. The river flows in the Atlantic Ocean and plays an important role in preserving the biodiversity, containing one of the most fertile and productive agricultural lands in the country.

The main objective in this study is to analyze the performance of the new BC-FVE method to handle complex geometries in coastal zones for long-term simulations. Therefore, we address the solution to the transport Eq. (1) within the computational domain defined by the Loukkos river and Larache coastal zone as illustrated in Fig. 7. An incompressible Navier-Stokes code implemented in FEniCS software is used to generate the velocity field and the results are depicted in Fig. 7. The problem model is subject to given inflow conditions $\mathbf{u}_\infty = 0.65\,\text{m/s}$

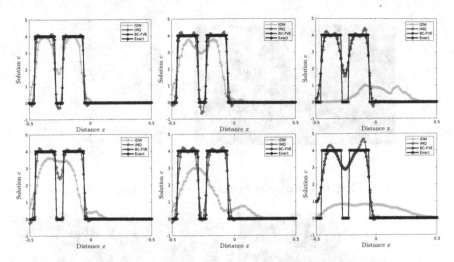

Fig. 5. Cross sections along $y = 0$. From left to right: CFL = 10.5, CFL = 4.63, CFL = 0.87. First row: 1 revolution. Second row: 4 revolutions.

Table 1. Computational results for Example 4.2 at time $t = 1$ obtained as a comparative study between the three discussed methods. The CPU times are given in seconds.

	n_v	$Re = 10^4$					$Re = 20$				
		Min(u)	Max(u)	L^2-error	Rate	CPU	Min(u)	Max(u)	L^2-error	Rate	CPU
IDW	257	1.452E−10	0.851	3.023E−01	—	1.18	2.929E−02	0.817	3.518E−01	—	1.38
	622	3.294E−14	0.884	1.898E−01	0.67	3.14	5.403E−18	0.870	1.986E−01	0.82	3.73
	1422	6.233E−16	0.894	9.421E−02	1.01	11.1	1.120E−57	0.891	8.986E−02	1.14	12.1
	2945	9.638E−18	0.938	4.331E−02	1.12	47.5	1.213E−17	0.914	3.986E−02	1.17	64.3
IMQ	257	1.452E−10	1.042	2.953E−02	—	1.58	−1.052E−03	1.432	1.946E−01	—	1.85
	622	1.144E−14	1.034	5.862E−03	2.33	4.17	−1.345E−03	1.303	4.953E−02	1.97	4.96
	1422	4.400E−15	1.026	1.101E−03	2.41	15.0	−1.293E−03	1.217	1.196E−02	2.05	16.3
	2945	1.213E−17	1.019	2.213E−04	2.31	63.2	1.213E−17	1.135	2.696E−03	2.14	85.6
BC-FVE	257	2.442E−27	0.998	2.942E−02	—	2.63	2.752E−16	0.999	1.836E−01	—	3.06
	622	3.534E−30	1.000	5.712E−03	2.36	6.95	1.425E−19	1.000	4.986E−02	1.88	8.27
	1422	4.218E−32	1.000	1.173E−03	2.28	25.2	1.293E−23	1.000	1.011E−02	2.30	27.3
	2945	1.423E−33	1.000	2.305E−04	2.34	105	1.003E−23	1.000	2.381E−03	2.08	144

on the river entrance, nonslip condition on land boundaries, while the ocean boundaries remains free. The problem assumes a kinematic viscosity of $10\,\mathrm{m^2/s}$ and is subject to a release defined by a source term specified as follows:

$$f(x, y, t) = \exp\left(-\frac{(x - x_r)^2 + (y - y_r)^2}{\sigma^2}\right), \tag{10}$$

Fig. 6. Location of Loukkos river in the map

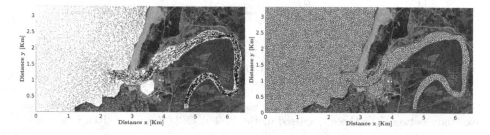

Fig. 7. Velocity field (left), computational domain and location of gauges (right).

where $\sigma = 0.2\,\mathrm{km}$ and $(x_r, y_r) = (4.69\,\mathrm{km},\ 0.15\,\mathrm{km})$ is the selected location for concentration release, see the release point R in Fig. 6. In our simulations, an unstructured triangular mesh with 5036 nodes for low order method and 19282 nodes for high order method, are used. In Fig. 8 we present the simulation of the concentration c at six distinct times: $t = 25\,\mathrm{min}$, $2\,\mathrm{h}$, $4\,\mathrm{h}$, $6\,\mathrm{h}$, $8\,\mathrm{h}$, and $10\,\mathrm{h}$. In the early stages of the simulation, the concentration fronts released in the river begins to evolve until it flows into the Atlantic Ocean. Henceforth, the advection of the sediment c into the ocean eventually condensates in the bay before it stagnates within it, forming large spot in the exit of the river. It should be stressed that despite the challenges posed by the complex geometry and the flow features, the novel BC-FVE method demonstrates high performance in capturing high gradients within the advancing plume. It is crucial to note that unlike prior cases, in which convection-diffusion problems were treated in basic geometries with known velocity fields, the current problem is addressed within a wide area with irregular geometry and complex flow patterns.

To further quantify the accuracy of the method with a reference solution, we monitor the concentration at three gauges, G1, G2, and G3, situated in the river at $(5.36\,\mathrm{km},\ 1.16\,\mathrm{km})$, $(4.64\,\mathrm{km},\ 2.05\,\mathrm{km})$, and $2.94\,\mathrm{km},\ 1.07\,\mathrm{km})$, respectively, as illustrated in Fig. 6. Keeping the same mesh resolution as before, we list in Fig. 9

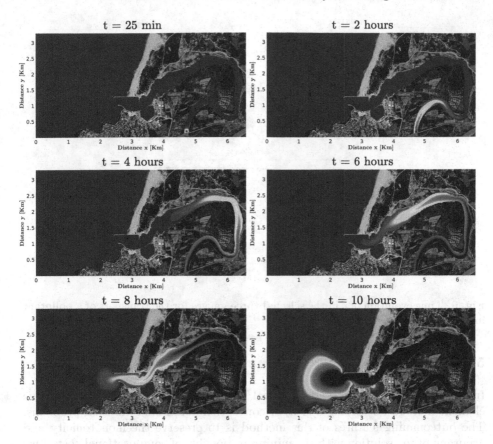

Fig. 8. Concentration snapshots for the transport problem in the Loukkos river using the proposed backward Petrov-Galerkin finite volume method (BC-FVE).

the time evolution of the concentration at the given gauges at time $t = 12\,h$. Note that the reference solution is computed on a reference mesh with 49429 elements and 100349 nodes. The clear indication from Fig. 9 is that the IMQ solution exhibits again oscillations and clearly steep layers appear in the regions of large gradients (compare Fig. 9 G 3). This is potentially attributed to complex feature of the bended geometry nearby G 1. From the same plot, the IDW solution exhibits numerical diffusion. By its turn, the BC-FVE solution results in a stable and monotone behavior in the selected gauges during time evolution. Moreover, a quantitative study confirms that the novel BC-FVE method results are in good agreement with the reference solution.

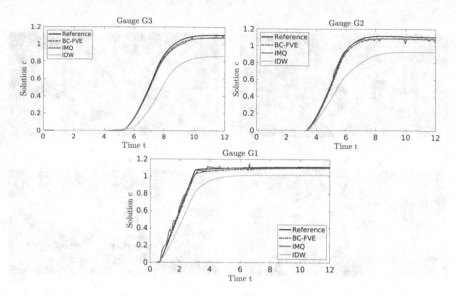

Fig. 9. Time evolution of c in the selected three gauges for the transport of a pollutant in the Loukkos river.

5   Conclusions

In this work we have presented a novel backward-characteristics finite volume element method for solving a class of convection-dominated diffusion problems. The outcomming features of this method is to preserve the monotonicity and accuracy of the solution with the minimum possible of computational cost. This is achieved by the virtue of backward-characteristics scheme where low and high order interpolation techniques are both used to balance the characteristics of the solution during time evolution. Indeed, the interpolated value remains within the largest and the smallest values of the solution in a set of points surrounding the feet of the characteristics. The diffusive term is approximated with a Petrov-Galerkin finite volume scheme where a finite partitioning set of control volumes is used to discretize the equation and a restriction of admissible functions of finite element space are used to discretize the solution. The obtained results demonstrate the ability of the new algorithm to accurately maintain the shape of the computed solutions in the presence of sharp gradients and shocks. **Future work will focus on the one hand, on extending the current approach to coupled flow-transport, where the flow is modeled by the incompressible Navier-Stokes equations. On the other hand, we aim to explore the time adaptation and construct an efficient adaptive BDF2 along the characteristics in order to minimize the computational cost.**

References

1. Bermejo, R.: Analysis of a class of quasi-monotone and conservative semi-Lagrangian advection schemes. Numer. Math. **87**(4), 597–623 (2001)
2. Crouseilles, N., Mehrenberger, M., Sonnendrücker, E.: Conservative semi-Lagrangian schemes for Vlasov equations. J. Comput. Phys. **229**(6), 1927–1953 (2010)
3. Després, B.: Polynomials with bounds and numerical approximation. Numer. Alg. **76**, 829–859 (2017)
4. Deuring, P., Mildner, M.: Stability of a combined finite element-finite volume discretization of convection-diffusion equations. Numer. Methods Partial Differ. Equ. **28**(2), 402–424 (2012)
5. Feistauer, M., Felcman, J., Lukácová-Medvid'ová, M., Warnecke, G.: Error estimates for a combined finite volume-finite element method for nonlinear convection-diffusion problems. SIAM J. Numer. Anal. **36**(5), 1528–1548 (1999)
6. Hairer, E., Nørsett, S.P., Wanner, G.: Solving Ordinary Differential Equations I. Computational Mathematics, 2nd edn. Springer, Heidelberg (1993). https://doi.org/10.1007/978-3-540-78862-1
7. Li, J., Zeng, J., Li, R.: An adaptive discontinuous finite volume element method for the Allen-Cahn equation. Adv. Comput. Math. **49**(4), 55 (2023)
8. Li, R., Gao, Y., Chen, J., Zhang, L., He, X., Chen, Z.: Discontinuous finite volume element method for a coupled Navier-Stokes-Cahn-Hilliard phase field model. Adv. Comput. Math. **46**, 1–35 (2020)
9. Lu, G.Y., Wong, D.W.: An adaptive inverse-distance weighting spatial interpolation technique. Comput. Geosci. **34**(9), 1044–1055 (2008)
10. Luo, Z., Li, H., Sun, P., An, J., Navon, I.M.: A reduced-order finite volume element formulation based on POD method and numerical simulation for two-dimensional solute transport problems. Math. Comput. Simul. **89**, 50–68 (2013)
11. Mongillo, M., et al.: Choosing basis functions and shape parameters for radial basis function methods. SIAM Undergraduate Res. Online **4**(190–209), 2–6 (2011)
12. Qiu, J.M., Shu, C.W.: Conservative semi-Lagrangian finite difference WENO formulations with applications to the Vlasov equation. Commun. Comput. Phys. **10**(4), 979–1000 (2011)
13. Sarra, S.A., Kansa, E.J.: Multiquadric radial basis function approximation methods for the numerical solution of partial differential equations. Adv. Comput. Mech. **2**(2), 220 (2009)
14. Xiong, T., Qiu, J.M., Xu, Z., Christlieb, A.: High order maximum principle preserving semi-Lagrangian finite difference WENO schemes for the Vlasov equation. J. Comput. Phys. **273**, 618–639 (2014)
15. Zhou, Y., Wu, J.: A unified analysis of a class of quadratic finite volume element schemes on triangular meshes. Adv. Comput. Math. **46**, 1–31 (2020)

A Three-Dimensional Fluid-Structure Interaction Model for Platelet Aggregates Based on Porosity-Dependent Neo-Hookean Material

Yue Hao[1(✉)], Alfons G. Hoekstra[1], and Gábor Závodszky[1,2]

[1] Computational Science Lab, Informatics Institute, Faculty of Science,
University of Amsterdam, Amsterdam, The Netherlands
y.hao@uva.nl
[2] Department of Hydrodynamic Systems,
Budapest University of Technology and Economics, Budapest, Hungary

Abstract. The stability of the initial platelet aggregates is relevant in both hemostasis and thrombosis. Understanding the structural stresses of such aggregates under different flow conditions is crucial to gaining insight into the role of platelet activation and binding in the more complex process of clot formation. In this work, a three-dimensional implicit partitioned fluid-structure interaction (FSI) model is presented to study the deformation and structural stress of platelet aggregates in specific blood flow environments. Platelet aggregates are considered as porous mediums in the model. The FSI model couples a fluid solver based on Navier-Stokes equations and a porosity-dependent compressible neo-Hookean material to capture the mechanical characteristics of the platelet aggregates. A parametric study is performed to explore the influence of porosity and applied body force on this material. Based on *in vitro* experimental data, the deformation and associated stress of a low shear aggregate and a high shear aggregate under different flow conditions are evaluated. This FSI framework offers a way to elucidate the complex interaction between blood flow and platelet aggregates and is applicable to a wider range of porous biomaterials in flow.

Keywords: Compressible neo-Hookean · Fluid-structure interaction · Porosity · Platelet aggregate

1 Introduction

Embolized thrombus portions can potentially block peripheral arteries, resulting in life-threatening cardiovascular diseases [3]. Platelet aggregation, the initial phase of thrombus formation, plays a vital role in thrombosis and hemostasis. The mechanical properties and morphology of platelet aggregates as well as their interaction with the flow of blood, significantly influence the further stages of aggregation and consequently the stability of the formed clot. Therefore, under-standing the mechanical properties of platelet aggregates allows insight into fur-ther details of the mechanical and chemical behaviour of thrombosis, that might contribute to the analysis and development of treatments for thrombus [6].

Computational modelling has been used to understand the interaction between blood flow and clots [13, 25]. In the earlier works clots were considered as a rigid non-porous material [24, 28]. The importance of the porous structure of clots has been revealed in more recent studies [23, 29]. In particular, Kadria et al. [16] examined the fluid-induced shear stress imposed on the surface of thrombi using a three-dimensional blood flow simulation. However, it's essential to note that most of the prior investigations focus primarily on hemodynamics, such as blood flow velocity and fluid shear stress instead of the mechanical properties of the clot.

Experimental and computational methods have been developed and applied to quantify the mechanical behaviour of clots, including compressibility and stress response under varying forces [5, 6, 22]. Notably, the nonlinear elastic behaviour of clots has been assumed to follow neo-Hookean materials [7, 24]. Our work focuses on platelet aggregates, that represent initial stages of blood clots, constituted entirely of platelets. Platelets themselves can be characterised by a neo-Hookean constitutive model, since their defining structural component, the internal marginal band, has the behaviour of a nearly incompressible hypere-lastic solid [12]. Therefore, a compressible neo-Hookean material [1] is employed to capture the mechanical properties of platelet aggregates, where the compress-ibility originates from the porous structure of aggregates and is dependent on porosity.

When a platelet aggregate is exposed to flow, an interaction force is exerted on the aggregates, leading to deformation. The resulting structural changes will in turn affect the fluid flow and finally reach a steady state under a constant flow condition. Generally, there are two schemes to address an FSI coupled problem, monolithic approaches or partitioned methods. With the monolithic approach, the governing equations for fluid and solid are solved simultaneously and cou-pling conditions at the interface are implicit in the solution procedure [4, 21]. While the partitioned method viewed fluid and solid as two subproblems indi-vidually [9, 10]. The partitioned approach can be further categorized into explicit and implicit strategies [11]. Explicit couplings solve flow and structural equations once per time step which may cause instability issues since the exact equilibrium conditions on the interface are not satisfied. Implicit techniques, on the other hand, enforce equilibrium at each time step by a coupling iteration [8]. Cur-rently, Gauss-Seidel and Jacobi coupling schemes are the most common implicit partitioned methods [20, 30].

In this work, a partitioned FSI model is proposed to simulate and analyze the deformations and associated stress distribution inside platelet aggregates subjected to specific flow conditions. The proposed FSI model incorporates the Navier-Stokes equations to simulate blood flow, exploring the interaction forces between platelet aggregates and the surrounding fluid. Simultaneously, a compressible neo-Hookean model is employed to assess the mechanical properties of the platelet aggregates, with its compressibility determined by the porosity distribution of the aggregate. The FSI model couples these two solvers by the Gaussi-Seidel scheme. A parametric study based on the compressible neo-Hookean model is carried out to evaluate the numerical stability of the method and investigate how the porosity and applied forces impact the deformation and stress of such materials. The FSI model is subsequently applied to study the deformation and stress distribution inside the platelet aggregates at steady state under different wall shear rates (WSRs).

2 Methodology

The FSI model consists of two submodels, a fluid dynamics model for hemodynamic simulations and a hyperelastic model to simulate the response of platelet aggregate under blood flow.

2.1 Computational Fluid Dynamics

In the fluid dynamics model, blood flow was modelled as an incompressible Newtonian fluid governed by the Navier-Stokes equations, coupling with the Brinkman term to introduce the influence of platelet aggregate with porous structure:

$$(\mathbf{u} \cdot \nabla)\mathbf{u} - \nu \nabla^2 \mathbf{u} = \nabla p - \frac{1}{\rho}\frac{\mu}{k}\mathbf{u},$$
$$\nabla \cdot \mathbf{u} = 0, \tag{1}$$

where $\mathbf{u}$, ν, p and μ are the flow velocity, blood kinematic viscosity, pressure and blood dynamic viscosity, respectively. In the last term on the right-hand side (Brinkman term), k represents the permeability of the platelet aggregates, which was estimated from the porosity using the Kozeny-Carman equation,

$$k = \Phi_s^2 \frac{\epsilon^3 D_p^2}{150 \left(1 - \epsilon\right)^2}. \tag{2}$$

Here, Φ_s is the sphericity of the platelets, ϵ is the porosity of the aggregates and D_p represents the platelet diameter.

Table 1. Parameter and parameter relations used in simulations.

Notation	Description	Value	Unit	Reference
ρ	density of fluid	1.025×10^{-3}	$g\ mm^{-3}$	[26]
μ	dynamic viscosity	3×10^{-3}	$g\ mm^{-1}\ s^{-1}$	[27]
ν	kinematic viscosity	μ/ρ	$mm^{-2}\ s^{-1}$	—
Φ_s	sphericity of activated platelet	0.71	—	[18]
D_p	platelet diameter	2×10^{-3}	mm	[2]

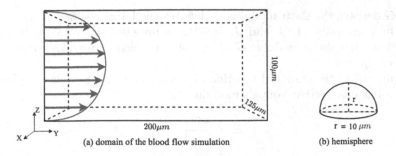

(a) domain of the blood flow simulation   (b) hemisphere

Fig. 1. Schematic diagram of the simulation domain. (a) Domain of the blood flow simulation. (b) Hemisphere used for compressible neo-Hookean materials parametric study.

The parameters used in the flow simulation and their literature sources are listed in Table 1. A cuboid domain was created to represent a blood flow channel, as shown in Fig. 1(a). The platelet aggregate was positioned on the bottom wall of the domain. The blood flowed through the domain along the positive y direction. As the blood flow domain represented a small part of the experimental channel in the x direction, consistent velocity profiles were assigned in the x direction, while parabolic profiles were designated in the z direction at the inlet. At the outlet, a constant pressure condition was prescribed. No-slip boundary conditions were imposed at the top and bottom walls.

2.2 Porosity-Dependent Compressible Neo-Hookean Materials

Platelet aggregates are not rigid bodies and can deform when subjected to the forces of blood flow. The force exerted on the porous structure in flow can be described using permeability as:

$$f = \frac{\mu}{k}\mathbf{u}. \tag{3}$$

This force can subsequently be interpreted as external body force for the mechanical model of the platelet aggregate to explore deformation and compression of the aggregate under flow.

We consider the platelet aggregates as a porous media characterized by compressible neo-Hookean model. Suppose that the aggregate is subjected to the displacement field $\mathbf{D}$, the deformation gradient can be written as $\mathbf{F} = \partial\mathbf{x}/\partial\mathbf{X}$ where $\mathbf{X}$ denotes the position of the aggregate in the reference configuration, while $\mathbf{x} = \mathbf{X} + \mathbf{D}$ is the corresponding deformed configuration of such aggregate. The strain energy density function of a compressible neo-Hookean material, W, is a scalar function of the right Cauchy-Green strain tensor $\mathbf{C} = \mathbf{F}^T\mathbf{F}$:

$$W = \frac{G}{2}(I_1 - 3) - G\ln[J] + \frac{\lambda}{2}(\ln[J])^2, \tag{4}$$

where G denotes the shear modulus, λ denotes the Láme coefficient, $I_1 = tr(\mathbf{C})$ is the first invariant of $\mathbf{C}$, and $J := det\mathbf{F}$ represents the volume change. In simulations, the shear modulus G of the platelet aggregates was chosen to be $1000\,\mathrm{Pa}$ [15].

Compared to the standard neo-Hookean materials, this porous neo-Hookean material has an effective Young's modulus E,

$$E = \frac{3G}{1 + \frac{1}{2}\epsilon^2}, \tag{5}$$

and an effective Poisson's ratio v,

$$v = \frac{1 - \epsilon^2}{2 + \epsilon^2}. \tag{6}$$

Then the Láme coefficient λ is given by:

$$\lambda = \frac{Ev}{(1 + v)(1 - 2v)}. \tag{7}$$

These material properties depend on the porosity ϵ of the material, which introduces the porosity-dependent compressibility to the porous structures [19]. Notably, a highly porous material exhibits an effective Poisson ratio approaching zero, as well as an effective Young's modulus tending toward $2G$. Conversely, a material with low porosity has $v \to \frac{1}{2}$ and $E \to 3G$, aligning with the expected behaviour of an incompressible neo-Hookean solid.

After a single run of the compressible neo-Hookean material simulation, the material undergoes deformation and compression, resulting in consequential changes to its porosity. The change of the porosity due to deformation is quantified by the volumetric strain, $\mathrm{div}(\mathbf{d})$, where $\mathbf{d}$ denotes the displacement. The updated porosity is then computed by

$$\epsilon' = \frac{\mathrm{div}(\mathbf{d}) + \epsilon}{\mathrm{div}(\mathbf{d}) + 1}. \tag{8}$$

Such variation of porosity may lead to changes in the local mechanical properties and result in further deformation. Therefore, we iteratively update the corresponding porosity and repeat the simulation until the material configuration reaches a steady state.

To quantify the stress distribution inside the aggregates, the Von Mises stress σ_v, which can be derived by the Cauchy stress tensor σ, is calculated:

$$\sigma_v = \sqrt{\frac{1}{2}\left[(\sigma_{11} - \sigma_{22})^2 + (\sigma_{22} - \sigma_{33})^2 + (\sigma_{33} - \sigma_{11})^2\right] + 3\left(\sigma_{12}^2 + \sigma_{23}^2 + \sigma_{31}^2\right)}. \tag{9}$$

where σ_{ij} denotes the element of Cauchy stress tensor.

2.3   The Fluid-Structure Coupling

An implicit partitioned FSI method is applied to couple the flow solver (Sect. 2.1) and the structural solver (Sect. 2.2). Platelet aggregate are considered as a porous medium attached to the channel wall. The aggregate will deform under the influence of hemodynamics, and correspondingly, the deformed configurations of the aggregate will result in a change in hemodynamics. Applying an implicit partitioned approach, an equilibrium state between fluid and structure should be reached. Note that, different from traditional FSI problem, there is no explicit interface between flow and structure but the entire structure is immersed in the fluid and exposed to the stresses. The flow solver can be viewed as a function of the configuration of the platelet aggregate,

$$f = \mathcal{F}(\mathbf{x}), \tag{10}$$

where f denotes the corresponding interaction kinetic force the platelet aggregate subjected to. Similarly, the structural solver is considered as a position function of the interaction force:

$$\mathbf{x} = \mathcal{S}(f). \tag{11}$$

The FSI problem therefore can be formulated as a root-finding problem such that the difference between the presumed configuration of the solid and the resulting one after FSI computation is minimised:

$$\underset{\mathbf{x}}{\arg\min} \, ||\mathcal{S} \circ \mathcal{F}(\mathbf{x}) - \mathbf{x}||. \tag{12}$$

The solution to the root-finding problem leads the presumed configuration to converge to the desired configuration in the equilibrium state. In our cases, $\mathcal{S} \circ \mathcal{F}(\mathbf{X})$ was employed as the initial guess of the presumed configuration. Given this initial guess, a Gauss-Seidel iteration scheme [20] updates the presumed configuration every step by minimizing the objective function Eq. 12 and the Aitken relaxation factor ω is adopted to accelerate the convergence [17], as shown in Algorithm 1.

It is noteworthy that due to the deformation of the platelet aggregate, the corresponding porosity and its position will be changed as well in the fluid domain. Therefore extra interpolations between the solid domains and the fluid domain are required. We present the detailed schematic workflow of one iteration (step 4 to 11 in Algorithm 1) of this FSI model in Fig. 2. The updated presumed position computed every iteration will affect the flow solver from two aspects: new geometry of the platelet aggregate and changes in porosity due to deformation and compression. First, the porosity position in the fluid domain has to be updated as the configuration of the platelet aggregate alters. Second, the deformation and compression of the geometries will result in the variation of porosity and subsequently influence the corresponding permeability. Therefore the new permeability information on the flow domain has to be interpolated from a deformed solid model based on the updated presumed configuration before simulation. The flow solver estimates the kinetic force applied on the porous medium and hands

---

Algorithm 1. The Gauss-Seidel iteration scheme with Aitken relaxation factor

---

Input: Flow solver $\mathcal{F}$, structural solver $\mathcal{S}$, stopping criteria ϱ, initial presumed configuration $\mathbf{x}^0$, initial Aitken relaxation factor ω^0.

Output: Converged configuration $\mathbf{x}^*$

1: $n = 0$
2: $\tilde{\mathbf{x}}^0 = \mathcal{S} \circ \mathcal{F}\left(\mathbf{x}^0\right)$
3: $r^0 = \tilde{\mathbf{x}}^0 - \mathbf{x}^0$
4: **while** $\|r^n\|_2 > \varrho$ **do**
5:     $n++$
6:     $\mathbf{x}^n = \mathbf{x}^{n-1} + \omega^{n-1} r^{n-1}$
7:     $\tilde{\mathbf{x}}^n = \mathcal{S} \circ \mathcal{F}\left(\mathbf{x}^n\right)$
8:     $r^n = \tilde{\mathbf{x}}^n - \mathbf{x}^n$
9:     $\omega^n = -\omega^{n-1} \frac{(r^{n-1})^T (r^n - r^{n-1})}{(r^n - r^{n-1})^T (r^n - r^{n-1})}$
10: **end while**
11: $\mathbf{x}^* = \mathbf{x}^n$

---

the information to the solid solver, considering it as body force. The solid solver resolves the corresponding displacement induced by this body force and quits the loop if the residual between the presumed and actual configuration is small enough.

Tetrahedral meshes were employed for both CFD and platelet aggregate meshes. The resolution of CFD meshes was refined to enhance accuracy specifically in the region surrounding the platelet aggregate. The model was implemented via FreeFEM [14] and performed on the Dutch national supercomputer Snelllius (SURF, The Netherlands) on 64 AMD Rome 7H12 CPU ×2 cores.

3   Results

3.1   Compressible Neo-Hookean Parametric Study

Before we implement the FSI model, a parametric study of porosity-dependent neo-Hookean materials was performed to investigate the influence of the porosity and the applied force on the deformation stress of materials. A hemisphere with a radius of $10\,\mu$m, was employed as a simplified configuration of aggregates. Homogeneous porosity was assigned to the hemisphere and a constant body force in the y-direction was applied.

The results of the parametric study are demonstrated in Fig. 3. The increase in magnitude of the body force leads to a larger average and maximum displacement and volumetric strain. Similarly, with the increase of the porosity, the average displacement also increases. However, compared with the influence of body force on the displacement, the effect of porosity on it is less significant. An interesting phenomenon can be observed from the change of the porosity in Fig. 3(c). The porosity change demonstrates a non-monotonic trend across various initial porosity values, with the largest occurring when the initial porosity equals 0.5. A potential reason for this behaviour is elaborated in the *Discussion*

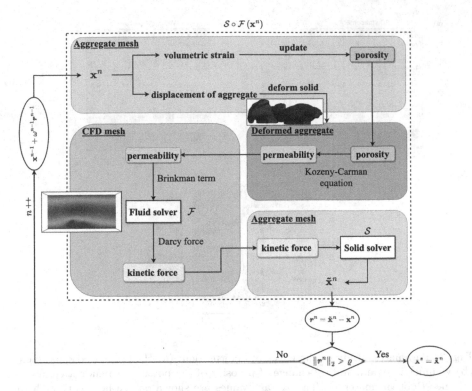

Fig. 2. Flow diagram for one iteration (step 4 to 11 in Algorithm 1) of the FSI model. The blocks in red, light blue, darker blue color represent the computational fluid dynamic (CFD) mesh, reference aggregate mesh and deformed aggregate mesh, respectively. The information shown on the blocks includes the data saved on this mesh and the operation performed based on this mesh. (Color figure online)

section. Notably, although the average change of porosity is small, the maximum porosity change among all cases reaches 0.152, approximately 30% of the corresponding initial porosity value.

Figure 4 shows the distribution of von Mises stress obtained from varying body forces and different porosity. An increase in body force leads to higher stress levels within the hemisphere. In the lower porosity cases, the object tends to behave more like an incompressible solid, therefore the ratio of high stress inside the hemisphere increases.

The deformation, porosity before and after deformation and stress distribution within the hemisphere subjected to a $5 \times 10^{-5}\,\mathrm{N/\mu m^3}$ body force and an initial porosity of 0.9 (case with the most significant displacement) are presented in Fig. 5. Since the applied body force acts in the y-direction, it is observed that the hemisphere undergoes a noticeable deformation towards the positive y-direction. This deformation leads to stretching around the left bottom edge of the hemisphere and compression around the right bottom edge. Consequently, the porosity of the left bottom part increases, while the bottom right part of

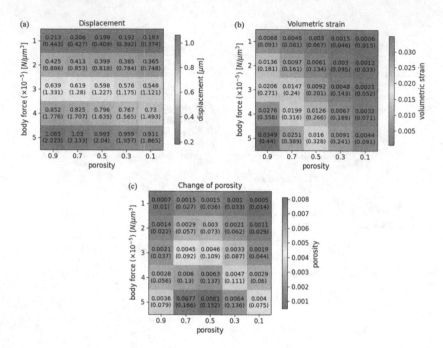

Fig. 3. Parametric study of the effect of deformation. (a) Heatmap of displacement, (b) volumetric strain, and (c) change of porosity of the porosity-dependent compressible neo-Hookean materials. The average values are shown on the heatmap, with the maximum values shown in brackets. The color corresponds to the magnitude of the average value. (Color figure online)

the hemisphere becomes denser. On the right bottom edge, the porosity has decreased to 0.82. Moreover, Fig. 5(d) demonstrates the von Mises stress distribution inside the hemisphere on a cross-section. The high stress region is distributed at the lower part of the hemisphere close to the immovable base.

3.2   FSI Simulation

We applied the proposed FSI model to both low shear and high shear aggregate data as obtained in earlier experiments [13]. In the low shear aggregate case the maximum inlet flow velocity was set to $20 \, \text{mm s}^{-1}$, corresponding to the flow condition of 800^{-1} WSR. For the high shear cases the maximum inlet flow velocities of $40 \, \text{mm s}^{-1}$ and $100 \, \text{mm s}^{-1}$ were prescribed. These velocities correspond to WSRs of $1600 \, \text{s}^{-1}$ and $4000 \, \text{s}^{-1}$ flow environments, respectively.

The original configuration, the deformed configuration, and the comparison of them under $\text{WSR} = 4000 \, \text{s}^{-1}$ flow are presented in Fig. 6(a). The aggregate deformed visibly along the direction of blood flow (positive y-direction). A noticeable difference between the reference and deformed configurations can be observed in the comparison figure. Furthermore, it is important to evaluate the

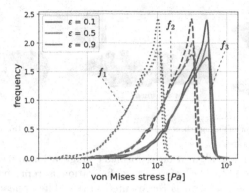

Fig. 4. Distribution of von Mises stress inside the hemispheres with a porosity of 0.1, 0.5 and 0.9 in log-scale. $f_1 = 1 \times 10^{-5}\,\mathrm{N/\mu m^3}$, $f_2 = 3 \times 10^{-5}\,\mathrm{N/\mu m^3}$ and $f_3 = 5 \times 10^{-5}\,\mathrm{N/\mu m^3}$.

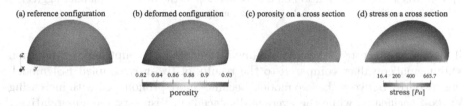

Fig. 5. (a), (b) - Reference configuration and deformed configuration of the hemisphere. The color represents the porosity on it. (c), (d) - Porosity and von Mises stress on a cross-section of the deformed hemisphere.

stress within the platelet aggregate, since it might be connected to the platelet activation and the stability of such aggregate. Figure 6(b) demonstrates the von Mises stress distribution on a cross-section of the deformed configuration of the same aggregate. As expected, the stress close to the attachment region (bottom) of the aggregates is higher than that on the top part.

Effect of Increasing WSR. With the increase of WSR, both the strain and the vom Mises stress of the aggregates becomes more significant (see Table 2, two-way). The maximum displacement of the low shear aggregate is 0.145 μm, while that for high shear aggregate under a flow condition of 4000 s⁻¹ WSR reaches 3.91 μm. This is a result of the larger forces induced by the high flow velocity. Moreover, the maximum von Mises stress is almost two orders of magnitude greater under high WSR compared to values of low WSR flow conditions, increasing from 43.46 Pa to 1263.3 Pa.

Comparison of Two-Way Coupled FSI with One-Way Coupled FSI Simulation. A comparison between the results of two-way coupled FSI and one-way coupled FSI simulation is presented in Table 2. It is observed that across

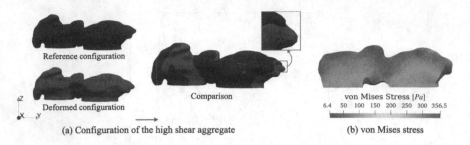

(a) Configuration of the high shear aggregate   (b) von Mises stress

Fig. 6. (a) The reference configuration, deformed configuration and comparison of them under a WSR $= 4000\,\mathrm{s}^{-1}$. The reference configuration is represented in blue color, while the deformed configuration is represented in red. The appearance of red in the comparison figure signifies a difference of the aggregate surfaces due to deformation. The small box in the comparison figure is a zoom-in figure showing the differences between the two configurations. The red arrow indicates the direction of blood flow. (b) Von Mises stress distribution on a cross-section of the deformed platelet aggregate under WSR $= 4000\,\mathrm{s}^{-1}$ flow conditions. (Color figure online)

all factors of interest, the results obtained from one-way coupled FSI simulation exhibit smaller values compared to the results of two-way coupled FSI model. This disparity between the two models becomes more pronounced with increasing WSRs. Specifically, while the average displacement disparity remains relatively minimal at low shear conditions, approximately $0.01\,\mu\mathrm{m}$, it increases to $0.31\,\mu\mathrm{m}$ under a WSR of $4000\,\mathrm{s}^{-1}$. Similar trends can be observed in stress and porosity variations. In the scenario of high shear aggregate under a flow condition of $4000\,\mathrm{s}^{-1}$ WSR, the discrepancy observed in maximum stress for the two models amounts to approximately $200\,\mathrm{Pa}$. Conversely, under a low WSR condition, this difference diminishes significantly to one-twentieth of its counterpart in the high WSR scenario.

Furthermore, it is noteworthy that the maximum porosity variation observed for the low shear aggregate and the high shear aggregate under a shear rate of $1600\,\mathrm{s}^{-1}$ remains below 0.1 for both two-way and one-way coupled FSI models, with an average value of less than 0.01. Such minor changes suggest that they are unlikely to exert a significant influence on the porous structure of the aggregate. In contrast, noticeable porosity changes are evident for the high shear aggregate subjected to a shear rate of $4000\,\mathrm{s}^{-1}$ in both two-way and one-way coupled FSI models, with a disparity of two models approximately 0.02.

4   Discussion

This work presents a novel methodology for modelling the interaction between platelet aggregates and hemodynamics under specific flow conditions. The model incorporates a fluid solver based on the Navier-Stokes equation to simulate hemodynamics and a porosity-dependent compressible neo-Hookean material to simulate aggregate deformation. Platelet aggregates are considered as porous mediums and the corresponding porosity varies along with volumetric strain under

Table 2. Comparison of two-way coupled FSI with one-way coupled FSI simulation results.

Low shear aggregate under WSR $= 800\,\mathrm{s}^{-1}$						
	d_{avg}	d_{max}	$\sigma_{v,\mathrm{avg}}$	$\sigma_{v,\mathrm{max}}$	$|\Delta\epsilon|_{\mathrm{avg}}$	$|\Delta\epsilon|_{\mathrm{max}}$
	$[\mu m]$	$[\mu m]$	$[Pa]$	$[Pa]$	$(\times 10^{-4})$	
two-way	0.042 ± 0.027	0.145	9.60 ± 3.22	43.46	3.3 ± 3.2	0.0061
one-way	0.031 ± 0.021	0.118	6.90 ± 2.78	33.48	2.5 ± 2.5	0.0047
High shear aggregate under WSR $= 1600\,\mathrm{s}^{-1}$						
two-way	0.556 ± 0.029	1.60	44.33 ± 25.22	474.5	2.20 ± 2.33	0.080
one-way	0.423 ± 0.232	1.351	33.28 ± 20.20	393.4	1.76 ± 1.76	0.075
High shear aggregate under WSR $= 4000\,\mathrm{s}^{-1}$						
two-way	1.388 ± 0.730	3.91	108.8 ± 63.95	1263.3	7.38 ± 6.38	0.185
one-way	1.079 ± 0.601	3.47	84.77 ± 51.89	1074.29	5.74 ± 4.75	0.165

Average displacement magnitude (d_{avg}), maximum displacement magnitude (d_{max}), average von Mises stress ($\sigma_{v,\mathrm{avg}}$), maximum von Mises stress ($\sigma_{v,\mathrm{max}}$), average poros-ity change ($|\Delta\epsilon|_{\mathrm{avg}}$), and maximum porosity change ($|\Delta\epsilon|_{\mathrm{max}}$) of low shear aggregate under a WSR of $800\,\mathrm{s}^{-1}$, high shear aggregate under a WSR of $1600\,\mathrm{s}^{-1}$ and high shear aggregate under a WSR of $4000\,\mathrm{s}^{-1}$ flow conditions.

deformation. The proposed FSI model provides a more precise and effective way to simulate the behavior of aggregates.

The parametric study of porosity-dependent compressible neo-Hookean materials elucidate the material behavior including the variation of porosity, stress distribution and deformation under different magnitude of body forces and initial porosity. Both the initial porosity and the magnitude of the body force have a positive correlation with material deformation. Varying the initial porosity also leads to different stress distributions within the materials. A non-monotonic relation was observed between the average change of the porosity and the magnitude of initial porosity (see Fig. 3(c), maximum at $\epsilon = 0.5$). An intuitive explanation to this phenomenon is that the change in porosity is not only a function of volumetric strain but also its initial porosity, e.g. $|\Delta\epsilon| = |(\epsilon - 1)\frac{\mathrm{div}(\mathbf{d})}{1+\mathrm{div}(\mathbf{d})}|$, therefore the change of porosity will be scaled by its initial porosity despite a monotonic tendency of volumetric strain observed along with the increase of the initial porosity.

The results of the FSI model reveal that noticeable deformation only occurs in very high WSR flow conditions. Within low WSR environments, the differ-ence in deformation between the two-way and one-way FSI models is negligible, which suggests that one-way FSI may be preferred over the two-way model for these cases considering the computational cost. Furthermore, the peak of the structural stress inside the aggregates increases with the increase of the WSR. This phenomenon can be attributed to two potential factors. On the one hand, faster flow will result in higher forces as shown in Eq. 3, and on the other hand, the geometry of the high shear aggregates tends to grow taller, protruding more into the flow and therefore will interact with the flow regions of higher velocity. However, it is shown in [15] that the shear modulus can change with the applied stress, and the accuracy of this shear modulus directly affects the magnitude of deformation. Therefore further investigations are required to evaluate precise material parameters under different flow conditions. By leveraging configuration data of aggregates measured from blood perfusion experiments with and without blood flow force, the shear modulus can be inferred by formulating an inverse problem. The obtained shear modulus can subsequently serve as a parameter in FSI model to simulate the behaviour of aggregates over time under different flow environments.

Although the proposed model demonstrates notable effectiveness, it is crucial to acknowledge its limitations. First, since an additional pressure term needs to be introduced to supplement the Cauchy stress in an incompressible material, setting $\epsilon = 0$ might result in numerical instabilities in the porosity-dependent compressible neo-Hookean materials. In this study, for numerical stability 0.01 was chosen as the minimum value for porosity which only introduces a negligible error. Moreover, since the FSI model is coupled by an implicit partitioned scheme, the fluid and solid solver are called repeatedly and may lead to high computational costs. To reduce the necessary computational resources, an adaptive mesh was generated for the CFD simulations. The mesh was refined at the position of aggregates to cope with the potential complexity of fluid dynamics while remaining relatively coarse away from the aggregate. This refinement enhances the accuracy of interaction forces derived from the CFD simulations, thereby improving the fidelity of input data for the neo-Hookean materials model. Furthermore, the interpolations between meshes in the Gauss-Seidel iteration will inevitably introduce discrepancies to the model and propagate. This error can be reduced by employing finer meshes at the cost of increased computational effort.

5   Conclusion

In this work, we present an FSI model to simulate and investigate platelet aggregate deformation and associated stress under three different flow conditions based on porosity-dependent compressible neo-Hookean materials. A parametric analysis of neo-Hookean materials highlights the importance of porosity on deformation and stress. The proposed FSI model incorporates the internal microstructure of the aggregates and facilitates a relatively precise prediction of the aggregate deformation stress in specific flow environments. The results suggests that employing a high-detail mechanical model is advantageous mostly under elevated flow conditions. Finally, this work lays a foundation for a better understanding of the deformability and mechanical properties of the platelet aggregates, which in turn could be used in the prediction of the stability of platelet aggregates or more complex porous biomaterials in the future.

References

1. Barsimantov, J., et al.: Poroelastic characterization and modeling of subcutaneous tissue under confined compression. Ann. Biomed. Eng. (2024)
2. Bath, P., Butterworth, R.: Platelet size: measurement, physiology and vascular disease. Blood Coagul. Fibrinol. Int. J. Haemostasis Thrombosis **7**(2), 157–161 (1996)
3. Bershadsky, E.S., Ermokhin, D.A., Kurattsev, V.A., Panteleev, M.A., Nechipurenko, D.Y.: Force balance ratio is a robust predictor of arterial thrombus stability. Biophys. J. (2024)
4. Blom, F.J.: A monolithical fluid-structure interaction algorithm applied to the piston problem. Comput. Methods Appl. Mech. Eng. **167**(3), 369–391 (1998)
5. Boodt, N., et al.: Mechanical characterization of thrombi retrieved with endovascular thrombectomy in patients with acute ischemic stroke. Stroke **52**(8), 2510–2517 (2021)
6. Cahalane, R.M.E., et al.: Tensile and compressive mechanical behaviour of human blood clot analogues. Ann. Biomed. Eng. **51**(8), 1759–1768 (2023)
7. van Dam, E.A., et al.: Non-linear viscoelastic behavior of abdominal aortic aneurysm thrombus. Biomech. Model. Mechanobiol. **7**(2), 127–137 (2008)
8. Degroote, J.: Partitioned simulation of fluid-structure interaction. Arch. Comput. Methods Eng. **20**(3), 185–238 (2013)
9. Degroote, J., Haelterman, R., Annerel, S., Vierendeels, J.: Coupling techniques for partitioned fluid-structure interaction simulations with black-box solvers. In: MpCCI User Forum, 10th, Proceedings, pp. 82–91. Fraunhofer Institute SCAI (2009)
10. Felippa, C.A., Park, K., Farhat, C.: Partitioned analysis of coupled mechanical systems. Comput. Methods Appl. Mech. Eng. **190**(24), 3247–3270 (2001). Adv. Comput. Methods Fluid-Struct. Interact
11. Fernández, M.A.: Coupling schemes for incompressible fluid-structure interaction: implicit, semi-implicit and explicit. SeMA J. **55**(1), 59–108 (2011)
12. Han, D., Zhang, J., He, G., Griffith, B.P., Wu, Z.J.: A prestressed intracellular biomechanical model for the platelet to capture the disc-to-sphere morphological change from resting to activated state. Int. J. Comput. Methods **19**(10), 2250021 (2022)
13. Hao, Y., Závodszky, G., Tersteeg, C., Barzegari, M., Hoekstra, A.G.: Image-based flow simulation of platelet aggregates under different shear rates. PLOS Comput. Biol. **19**(7), 1–20 (2023)
14. Hecht, F.: New development in FreeFem++. J. Numer. Math. **20**(3–4), 251–265 (2012)
15. Huang, C.C., Shih, C.C., Liu, T.Y., Lee, P.Y.: Assessing the viscoelastic properties of thrombus using a solid-sphere-based instantaneous force approach. Ultrasound Med. Biol. **37**(10), 1722–1733 (2011)
16. Kadri, O.E., Chandran, V.D., Surblyte, M., Voronov, R.S.: In vivo measurement of blood clot mechanics from computational fluid dynamics based on intravital microscopy images. Comput. Biol. Med. **106**, 1–11 (2019)
17. Küttler, U., Wall, W.A.: Fixed-point fluid-structure interaction solvers with dynamic relaxation. Comput. Mech. **43**(1), 61–72 (2008)
18. Lee, S., Jang, S., Park, Y.: Measuring three-dimensional dynamics of platelet activation using 3-d quantitative phase imaging. bioRxiv (2019)

19. Maas, S.A., Ellis, B.J., Ateshian, G.A., Weiss, J.A.: FEBio: finite elements for biomechanics. J. Biomech. Eng. **134**(1), 011005 (2012)
20. Matthies, H.G., Niekamp, R., Steindorf, J.: Algorithms for strong coupling procedures. Comput. Methods Appl. Mech. Eng. **195**(17), 2028–2049 (2006)
21. Rugonyi, S., Bathe, K.: On the analysis of fully-coupled fluid flows with structural interactions - a coupling and condensation procedure. Int. J. Comput. Civil Struct. Eng. **1**, 29–41 (2000)
22. Slaboch, C.L., Alber, M.S., Rosen, E.D., Ovaert, T.C.: Mechano-rheological properties of the murine thrombus determined via nanoindentation and finite element modeling. J. Mech. Behav. Biomed. Mater. **10**, 75–86 (2012)
23. Stalker, T.J., et al.: Hierarchical organization in the hemostatic response and its relationship to the platelet-signaling network. Blood **121**(10), 1875–1885 (2013)
24. Storti, F., van de Vosse, F.N.: A continuum model for platelet plug formation, growth and deformation. Int. J. Numer. Methods Biomed. Eng. **30**(12), 1541–1557 (2014)
25. Teeraratkul, C., Tomaiuolo, M., Stalker, T.J., Mukherjee, D.: Investigating clot-flow interactions by integrating intravital imaging with in silico modeling for analysis of flow, transport, and hemodynamic forces. Sci. Rep. **14**(1), 696 (2024)
26. Trudnowski, R.J., Rico, R.C.: Specific gravity of blood and plasma at 4 and 37°C. Clin. Chem. **20**(5), 615–616 (1974)
27. Voronov, R.S., Stalker, T.J., Brass, L.F., Diamond, S.L.: Simulation of intrathrombus fluid and solute transport using in vivo clot structures with single platelet resolution. Ann. Biomed. Eng. **41**(6), 1297–1307 (2013)
28. Wang, W., Lindsey, J.P., Chen, J., Diacovo, T.G., King, M.R.: Analysis of early thrombus dynamics in a humanized mouse laser injury model. Biorheology **51**(3–14), 1 (2014)
29. Welsh, J.D., et al.: A systems approach to hemostasis: 1. The interdependence of thrombus architecture and agonist movements in the gaps between platelets. Blood **124**(11), 1808–1815 (2014)
30. Winterstein, A., Lerch, C., Bletzinger, K.U., Wüchner, R.: Partitioned simulation strategies for fluid-structure-control interaction problems by Gauss-Seidel formulations. Adv. Model. Simul. Eng. Sci. **5**(1), 29 (2018)

Modeling of Turbulent Flow over 2D Backward-Facing Step Using Generalized Hydrodynamic Equations

Alex Fedoseyev[✉] [iD]

Ultra Quantum Inc., Huntsville, AL 35738, USA
af@ultraquantum.com
http://www.ultraquantum.com

Abstract. The Generalized Hydrodynamic Equations are being investigated for simulating turbulent flows. They were derived from the Generalized Boltzmann Equation by Alexeev (1994), which itself was obtained from first principles via a chain of Bogolubov kinetic equations and considers particles of finite dimensions. Compared to the Navier-Stokes equations, the Generalized Hydrodynamic Equations include new terms representing temporal and spatial fluctuations. These terms introduce a timescale multiplier denoted by τ, and the Generalized Hydrodynamic Equations reduce to the Navier-Stokes equations when τ equals zero. The nondimensional τ is calculated as the product of the Reynolds number and the squared ratio of length scales, $\tau = Re \times (l/L)^2$, where l represents the apparent Kolmogorov length scale and L denotes a hydrodynamic length scale.

In this study, 2D turbulent flow over a Backward-Facing Step (BFS) with a step height of $H = L/3$ (where L is the channel height) at Reynolds number Re $= 132000$ was investigated using finite-element solutions of the GHE. The results were compared to Direct Numerical Simulations (DNS) utilizing the Navier-Stokes equations, and to a $k - \varepsilon$ turbulence model, as well as experimental data. The comparison encompassed velocity profiles, recirculation zone length, and the velocity flow field. The obtained data confirm that the GHE results are in good agreement with the experimental findings, while other approaches diverge significantly from the experimental data.

Keywords: Turbulent flow · Generalized Hydrodynamic Equations · DNS · Navier-Stokes equations · $k - \varepsilon$ turbulence model · Numerical solution · Comparison with experimental data

1 Introduction

The Generalized Hydrodynamic Equations (GHE) were derived by Alexeev in 1994 from the Generalized Boltzmann Equation. The Generalized Boltzmann Equation itself was obtained from first principles through the Bogolubov kinetic equations chain, taking into account particles with finite dimensions [1].

© The Author(s), under exclusive license to Springer Nature Switzerland AG 2024
L. Franco et al. (Eds.): ICCS 2024, LNCS 14838, pp. 63–69, 2024.
https://doi.org/10.1007/978-3-031-63783-4_6

1.1   Generalized Boltzmann Transport Equation (GBE)

The kinetic theory of gases is founded upon the solution to the Boltzmann transport equation, governing the space-time evolution of the particle velocity distribution function, f, expressed as

$$\frac{Df}{Dt} = J, \tag{1}$$

where

$$\frac{D}{Dt} = \frac{\partial}{\partial t} + \mathbf{v} \cdot \frac{\partial}{\partial \mathbf{r}} + \mathbf{F} \cdot \frac{\partial}{\partial \mathbf{v}} \tag{2}$$

represents material derivative in space, velocity space and time. J denotes the collision integral ([2], p.11), where $\mathbf{v}$ and $\mathbf{r}$ are the velocity and the radius-vector of the particle, respectively, and $\mathbf{F}$ is the force acting on the particle.

The standard Boltzmann transport equation takes into account the changes in distribution function f on hydrodynamic and mean time between collision scales of infinitesimal particles. Accounting for a third time scale, associated with finite dimensions of interacting particles, gives rise to an additional term in the Boltzmann transport equation resulting in a generalized form given by

$$\frac{Df}{Dt} - \frac{D}{Dt}(\tau \frac{Df}{Dt}) = J, \tag{3}$$

where τ is the timescale, a material property, that Alexeev (1994) related to the mean time between particle collisions. The new term is thermodynamically consistent; more details on the GBE are provided in Alexeev's book [2].

2   Generalized Hydrodynamic Equations as Governing Equations

Hydrodynamic equations can be obtained from Eq. (3) by multiplying the latter by the standard collision invariants (mass, momentum, energy) and integrating the result in the velocity space. These equations are valid for incompressible viscous flow. They have the following form and were originally presented in [5]:

$$\frac{\partial \mathbf{V}}{\partial t} + (\mathbf{V}\nabla)\mathbf{V} - Re^{-1}\nabla^2\mathbf{V} + \nabla p - \mathbf{F} = \tau \left\{ 2\frac{\partial}{\partial t}(\nabla p) + \nabla^2(p\mathbf{V}) + \nabla(\nabla \cdot (p\mathbf{V})) \right\} \tag{4}$$

while continuity equation is

$$\nabla \cdot \mathbf{V} = \tau \left\{ 2\frac{\partial}{\partial t}(\nabla \cdot \mathbf{V}) + \nabla \cdot (\mathbf{V}\nabla)\mathbf{V} + \nabla^2 p - \nabla \cdot \mathbf{F} \right\} \tag{5}$$

where $\mathbf{V}$ and p are nondimensional velocity and pressure, $Re = V_0 L_0/\nu$ - the Reynolds number, V_0 - velocity scale, L_0 - hydrodynamic length scale, ν -

kinematic viscosity, $\mathbf{F}$ is a body force and a nondimensional $\tau = \tau^* L_0^{-1} V_0 = \tau^* \nu / L^2 Re$. The terms containing τ are called the fluctuations (temporal and spatial) [1].

We made the following assumptions deriving Eq. (4, 5): (i) τ is assumed to be constant, (ii) the nonlinear terms of the third order in the fluctuations, and the terms of the order τ / Re, are neglected, so the focus is on large Re numbers, (iii) slow flow variation is assumed, neglecting second derivatives in time.

Additional boundary conditions on walls require fluctuations to be zero. The following boundary condition for pressure on walls is expressed as:

$$(\nabla p - \mathbf{F}) \cdot \mathbf{n} = 0, \tag{6}$$

where $\mathbf{n}$ is a wall normal.

The Eqs. (4) together with equation (6), and boundary and initial conditions for the velocities, constitute the governing equations that we are going to solve for turbulent flows. The Generalized Hydrodynamic Equations become the Navier-Stokes equations when the timescale τ is zero. No additional turbulent model is involved or used to obtain the solution for turbulent flow.

Recently, an analytical solution of GHE for turbulent flow in channel has been obtained, which compares well with a number of turbulent channel flow experiments for Reynolds number from $Re = 2970$ to $Re = 970000$ [6].

3   Backward-Facing Step Flow Problem

The Backward-Facing Step (BFS) flow is illustrated in Fig. 1. The flow progresses from left to right over a backward-facing step of height H. The entrance channel width is $W = 2 \cdot H$, as depicted in Fig. 1. The Reynolds number for the channel exit width is Re = 132000, as in the experiment [11], or the Reynolds number calculated for the step height H is Re = 44000. The inlet velocity is U=1.

3.1   Numerical Simulations Results

GHE Model. The averaged flow field, the horizontal velocity contours, obtained with the GHE model are shown in Fig. 1(a). We used a finite element mesh refined near the walls for GHE consisting of 27700 nodes (triangular elements with linear approximation for all variables). The non-stationary problem described by Eq. (4) and (5) was been solved until it reached a nearly quasi steady state, and then the averaging of the velocity field was performed over the time interval $t = [100, 200]$.

The parameter τ^* in the expression $\tau = \tau^* \nu / L^2 Re$ is a material property and is not currently known for the air, which was used in the experiment. We varied the parameter τ to fit the velocity profile as shown in Fig. 2. By fitting the velocity profile, we also obtained an excellent recirculation zone length, which compared well with the experiment. A similar procedure was performed in our work [4] for water, where fitting of one velocity profile by varying τ^* resulted

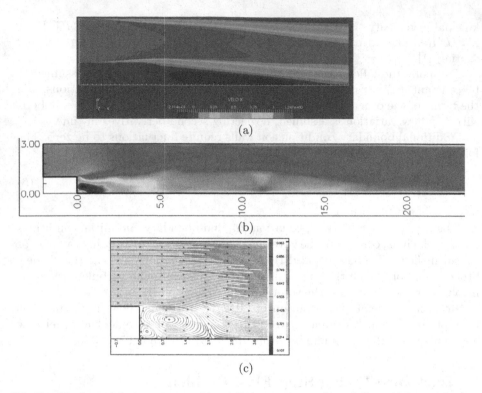

Fig. 1. Backward-facing step problem: flow progresses from left to right, with the entrance channel width denoted as W and the step height as H = W/3. (a) Contours of the averaged horizontal velocity obtained with the solution GHE by finite-element method at Re = 132000, with a mesh of 27700 nodes, (b) Contours of the averaged horizontal velocity obtained with the DNS (Navier-Stokes) solution at Re = 132000 with a mesh of 1.1 million, (c) Zoomed-in of streamlines for the DNS (Navier-Stokes) solution.

in excellent agreement for all the velocity profiles across different experiments conducted at various Reynolds numbers. The determined value of τ^* for distilled water remained consistent across different experiments where distilled water was also used.

DNS (Navier-Stokes Model). We can estimate how fine the mesh must be to perform a resolved DNS. The smallest scale of the flow is [10]

$$\lambda = \sqrt{\frac{\nu}{|\nabla \mathbf{V}|}}. \tag{7}$$

If we assume that the non-dimensional vorticity is on the order of $Re^{1/2} = 363$ and use $|\nabla \mathbf{V}| = 100$, then we arrive at the estimate $\lambda = 3 \cdot 10^{-4}$. Thus, the grid spacing in the boundary layer (where the vorticity is largest) should be roughly $\Delta s = 3 \cdot 10^{-4}$, and the total mesh size becomes approximately 1.1M grid points.

Figure 1(b) shows the time-averaged solution when averaged over the time interval $t = [50, 200]$, illustrating the averaged flow field, the horizontal velocity contours. The streamlines of the time-averaged solution, depicted in Fig. 1(c), reveal a small recirculation bubble that re-attaches at approximately $x = 2.1H$.

The Overture Cgins module was used for the DNS (Navier-Stokes model) [8,9].

$K - \epsilon$ Model. The solution with the $k - \epsilon$ turbulence model was obtained using the commercial CFD2000 software with a mesh of about 45000 nodes refined at a boundary.

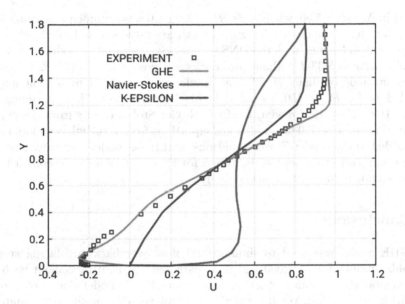

Fig. 2. Flow over a backward facing step, $Re = 132,000$, comparison of the averaged horizontal velocity for the GHE solution, DNS (Navier-Stokes), $k - \epsilon$ model [13], and experimental data (squares) [11] at x = 5.33H.

Comparison of Velocity Profiles. The flow patterns in Fig. 1 are quite distinct. Figure 2 presents the computed velocity profiles at $x = 5.33H$ at the end of the recirculation zone ($x = 0$ at the edge of the step) for different models, along with experimental mean velocity measurements [11].

The solution with a standard $k - \epsilon$ model shows the velocity profile at $x = 5.33H$ that has no backward flow. A standard $k - \epsilon$ model underpredicts the recirculation zone length $X_r = 5.5H$ by a substantial amount, 20–25% according to [12], where more sophisticated turbulence models have been proposed for this problem.

The GHE model output satisfactorily agrees with the experimental data for both the velocity profile and the recirculation zone length X_r. The GHE model

Table 1. Comparison of the recirculation zone length for different models and experiment.

Model	Recirculation zone length/H
Navier-Stokes	2.1
$k - \varepsilon$	5.5
GHE	7.5
Experiment [11]	7.0 ± 1.0

resulted in $X_r/H = 7.50$, while $X_r^{exp}/H = 7.0 \pm 1.0$ was obtained experimentally. All the data for the recirculation zone length are provided in Table 1.

The results obtained with the DNS (Navier-Stokes equations) show the velocity profile at $x = 5.33H$ without any backward flow, and $X_r/H = 2.1$. Similar conclusion made by Jiang (1993), the Navier-Stokes cannot match the experimental data for $Re \geq 450$, as noted in his NASA report [7]. Experiments by Amaly (1996) show that deviation of the Navier-Stokes results from the experiment begins at $Re = 350$ [3]. Additional equations from a turbulence model need to be added to the Navier-Stokes equations, and these models typically have two to five parameters that need to be tuned for specific problems. In contrast, the GHE does not require a turbulence model.

4   Conclusions

The GHE model was used to simulate 2D flow over backward-facing step at Reynolds number $Re = 132000$ and demonstrated excellent agreement with the experimental data. Unlike other methods, the GHE model does not require any turbulence models, yet it yielded turbulent velocity profiles that matched well with the experimental data (Kim, 1980). This outcome marks a significant improvement over our previous results presented at the ICCS 2010 conference for this type of flow [4], achieved by utilizing a larger number of nodes in the finite element mesh. In contrast, both DNS (Navier-Stokes equations) and the $k - \varepsilon$ turbulence model results for this turbulent flow deviated considerably from the experimental data.

Acknowledgments. Author would like to extend the sincere thanks to W.D. Henshaw for the DNS (Navier-Stokes equations) solution using his Overture Cgins software.

References

1. Alexeev, B.V.: The generalized Boltzmann equation, generalized hydrodynamic equations and their applications. Phil. Trans. Roy. Soc. London. A. **349**, 417–443 (1994)
2. Alexeev B.V., Generalized Boltzmann Physical Kinetics, Elsevier, Amsterdam (2004)
3. Armaly, B.F., Durst, F., Pereira, J.C.F., Schonung, B.: Experimental and theoretical investigation of backward-facing step flow. J. Fluid Mech. **127**, 473–496 (1983)
4. Fedoseyev, A.I., Alexeev, B.V.: Simulation of viscous flows with boundary layers within multiscale model using generalized hydrodynamics equations. Procedia Comput. Sci. **1**, 665–672 (2010)
5. Fedoseyev, A., Alexeev, B.V.: Generalized hydrodynamic equations for viscous flows-simulation versus experimental data, in AMiTaNS-12. Am. Inst. Phys. AIP CP **1487**, 241–247 (2012)
6. Fedoseyev, A.: Approximate analytical solution for turbulent flow in channel. J. Phys. Conf. Ser. **2675**, 012011 (2023)
7. Jiang, B.-N., Hou, L.-J., Lin, T.-L.: Least-squares finite element solutions for three-dimensional backward-facing step flow. In: Fifth International Symposium on Computational Fluid Dynamics, NASA Technical Memorandum ICOMP-93-31 10635, ICOMP-93-31, Sendai, Japan, August 31-September 3, 1993
8. Henshaw, W.D.: A fourth-order accurate method for the incompressible Navier-Stokes equations on overlapping grids. J. Comput. Phys. **113**, 13–5 (1994)
9. Henshaw, W.D.: Cgins user guide: an overture solver for the incompressible Navier-Stokes equations on composite overlapping grids, Software Manual LLNL-SM-455851, Lawrence Livermore National Laboratory (2010)
10. Henshaw, W.D., Kr eiss, H.-O., Reyna, L.G.M.: On the smallest scale for the incompressible Navier-Stokes equations. Theoret. Comput. Fluid Dyn. **1**, 65–95 (1989)
11. Kim, J., Kline, S.J., Johnston, J.P.: Investigation of a reattaching turbulent shear layer: flow over a backward-facing step. ASME J. Fluids Engng. **102**, 302–308 (1980)
12. Thangam, S., Speziale, C.G.: ICASE Report 91–23. NASA Langley, Hampton, Virginia (1991)
13. Zijlema, M., Segal, A., Wesseling, P.: Report DUT-TWI-94-24, The Netherlands (1994)

Simulation of Droplet Dispersion from Coughing with Consideration of Face Mask Motion

Ayato Takii[1,4]([✉]), Tatsuya Miyoshi[2], Masashi Yamakawa[2], Yusei Kobayashi[2], Shinichi Asao[3], Seiichi Takeuchi[3], and Makoto Tsubokura[1,4]

[1] RIKEN Center for Computational Science, Kobe 650-0047, Hyogo, Japan
Ayato.takii@riken.jp
[2] Kyoto Institute of Technology, Sakyo-ku 606-8585, Kyoto, Japan
[3] College of Industrial Technology, Amagasaki 661-0047, Hyogo, Japan
[4] Kobe University, Kobe 657-0013, Hyogo, Japan

Abstract. Wearing a face mask is widely acknowledged as a critical defense against the transmission of the novel coronavirus (COVID-19) and influenza. This research focuses on the deformation of face masks during a cough and uses fluid dynamics simulations to more precisely predict the trajectory of virus-laden droplets. By employing motion capture technology, we measured the mask's displacement, which reaches up to 6 mm during a cough. Moreover, this paper delves into how the mask's deformation influences the movement of these droplets. We created a model for a small, spherical droplet and analyzed its dispersion by solving its motion equation, factoring in the cough's flow rate, droplet size distribution, and evaporation, all while considering the mask's deformation. Our findings reveal that mask deformation leads to a 7% reduction in average flow velocity compared to analyses using a non-deforming mask. Additionally, the distance droplets disperse increases over time when mask deformation is considered. As a practical application, we analyzed droplet dispersion in a scenario where a wheelchair is being pushed, utilizing airflow data that accounts for mask deformation. This analysis indicated that pushing a wheelchair at a speed of 0.5 m/s significantly raises the infection risk for individuals behind it.

Keywords: COVID-19 · Face Mask · Computational Fluid Dynamics

1 Introduction

In 2019, the emergence of a novel coronavirus infection (COVID-19), caused by the SARS-CoV-2 virus, was confirmed [1]. Subsequently, COVID-19 proliferated globally at an explosive rate. Presently, while the trepidation towards the novel coronavirus has diminished among people, it remains important to strategize against potential unknown infectious diseases and mutations of the coronavirus that may arise in the future. Recommended practices for protecting oneself from viral droplets include maintaining social distancing, wearing masks and ensuring adequate ventilation. Due to the challenges in reproducing studies on virus droplets and the complexity of experiments, an engineering approach is efficacious. Notably, research employing Computational Fluid Dynamics (CFD) for numerical simulations has garnered significant attention. Bale et al. [2]

L. Franco et al. (Eds.): ICCS 2024, LNCS 14838, pp. 70–84, 2024.
https://doi.org/10.1007/978-3-031-63783-4_7

developed a framework for quantifying the risk of airborne infections like COVID-19, demonstrating that infection risk decreases as distance from an infected person increases. Yamakawa et al. [3] used a classroom model with ventilation to show that droplets can remain airborne for extended periods, thereby increasing the risk of infection. Bale et al. [4] analyzed the spread of virus droplets during conversations by changing the direction of the wind. In addition, numerical simulations in environments such as classrooms, trains [5], and escalators [6] have been conducted as simulations in applied situations, demonstrating the usefulness of simulating droplet behavior using CFD. Regarding social distancing, Dbouk et al. [7] points out that 2 m is not a sufficient distance. Moreover, Bourouiba et al. [8] revealed that droplets can be dispersed up to 6 to 7 m indoors. Furthermore, Blocken et al. [9] showed that the range in which droplets fly changes greatly depending on the surrounding airflow environment. Therefore, social distancing alone is insufficient to prevent infection, and it is recommended to wear a mask to reduce the risk of infection. Focusing on masks, Bagchi et al. [10] clarified the behavior of droplets passing through a mask in experiment. In addition, Bourrianne et al. [11] actually filmed people breathing while wearing a mask and showed that the mask moderates the flow rate of exhaled air. In a study using numerical simulations, Pender et al. [12] suggested that social distancing of at least 4 m is required when sneezing while wearing a mask. Additionally, Dbouk et al. [13] revealed that wearing a mask reduces the distance that droplets are spread. As described, studies using numerical simulations regarding masks have revealed the effectiveness of wearing masks in preventing infection and reducing the spread distance of droplets. These studies, however, have adopted static masks that do not deform during coughing or breathing for simplification. If the mask move, it's possible that area of the gap between a mask and face increases and the speed of air leaking decreases. As a result, the traveling distance of viral droplets may be affected. Though many previous literatures have ignored the effect of mask movement on virus droplets, no research has yet verified this. Therefore, we conduct research taking into account the displacement and deformation of masks during coughing to more precisely evaluate the effectiveness of masks in preventing droplet spread. In this paper, we evaluate the influence of mask displacement and deformation due to coughing on the movement of virus droplets. In order to trace the movement of the mask, the displacements of the mask during coughing are measured in experiment. This allows to predict the behavior of virus droplets under conditions close to actual phenomena. Furthermore, numerical simulations with static mask are performed to analyze the effect of considering mask displacement and deformation. Finally, we perform applied calculations to utilize the dynamic mask calculation results.

2 Numerical Approach

2.1 Governing Equations for Fluid Flow

We first calculated fluid flow by coughing, and then analyzed the movement of virus-laden droplets by using the fluid flow data. The continuity equation and the three-dimensional incompressible Navier-Stokes equation used to compute the airflow are as follows:

$$\frac{\partial u}{\partial x} + \frac{\partial v}{\partial y} + \frac{\partial w}{\partial z} = 0, \tag{1}$$

$$\frac{\partial \boldsymbol{q}}{\partial t} + \frac{\partial \boldsymbol{E}}{\partial x} + \frac{\partial \boldsymbol{F}}{\partial y} + \frac{\partial \boldsymbol{G}}{\partial z} = \left(\frac{\partial \boldsymbol{E_V}}{\partial x} + \frac{\partial \boldsymbol{F_V}}{\partial y} + \frac{\partial \boldsymbol{G_V}}{\partial z} \right),$$ (2)

where

$$\boldsymbol{q} = \begin{bmatrix} u \\ v \\ w \end{bmatrix}, \boldsymbol{E} = \begin{bmatrix} u^2 + p \\ uv \\ uw \end{bmatrix}, \boldsymbol{F} = \begin{bmatrix} uv \\ v^2 + p \\ vw \end{bmatrix}, \boldsymbol{G} = \begin{bmatrix} uw \\ vw \\ w^2 + p \end{bmatrix}$$

$$\boldsymbol{E_V} = \frac{1}{Re} \begin{bmatrix} \partial u / \partial x \\ \partial v / \partial x \\ \partial w / \partial x \end{bmatrix}, \boldsymbol{F_V} = \frac{1}{Re} \begin{bmatrix} \partial u / \partial y \\ \partial v / \partial y \\ \partial w / \partial y \end{bmatrix}, \boldsymbol{G_V} = \frac{1}{Re} \begin{bmatrix} \partial u / \partial z \\ \partial v / \partial z \\ \partial w / \partial z \end{bmatrix}$$ (3)

u, v, w are the corresponding velocity components in the x, y, z direction, respectively, t is time, p is pressure and $Re(\approx 1.6 \times 10^4)$ is the Reynolds number with the representative length that is half width of mouth (=0.02 m) and the representative velocity (=12.0 m/s). Additionally, $\boldsymbol{E}, \boldsymbol{F}, \boldsymbol{G}$ are the advection vectors, $\boldsymbol{E_V}, \boldsymbol{F_V}, \boldsymbol{G_V}$ are the corresponding viscous stress vectors. Physical variables are defined at cell centers. The governing equations are discretized using the Moving-Grid Finite-Volume Method with an unstructured grid, which is capable of handling moving geometries. This method ensures conservation laws are satisfied even when the computational grid undergoes deformation, such as when the face mask moves and deforms. The specific details are documented in the literature [14]. To solve for incompressible flow, the fractional step method is employed, utilizing the LU-SGS (Lower-Upper Symmetric-Gauss-Seidel) method [15] for the first step and the SOR (Successive Over Relaxation) method for the second step.

2.2 Equation of Motion for Droplets

For calculation of droplets motion, we employed the Lagrangian tracking method [3], assuming that the diameter of the droplets is sufficiently small to consider the effect of the droplets on the fluid negligible. Therefore, we first calculated the fluid flow from coughing, and then simulated the motion of droplets using the flow velocity. Here, droplets were treated as un-deformable spherical droplets, and forces between droplets were ignored. The equation for the motion of droplets is given by:

$$\rho_d V_d \frac{d\boldsymbol{q}_d}{dt} + \rho_d V_d \boldsymbol{g} + \frac{1}{2} C_D \rho_a S_d |\boldsymbol{q}_a - \boldsymbol{q}_d| (\boldsymbol{q}_a - \boldsymbol{q}_d) = 0$$ (4)

where the suffixes d and a indicate values pertaining to the droplet and air, respectively. Here, ρ represents density, V is volume, $\boldsymbol{g}$ is gravitational acceleration, $\boldsymbol{q}$ is velocity, C_D is drag coefficient of the droplet. S_d is the projection area of the droplet. The shape of the droplet is assumed to be spherical.

3 Measurement of Face Mask Deformation

3.1 Measuring Instrument

Face mask

We used a disposal face mask for the measurement. The size of the face mask is 170 mm

× 95 mm. Figure 1 shows a face mask and infrared markers (5 rows and 5 columns a total 25). The diameter of infrared markers was 1.5 mm. They were glued with glue tape, which has few effects on the deformation of the face mask. Here, we assign names to the markers, ranging from I-1 to V-5. Furthermore, some reference markers were placed on face of test subject. By measuring the relative displacement of the reference marker and the marker on the face mask, the displacement of face by coughing can be neglected.

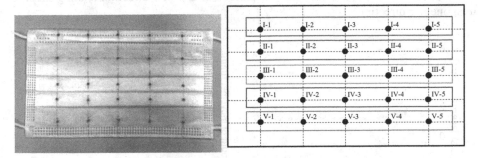

Fig. 1. Markers position on the disposal face mask

Motion capture system

To measure the motion of the face mask in three-dimension, we used a motion capture system (OptiTrack motion capture system, Acuity Inc). Figure 2 shows a schematic diagram of the coordinate and configuration of the motion capture system. 6 cameras were placed at a position of about 2 m from the test subject wearing the face mask. To capture all markers from each camera, the cameras were placed at right position. The measurement uncertainty was about 0.015 mm.

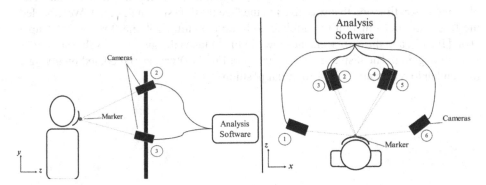

Fig. 2. Schematic diagram of motion capture system

3.2 Measurement Results

Figure 3 (a) shows the time series of displacement on marker III-2 in first experimental run. It can be seen that the mask reached the maximum displacement of 5.5 mm in the z direction (front of the human body) at ~0.1 s after coughing. After coughing ($t \geq 1.0$ s), forward the y and z displacement did not return to the original position because the position of the face mask was displaced by coughing. We considered the deformation in the z direction, since the maximum displacement in x or y direction relating smaller than one in z direction.

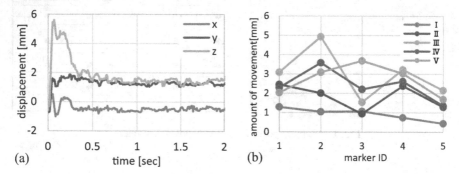

Fig. 3. (a) Displacement in each axial direction on marker III-2 and (b) average amount of movement for each marker in forward direction of the face.

Figure 3(b) displays the ensemble average of the maximum displacement for each marker in the z-direction, based on five experimental runs. We modeled the average displacement by averaging left and right markers to eliminate the asymmetry due to the test subject. The displacement between each marker was calculated by the Lagrange interpolation in x direction and linear interpolation in y direction. The maximum displacement in z direction that was given to the face mask is shown in Fig. 4. We modeled the face mask deformation to correlate with the coughing volume flow rate of Gupta et al. [16, 17], and the displacement shown in Fig. 4 takes the maximum value at $t = 0.1$ s after coughing. The displacement tendency in Fig. 3(a) was also modeled by giving a movement to return about 70% to initial position.

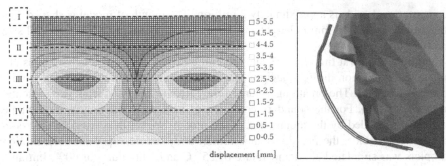

Fig. 4. Visualization of maximum displacement (left) and a cross-sectional view of the mask (right): the blue line represents the original position, while the red line indicates the position of the mask at its maximum displacement. (Color figure online)

4 Simulation of Human Model Wearing Mask

4.1 Computational Mesh and Conditions

Figure 5 shows the computational domain for our simulations, with dimensions of 6.25 × 6.25 × 3.75 m. To simplify the model and reduce computational complexity, the human body model excludes the lower body and both arms, standing 0.9 m tall. The face mask model in the simulation is 2.5 mm thick to allow for the creation of at least three computational grids along its thickness. Initially, the gap between the face mask and the nose is set at 3.5 mm [18, 19]. The coordinate system used is defined as follows: x represents the lateral direction relative to the human model, y is the vertical direction, and z is the direction facing forward from the human model.

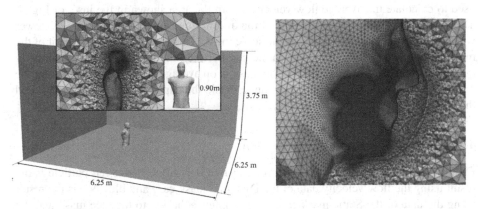

Fig. 5. Overview of the computational domain and the grid surrounding the face mask

Computational grids were generated using MEGG3D software. The mesh distribution is detailed in the cross-sectional views of the computational grid at the center of the human body. The grids are more refined around the face mask. The total number of computational grids is approximately 2 million. The mouth of model is shaped as a rectangle, measuring 10 mm in height and 40 mm in width, with the smallest cell size being 0.5 mm.

Coughing begins at time $t = 0$ s and breathing begins after the coughing ends. To simulate the exhalation during coughing, we used volume flow rates recorded from a spirometer test [16]. The breathing volume flow rate was obtained from experimental measurements [20]. For the boundary condition of velocity at the mouth, the volume flow rate was divided by the mouth area of the model. We set the ceiling of the room as an outflow surface, the face mask and human body are no-slip walls. The motion of droplets was simulated at a temperature of 25 °C and a humidity of 60%. Initially, droplets were positioned in front of the mouth, with a total count of 10,000 droplets. To model realistic droplet behavior, we utilized the frequency distribution of droplet diameters expelled during coughing [21, 22]. Droplets were considered adhered upon crossing any boundary. The motion of the droplets was calculated using pre-calculated unsteady flow velocity data, which was updated every 1/600 of a second.

4.2   Result of Fluid Flow Simulation

Figure 6 shows the cross-sectional velocity distribution of the human body center at 0.1, 0.3 and 0.4 s. From the figure, it is observed that 0.1 s after the start of a cough, the mask moves forward, widening the airflow path between the mask and the human body. As a result, the flow velocity between the nose and the mask is reduced in the dynamic mask scenario.

At 0.4 s, the flow tends to spread backward with the dynamic mask, while with the static mask, the tendency is for the flow to spread upward over the human body. Next, the average flow velocity in the gap between the mask and the human body model is shown in Fig. 7. The elements of the gap between the mask and the human body model used to calculate the average flow velocity are in the area shown in the inset of Fig. 7. The average flow velocity, at the time of maximum flow velocity, was about 7% lower with the dynamic mask compared to the static mask. This is due to the movement of the mask, which widened the flow path in the gap between the mask and the human body model. In the case of the static mask, the mask is on average about 3 mm away from the human body model, while with the dynamic mask, the gap area between the mask and the body increased by up to about 6%.

4.3   Result of Droplet Motion Simulation

Figure 8(a) shows the positions of droplets at $t = 0.4, 4.0$ s, where red particle is the result using the flow velocity data of the Dynamic mask case and blue one is the result using the data of the Static mask case. The droplets adhering to the face mask and the human body are not drawn, and thus we can observe only the floating droplets shown at each time. In the Static mask case, the dispersion distance of droplets in the positive y direction was greater than Dynamic mask case. In addition, the average direct distance from the initial position of droplets was about 12% smaller with the Dynamic mask case than with the Static mask case, because of the flow velocity decreasing written in the last

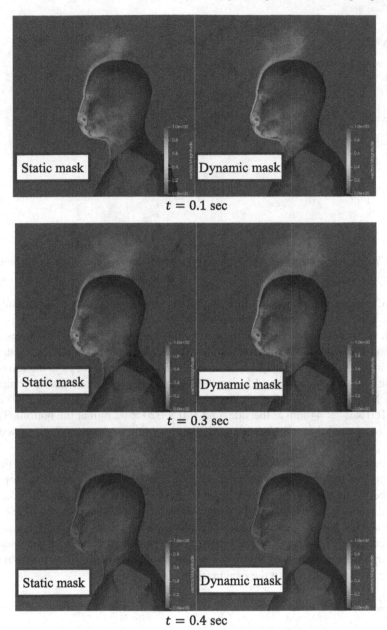

Fig. 6. Comparison between static and dynamic mask with velocity distribution

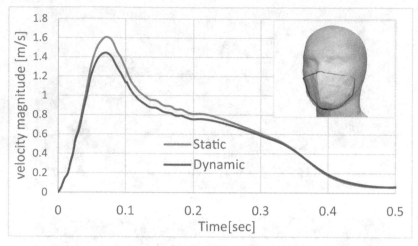

Fig. 7. Average velocity magnitude of cells between mask and face during coughing

section. Consequently, the Dynamic mask was effective comparing to the Static mask to suppress the moving distance of droplets.

Next, we set some boxes around the human model as Bale et al. [4], and investigated the relationship between the number of droplets reaching each box and the total droplets volume, in 0.0–0.5 s. To identify where droplets leaked the most through the top of face mask, the side of face mask, and lower of face mask, boxes were sized to completely cover each gap illustrated in Fig. 8. The size of the boxes was 140 × 120 × 20 mm at the top, 160 × 24 × 44 mm at the side, and 140 × 320 × 52 mm at the bottom. Table 1 summarizes the number of droplets and the total volume of droplets that reached each box.

We found that the number of leaked droplets were much larger on the top of the face mask compared to the sides and lower of the face mask. Next, we focus on the top parts, the number of droplets in the Dynamic mask case reduced by about 5.8% compared to the static mask case. However, the total volume of droplets increased by 18% in the dynamic mask. Large-diameter droplets tended to leak out from the top by considering the displacement of the mask. It is considered that this trend is because the wider gap caused large-diameter droplets, which has adhered to the Static mask, to flow out of the face mask without adhering to the face mask.

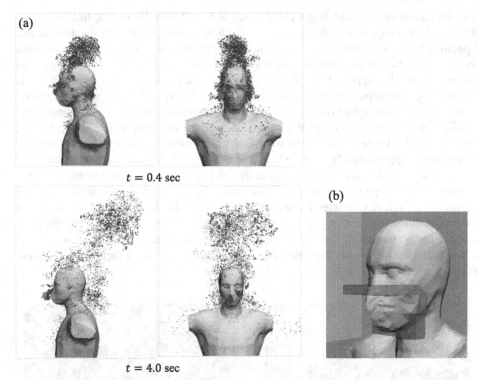

Fig. 8. (a) Distribution of virus-laden droplets: Blue droplets result from using a static mask, while red droplets are from a dynamic mask and (b) position of boxes counting droplets. (Color figure online)

Table 1. Number of droplets and volume in each box

	Number of droplet		Droplet volume [$\times 10^{-6}$ ml]	
	Static	Dynamic	Static	Dynamic
Top box	2997	2823	4.5	5.3
Side box(right)	216	226	5.2×10^{-2}	5.9×10^{-2}
Side box (left)	188	195	6.1×10^{-2}	5.7×10^{-2}
Lower box	334	323	6.3×10^{-2}	6.1×10^{-2}

5 Applicational Computation

5.1 Transplant Calculation for Dynamic Mask

The above computation considering deforming a mask is high cost because very fine mesh is required around in/around the mask. To reduce computational costs, calculations within the mask are omitted. Instead, a model that fills the gap between the mask and the human body model, highlighted in red in Fig. 9(left), is created. Inflow boundary condition is applied to this gap, using the flow velocity from the dynamic mask. Specifically, the flow velocity for the transplant calculation references the value from the cell whose element centroid is closest to the centroid of the boundary surface on the new gap. This approach is referred to as the transplant calculation in this study. Figure 9(right) compares the cross-sectional meshes for transplanted computation with those from a full computation in a dynamic mask, demonstrating that the area inside the mask, which previously occupied a significant portion of the computational grid, has been reduced. The new grid for the transplant calculation maintains the same number of elements (approximately 2 million) as that used in the dynamic mask calculation to facilitate a fair comparison of calculation times. The transplant calculations introduce a minor modification to the droplet calculations. The initial conditions for the droplets (coordinates, velocity, droplet diameter) in the transplant calculations are derived from their state at the moment they were ejected from the mask in the dynamic calculations. Consequently, the initial number of droplets for the transplant calculation was set at 3248, matching the count of droplets that emerged from the mask gaps in the dynamic mask droplet calculation.

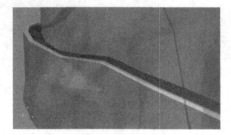 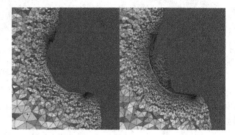

Fig. 9. Gap-filled mask model (left), and a comparison of cross-sectional meshes for transplanted computation versus full computation in a dynamic mask (right) (Color figure online)

The results of the transplant calculation, compared to those of the dynamic mask, are shown in Fig. 10. Figure 10(a) and (b) display the velocity distribution in a cross-section at the center of the human body at 0.1 s, representing the results of the dynamic and the transplant calculations, respectively. In the transplant calculation, since the inside of the mask is outside the computational domain, there is no data inside the mask. It can be seen that the tendency for the flow to spread backward is also reproduced in the transplant calculation. Moreover, by transplanting data from the dynamic mask, it is possible to replicate the reduction in flow velocity in the gap between the mask and the human body. The results of the droplet calculation using the airflow data from the transplant calculation are shown in Fig. 10(c). The blue droplets represent the results of the transplant calculation, and the red droplets represent the results of the dynamic calculation. Although there are some differences in the motion of the droplets, it is evident that the transplant calculation can replicate the trend of droplets flowing backward, as seen in the dynamic mask calculation. The transplant calculation has reduced the computation time by approximately 56% compared to the calculation considering the movement of the mask.

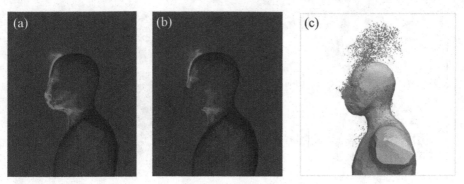

Fig. 10. Distribution of flow velocity magnitude at 0.1 s in (a) the dynamic mask model and (b) the transplant model. (c) shows a comparison of droplet dispersion at 0.4 s. (Color figure online)

5.2 Setting up the Pushing Wheelchair Scenario

The transplant calculation significantly reduced computational costs while retaining the characteristics of the dynamic mask. From the calculations so far, it has been observed that droplets scatter further backward when using a dynamic mask on a human model compared to a model with a static mask. This indicates an increased risk of infection behind a mask-wearing individual. Therefore, as an applied calculation, we consider the scenario of pushing a wheelchair, where the infection probability increases behind the mask wearer. The computational model is shown in Fig. 11. It is assumed that the wheelchair does not significantly affect the movement of droplets scattered near the rear of the human body, and it is omitted to reduce the number of grids and simplify the model. The size of the whole computational domain is set to 8.0 × 8.0 × 6.0 m. The grid width becomes finer around the mask and between the human models through which the droplets pass. The minimum grid width is about 0.5 mm. Transplant calculations using the airflow from the dynamic mask are applied to the human model wearing the mask. Furthermore, assuming a state of moving forward while pushing a wheelchair, a uniform flow is applied from the front. The propulsion speed of the wheelchair in straight-line motion is reported to be approximately 1 m/s, and it is noted that the sense of anxiety and fear of the occupant increases with speed [23]. Therefore, inflow speeds of 0.0, 0.5, 1.0, 1.5 m/s are set, and the number and volume of droplets passing around the respiratory area of the human model behind the infected person are investigated at each speed. Under the condition of providing inflow from the front, pre-calculations are performed until the flow develops to stabilize the flow caused by the uniform stream from the front, and then the cough velocity are transplanted. The boundary conditions are set as outflow for the ceiling, sides, and rear wall, with the mask and human body being no-slip wall.

5.3 Results of the Pushing Wheelchair Scenario

To assess the risk of infection, a box is placed near the mouth of an infected individual, and the relationship between the number of droplets reaching the box and the total volume of droplets is investigated from 0.0 to 3.0 s after the start of a cough. The size of

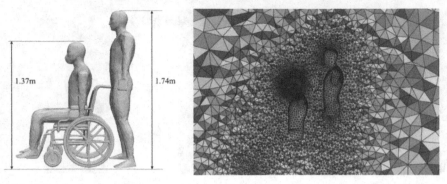

Fig. 11. Model of a seated human wearing a face mask and a standing human (left), and computational mesh (right).

the box is determined based on the volume of air a typical adult inhale at once, set to 10 × 15 × 10 cm³, following Bale et al. [2]. The number of droplets reaching the box and their total volume at each forward speed are shown in Fig. 12. Additionally, the state of droplets at 0.1, 0.55 and 1.0 s after the start of the cough is depicted in Fig. 13.

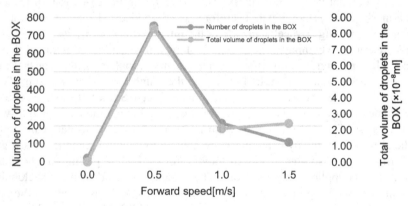

Fig. 12. Number of droplets and total droplet volume in the BOX, counted at each wheelchair pushing speed

When the wheelchair is stationary, droplets hardly reach the box placed near the respiratory area of the rear human model. When the wheelchair is pushed at a speed of 0.5 m/s, the highest number of droplets reach the box, and the total volume of droplets is also the largest. As the speed increases to 1.0 m/s and 1.5 m/s, the number of droplets reaching the box decreases, but the total volume of droplets reaching the box at 1.5 m/s is larger than at 1.0 m/s. This is believed to be because larger droplets are carried further backward due to the increased inflow from the front. Therefore, under the conditions of this study, the result shows that moving forward at a speed of about 0.5 m/s while pushing a wheelchair with an infected mask-wearer seated in it presents the highest risk of infection.

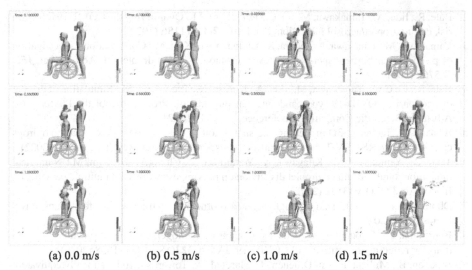

| (a) 0.0 m/s | (b) 0.5 m/s | (c) 1.0 m/s | (d) 1.5 m/s |

Fig. 13. Droplets dispertion at each forward speed at 0.1, 0.55 and 1.0 s after the start of the cough

6 Conclusion

To investigate the impact of face mask deformation caused by coughing on droplet behavior, we conducted a numerical simulation that integrated fluid flow and droplet motion. By measuring the displacement of the mask and applying dynamic fluid calculations that accounted for these displacements, we captured changes in droplet trajectories. Our findings revealed that coughing while wearing a disposable mask can result in approximately 6mm of mask displacement, accompanied by puffing out of the mask. In scenarios where the mask's dynamic behavior was considered, there was a 7% decrease in the average flow velocity in the gap between the mask and the human body, compared to using a static mask. Initially, droplets dispersed more rapidly when using a static mask; however, over time, the dynamic mask scenario allowed for a greater dispersion distance, with droplets spreading further backward. Additionally, we applied our findings to practical scenarios, such as assessing infection risk when pushing a wheelchair from behind. The simulation indicated that a caregiver pushing an infected individual who is wearing a mask at a speed of about 0.5 m/s faces the highest infection risk. It's important to note that these calculations were conducted assuming no obstructions at the head of the human model. This suggests that implementing barriers, such as wearing a hat, might obstruct airflow from the nose towards the head and potentially reduce the risk of infection.

Acknowledgements. This work was partially supported by JST CREST Grant Number JPMJCR20H7 and by JSPS KAKENHI Grant Number 21K03856.

References

1. World Health Organization. https://www.who.int/emergencies/diseases/novel-coronavirus-2019/question-and-answers-hub. Accessed 29 Feb 2024

2. Bale, R., Iida, A., Yamakawa, M., Li, C., Tsubokura, M.: Quantifying the COVID19 infection risk due to droplet/aerosol inhalation. Sci. Rep. **12**(1), 11186 (2022)
3. Yamakawa, M., Kitagawa, A., Ogura, K., Chung, Y.M., Kim, M.: Computational investigation of prolonged airborne dispersion of novel coronavirus-laden droplets. J. Aerosol Sci. **155**, 105769 (2021)
4. Bale, R., Li, C.G., Yamakawa, M., Iida, A., Kurose, R., Tsubokura, M.: Simulation of droplet dispersion in COVID-19 type pandemics on Fugaku. In: Proceedings of the Platform for Advanced Scientific Computing Conference, pp. 1–11 (2021)
5. Armand, P., Tâche, J.: 3D modelling and simulation of the dispersion of droplets and drops carrying the SARS-CoV-2 virus in a railway transport coach. Sci. Rep. **12**(1), 4025 (2022)
6. Takii, A., Yamakawa, M., Kitagawa, A., Watamura, T., Chung, Y.M., Kim, M.: Numerical model for cough-generated droplet dispersion on moving escalator with multiple passengers. Indoor Air **32**(11), e13131 (2022)
7. Dbouk, T. and Drikakis, D.: On coughing and airborne droplet transmission to humans. Phys. Fluids **32**(5) (2020)
8. Bourouiba, L.: Turbulent gas clouds and respiratory pathogen emissions: potential implications for reducing transmission of COVID-19. JAMA **323**(18), 1837–1838 (2020)
9. Blocken, B., Malizia, F., van Druenen, T., Marchal, T.: Towards aerodynamically equivalent COVID19 1.5 m social distancing for walking and running. Preprint (1) (2020)
10. Bagchi, S., Basu, S., Chaudhuri, S., Saha, A.: Penetration and secondary atomization of droplets impacted on wet facemasks. Phys. Rev. Fluids **6**(11), 110510 (2021)
11. Bourrianne, P., Xue, N., Nunes, J., Abkarian, M., Stone, H.A.: Quantifying the effect of a mask on expiratory flows. Phys. Rev. Fluids **6**(11), 110511 (2021)
12. Pendar, M.R., Páscoa, J.C.: Numerical modeling of the distribution of virus carrying saliva droplets during sneeze and cough. Phys. Fluids **32**(8) (2020)
13. Dbouk, T., Drikakis, D.: On respiratory droplets and face masks. Phys. Fluids **32**(6) (2020)
14. Takii, A., Yamakawa, M., Asao, S., Tajiri, K.: Six degrees of freedom flight simulation of tilt-rotor aircraft with nacelle conversion. J. Comput. Sci. **44**, 101164 (2020)
15. Yoon, S.: Lower-upper Symmetric-Gauss-Seidel method for the euler and navier-stokes equations. AIAA J. **26**(9), 1025–1026 (1988)
16. Gupta, J.K., Lin, C.-H., Chen, Q.: Flow dynamics and characterization of a cough. Indoor Air **19**, 517–525 (2010)
17. Gupta, J.K., Lin, C.-H., Chen, Q.: Characterizing exhaled airflow from breathing and talking. Indoor Air **20**, 31–39 (2010)
18. Khosronejad, A., et al.: Fluid dynamics simulations show that facial masks can suppress the spread of COVID-19 in indoor environments. AIP Adv. **10**(12) (2020)
19. Dbouk, T., Drikakis, D.: On respiratory droplets and face masks. Phys. Fluids **33**, 073315 (2021)
20. Fenn, W.O., Rahn, H.: Handbook of Physiology. Section3: Respiration. American Physiological Society, Washington DC (1965)
21. Duguid, J.P., et al.: The size and the duration of air-carriage of respiratory droplets and droplet-nuclei. Epidemiol. Infect. **44**(6), 471–479 (1946)
22. Yang, S., Lee, G.W.M., Chen, C.-M., et al.: The size and concentration of droplets generated by coughing in human subjects. J. Aerosol Med. **20**(4), 484–494 (2007)
23. Noto, H., Saito, S., Muraki, S.: Influence of wheelchair pushing speed on riding comfort and helpers physical strain on helpers. Japan. J. Nurs. Art Sci. **8**(2), 37–45 (2009)

Implementation of the QGD Algorithm Using AMR Technology and GPU Parallel Computing

Ivan But[1,2]([✉])[iD], Andrey Epikhin[1,2][iD], Maria Kirushina[2][iD], and Tatiana Elizarova[2][iD]

[1] Ivannikov Institute for System Programming of the RAS, Moscow 109004, Russia
[2] Keldysh Institute of Applied Mathematics of the RAS, Moscow 125047, Russia
ivan.but@ispras.ru

Abstract. The paper presents an algorithm based on the quasi-gasdynamic approach for the solution of unsteady compressible flows over a wide range of Mach numbers. It is implemented on the AMReX open platform, which uses adaptive mesh refinement technology to facilitate parallelization of computations on GPU architectures. To validate its effectiveness, the developed solver is applied to the numerical simulation of the shock-vortex interaction problem with flow parameter values of $M_v = 0.9$ and $M_s = 1.5$. Cross-validation to assess its performance is conducted with OpenFOAM-based solvers, specifically rhoCentralFoam and QGDFoam. Schlieren fields are used to evaluate oscillations of the numerical schemes and algorithms, while resolution capabilities of the algorithm are assessed by comparing density fields in five cross-sections with the reference values.

Keywords: Shock-vortex interactions · Compressible flow · Quasi-gas dynamic equations · OpenFOAM · AMReX

1 Introduction

The increasing complexity of physical processes modelling has made parallelization of computations on GPUs and integration of adaptive mesh refinement (AMR) technology essential [1]. GPUs, capable to solve massively parallel problems, offer a significant advantage in accelerating the computation of complex aerohydrodynamic problems. At the same time, AMR has emerged as a promising method to solve problems with complex and dynamic nature of aerohydrodynamic flows. The ability to automatically adjust mesh resolution in regions of interest not only improves simulation accuracy, but also optimises computational resources by allocating higher resolution only where necessary.

The Quasi-Gasdynamic (QGD) algorithm based on the regularised equations [2] allows modelling of compressible ideal gas flows over a wide range of Mach numbers, from subsonic to supersonic. The QGD equations, which differ from the Navier-Stokes equations, include additional terms proportional to the small parameter τ, which depends on the size of the computational cell and local

L. Franco et al. (Eds.): ICCS 2024, LNCS 14838, pp. 85–99, 2024.
https://doi.org/10.1007/978-3-031-63783-4_8

sound velocity. In particular, the universality of the algorithm for all types of flows distinguishes it. This numerical algorithm was implemented in OpenFOAM [3–6] and showed high efficiency in modelling complex unsteady flows and flows with strong discontinuities [7,8]. While this algorithm has shown high efficiency in modelling unsteady flows and flows with strong discontinuities, it tends to be slower than the alternatives due to its numerical specificity.

To address this challenge, the use of AMR technology and the potential for GPU computations parallelization offers a solution to speed up calculations and reduce expended resources. Currently, the most suitable free platform to achieve this goal is AMReX [9,10], which has demonstrated higher computational efficiency compared to OpenFOAM [11]. Moreover, as of 2023, AMReX is part of the Linux Foundation's High-Performance Software Foundation [12], which means it will receive significant support and development.

The rest of the paper is structured as follows: Sect. 2 describes the mathematical model underlying QGD. Section 3 describes the implementation of QGD in AMReX by demonstrating the numerical algorithm and the structure of the AmrQGD solver. Section 4 presents the formulation of the shock-vortex interaction problem. Section 5 shows the computational results in AmrQGD and OpenFOAM-based cross-validation with three solvers. The performance study of the algorithms is presented in Sect. 5.3. Section 6 contains a paper with the main conclusions.

2   Mathematical Model

The regularized gas dynamics equations in the form of continuity, momentum, total energy and state equations are used for the implementation of the AmrQGD solver in AMReX:

$$\frac{\partial \rho}{\partial t} + \nabla \cdot \mathbf{j_m} = 0, \tag{1}$$

$$\frac{\partial (\rho \mathbf{u})}{\partial t} + \nabla \cdot (\mathbf{j_m} \otimes \mathbf{u}) + \nabla p = \nabla \cdot \hat{\sigma}, \tag{2}$$

$$\frac{\partial (\rho E)}{\partial t} + \nabla \cdot (\mathbf{j_m} E) + \nabla \cdot \mathbf{q} = \nabla \cdot (\hat{\sigma} \mathbf{u}), \tag{3}$$

$$p = \rho R T, \tag{4}$$

where ρ is density; $\mathbf{j_m}$ is mass flux density; $\mathbf{u}$ is the velocity vector; p is pressure; $E = e + |\mathbf{u}|/2$ is total energy, e is the specific internal gas energy; $\hat{\sigma}$ is the viscous stress tensor, $\mathbf{q}$ is heat flux; $\otimes$ is the direct tensor product.

The presence of additional QGD terms $\mathbf{w}$, $\hat{\sigma}_{QGD}$, $\mathbf{q}_{QGD}$ proportional to the small parameter τ is the main difference of quasi-gasdynamic equations:

$$\mathbf{j_m} = \rho(\mathbf{u} - \mathbf{w}), \quad \mathbf{q} = \mathbf{q}_{NS} + \mathbf{q}_{QGD}, \quad \hat{\sigma} = \hat{\sigma}_{NS} + \hat{\sigma}_{QGD} \tag{5}$$

$$\mathbf{w} = \frac{\tau}{\rho} \left(\text{div} \left(\rho \mathbf{u} \otimes \mathbf{u} \right) + \nabla p \right), \quad \mathbf{q}_{QGD} = -\tau \rho \mathbf{u} ((\mathbf{u} \cdot \nabla) e + p(\mathbf{u} \cdot \nabla)/\rho)$$

$$\hat{\sigma}_{QGD} = \tau \mathbf{u} \otimes (\rho(\mathbf{u} \cdot \nabla)\mathbf{u} + \nabla p) + \tau \hat{I}((\mathbf{u} \cdot \nabla)p + \gamma p \nabla \mathbf{u})$$

$$\tau = \alpha \frac{\Delta_h}{c} + \frac{\mu_d}{p}; \ c = \sqrt{\gamma RT} \tag{6}$$

where Δ_h is the local step size of the spatial mesh; c is the local speed of sound; α is the tuning parameter of the algorithm; μ_d is the dynamic viscosity coefficient; R is the specific gas constant. Viscosity coefficient $\mu = \mu_d + \tau p Sc_{QGD}$, where α and Sc_{QGD} are the tuning parameter of the related numerical algorithm.

A detailed description of the derivation and characteristics of these equations is given in, e.g. [2,5].

3   QGD Implementation in AMReX

The AMReX software package allows us to use ready-made logic to refine mesh into levels with a corresponding change in time step (Fig. 1a,b), and also provides the possibility of transferring the computations to the GPU, which makes the computations much less time-consuming compared to OpenFOAM. On this basis, the implementation of the numerical algorithm based on the QGD equations in AMReX will significantly reduce the computational cost and increase the computational speed for problems in a wide range of Mach numbers.

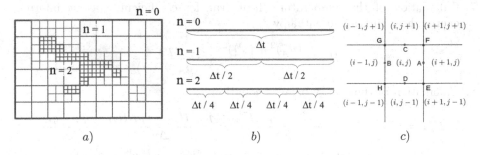

Fig. 1. QGD implementation in AMReX: a) Scheme of adaptive mesh refinement by levels; b) Scheme of adaptive time step refinement at each level of the computational mesh; c) Scheme of the numerical QGD template implemented in AmrQGD.

3.1   Numerical Algorithm Based on QGD in AMReX

Figure 1c shows the structure of the template for the QGD equations, which, unlike classical approaches, requires all surrounding cells. A two-dimensional (2D) numerical algorithm is currently implemented:

1. Discretization of ρ, u_x, v_y, p is carried out by the central differences method:

$$\rho_A = \frac{\rho_{i,j} + \rho_{i+1,j}}{2}, \ \rho_E = \frac{\rho_{i,j} + \rho_{i+1,j} + \rho_{i,j-1} + \rho_{i+1,j-1}}{4} \tag{7}$$

2. Calculation of the sound speed, the parameter τ, the QGD part w;
3. Calculation of mass flux density j_m;
4. The continuity equation is solved (the parameter with the hat is the variable value at the new time step):

$$\hat{\rho}_{i,j} = \rho_{i,j} - \frac{\Delta t}{\Delta x}(j_{mA} - j_{mB}) - \frac{\Delta t}{\Delta y}(j_{mC} - j_{mD}) \qquad (8)$$

5. The viscosity coefficient, the Reynolds viscous stress tensor and its QGD analogue are considered;
6. Calculation of the momentum equations:

$$\hat{u}_{i,j}^x = \rho_{i,j}u_{i,j}^x - \Delta t \left(\frac{j_{mA}u_A^x - j_{mB}u_B^x}{\Delta x} + \frac{j_{mC}u_C^x - j_{mD}u_D^x}{\Delta y} + \frac{p_A - p_B}{\Delta x} \right)$$
$$+ \Delta t \left(\frac{\sigma_A^{xx} - \sigma_B^{xx}}{dx} + \frac{\sigma_C^{yx} - \sigma_D^{yx}}{\Delta y} \right) \qquad (9)$$

$$\hat{u}_{i,j}^y = \rho_{i,j}u_{i,j}^y - \Delta t \left(\frac{j_{mA}u_A^y - j_{mB}u_B^y}{\Delta x} + \frac{j_{mC}u_C^y - j_{mD}u_D^y}{\Delta y} + \frac{p_C - p_D}{\Delta y} \right)$$
$$+ \Delta t \left(\frac{\sigma_C^{yy} - \sigma_D^{yy}}{\Delta y} + \frac{\sigma_A^{xy} - \sigma_B^{xy}}{\Delta x} \right) \qquad (10)$$

7. Calculation of the temperature, heat transfer coefficient, specific internal energy, heat fluxes and enthalpy

$$H_A = \frac{(u_A^x)^2 + (u_A^y)^2}{2} + \gamma \frac{p_A}{\rho_A(\gamma - 1)}$$

8. Calculation of the energy balance equation:

$$\hat{E}_{i,j} = E_{i,j} - \Delta t \left(\frac{j_{mA}H_A - j_{mB}H_B}{\Delta x} + \frac{j_{mC}H_C - j_{mD}H_D}{\Delta y} + \frac{q_A - q_B}{\Delta x} \right.$$
$$+ \frac{q_C - q_D}{\Delta y} \left. \right) + \Delta t \left(\frac{\sigma_A^{xx}u_A^x - \sigma_B^{xx}u_B^x}{\Delta x} + \frac{\sigma_C^{yy}u_C^y - \sigma_D^{yy}u_D^y}{\Delta y} \right.$$
$$+ \frac{\sigma_A^{xy}u_A^y - \sigma_B^{xy}u_B^y}{\Delta x} + \frac{\sigma_C^{yx}u_C^x - \sigma_D^{yx}u_D^x}{\Delta y} \left. \right) \qquad (11)$$

9. Pressure definition:

$$\hat{p}_{i,j} = (\gamma - 1) \left(\hat{E}_{i,j} - \hat{\rho}_{i,j} \frac{(\hat{u}_{i,j}^x)^2 + (\hat{u}_{i,j}^y)^2}{2} \right) \qquad (12)$$

3.2   Solver Structure

Figure 2 shows the structure of the developed QGD numerical algorithm in AMReX.

The **QGD/Source** folder contains the solver sources:

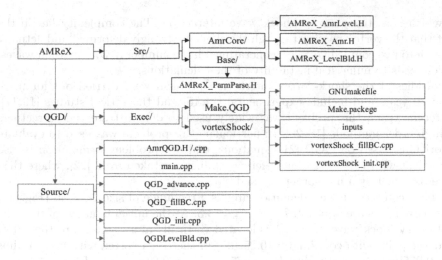

Fig. 2. Structure of the AmrQGD solver in AMReX. Black colour - AMReX kernel; Blue - case location folder including its initial and boundary conditions; Green - solver sources location. (Color figure online)

- **AmrQGD.H** - the main solver class AmrQGD is declared in the file. It inherits from the AmrLevel class defined in the AMReX core;
- **AmrQGD.cpp** - the main solver class AmrQGD is implemented in this file;
- **main.cpp** - numerical algorithm;
- **QGD_advance.cpp** - quasi-gasdynamic equations are implemented in this file;
- **QGD_fillBC.cpp** - the file describes the boundary conditions;
- **QGD_init.cpp** - initial conditions are set in this file;
- **QGDLevelBld.cpp** - the description of mesh refining.

The **QGD/Exec** folder contains statements of calculation tasks:

- **inputs** - the file where the calculation parameters are set;
- **vortexShock_fillBC.cpp** - a file containing the boundary conditions;
- **vortexShock_init.cpp** - file containing the initial conditions.

It is also worth noting that at each level it is not necessary to refine the whole mesh, but only some parts of it (Fig. 1a,b). For this purpose, the errorEst method is defined in the **AmrQGD.cpp** file. This method takes a reference to the tags instance of the TagBoxArray container. Each mesh cell is defined, which is marked for partitioning if it meets some condition (criterion). The remaining cells that do not meet the condition are not subject to partitioning.

Solver is available on GitHub [13].

4   Problem Statement

To demonstrate features and advantages of the developed AmrQGD solver, it has been decided to choose a problem with complex unsteady flow, a good example of which is a strong vortex-shock wave interaction. The complexity lies in the fact that the vortex passing through the shock is strongly deformed and actually splits into two coupled vortices, generating a large number of compression waves, which leads to numerical instability of the calculations.

Studies of the shock-vortex interaction problem were carried out for more than 70 years and include experimental [14,15] and theoretical studies [16,17]. In recent years, this problem was widely used to demonstrate the correctness of high-order methods [18–20]. In particular, this problem was used to evaluate algorithms based on QGD [21] equations, and our problem formulation is also consistent with this work. The reference values are taken from [22], where they were obtained by a high-order method.

The geometry of the calculation area is a rectangle of size 2×1 m (Fig. 3a). Gas parameters are: $\gamma = 1.4$, $R = 1$, $C_p = 3.5$. At the initial moment of time the stationary shock wave $M_s = 1.5$ is located vertically at $x = 0.5$ m, to the left of it at the point with coordinates $(0.25, 0.5)$ m there is a vortex with inner radius $a = 0.075$ m and outer radius $b = 0.175$ m, $M_v = 0.9$, $v_{max} = M_v\sqrt{\gamma}$.

Flow conditions before the shock ($x < 0.5$):

$$(\rho, u_x, v_y, p, T)_{\text{left}} = (1,\ M_s\sqrt{\gamma},\ 0,\ 1,\ p/(\rho R)) \tag{13}$$

Stationary shock conditions:

$$\frac{\rho_{\text{left}}}{\rho_{\text{right}}} = \frac{u_{\text{right}}}{u_{\text{left}}} = \frac{2 + (\gamma - 1)M_s^2}{(\gamma + 1)M_s^2}, \quad \frac{p_{\text{left}}}{p_{\text{right}}} = 1 + \frac{2\gamma}{\gamma + 1}(M_s^2 - 1), \quad v_{\text{right}} = 0 \tag{14}$$

The domain's initial conditions:

$$(\rho, u_x, v_y, p, T) = \begin{cases} (1,\ 1.77482,\ 0,\ 1,\ 1) & x < 0.5 \\ (1.862,\ 0.953146,\ 0,\ 2.45833,\ 1.32022) & x \geq 0.5 \end{cases} \tag{15}$$

The initial conditions in the vortex zone are shown in Fig. 3b. The condition for the angular velocity of the vortex is calculated by the formula for vortex velocity v_θ:

$$v_\theta(r) = \begin{cases} v_m \dfrac{r}{a} & r \leq a \\ v_m \dfrac{a}{a^2 - b^2}\left(r - \dfrac{b^2}{r}\right) & a < r \leq b \\ 0 & r > b \end{cases} \tag{16}$$

where $r = \sqrt{(x - 0.25)^2 + (y - 0.5)^2}$ - radius from the vortex center. Then in the projection on the OX and OY axes: $u_x(r) = u_{\text{left}} - v_\theta(r)\sin(\theta)$, $v_y(r) = v_{\text{left}} + v_\theta(r)\cos(\theta)$

The temperature inside the vortex is set as:

$$T(r) = \begin{cases} T(b) - \dfrac{\gamma - 1}{R\gamma}\dfrac{v_m^2}{a^2}\left(\dfrac{a^2 - r^2}{2}\right) - \\ \quad - \dfrac{\gamma - 1}{R\gamma}v_m^2\dfrac{a^2}{\sqrt{a^2 - b^2}}\left(\dfrac{b^2 - a^2}{2} - 2b^2\ln\dfrac{b}{a} - \dfrac{b^4}{2}\left(\dfrac{1}{b^2} - \dfrac{1}{a^2}\right)\right) & r \leq a \\ T(b) - \dfrac{\gamma - 1}{R\gamma}v_m^2\dfrac{a^2}{\sqrt{a^2 - b^2}}\left(\dfrac{b^2 - r^2}{2} - 2b^2\ln\dfrac{b}{r} - \dfrac{b^4}{2}\left(\dfrac{1}{b^2} - \dfrac{1}{r^2}\right)\right) & a < r \leq b \\ 0 & r > b \end{cases} \tag{17}$$

Density and pressure inside the vortex are:

$$\rho(r) = \rho_{\text{left}} \left(\frac{T(r)}{T_{\text{left}}} \right)^{\frac{1}{\gamma-1}}, \ p(r) = p_{\text{left}} \left(\frac{T(r)}{T_{\text{left}}} \right)^{\frac{\gamma}{\gamma-1}} \tag{18}$$

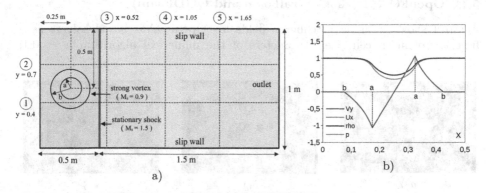

Fig. 3. Problem statement: a) Geometry of the computational domain; b) Initial conditions on the line $Y = 0.5$ m in the vortex location region ($a = 0.075$ m, $b = 0.175$ m).

Boundary conditions:

- Smooth wall conditions are placed at the top and bottom (zero gradient for pressure and density, slip condition for velocity).
- Left (inlet): Conditions correspond to the initial parameters for $x < 0.5$.
- Right (outlet): Smooth boundary conditions (zero gradient).

The numerical Schlieren value fields are used to visualize the calculation Sch $\subset (0.05; 2.4)$: Sch $= \frac{\ln(1+\nabla\rho)}{\ln(10)}$

The density values on five lines are also compared (Fig. 3a), the reference values are taken from [22], obtained in this work by the higher order method on a mesh of 12800×6400.

5   Results and Discussion

The vortex-shock wave interaction problem is numerically modelled on $\frac{1}{400}, \frac{1}{800}, \frac{1}{1600}$ meshes along the OY axis and different Courant numbers. The results of the calculations are presented for the moment $t_{end} = 0.7$.

The following software packages and solvers are used for further cross-validation:

- rhoCentralFoam solver for compressible flows based on the central Kurganov-Tadmor schemes [23], implemented in OpenFOAM. The upwind [24] (1st order) and Van Leer [25] (2nd order) numerical schemes are used.

- QGDFoam [4] solver for flows over a wide range of Mach numbers, implemented in OpenFOAM and based on the QGD equations. [2]
- A new QGD-based equation solver AmrQGD implemented on AMReX software.

5.1 OpenFOAM (rhoCentralFoam and QGDFoam)

Figure 4 shows numerical Schlieren fields for OpenFOAM based solvers as a function of mesh cell size. Parameters of the numerical algorithm $\alpha = 0.1$,

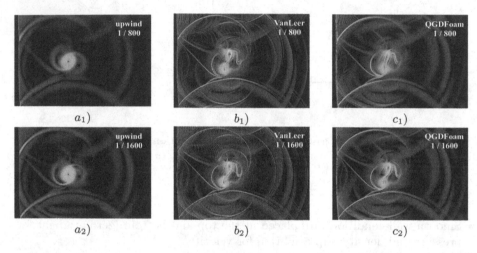

Fig. 4. Numerical Schlieren with $\Delta t = 10^{-5}$s and $t_{end} = 0.7$s.

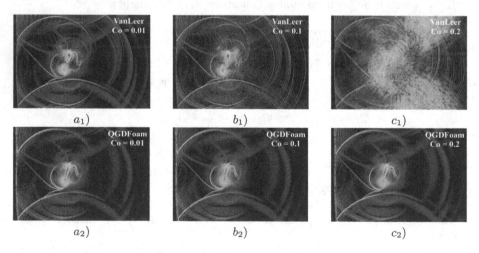

Fig. 5. Numerical Schlieren on a 1/800 mesh. (a_1, b_1, c_1) - Van Leer; (a_2, b_2, c_2) - QGDFoam. (a_1, a_2)-$Co = 0.01$ ($\Delta t = 0.5 \cdot 10^{-5}$s); (b_1, b_2) - $Co = 0.1$; (c_1, c_2) - $Co = 0.2$.

$Sc_{QGD} = 0.1$ are used for the QGDFoam, these parameters are defined in the paper [21] where this problem is considered.

It can be seen that the upwind scheme does give an unacceptable solution even on a $1/1600$ mesh, but the Van Leer and QGD schemes resolve the flow most correctly. However, when investigating different values of the time step (Fig. 5), it is found that the Van Leer scheme is oscillatory, the QGD approach gives an acceptable solution at the Courant number $Co = 0.2$, while to obtain a non-oscillatory solution using the Van Leer scheme, the Courant number $Co = 0.01$ is required.

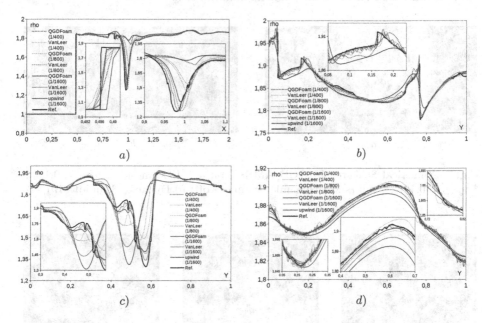

Fig. 6. Comparison of the density field plots on the lines (Fig. 3): a) Line 1; b) Line 3; c) Line 4; d) Line 5. $t_{end} = 0.7$s.

The density plots in Fig. 6 show that the upwind scheme does not resolve the vortex structure, making it inapplicable in the context of the problem, while the Van Leer and QGD schemes correctly resolve the vortex structure and its features.

5.2   AmrQGD Features

The QGD algorithm has two tuning parameters α and Sc_{QGD}, in [21] it was noted that the best choice for this problem is $\alpha = 0.1$ and $Sc_{QGD} = 0.1$, however these recommendations were given for numerical implementation based on the OpenFOAM package, so additional study is required for the newly developed solver. Figure 7 a gives the initial simulation result, which shows that the solution is free from numerical oscillations only at $Sc_{QGD} = 0.5$, but the solution is viscous due to the large value of Sc_{QGD}. In order to reduce oscillations, a local variation of the Sc_{QGD} number (variable varSc) is introduced, which is equal to 1 on the stationary shock when varSc = true. Figure 7 b shows the effect of the additional viscosity on the shock. It can be seen that this approach significantly reduced numerical oscillations after the vortex passes the shock, and an acceptable solution is obtained when $\alpha = 0.1$, $Sc_{QGD} = 0.1$, which is similar to the numerical algorithm implemented in OpenFOAM. It is likely that the numerical implementation of the QGD algorithm in OpenFOAM automatically includes numerical limiters, resulting in less oscillations, while the numerical implementation in AMReX does not include limiters.

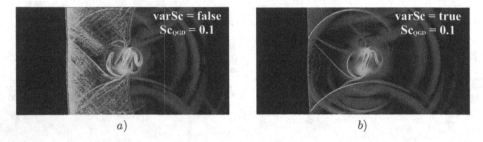

Fig. 7. Numerical Schlieren in QGD AMReX on a 1/1600 mesh with $\alpha = 0.1$ and $Sc_{QGD} = 0.1$, $t_{end} = 0.7$s : a) Dynamic Schmidt - **off**; b) Dynamic Schmidt - **on**,

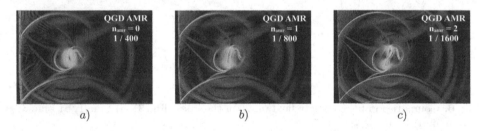

Fig. 8. Numerical Schlieren in QGD AMReX with $\alpha = 0.1$, $Sc_{QGD} = 0.1$: a) 1/400; b) 1/800; c) 1/1600.

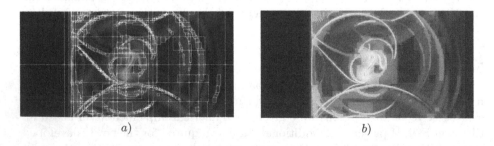

Fig. 9. Example of adaptive mesh refinement: a) by blocks; b) by cells.

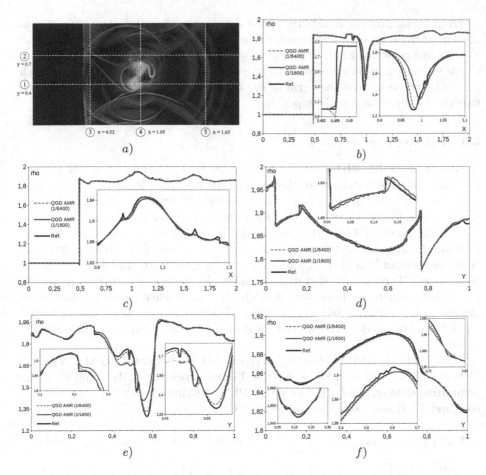

Fig. 10. Comparison of density values on five reference lines: a) Numerical Schlieren field and five reference lines; b) Line 1; c) Line 2; d) Line 3; e) Line 3; d) Line 4; f) Line 5; $t_{end} = 0.7\,\mathrm{s}$.

The influence of the adaptive mesh refinement function (number of levels n_{amr}) on the results obtained in AmrQGD is shown in Figs. (8–10). The adaptation is performed on the density gradient, i.e. if the gradient (the difference of the density values in the centers of neighboring cells divided by the distance between them) is greater than 0.0003, the mesh cell is split into 4 and the time step in each of them is reduced by half compared to the time step on the previous equation of the computational cell. Figure 8a shows the flow structure without splitting, on a mesh of 1/400. Figure 8b has a level of adaptation ($n_{amr} = 1$), i.e. in a given region the mesh is equivalent to 1/800. In Fig. 8c there are 2 levels of adaptation ($n_{amr} = 2$), corresponding to the mesh 1/1600 in the vortex resolution zone and reflected shock waves. Figure 9 shows the partitioning of the computational domain into blocks according to the density gradient criterion (Fig. 9a) and the refinement of the computational mesh in these blocks (Fig. 9b).

To demonstrate the capability of the developed numerical algorithm, a simulation of the flow with local mesh refinement corresponding to a mesh size of 1/6400 was performed (Fig. 10). The use of AMR technology allows the calculation for such a mesh refinement to be performed within 6 h on 24 cores. Such a calculation on a stationary grid in the OpenFOAM package would have been mush more computationally expensive.

5.3   Solving Performance

A cluster of 12 Intel(R) Xeon(R) CPU 5160 @ 3.00 GHz core nodes is used to evaluate the performance of the selected algorithms. Measurements are performed on an orthogonal mesh of 1/1600 on the OY axis with $\Delta t \doteq 10^{-5}$ s. and $t_{end} = 0.01$ s. (Table 1). Where AmrQGD (off) means that there is no mesh adaptation, and AmrQGD (on) that two levels of adaptation are on. It can be seen that on 12 cores, AmrQGD ($n_{amr} = 2$) computes 9 times faster than upwind, 15 times faster than Van Leer, 32 times faster than QGDFoam and 4.5 times faster than AmrQGD on a 1/1600 stationary mesh. As the number of cores increases, the speedup difference decreases, due to both the non-ideal parallelization of the algorithm in AmrQGD and the fact that the algorithm reaches its lower bound on the number of cells per core (which is about twenty thousand cells per CPU core).

The study of GPU computing performance is conducted on the NVIDIA GeForce RTX 2060 card based on the TU106 processor. It is found that the computing time on one CPU core plus GPU is 4.8 times faster than on one CPU core and 1.4 times faster than on four CPU cores.

Table 1. Calculation time on the 1/1600 mesh is $\Delta t = 10^{-5}$ s and $t_{end} = 0.01$ s.

CPU	Cell per CPU	upwind	vanLeer	QGDFoam	AmrQGD (off)	AmrQGD (on)
12	426.7k	1317.23	2124.01	4597.02	636	143
24	213.3k	642.82	1060.39	2297.98	422	88
48	106.7k	338.25	544.53	1160.61	217	62
96	53.3k	173.05	286.46	595.18	127	63
192	26.7k	86.74	130.16	318.27	81	65

Table 2. Algorithms parallelization efficiency [%]. Computation mesh 1/400 $\Delta t = 10^{-5}$ s and $t_{end} = 0.01$ s

CPU	Cell per CPU	upwind	vanLeer	QGDFoam	AmrQGD (off)	AmrQGD (on)
1	320k	–	–	–	–	–
2	160k	83	87	89	81	60
4	80k	78	81	86	79	47
8	40k	64	67	67	61	45

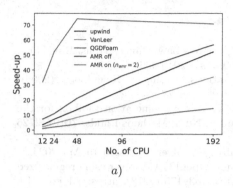
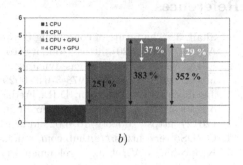

Fig. 11. Solving performance: a) Speedup of various algorithms on CPU; b) relative AMR speedup on GPU.

6 Conclusion

A new two-dimensional quasi-gasdynamic solver AmrQGD based on AMReX with the possibility of adaptive mesh refinement and parallelization of computations to graphics kernels is developed and described.

Cross-validation of the OpenFOAM and AMReX software packages, as well as the different upwind, Van Leer and QGD algorithms, the latter being implemented in two software packages simultaneously. It is found that the upwind algorithm does not give an acceptable solution on a reasonable grid and time step. The Van Leer algorithm gives an acceptable solution on a grid of $1/1600$ and a time step of $\Delta t = 10^{-5}$ s, corresponding to a Courant number of $Co = 0.03$. A comparable result is given by the QGD algorithm implemented in QGDFoam and AmrQGD, with a Courant number of $Co = 0.2$.

When the performance of the algorithms is examined on an orthogonal grid of $1/1600$ on the axis of the OY and with a time step $\Delta t = 10^{-5}$ s, it is found that AmrQGD on 12 cores with two levels of adaptation is 15 times faster than Van Leer, 32 times faster than QGDFoam and 4.5 times faster than AmrQGD on a stationary mesh. However, because AMReX is designed to handle large numbers of computational cells, the parallelization efficiency of QGD in AMReX decreases as the number of cells per core decreases. It is found that adding a GPU to the computation in AmrQGD can speed up the computation by up to 4.8 times.

It is planned to extend the presented AMReX QGD solver to three-dimensional flows, to optimize the parallelization of the numerical algorithm and to test it on a wide range of validation tasks in the future.

Acknowledgments. This work was supported by Moscow Center of Fundamental and Applied Mathematics, Agreement with the Ministry of Science and Higher Education of the Russian Federation, No. 075-15-2022-283.

Disclosure of Interests. The authors have no competing interests to declare that are relevant to the content of this article.

References

1. Shalf, J.: The future of computing beyond Moore's Law. Philos. Trans. Royal Soc. A **378**, 2166 (2020). https://doi.org/10.1098/rsta.2019.0061
2. Elizarova, T.G.: Quasi-gas-dynamic Equations. Springer, Heidelberg (2009). https://doi.org/10.1007/978-3-642-00292-2
3. Jacobsen, N.G., Fuhrman, D.R., Fredsøe, J.: A wave generation toolbox for the open-source CFD library: OpenFoam®. Int. J. Num. Methods Fluids. **70**(9), 2726 (2012). https://doi.org/10.1002/fld.2726
4. QGDSolvers. https://github.com/unicfdlab/QGDsolver. Accessed 19 Apr 2024
5. Kraposhin, M.V., et al.: Development of a new OpenFOAM solver using regularized gas dynamic equations. Comput. Fluids **166**, 163–175 (2018). https://doi.org/10.1016/j.compfluid.2018.02.010
6. Kraposhin, M.V., Ryazanov, D.A., Elizarova, T.G.: Numerical algorithm based on regularized equations for incompressible flow modeling and its implementation in OpenFOAM. Comput. Phys. Commun. **271**, 108216 (2022). https://doi.org/10.1016/j.cpc.2021.108216
7. Epikhin, A.S., Elizarova, T.G.: Numerical simulation of underexpanded supersonic jets impingement on an inclined flat plate. Thermophys. Aeromech. **28**, 479–486 (2021). https://doi.org/10.1134/S0869864321040028
8. Melnikova, V.G., Epikhin, A.S., Kraposhin, M.V.: The Eulerian-Lagrangian approach for the numerical investigation of an acoustic field generated by a high-speed gas-droplet flow. Fluids. **6**(8), 274 (2021). https://doi.org/10.3390/fluids6080274
9. Zhang, W., et al.: AMReX: a framework for block-structured adaptive mesh refinement. J. Open Source Softw. **4**(37), 1370 (2019). https://doi.org/10.21105/joss.01370
10. Zhang, W., et al.: AMReX: block-structured adaptive mesh refinement for multiphysics applications. Int. J. High Perform. Comput. App. **35**(6), 508–526 (2021). https://doi.org/10.1177/10943420211022811
11. Epikhin, A., But, I.: Numerical simulation of supersonic jet noise using open source software. In: Mikyska, J., de Mulatier, C., Paszynski, M., Krzhizhanovskaya, V.V., Dongarra, J.J., Sloot, P.M. (eds.) Computational Science – ICCS 2023. ICCS 2023. LNCS, vol. 14077 pp. 292–302. Springer, Cham (2023). https://doi.org/10.1007/978-3-031-36030-5_24
12. TheLinuxFoundation. https://www.linuxfoundation.org/press. Accessed 19 Apr 2024
13. AmrQGDSolvers. https://github.com/unicfdlab/AmrQGDSolvers. Accessed 19 Apr 2024
14. Hollingsworth, W.A.: A Schlieren study of the interaction between a vortex and a shock wave in a shock tube. Br. Aeronaut. Res. Council Rept. **17**, 985 (1955)
15. Dosanjh, D.S., Weeks, T.M.: Interaction of a starting vortex as well as a vortex street with a traveling shock wave. AIAA J. **3**(2), 216–223 (1965). https://doi.org/10.2514/3.2833
16. Ribner H.S.: The sound generated by interaction of a single vortex with a shock wave. University of Toronto (1959)
17. Dosanjh, D.S., Weeks, T.M.: Sound generation by shock-vortex interaction. AIAA J. **5**(4), 660–669 (1967). https://doi.org/10.2514/3.4045
18. Rault, A., Chiavassa, G., Donat, R.: Shock-vortex interactions at high Mach numbers. J. Sci. Comput. **19**, 347–371 (2003). https://doi.org/10.1023/A:1025316311633

19. Rodionov, A.V.: Simplified artificial viscosity approach for curing the shock instability. Comput. Fluids **219**, 104873 (2021). https://doi.org/10.1016/j.compfluid.2021.104873
20. Abhishek, K., Biswas, G.: Analysis of multipolar vortices in the interaction of a shock with a strong moving vortex. Comput. Fluids **248**, 105686 (2022). https://doi.org/10.1016/j.compfluid.2022.105686
21. Kirushina, M.A., Elizarova, T.G., Epikhin, A.S.: Simulation of vortex interaction with a shock wave for testing numerical algorithms. Math. Models Comput. Simul. **15**(2), 277–288 (2023). https://doi.org/10.1134/s2070048223020072
22. 5th International workshop on hight-order CFD methods. https://how5.cenaero.be/. Accessed 29 Feb 2024
23. Kurganov, A., Tadmor, E.: New high-resolution central schemes for nonlinear conservation laws and convection-diffusion equations. J. Comput. Phys. **160**(1), 241–282 (2000). https://doi.org/10.1006/jcph.2000.6459
24. Swanson, R.C., Turkel, E.: On central-difference and upwind schemes. J. Comput. Phys. **101**(2), 292–306 (1992). https://doi.org/10.1016/0021-9991(92)90007-L
25. Van-Leer, B.: Towards the ultimate conservative difference scheme. II. Monotonicity and conservation combined in a second-order scheme. J. Comput. Phys. **14**(4), 361–371 (1974). https://doi.org/10.1016/0021-9991(74)90019-9

Smart Systems: Bringing Together Computer Vision, Sensor Networks, and Artificial Intelligence

Deep Learning Residential Building Segmentation for Evaluation of Suburban Areas Development

Agnieszka Łysak[1]([envelope])[iD] and Marcin Luckner[2][iD]

[1] Institute of Physics, Jan Kochanowski University,
ul. Uniwersytecka 7, 25-406 Kielce, Poland
agnieszka.lysak@ujk.edu.pl
[2] Faculty of Mathematics and Information Science, Warsaw University of Technology,
ul. Koszykowa 75, 00-662 Warsaw, Poland
marcin.luckner@pw.edu.pl

Abstract. Deep neural network models are commonly used in computer vision problems, e.g., image segmentation. Convolutional neural networks have been state-of-the-art methods in image processing, but new architectures, such as Transformer-based approaches, have started outperforming previous techniques in many applications. However, those techniques are still not commonly used in urban analyses, mostly performed manually. This paper presents a framework for the residential building semantic segmentation architecture as a tool for automatic urban phenomena monitoring. The method could improve urban decision-making processes with automatic city analysis, which is predisposed to be faster and even more accurate than those made by human researchers. The study compares the application of new deep network architectures with state-of-the-art solutions. The analysed problem is urban functional zone segmentation for the urban sprawl evaluation using targeted land cover map construction. The proposed method monitors the expansion of the city, which, uncontrolled, can cause adverse effects. The method was tested on photos from three residential districts. The first district has been manually segmented by functional zones and used for model training and evaluation. The other two districts have been used for automated segmentation by models' inference to test the robustness of the methodology. The test resulted in 98.2% accuracy.

Keywords: Transformers · SegFormer · Deep learning · Semantic segmentation · Computer vision

1 Introduction

Suburbanisation, the process through which urban areas expand and evolve, is often seen as a positive outcome of population growth, rising incomes, advances in transportation technology, and the decentralisation of jobs. However, a byproduct of this process, known as urban sprawl, poses several challenges [6, 22]. Urban sprawl, a term encompassing various phenomena such as

L. Franco et al. (Eds.): ICCS 2024, LNCS 14838, pp. 103–117, 2024.
https://doi.org/10.1007/978-3-031-63783-4_9

(a) *Wietrznia*, Kielce   (b) *Pod Telegrafem*, Kielce   (c) *Marymont*, Warsaw

Fig. 1. Analysed residential zones in Kielce and Warsaw

the expansion of urban boundaries, land use practices, and their consequences, leads to issues like spatial mismatch between housing and employment zones, over-reliance on automobiles, fragmented local governance, and inefficient spatial planning [13]. These issues impact various dimensions, including economic (higher infrastructure and public service cost, increasing unemployment), policy (unplanned growth, un-coordinated development), environmental (loss of vegetation, increased pollution) and social (car dependency, poverty, reduction in social interaction) [4].

Traditional methods of urban analysis, often manual and time-consuming, are increasingly supplemented by advanced technologies. There are some attempts, including urban sprawl monitoring using artificial neural networks [25] and convolutional neural networks [16,30]. There are also approaches based on Seg-Former architecture, described in Sect. 3.1, for the problem of semantic segmentation of roads for sustainable mobility development [23], or buildings [12,21].

Our work examines the possibility of the application of deep-learning approaches to the problem of urban sprawl monitoring. We compared four deep learning architectures (with a particular interest in SegFormer architecture [28]) and analysed their generalisation ability. In this context, we examined the City of Kielce, experiencing significant urban sprawl [20]. The occurrence is growing [13] and is particularly evident in suburban residential districts like *Wietrznia* area (Fig. 1a) and *Pod Telegrafem* area (Fig. 1b). In contrast, the City of Warsaw, particularly the densely developed *Marymont* district (Fig. 1c), presents a different scenario, with lower levels of urban sprawl [20].

Our research and obtained results show that the Segformed architecture, trained initially for scene segmentation [34], can be applied – using manually labelled aerial photos and AdamW optimising algorithm [15] – for urban area segmentation. The proposed model achieved over 91 per cent accuracy on the testing data, which was not obtained by several other architectures.

After segmentation, a vector representation of the functional area was created, which can be converted to a segmentation map, substantively the same as the land use map, within the selected functional zones. This map can be used for further analysis, which is necessary in many fields, such as urban, economic, and spatial planning. If this methodology were standard, there would be no problem calculating such a map on the fly when new data is derived and monitoring the growth of the residential area with streamed new data.

The rest of this paper is structured as follows. Section 2 summarises related works in the segmentation area and deep-learning techniques. Section 3 presents the workflow of the proposed solution and the deep-learning model – the Seg-Former – used in the tests. Section 4 presents the obtained results on three different data sets and compares our framework with other state-of-the-art methods. Section 5 concludes the paper and presents possible further research.

2   Related Works

A typical urban zone segmentation approach is image data for vector conversion, often with the addition of socioeconomic data and then pixel-wise classification with the use of algorithms like SVM [7], K-means [32], and XGBoost [2]. Also, convolutional neural networks (CNN) are introduced as models, which operate on pure image data only and extract features from them as a part of the learning process, like AlexNet and ResNet [9], custom CNN [35], DFCNN [1], SegNet [24], DeepLab family [27], and attention-based systems [17]. Though semantic segmentation of functional zones is complex and requires understanding crucial image features and capturing context dependencies, the transformer-based methodology seems the best intuition for the problem.

Vision Transformers (ViT) started from a concept of context dependencies extended on the whole image [8]. A disadvantage of the ViT architecture is the impossibility of solving more complex computer vision tasks, like detection or segmentation. The first attempt to overcome this drawback was Pyramid Vision Transformer (PVT) [26], a Transformer backbone for convolution block replacement, not a complete architecture. Thanks to the pyramid feature generation, it can work on image classification, detection and segmentation.

Transformer-based architectures have been invented for computer vision tasks with different purposes like Swin Transformer [14] to improve ViT, networks for image detection [5], or networks for image segmentation [5,33], especially SegFormer [28], the model chosen for this work.

Several works discussed urban segmentation – using deep learning – before. Pan et al. proposed using the U-net deep learning architecture to detect unplanned urban settlements [18]. The proposed method led to an overall accuracy of over 86% for the building segmentation, which can probably be improved using newer network architecture.

Zhang et al. proposed a deep learning-based framework called RFCNet [31]. The solution did not work on aerial photos but fused multiple views and generated plausible and complete structures. The solution could be used in urban sprawl observation because the authors presented that their solution can construct roof structures from photos

Finally, Yi et al. created UAVformer, a composite transformer network for urban scene segmentation of unmanned aerial vehicle (UAV) images [29]. The system works on various kinds of photos and performs more complex scenery segmentation than in our work. The obtained building recognition varies from 88.5% to 95.2% according to a data set. However, because the types of photos and the set issues are slightly different in our case, it is hard to compare the results directly.

3   Methods

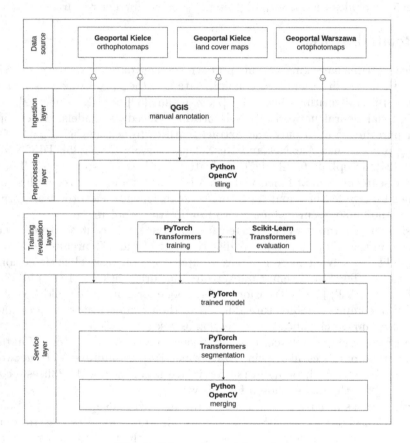

Fig. 2. Schematic workflow chart

In our work, we used maps from Kielce Geoportal (aerial photos and land cover maps) from 2019 and Warsaw Geoportal (aerial photos) from 2022. Photos from Kielce were used in the learning stage. Due to shifts and inconsistencies in combination with the photos, the land cover map has been rejected as a ground truth-functional zones segmentation map. Hence, data was labelled manually through geographic information system (GIS) software based on information about residential zone locations from the land cover maps. Then, the dataset was tiled and split into train and test parts.

Workflow (Fig. 2) takes a tiled image and produces a segmented image to classify a residential area. Although the dataset is limited to only three districts from two Polish cities, it was selected based on availability and relevance to the specific urban areas. The intent was to focus on these regions to develop and validate our deep-learning framework under controlled and consistent conditions. This decision was also influenced by the constraints in data accessibility and the computational resources required for processing more extensive datasets.

Construction of the Dataset. Data for SegFormer training and evaluation was taken from Kielce's open-access maps. The photo was from April 2019, and it was constructed from a 5×5 cm from a plane flying 700 m over ground height. It was downloaded from GIS Kielce open-access Geoportal via Web Map Service (WMS). The orthophoto was downloaded as a 1:500 scale map, converted to 1200 dpi Portable Graphics Format (PNG). The map was labelled in QGIS open-source software based on residential zone localisation information from the land cover map obtained from Kielce Geoportal for the same scope as the photo. Substantively, there were two classes: buildings and ground (building surroundings). Later, the image from the photo was converted to an RGB JPG image, and the annotated image with labels was converted to an 8-bit GRAY JPG image. Next, both images were tiled to 512×512 pixels tiles. The class information was coded according to the approach applied in the ADE20K semantic segmentation dataset, which contains over 20K images annotated with pixel-level objects [34]. Because we were implementing SegFormer architecture from the Huggingface Transformers library, where class info is inherited from the ADE20K dataset by default, we followed this behaviour and used the ADE20K *building* class for houses and *grass* class for building surroundings.

Dataset Preprocessing. Training $29,113 \times 15,938$ pixels image from Kielce *Wietrznia* residential district was tiled via OpenCV in Python programming language. The final training dataset contained three-channel (RGB) 512×512 tiles. The supporting segmentation masks were delivered as one channel (GRAY) 512×512 tiles. We removed empty tiles and tiles with 80% background and no building class existence. So, the training dataset finally counted 812 images with corresponding annotations.

Testing Stages. SegFormer was evaluated three times, presented in Fig. 3. During the training, we checked the models' performance on one house image from *Wietrznia* district, which was from the same location as the training set but was not a part of the training data. The model has not seen this exact house. However, it was extracted from *Wietrznia* district, where the rest of the area was used for the models' training (Fig. 3a). The image was 1024×1024 RGB JPG, which was tiled into four 512×512 tiles.

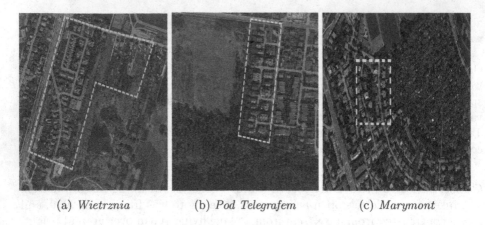

(a) *Wietrznia*   (b) *Pod Telegrafem*   (c) *Marymont*

Fig. 3. Testing datasets from residential districts. The area contoured in blue was the training data for SegFormer. The yellow area (at the lower-left corner) was the local part of the testing data. The areas contoured in magenta were the remote parts of the testing data. (Color figure online)

After the experiment was done, the model was tested on the second *Pod Telegrafem* dataset (Fig. 3b), which was a district 4 km away from the localisation of the training dataset and, of course, it also has not been seen by the model during the learning phase. Photo was also RBG JPG image, with 3072 × 2560 resolution, divided into 30 512 × 512 tiles. The last test was performed on the residential district *Marymont* from Warsaw (Fig. 3c). Data for the SegFormer test was the Warsaw open-access map. The photo was from April and May 2022, and it was constructed from a 5 × 5 cm vertical photo from a plane flying at 1600 m over ground height, taken with a camera for vertical photographs. It was downloaded from GIS Warszawa open-access Geoportal via WMS. The photo was 2048 × 2048 RGB JPG, divided into 16 512 × 512 tiles.

To sum up, the testing stage included the same area but a different house, from training data, data from the same city with similar building density and architecture (mostly detached houses), and data from a different city, from more densely built regions and a little more condensed architecture (more semi-detached and terrace houses).

3.1   Deep Learning Model

In the proposed framework, we used SegFormer b5 [28], pre-trained on the ADE20K dataset. We experimented with different versions of SegFormer, different epochs, and learning rates. The best combination was on the b5 version, after 28 training epochs and a 0.0006 learning rate.

SegFormer Architecture. The applied model combines Transformer and Multi-Layer Perceptron (MLP) architectures [26]. It consists of the encoder part

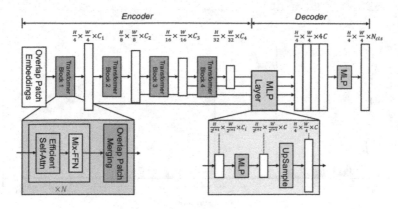

Fig. 4. SegFormer architecture [26]

for feature extraction and the decoder part for upsampling and segmentation mask prediction (Fig. 4).

The encoder's input image is divided into patches of 4×4 pixels. In contrast to ViT, smaller patches work better in detailed classification, like semantic or instance segmentation. The transformer block – an equivalent to the convolution block in CNN – performs feature extraction. This Transformer block works without a positional encoding module. In semantic segmentation, the input image should be arbitrarily shaped. Therefore, classical rigid positional encoding is not possible to deploy. Interpolation for positional encoding in SegFormer is replaced by Mix-Feed Forward Network (Mix-FFN) with local positional information share.

The encoder consists of Efficient Self-Attention, Mix-FFN and Overlap Patch Merging. Self-Attention is being computed as standard, but with efficiency improvement, thanks to reducing the density of one of the attention mechanism formula components. Mix-FFN consists of a convolutional layer and Multi Layer Perceptron (MLP) for data-driven, flexible positional encoding, which is not fixed, like in a typical Transformer. Overlap Patch Merging enables feature size reduction. In the decoder, which is very lightweight, there is MLP, which takes features from the encoder and fuzzes them together to unify channel dimensions. Then, features are upsampled and concatenated together. Second, MLP fuses concatenated features and predicts a segmentation map.

Training. The SegFormer model, by default, employs the cross-entropy loss function for optimisation. Our implementation utilised the PyTorch and Huggingface libraries, enabling us to choose from Huggingface-supported PyTorch optimisers. We opted for the AdamW optimising algorithm [15]. Our choice to utilise the AdamW optimiser was driven by its distinct advantages in enhancing model generalisation. A key feature of AdamW, an extension of the Adam optimizer [11], is its implementation of weight decay regularisation. This approach is particularly effective in minimising the loss function by selectively adjusting

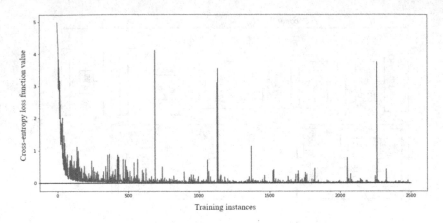

Fig. 5. Cross-entropy loss function during the learning phase plot

smaller weight values. AdamW minimises loss function by finding small weight values, which helps to overfit less and generalise better due to eliminating irrelevant components and suppressing static noise on the target.

Our choice of a learning rate of 0.0006 was based on empirical testing, yielding the most favourable results regarding model performance. During training, images were batch-processed, each batch containing two images of dimensions 512×512 pixels in RGB colour mode. The optimisation process minimises the loss function by comparing the cross-entropy between the target and predicted pixel classes in logit mode.

Figure 5 presents the loss function. The function decreases quickly and stabilises at a low level, which is a proper and expected behaviour. Sporadic peaks in its value could testify to outlying image tiles, in which models' predictions were inaccurate.

The model's parameters, which facilitated this minimisation, were adjusted following the selected learning rate. The training was concluded once the model performance metrics reached satisfactory levels, after which the model's parameters were saved for future inference purposes. Notably, an automatic hyperparameter tuning stage was not incorporated into this research.

Evaluation. A two-step evaluation process assessed pixel-wise accuracy and the mean intersection over union metrics. Pixel-wise accuracy represents the percentage of pixels correctly classified concerning the target mask. A second metric – mean Intersection over Union (mIoU) – was introduced because accuracy could be misleading in cases where many pixels belong to the ground and fewer to residential buildings (imbalanced classes). mIoU calculates the overlap between the predicted and target masks, divided by the combined area of both masks. Figure 6 presents metrics in the training stage using *Wietrznia* dataset. The trained model achieved an average pixel-wise accuracy of 0.93 (Fig. 6a) and 0.86 mIoU (Fig. 6b).

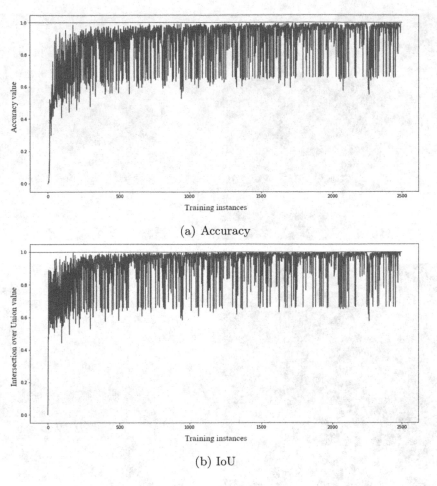

(a) Accuracy

(b) IoU

Fig. 6. Metric plots for the learning phase

4   Results

4.1   SegFormer Tests on Various Data Sets

The model was evaluated on data marked with yellow and magenta in Fig. 3. Figure 7 presents the qualitative results for these data sets.

In the testing phase, the model's inferencing capability was initially tested on a single house from the *Wietrznia* dataset. The model's proficiency in processing images of varying dimensions was evaluated using a 2048 × 2048 pixels map tile. In this test, the model successfully segmented the building area, achieving a 96.2% mIoU and 99.6% accuracy. The qualitative results (Fig. 7a) were highly encouraging regarding qualitative performance.

Subsequently, we extended our evaluation to include two additional datasets: *Pod Telegrafem* and *Marymont*. For these tests, the input tiles maintained the

(a) *Wietrznia*

(b) *Pod Telegrafem*

(c) *Marymont*

Fig. 7. SegFormer results

exact dimensions as those used in the training phase (512 × 512 pixels). The SegFormer model demonstrated robust segmentation capabilities in urban areas, achieving an 86.7% mIoU and 98.2% accuracy for *Pod Telegrafem* district (the results are shown in Fig. 7b) and 50.3% mIoU and 91.4% accuracy for *Marymont* district (the results are shown in Fig. 7c). These results validate the model's effectiveness in diverse urban environments.

4.2   SegFormer vs. Other Methods

The SegFormer results were compared against several pre-trained state-of-the-art algorithms to evaluate their effectiveness. That includes the vision transformers-based solution DPT [19] published in 2021, the mask transformer-based method Mask2Former [3] published in 2022, and the latest version of the acclaimed real-time object detection and image segmentation model YOLOV8 [10] published in 2023.

All algorithms were subjected to the same testing procedure to ensure a fair and consistent comparison. The *Wietrznia* dataset served as the basis for this evaluation. Each algorithm was applied to segment a single building within this district and then extended the testing to include the *Pod Telegrafem* and *Marymont* districts. The outcomes of these comparative tests have been presented in Table 1.

Table 1. Comparison of the results of our framework and state-of-the-art methods for image semantic segmentation

Algorithm	Dataset	IoU [%]	Accuracy [%]
DPT	*Wietrznia*	84.0	97.9
Mask2Former	*Wietrznia*	76.2	96.5
SegFormer	*Wietrznia*	96.1	99.5
YOLOV8	*Wietrznia*	96.5	99.6
DPT	*Pod Telegrafem*	20.1	87.0
Mask2Former	*Pod Telegrafem*	38.7	82.2
SegFormer	*Pod Telegrafem*	86.7	98.2
YOLOV8	*Pod Telegrafem*	34.9	87.4
DPT	*Marymont*	37.3	71.6
Mask2Former	*Marymont*	56.6	87.4
SegFormer	*Marymont*	50.3	91.4
YOLOV8	*Marymont*	62.2	91.6

The proposed SegFormer-based approach outperformed most of the other methods in the tests. However, it is notable that the YOLOV8 algorithm achieved a higher mIoU on *Wietrznia* and *Marymont* test sets. Additionally, YOLOV8 demonstrated slightly better accuracy in these datasets.

It is worth noticing that any reference method could not obtain the IoU close to the proposed framework on a separate *Pod Telegrafem* data set. This finding underscores the impressive generalisation capabilities of the SegFormer framework, particularly in diverse urban settings.

5   Conclusions

Our study demonstrates the effective application of the SegFormer model, which was fine-tuned rather than pre-trained, for semantic segmentation in residential zones. This approach yielded promising results, both qualitatively and quantitatively.

Comparison with the state-of-the-art methods, in their pre-trained versions, showed that in many cases, SegFormer and the workflow proposed in this research achieved the best results on the test sets from *Wietrznia*. In the second and third testing districts, SegFormer architecture performed very well, being outperformed by YOLOV8 in two testing sets.

Notably, the SegFormer model exhibited strong generalisation capabilities, particularly in the additional tests using data from another city. Despite a decrease in performance compared to the original city (Kielce), which was anticipated due to the model not being trained on data from other cities, the results were still robust and satisfactory. The model had not been trained on data from any other city; therefore, urban aesthetics from another region can cause the outcome to deteriorate. A solution for problems like that is additional training of the actual model on data from other metropolises of interest.

The main contribution is the novel application of the known SegFormer architecture to urban sprawl monitoring, a relatively underexplored area. Also, analysis of the custom dataset from Kielce and Warsaw provides new insights specific to these urban areas.

We acknowledge certain limitations in our research, primarily related to the dataset's size and diversity. These limitations could affect the model's generalizability across different urban settings. To address this, we suggest expanding the dataset to include various photos from diverse areas. Such an approach would likely enhance the model's ability to account for variations in urban characteristics, including different residential area locations and roofing materials, thereby bolstering the overall robustness and generality of the method.

Likewise, more functional zone types and data from a more comprehensive time range could be included to broaden the methodology. Also, exploring the integration of historical data to track and predict urban sprawl over time presents a promising direction. Additionally, fine-tuning the SegFormer configuration's hyperparameters could yield improved results for specific applications, such as photo segmentation. Finally, augmenting our dataset is another strategy that could further refine our outcomes.

In summary, while our study demonstrates the potential of using advanced deep learning models like SegFormer in urban sprawl monitoring, it also opens up several avenues for further exploration and improvement. The insights gained from this research contribute to deep learning, computer vision, urban planning, and sustainable development. The practical implications of this research for urban planners and policymakers are significant. The ability to accurately and efficiently monitor urban sprawl can inform more sustainable urban development practices, help resource allocation and support environmental conservation efforts.

References

1. Bao, H., Ming, D., Guo, Y., Zhang, K., Zhou, K., Du, S.: DFCNN-based semantic recognition of urban functional zones by integrating remote sensing data and POI data. Remote Sens. **12**(7), 1088 (2020). https://doi.org/10.3390/rs12071088
2. Chen, S., Zhang, H., Yang, H.: Urban functional zone recognition integrating multisource geographic data. Remote Sens. **13**(23) (2021).https://doi.org/10.3390/rs13234732
3. Cheng, B., Misra, I., Schwing, A.G., Kirillov, A., Girdhar, R.: Masked-attention mask transformer for universal image segmentation. In: Proceedings of the IEEE Computer Society Conference on Computer Vision and Pattern Recognition June 2022, pp. 1280–1289 (2022). https://doi.org/10.1109/CVPR52688.2022.00135
4. Chiguvi, D., Kgathi-Thite, D.: Analysis of the positive and negative effects of urban sprawl and dwelling transformation in urban cities: case study of Tati Siding Village in Botswana. J. Legal Ethical Regul. Issues **25**(S2), 1–13 (2022)
5. Chu, X., et al.: Twins: revisiting the design of spatial attention in vision transformers. In: Advances in Neural Information Processing Systems, vol. 12(NeurIPS), pp. 9355–9366 (2021)
6. Cocheci, R.M., Petrisor, A.I.: Assessing the negative effects of suburbanization: the urban sprawl restrictiveness index in Romania's metropolitan areas. Land **12**(5) (2023) https://doi.org/10.3390/land12050966
7. Deng, Y., He, R.: Refined urban functional zone mapping by integrating open-source data. ISPRS Int. J. Geo-Inf. **11**(8) (2022) https://doi.org/10.3390/ijgi11080421
8. Dosovitskiy, A., et al.: An image is worth 16x16 words: transformers for image recognition at scale. In: International Conference on Learning Representations (2021). https://openreview.net/forum?id=YicbFdNTTy
9. Izzo, S., Prezioso, E., Giampaolo, F., Mele, V., Di Somma, V., Mei, G.: Classification of urban functional zones through deep learning. Neural Comput. Appl. **34**(9), 6973–6990 (2022). https://doi.org/10.1007/s00521-021-06822-w
10. Jocher, G., Chaurasia, A., Qiu, J.: YOLO by Ultralytics (2023). https://github.com/ultralytics/ultralytics
11. Kingma, D.P., Ba, J.L.: Adam: a method for stochastic optimization. In: 3rd International Conference on Learning Representations. ICLR 2015 - Conference Track Proceedings, pp. 1–15 (2015)
12. Li, M., et al.: Method of building detection in optical remote sensing images based on SegFormer. Sensors **23**(3) (2023). https://doi.org/10.3390/s23031258
13. Lityński, P.: The intensity of urban sprawl in Poland. ISPRS Int. J. Geo-Inf. **10**(2) (2021). https://doi.org/10.3390/ijgi10020095
14. Liu, Z., et al.: Swin transformer: hierarchical vision transformer using shifted windows. In: Proceedings of the IEEE International Conference on Computer Vision, pp. 9992–10002 (2021). https://doi.org/10.1109/ICCV48922.2021.00986

15. Loshchilov, I., Hutter, F.: Decoupled weight decay regularization. In: 7th International Conference on Learning Representations. ICLR 2019 (2019)

16. Mansour, D., Souiah, S.A., El Amin Larabi, M.: Built-up area extraction through deep learning. In: 2021 IEEE International Geoscience and Remote Sensing Symposium. IGARSS, pp. 6805–6808 (2021). https://doi.org/10.1109/IGARSS47720.2021.9554694

17. Niu, R., Sun, X., Tian, Y., Diao, W., Chen, K., Fu, K.: Hybrid multiple attention network for semantic segmentation in aerial images. IEEE Trans. Geosci. Remote Sens. **60**, 1–18 (2022). https://doi.org/10.1109/TGRS.2021.3065112

18. Pan, Z., Xu, J., Guo, Y., Hu, Y., Wang, G.: Deep learning segmentation and classification for urban village using a worldview satellite image based on U-net. Remote Sens. **12**(10), 1–17 (2020). https://doi.org/10.3390/rs12101574

19. Ranftl, R., Bochkovskiy, A., Koltun, V.: Vision transformers for dense prediction. In: Proceedings of the IEEE International Conference on Computer Vision, pp. 12159–12168 (2021). https://doi.org/10.1109/ICCV48922.2021.01196

20. Renata, R.C., Barbara, C., Andrzej, S.: Which polish cities sprawl the most. Land **10**(12) (2021). https://doi.org/10.3390/land10121291

21. Song, J., Zhu, A.X., Zhu, Y.: Transformer-based semantic segmentation for extraction of building footprints from very-high-resolution images. Sensors **23**(11) (2023). https://doi.org/10.3390/s23115166

22. Spirkova, D., Adamuscin, A., Golej, J., Panik, M.: Negative effects of urban sprawl. In: Charytonowicz, J. (ed.) AHFE 2020. AISC, vol. 1214, pp. 222–228. Springer, Cham (2020). https://doi.org/10.1007/978-3-030-51566-9_30

23. Tao, J., et al.: Seg-road: a segmentation network for road extraction based on transformer and CNN with connectivity structures. Remote Sens. **15**(6) (2023). https://doi.org/10.3390/rs15061602

24. Tian, T., Chu, Z., Hu, Q., Ma, L.: Class-wise fully convolutional network for semantic segmentation of remote sensing images. Remote Sens. **13**(16), 200–215 (2021). https://doi.org/10.3390/rs13163211

25. Tsagkis, P., Bakogiannis, E., Nikitas, A.: Analysing urban growth using machine learning and open data: an artificial neural network modelled case study of five Greek cities. Sustain. Cities Soc. **89**, 104337 (2023). https://doi.org/10.1016/j.scs.2022.104337

26. Wang, W., et al.: Pyramid vision transformer: a versatile backbone for dense prediction without convolutions. In: Proceedings of the IEEE International Conference on Computer Vision, pp. 548–558 (2021). https://doi.org/10.1109/ICCV48922.2021.00061

27. Wang, Y., et al.: Mask DeepLab: end-to-end image segmentation for change detection in high-resolution remote sensing images. Int. J. Appl. Earth Obs. Geoinf. **104**, 102582 (2021). https://doi.org/10.1016/j.jag.2021.102582

28. Xie, E., Wang, W., Yu, Z., Anandkumar, A., Alvarez, J.M., Luo, P.: SegFormer: simple and efficient design for semantic segmentation with transformers. In: Advances in Neural Information Processing Systems, vol. 15, 12077–12090. NeurIPS (2021)

29. Yi, S., Liu, X., Li, J., Chen, L.: UAVformer: a composite transformer network for urban scene segmentation of UAV images. Pattern Recogn. **133** (2023). https://doi.org/10.1016/j.patcog.2022.109019

30. Yin, B., et al.: How to accurately extract large-scale urban land? Establishment of an improved fully convolutional neural network model. Front. Earth Sci. **16**(4) (2022). https://doi.org/10.1007/s11707-022-0985-2

31. Zhang, X., Aliaga, D.: RFCNet: enhancing urban segmentation using regularization, fusion, and completion. Comput. Vis. Image Underst. **220**(April), 103435 (2022). https://doi.org/10.1016/j.cviu.2022.103435
32. Zhang, X., Li, W., Zhang, F., Liu, R., Du, Z.: Identifying urban functional zones using public bicycle rental records and point-of-interest data. ISPRS Int. J. Geo-Inf. **7**(12) (2018). https://doi.org/10.3390/ijgi7120459
33. Zheng, S., et al.: Rethinking semantic segmentation from a sequence-to-sequence perspective with transformers. In: Proceedings of the IEEE Computer Society Conference on Computer Vision and Pattern Recognition, pp. 6877–6886 (2021). https://doi.org/10.1109/CVPR46437.2021.00681
34. Zhou, B., et al.: Semantic understanding of scenes through the ADE20K dataset. Int. J. Comput. Vision **127**(3), 302–321 (2019). https://doi.org/10.1007/s11263-018-1140-0
35. Zhou, W., Ming, D., Lv, X., Zhou, K., Bao, H., Hong, Z.: SO–CNN based urban functional zone fine division with VHR remote sensing image. Remote Sens. Environ. **236**(November 2019), 111458 (2020).https://doi.org/10.1016/j.rse.2019.111458

Analysing Urban Transport Using Synthetic Journeys

Marcin Luckner[1] , Przemysław Wrona[1] , Maciej Grzenda[1](✉) ,
and Agnieszka Łysak[2]

[1] Faculty of Mathematics and Information Science, Warsaw University of Technology,
ul. Koszykowa 75, 00-662 Warsaw, Poland
{marcin.luckner,przemyslaw.wrona,maciej.grzenda}@pw.edu.pl
[2] Institute of Physics, Jan Kochanowski University,
ul. Uniwersytecka 7, 25-406 Kielce, Poland
agnieszka.lysak@ujk.edu.pl

Abstract. Travel mode choice models make it possible to learn under what conditions people decide to use different means of transport. Typically, such models are based on real trip records provided by respondents, e.g. city inhabitants. However, the question arises of how to scale the insights from an inevitably limited number of trips described in their travel diaries to entire cities.

To address the limited availability of real trip records, we propose the Urban Journey System integrating big data platforms, analytic engines, and synthetic data generators for urban transport analysis. First of all, the system makes it possible to generate random synthetic journeys linking origin and destination pairs by producing location pairs using an input probability distribution. For each synthetic journey, the system calculates candidate routes for different travel modes (car, public transport (PT), cycling, and walking). Next, the system calculates Level of Service (LOS) attributes such as travel duration, waiting time and distances involved, assuming both planned and real behaviour of the transport system. This allows us to compare travel parameters for planned and real transits.

We validate the system with spatial, schedule and GPS data from the City of Warsaw. We analyse LOS attributes and underlying vehicle trajectories over time to estimate spatio-temporal distributions of features such as travel duration, and number of transfers. We extend this analysis by referring to the travel mode choice model developed for the city.

Keywords: Travel mode choice · synthetic journeys · public transport

1 Introduction

Modern cities have typically been designed and built with the primary focus on the needs of car drivers [10]. Planning concepts, such as the 15-minute city, aim to minimise car usage by ensuring access to critical urban facilities within walking distance. Another approach is to promote less energy consumption and pollution-emitting means of transport [2]. However, the proposed solutions can be difficult

L. Franco et al. (Eds.): ICCS 2024, LNCS 14838, pp. 118–132, 2024.
https://doi.org/10.1007/978-3-031-63783-4_10

to implement in the existing urban infrastructure [8]. Therefore, it is necessary to develop tools that enable data-driven decisions for urban development. These include vehicle trajectory and LOS analysis tools that can give insights into traffic and street congestion which are useful for routing and transportation planning [14].

Travel mode choice (TMC) modelling [4–6] is vital to understanding under what conditions people decide to use different means of transport. In particular, it helps us to understand what makes people use (or not use) public transport. However, while TMC models predict whether for a trip of interest a car or another means of transport is likely to be used, identifying the spatial distribution of mode choices in urban areas is difficult. This is because the number of trips collected in surveys is inevitably limited, and increasing this number is expensive. Moreover, as not everyone is equally likely to share their data, increasing the size of the representative sample of real journeys in the areas of interest is additionally difficult.

Motivated by these needs, we propose the Urban Journey System (UJS) integrating Apache big data platforms with spatial data analytic engines based on OpenTripPlanner (OTP). The system combines open-source platforms with the newly proposed JourneyGenerator, JourneyDescriber, and JourneyAnalyser modules. JourneyGenerator generates synthetic journeys which are used to create public and individual transport trajectories. In this way, an arbitrarily large number of journeys can be obtained. To obtain representative journey origin and destination pairs, locations are generated using an input probability distribution, such as the probability distribution of journey endpoints based on time-dependent transport model demand matrices.

Next, JourneyDescriber calculates candidate routes for each synthetic journey. These include routes for car, walking, cycling, and PT using OTP instances provided with planned and real timetables in the form of General Transit Feed Specification (GTFS) files. While planned GTFS files are obtained from transport authorities, real GTFS files are developed by our system using a real-time location stream of public transport vehicles processed inter alia by a module based on Apache Flink. JourneyDescriber also calculates LOS attributes for various travel modes. Finally, the JourneyAnalyser produces an interactive HTML report on the generated journeys and compares LOS attributes.

We validate the system with the results obtained for the City of Warsaw, Poland. The GTFS feed and real-time PT location stream were processed to calculate scheduled and real PT networks. Hence, we use a selection of 399 planned and real daily public transport schedules already collected by the system. Journeys are generated based on the origin-destination hourly demand matrices of the transport model. In our analysis, we pay particular attention to journey attributes having a key impact on travel mode choices.

The remainder of this work is organised as follows. In Sect. 2 we analyse related works. This is followed by a summary of the system in Sect. 3, including an overview of its implementation. Next, results obtained for the City of Warsaw are analysed in Sect. 4. Finally, conclusions are made in Sect. 5.

2   Related Works

Several other works have focused on the spatial distribution of the choice of means of transport and its LOS. The applied approach to estimating their spatial distribution depended on the available data. Chia et al. explored the relationship between the spatial distribution of transfer location and the transit service's attractiveness in Brisbane [3]. The study was limited to services operated with a passenger card because the analysis was partially based on data the card operator system collected. The rest of the data came from a travel survey.

Yousefzadeh Barri et al. explored data from an extensive household travel survey in the Toronto region, including a one-day household travel diary [15]. They used statistical and machine learning (ML) models to predict newly generated transit trips by a low-income carless group after improving job accessibility. The study was limited to the use of public transport. Rocha et al. combined genetic algorithms and geostatistical methods to forecast travel demand variables and, as a result, the distribution of car trip rates in the São Paulo Metropolitan Area [11]. The data used in the study came from the Origin-Destination (OD) survey. Regarding means of transport, the study was limited to cars.

Tenkanen et al. analyzed different travel modes in the Helsinki region [13]. The analysis included door-to-door walking, cycling, driving and transit journeys. Centroids of statistical grid cells were used as origin and destination points for the calculation of the journeys. Their work is focused on distance and duration. Other LOS attributes were not analyzed. However, some of them were estimated to obtain the total travel time, e.g. PT walking times to and from the nearest stops were estimated based on Euclidean distances.

In [1], public transport time inaccuracy and variability were addressed. The duration of door-to-door PT journey and its components e.g. waiting and in-vehicle time were estimated using both scheduled timetables and actual timetables determined based on past GPS traces of PT vehicles. Both the origin and destination of each generated journey were based on centroids of the hexagonal spatial index developed by Uber with the area of each hexagon of $0.1\,km^2$. Geolocated jobs were used to enable spatial analysis of employment accessibility. LOS attributes of other than PT transport modes were not considered. Demand matrices were not used to vary demand for transportation between zones. Travel time reliability was recently addressed also in [16], where a proposal for formulas quantifying PT competitiveness compared to cars based on maximising entropy value to obtain more dispersed competitiveness was made. Maps of areas of Hangzhou, China with low and high competitiveness according to these formulas were obtained.

As observed in [6], travel mode choice models most frequently rely on survey data only. This is even though the LOS attributes documenting the choices faced by a traveller are also considered to be important [5]. Among the attempts to calculate features quantifying trip characteristics under different travel modes, considering exact point coordinates, the study developed for London can be mentioned [6]. The LOS attributes in the work included durations of walking, cycling, interchanges and the whole PT route, and were calculated for real trips only. No attempts to generate trip endpoints of representative trips were made. Furthermore, the spatial variability of the values of LOS features was not considered [6]. The inevitably limited number of real trips, compared to the London city area, illustrates the challenges caused by the use of real trip data.

In this work, we aim to go beyond these studies by generating and analysing an arbitrary number of detailed trips in urban areas, linking points generated based on probability distributions calculated with data from a transport model. Our system enables spatial analysis of multiple trip features under four different travel modes rather than the use of PT and car only. We consider inter alia features found relevant for TMC models built with ML methods.

3   System Overview

The objective of the UJS system is to enable large-scale analysis of journeys in urban areas. The high-level architecture of the system including its core components is presented in Fig. 1. Let us note that both JourneyGenerator and JourneyDescriber were designed to be a part of the Use4IoT architecture [7].

The PT schedule data in GTFS format for each day is developed based on data downloaded from transport operators. This provides planned daily timetables. However, disruptions such as delays and cancellations sometimes occur in public transport. Hence, real GTFS is created using a real-time location stream of public transport vehicles obtained from the API of public transport entities and processed inter alia by Apache Flink. In this way, real daily timetables in GTFS format are obtained. This makes it possible to construct a real transit network. Hence, for example, delays causing possibly missed transfers can be considered when calculating LOS attributes for a day of interest, based on the real behaviour of the PT system. This enables analysis of the actual experience of using PT.

3.1   JourneyGenerator

Planning an urban transport system is a complex task that relies on travel demand patterns. To model such demand, the urban area is frequently divided into disjoint transport zones. A travel demand (TD) matrix is a square matrix which shows the expected number of passengers moving between each combination of zones. The columns represent the origin zones, while the rows represent the destination zones of the city. In our approach, we use a potentially different matrix for each hour of the day. Hence, we consider up to 24 demand matrices for a demand scenario. One scenario can denote, e.g. real demand observed currently or hypothetical demand expected in 5 years during working days. Each matrix can come from the transport model of the city or be generated to reflect, e.g. the number and location of children travelling to schools.

TD matrices are used in JourneyGenerator to generate synthetic journeys, as shown in Algorithm 1. We use a demand matrix to generate a random zone pair in line 4 of the algorithm. Although any zone combination is possible, the

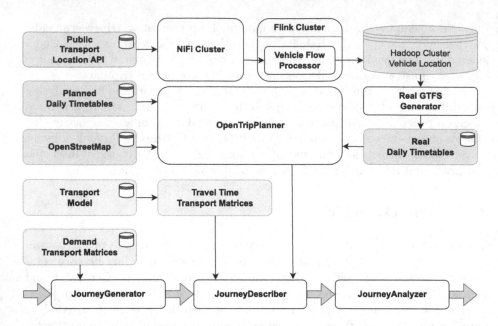

Fig. 1. High-level architecture of UJS system

likelihood of selecting each zone varies. A zone pair and an hour with higher passenger traffic between the zones are more likely to be selected. After randomly selecting an origin zone and a destination zone, in lines 5 and 6 we select random addresses from the lists of address points in each zone area. The probability of selecting an address is proportional to its weight. This makes it possible, e.g. to make journeys from buildings populated by a large number of residents more likely to be generated. Finally, an exact time during the one hour is randomly generated.

3.2   JourneyDescriber

For each synthetic journey produced by JourneyGenerator, JourneyDescriber calculates trajectories likely to be used with different travel modes and LOS attributes for each travel mode. Estimating a travel path and calculation of its LOS attributes such as duration and distance is carried out using OTP[1], a multimodal trip planning module relying on OpenStreetMap[2] (OSM) [9]. OSM provides information on transport infrastructure such as the street network and the location of public transport stops.

OTP calculates private car routes and estimates trip duration using OSM data about the street network, taking into consideration traffic regulations and

---

[1] https://www.opentripplanner.org.
[2] Map data copyrighted by OpenStreetMap contributors and available from https://www.openstreetmap.org.

Algorithm 1. The generation of journeys

Input:

int N - the number of journeys to generate

$int[][][]$ $matrices$ - a table of H two-dimensional matrices representing the number of passengers moving between zones $(i,j), i,j \in \{1,\ldots,Z\}$, one matrix per one hour time slot $h \in \{h_1,\ldots,h_H\}$.

$A[]$ $addresses$ - list of address points to consider in each zone $i \in \{1,\ldots,Z\}$

1: **procedure** GENERATERANDOMJOURNEYS(N, matrix, addresses)
2:     $journeys \leftarrow emptyList()$
3:     **for** $k \leftarrow 1$ to N **do**
4:         $(C_{k_{start}}, C_{k_{end}}, h) \leftarrow GetRandomZonePair(matrices)$
5:         $A_{k_{start}} \leftarrow GetRandomAddress(C_{k_{start}}, addresses)$
6:         $A_{k_{end}} \leftarrow GetRandomAddress(C_{k_{end}}, addresses)$
7:         $journeys.add(A_{k_{start}}, A_{k_{end}}, randomTime(h))$
8:     **end for**
9:     $return(journeys)$
10: **end procedure**

---

obstacles such as traffic lights, road types, and crossroads limiting the estimated car speed. Based on travel time matrices from a transport model, JourneyDescriber also calculates the travel time by car under the expected street congestion for a given zone pair and time of the day.

In the case of public transport, the routes are calculated between public transport stops, and the waiting and walking times are added to the total trip duration. The calculated distance includes walking from the trip origin to the first stop, from the last stop to the destination, and the potential distance covered during transfers in multimodal travel.

For a private car, a single route is calculated. For PT, a set of routes is created that consists of all connections that start up to 5 min before and 10 min after the given journey starting time. Sample car and PT routes calculated for the same input synthetic journey are presented in Fig. 2.

For every synthetic journey, JourneyDescriber makes requests to the OTP instance(s). Each instance is configured with planned or real daily timetables. OTP provides JourneyDescriber with data about possible routes for car, PT, walking and cycling, and LOS attributes such as travel duration for each of these modes. Importantly, LOS features can be calculated using both planned timetables and real timetables developed by the UJS system based on GPS traces of public transport vehicles. Thus, LOS features documenting planned connections can be compared with LOS features quantifying connections which were feasible in the past. In particular, the impact of delays on missed connections is reflected in the LOS features developed based on real timetables.

In the OTP responses, for requests to OTP instances based on both planned and real timetables, we also receive information inter alia on potential PT connections within a 15-minute timeframe, the duration of each connection, and the number of transfers required by each of them. Data on possibly many PT

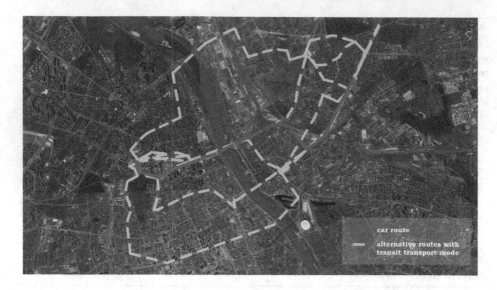

Fig. 2. Example car route and its PT alternatives

connections per journey is aggregated by JourneyDescriber to determine LOS attribute values such as minimum and average travel time and the number of possible connections within a 15-minute window from the start of the journey. Finally, every journey with its LOS attributes and trajectories is included in the output list of synthetic journeys.

3.3 JourneyAnalyser

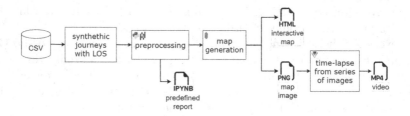

Fig. 3. The workflow of JourneyAnalyser

Figure 3 shows the workflow of JourneyAnalyser – a module implemented in Python that generates the final analysis. Input journey records in CSV format with embedded trajectories are filtered to develop maps and plots, e.g. for specific hours and means of transport. Data is also transformed to extract individual car and PT trajectories for each journey. The Folium library generates interactive maps from preprocessed data and OSM. The maps are exported in PNG format for coarse analysis and in HTML format for more geospatial insight. Figure 4 presents sample maps for car (Fig. 4a) and PT (Fig. 4b) connections. A time-lapse video from the PNG maps can be generated for the period of interest to better understand the geospatial-temporal data. The preprocessed data is also the source for predefined reports implemented as a Jupyter Notebook. The reports compare global LOS for means of transport, timetables or hours, e.g. by generating empirical cumulative distribution function (ECDF) plots for LOS attributes such as the ones shown in Fig. 5 and Fig. 6.

(a) Routes of cars. 6 am to 12 pm   (b) Connections by PT. 6 am to 12 pm

Fig. 4. Sample interactive maps produced by JourneyAnalyser. `SYNTH_WAW_EQW` data. Background: [9].

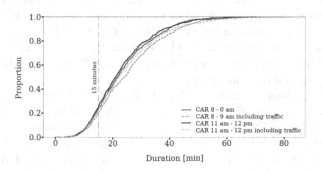

Fig. 5. ECDFs of car trip duration not considering street congestion and considering congestion. Selected hours. `SYNTH_WAW_EQW` data.

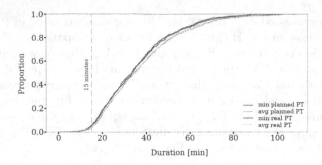

Fig. 6. ECDFs of PT trip duration, including minimum and average travel time under planned and real timetables. All considered hours. SYNTH_WAW_EQW data.

3.4 Implementation

The system was implemented at a central unit with 2xLenovo P/N BF78 2.65 GHz and 48 cores, 1536 GB RAM and 64 TB mass storage, which provided the basis for Apache NiFi, Apache Hadoop, Apache Flink, six OTP instances, and R and Python environments. It serves UJS needs by inter alia collecting planned timetables and GPS traces of PT vehicles, and calculating real timetables.

4 Results

4.1 Aggregation of Journey Features into Frequent Routes and Distribution Functions

The system was validated using real data from the City of Warsaw. First, public transport APIs were used to download raw data and prepare planned and real daily timetables for buses and trams. For the metro, GTFS files were calculated using metro frequency information. The Warsaw transport model was used to provide travel time transport matrices for cars and OTP was fed with the Warsaw infrastructure spatial data from OSM.

Two synthetic data sets were developed with UJS. In both cases, real demand matrices for working days from the transport model were used as input for JourneyGenerator. In the first case, the addresses considered included all available addresses in the City of Warsaw, sampled with equal weights. 4,000 journeys in the period of 6 AM to 12 PM based on these weights were generated with Algorithm 1. LOS attributes were calculated using both planned schedules and real schedules. This was done by submitting requests to two different OTP instances in which planned and real Warsaw transport networks were configured. In this way, the SYNTH_WAW_EQW data set was developed.

Data sets such as the SYNTH_WAW_EQW data set can be used for both calculating distributions of individual features and visualising routes of synthetic trips on maps. Examples of interactive maps generated by JourneyAnalyser are given in Fig. 4. Figure 4a identifies segments of traffic infrastructure heavily used by individual means of transport in morning hours (between 6 am and 12 pm). Traffic congestion is visible on the east bank of the Vistula River in the south part of the city. Studies on air pollution [12] can be complemented by this kind of analysis.

Similarly, Fig. 4b shows the quickest PT connections which could be used for journeys present in the SYNTH_WAW_EQW data set. Public transport in Warsaw mainly consists of low-emission vehicles (buses and electric trams). Therefore, this visualisation can be used to identify areas without good direct connections with the city centre. Such areas are observed in the north-west outskirts of the city along the Vistula river, splitting the city into two parts. In this area, there are more direct car routes than PT connections.

These results can also be analysed statistically, e.g. aggregated using an empirical cumulative distribution function (ECDF). Figure 5 and Fig. 6 show the distribution of travel duration. Figure 5 illustrates the differences in the distribution of car travel duration during and after morning rush hours. The plots compare the duration calculated with and without considering traffic congestion. Let us note that not only congestion but also travel origin-destination patterns are different depending on the time of the day.

Similarly, Fig. 6 compares PT travel duration calculated using planned and real daily timetables. While the planned travel duration is calculated using planned timetables, the real travel duration is calculated using data from GPS sensors in the vehicles. Because the duration for PT is calculated as a statistic from available connections for each trip, minimal and average travel times are compared. The plots show that the duration difference between planned and real connections is higher for longer journeys, which frequently include transfers between lines and are more susceptible to delays of individual vehicles. Both Figs. 5 and 6 illustrate how data produced by UJS can be used for in-depth analysis of LOS attributes under different conditions.

4.2   Analysing Spatial Distribution of the Level of Service Features

The SYNTH_WAW_EQW data set shows that aggregates of trips and ECDFs can be developed with moderately sized data sets. However, a few thousand trips are not sufficient to generate high-density plots showing spatial distribution of LOS.

Hence, the second data set used as an illustration in this study was populated with 100,000 journey records. This time the addresses considered included all available addresses in the City of Warsaw, sampled with weights proportional to the number of inhabitants of an address. As before, the journeys were randomly generated with Algorithm 1. LOS attributes were calculated using planned schedules. In this way, the SYNTH_WAW_INHW data set was developed. Trips were generated for random times for hours between 6 AM and 8 PM. As discussed in Sect. 3.1, the probability of selecting an hour depended on the overall number of journeys for this hour according to the demand matrices. In this way, journeys during rush hours were more likely to be produced.

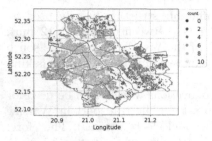

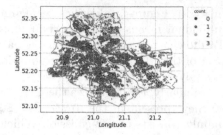

(a) Number of available PT connections

(b) Minimum number of required transfers

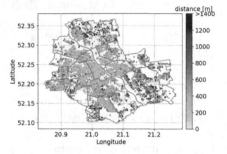

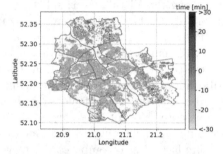

(c) Minimum walking distances required to use PT connections

(d) Difference of travel duration between bicycle and PT

Fig. 7. Distributions of sample LOS values. `SYNTH_WAW_INHW` data.

Maps showing the distribution of selected LOS features are provided in Fig. 7. Each point represents the location of the origin of the synthetic trips present in the data. It can be observed that the density of points varies greatly, which corresponds to the fact that trips from some locations, such as forest areas, are far less likely during the working days for which the data set was generated.

Figure 7a shows the number of available PT connections to a destination within 15 min i.e. in the period $[t(j) - 5\,\text{min}, t(j) + 10\,\text{min}]$, where $t(j)$ denotes the randomly assigned start time of journey j. In the city centre, 10 or more connections are frequently available during such periods, as travellers may rely on multiple trams and buses travelling through major transportation hubs. However, it can be observed that for some areas from which many trips are likely to be initiated such as the eastern part of the city, the number of available connections is significantly lower. Similarly, in such areas, as shown in Fig. 7b, even the best connection may require one or more transfers.

Figure 7c shows that the overall walking distance needed by travellers in some areas of the city to use a suitable PT connection varies greatly. Importantly, when generating LOS features, UJS considers feasible routes, i.e. in this case walking paths resulting from street and pavement networks rather than Euclidean distances. In less densely populated areas the density of feasible walking paths is likely to be much lower, resulting in a major difference between the Euclidean distance and the actual length of the walking route. Finally, Fig. 7d shows the values calculated by subtracting from travel duration by bicycle the duration of the quickest connection with public transport.

4.3   Adding Spatial Aspects to Travel Mode Choice Modelling

Let us note one more use for the data generated by the system. Municipalities struggle to reduce air pollution and foster sustainable mobility. Hence, the question arises of how to analyse the potential for environmentally-friendly travel mode choices. In Fig. 8, we provide a TMC model developed for the City of Warsaw, taking the form of a decision tree and pruned to include key top-level splits only[3]. Importantly, the model was developed using both real survey data including trip diaries and an extensive set of LOS features, as suggested in [6].

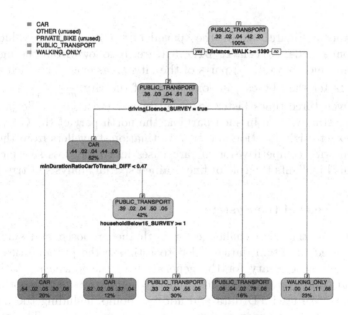

Fig. 8. Simplified TMC model for the City of Warsaw. Decision tree.

The model illustrated in Fig. 8 reveals that people tend to walk if the walking distance is no longer than 1390 m. Otherwise, and assuming they have a driving license, people tend to travel by car when $\frac{d_{CAR}(j)}{min(d_{PT}(j))} < 0.47$ i.e. when the travel duration required when relying on the quickest connection by PT is at least twice as large as the duration of travel by car under free vehicle flow conditions denoted by $d_{CAR}(j)$. Let us analyse for which journeys this is

---

[3] While it is not the objective of this work to describe the process of model development, let us note that some further details on TMC modelling for the City of Warsaw that we rely on in this work can be found in [4].

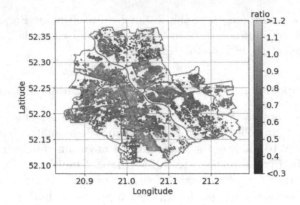

Fig. 9. Travel duration by car vs. public transport. `SYNTH_WAW_INHW` data.

likely to happen. Figure 9 shows the spatial distribution of the values of the `minDurationRatioCarToTransit_DIFF` feature. It follows from the figure that while in some (mostly western) parts of the city, areas exist for which travel by PT is even faster than by car, in some parts of the city $\frac{d_{\mathrm{CAR}}(j)}{min(d_{\mathrm{PT}}(j))} < 0.4$ and it may be even three times faster to travel by car (without traffic jams) than by PT. Interestingly, even in such parts as the north part of the City, depending on the exact origin location and trip destination, travellers from these areas are provided with competitive or in rare cases inevitably less satisfactory PT services. This highlights the role of fine-grained spatial analysis of trip features.

4.4 Data Needs of the System

Let us note that the main challenge for both the proposed and similar solutions is the need to obtain data needed to estimate the probabilities of exact travel parameters necessary for the generation of the journeys. Without TD matrices and address point data, it is hard to generate realistic coordinates of trip endpoints. We used TD data and address points including the number of house inhabitants for journey generation for our experiments. The number of house inhabitants can be used to estimate the probability of the trips from/to residential buildings. However, this is not the case for commercial buildings. For buildings such as shopping centers estimating at a city scale the number of arriving/departing persons and the origin and destination of their travel can be difficult. Still, journeys for zone combinations can be generated and if zones have moderate area the impact of the problems discussed above is limited.

5   Conclusions

Travellers in urban areas have to select travel modes for individual trips they make. While for short-distance trips, walking and cycling are most frequently used, sustainable mobility aims at the reduction of private car use. This means inter alia increasing the use of public transport. However, multiple factors influence the decisions of travellers. Furthermore, the overall travel duration depends on walking distances and waiting times such as waiting for transfers.

In this study, we propose a system combining Apache platforms, OpenTrip-Planner and transport model data to generate and analyse representative trips in urban areas. The system enables an in-depth understanding of the distribution of trips, their routes and LOS features. Once used for spatial analysis, these data enable the understanding of which areas benefit from short walking distances needed to use PT connections, direct connections and overall PT travel duration comparable to travel duration by car. This provides interesting opportunities for analysing travel mode choices across different city areas. In the future, we plan to further exploit the visualisation of synthetic trip data. An interesting challenge is the use of clustering techniques for these data, though the existence of spatially close points with substantially different feature values confirms the complexity of PT service level patterns.

Acknowledgments. The research leading to these results has received funding from the EEA/Norway Grants 2014–2021 through the National Centre for Research and Development. CoMobility benefits from a 2.05 million€ grant from Iceland, Liechtenstein and Norway through the EEA Grants. The aim of the project is to provide a package of tools and methods for the co-creation of sustainable mobility in urban spaces.

Data made public by the City of Warsaw including schedules and GPS traces from the City Open Data portal (http://api.um.warszawa.pl) acquired and processed in the years 2022–2023 were used to develop public transport schedules. Further details on these data can be found on the Open Data portal.

References

1. Braga, C.K.V., Loureiro, C.F.G., Pereira, R.H.: Evaluating the impact of public transport travel time inaccuracy and variability on socio-spatial inequalities in accessibility. J. Transp. Geography **109** (2023). https://doi.org/10.1016/j.jtrangeo.2023.103590

2. Brunner, H., Hirz, M., Hirschberg, W., Fallast, K.: Evaluation of various means of transport for urban areas. Energy Sustain. Soc. **8**(1), 9 (2018). https://doi.org/10.1186/s13705-018-0149-0

3. Chia, J., Lee, J., Han, H.: How does the location of transfer affect travellers and their choice of travel mode? A smart spatial analysis approach. Sensors (Switzerland) **20**(16), 1–17 (2020). https://doi.org/10.3390/s20164418

4. Grzenda, M., Luckner, M., Brzozowski, Ł: Quantifying parking difficulty with transport and prediction models for travel mode choice modelling. In: Mikyška, J., de Mulatier, C., Paszynski, M., Krzhizhanovskaya, V.V., Dongarra, J.J., Sloot, P.M. (eds.) ICCS 2023. LNCS, pp. 505–513. Springer, Cham (2023). https://doi.org/10.1007/978-3-031-36030-5_40

5. Hillel, T., Bierlaire, M., Elshafie, M.Z., Jin, Y.: A systematic review of machine learning classification methodologies for modelling passenger mode choice. J. Choice Model. **38**, 100221 (2021). https://doi.org/10.1016/j.jocm.2020.100221

6. Hillel, T., Elshafie, M.Z.E.B., Jin, Y.: Recreating passenger mode choice-sets for transport simulation: a case study of London, UK. In: Proceedings of the Institution of Civil Engineers - Smart Infrastructure and Construction, vol. 171, pp. 29–42 (2018). https://doi.org/10.1680/jsmic.17.00018

7. Luckner, M., Grzenda, M., Kunicki, R., Legierski, J.: IoT architecture for urban data-centric services and applications. ACM Trans. Internet Technol. **20**(3), 1–30 (2020)

8. Moreno, C., Allam, Z., Chabaud, D., Gall, C., Pratlong, F.: Introducing the "15-minute city": sustainability, resilience and place identity in future post-pandemic cities. Smart Cities **4**(1), 93–111 (2021)

9. OpenStreetMap contributors: Planet dump˙ (2017). https://planet.osm.org, https://www.openstreetmap.org

10. Redman, L., Friman, M., Gärling, T., Hartig, T.: Quality attributes of public transport that attract car users: a research review. Transp. Policy **25**, 119–127 (2013). https://doi.org/10.1016/j.tranpol.2012.11.005

11. Rocha, S.S., Pitombo, C.S., Costa, L.H.M., Marques, S.d.F.: Applying optimization algorithms for spatial estimation of travel demand variables. Transp. Res. Interdisc. Perspect. **10**(April), 100369 (2021). https://doi.org/10.1016/j.trip.2021.100369

12. Sousa Santos, G., et al.: Evaluation of traffic control measures in Oslo region and its effect on current air quality policies in Norway. Transp. Policy **99**(August), 251–261 (2020). https://doi.org/10.1016/j.tranpol.2020.08.025

13. Tenkanen, H., Toivonen, T.: Longitudinal spatial dataset on travel times and distances by different travel modes in Helsinki Region. Sci. Data **7**(1) (2020). https://doi.org/10.1038/s41597-020-0413-y

14. Waury, R., Dolog, P., Jensen, C.S., Torp, K.: Analyzing trajectories using a path-based API. In: Proceedings of the 16th International Symposium on Spatial and Temporal Databases. SSTD '19, pp. 198–201. Association for Computing Machinery, New York, NY, USA (2019). https://doi.org/10.1145/3340964.3340990

15. Yousefzadeh Barri, E., Farber, S., Jahanshahi, H., Beyazit, E.: Understanding transit ridership in an equity context through a comparison of statistical and machine learning algorithms. J. Transp. Geogr. **105**(August), 103482 (2022). https://doi.org/10.1016/j.jtrangeo.2022.103482

16. Zhang, G., Wang, D., Cai, Z., Zeng, J.: Competitiveness of public transit considering travel time reliability: a case study for commuter trips in Hangzhou, China. J. Transp. Geogr. **114** (2024). https://doi.org/10.1016/j.jtrangeo.2023.103768

LoRaWAN Infrastructure Design and Implementation for Soil Moisture Monitoring: A Real-World Practical Case

Erika Pamela Silva Gómez[1,2] and Sang Guun Yoo[1,2,3](✉) (iD)

[1] Departamento de Informática y Ciencias de la Computación, Escuela Politécnica Nacional, Quito, Ecuador
sang.yoo@epn.edu.ec

[2] Smart Lab, Escuela Politécnica Nacional, Quito, Ecuador

[3] Departamento de Ciencias de la Computación, Universidad de las Fuerzas Armadas ESPE, Sangolquí, Ecuador

Abstract. The application of Internet of Things technology in the agricultural sector has allowed to achieve a significant improvement in the process of growing and harvesting products. This has been possible since it allows obtaining a more exact control of the information in real time, thus allowing better decision-making in crop management and thereby improving their quality. Faced with this situation, this paper proposes the design and implementation of a soil moisture monitoring system for a strawberry crop using LoRaWAN technology to allow the farmer to improve the production of their crops, while maintaining low technological implementation costs. The system allows the visualization of the data in real time through a web application, which are obtained from the sensors installed in the ground and which are transmitted through the LoRaWAN network. Once the system was developed using different trending technological tools, its functionality could be verified with satisfactory results. The functionality of the application obtained an acceptance of 94% and an usability a score of 86.87, indicating that the system meets the expectations of the users. Additionally, in the coverage tests, it was possible to verify the long communication range of the installed LoRaWAN devices.

Keywords: LoRaWAN · Smart Farming · IoT · Soil Moisture Monitoring

1 Introduction

Year after year, the Internet of Things (IoT) technology is used more frequently in different areas such as health, transportation, communication, among others [1]. However, its use is still not very common in some areas in developing countries, such is the case with the agriculture and farming [2], especially with the small and medium size producers. The lack of knowledge and low budgets are the main reasons why they do not apply technology in their activities. This situation causes the progress of small and medium size producers to be delayed compared to the large producers who use technology more frequently creating an important gap in production levels.

© The Author(s), under exclusive license to Springer Nature Switzerland AG 2024
L. Franco et al. (Eds.): ICCS 2024, LNCS 14838, pp. 133–146, 2024.
https://doi.org/10.1007/978-3-031-63783-4_11

According to the Food and Agriculture Organization (FAO) of the United Nations, by 2050, the world should produce 70% more food than in previous years [3]. In order to promote the development of this sector and take it to another level, it is essential to use the latest technologies, looking for solutions for improving the level of production but avoiding high costs [4].

In the case of developed countries, technologies such as the Internet of Things (IoT) are used frequently in agriculture. This has allowed them to achieve important production improvement. Among the used technologies in this area, one of the most important one is the LPWAN networks. This type of technology allows farmers to access accurate information in real time, achieving better decision-making in crop management in order to improve its quality [5].

On the other hand, efficient production of agricultural crops is affected by different physical variables such as temperature, humidity, soil pH, among others; and depending on the good control of these variables, you can have a successful harvest, as well as have product losses [4].

In the specific case of strawberry cultivation, soil moisture represents one of the main factors influencing its production. Most of the process from planting, growing and maturing of the strawberries depends on this variable. If this variable is not adequately controlled, it can generate ineffective production and can even generate diseases in the plants [6].

With this background, in order to contribute to the sector of agricultural production, this work proposes a smart solution (design and implementation) for a strawberry crop based on LoRaWAN technology. The proposed solution has been implemented in a real farm located in Ecuador in order to verify its functionality.

The proposed solution is composed of a sensor infrastructure that accumulates soil moisture monitoring data, a LoRaWAN network which allows transmitting the gathered data to the server, and an web application which allows farmers to visualize the real time data in order to make decisions in their crops. Through the proposed system it is intended to obtain a more efficient cultivation process, while having a low-cost investment. This solution allows not only to improve soil moisture control but also helps to manage natural resources more efficiently, i.e. water [7].

This paper is organized as follows. Section 2 describes the state of the art of the usage of LoRaWAN technology in agriculture solutions. Then, in Sect. 3, the methodology used for the development of the proposed solution is explained. Later, Sect. 4 describes the details of the development process. Then, in Sect. 5, tests executed to the developed system are described. Finally, Sect. 6 concludes the present work.

2   State of the Art

In the present work, an analysis of several previous works has been carried out with the aim of knowing in depth how strawberry crops work and better understanding the problem farmers have. Additionally, the intention of analyzing previous works is to understand the used technologies and, at the same time, to understand their limitations with the objective of overcoming such restrictions in the present work.

One of the analyzed works presents a greenhouse prototype using LoRa technology with various sensors [6]. This work presents a technical solution, listing the different types of LPWAN technologies, as well as a more detailed explanation of the elaboration of the used sensors used. Although this work is quite interesting, it has certain limitations: e.g., a field test has not been carried out on an real crop; on the other hand, the authors do not analyze what are the problems that farmers present in their daily basis activities [8].

Another analyzed work is the evaluation of LPWAN technologies to guide the approach of long-range communication solutions for the agricultural sector of Ecuador [9]. This work is quite similar to the previous one. It also focuses only on the technical part, leaving aside an important part that is to know the problems that farmers live with their crops on a daily basis. Additionally, this work has not been implemented in a real crop to test its overall performance in real environments.

In the work called "Design of a low-cost LoRa network for monitoring the agricultural sector", the authors design a technological solution and make a study of its cost. However, the work does not carry out the implementation of the proposed solution in a real crop which could generate a deeper social contribution [10].

Additionally, a scientific article entitled "IoT agriculture system based on LoRaWAN", a work carried out in Greece in a grape field, presents an IoT model in an agricultural system using the LoRaWAN protocol in the data transmission between the sensors and the TTN network [11]. In this paper, the authors highlight the advantages of implementing LoRaWAN in different fields such as agriculture and food production.

On the other hand, the article "Internet of Things and LoRaWAN - Enabled Future Smart Farming" indicates the different use cases of IoT applications in the agricultural field [12]. This work mentions that LoRaWAN shows great potential to be applied in the agricultural and industrial sector. This paper also presents an analysis of a possible limitation of LoRaWAN i.e., the effect of packet collision that would occur in the case that the number of devices connected to the network increases significantly. However, there are already several investigations that seek to solve this future limitation, but it is important to keep this in mind when working with this technology.

As we can see, several previous works have been carried out on the application of IoT technologies in the agricultural sector. However, most of them focus on the development of a design or prototype and not in the implementation of the solution in a real case. We believe that it is important to carry out a real implementation, since new requirements from real users usually appear from such stage.

In this situation, in this work, in addition to carrying out an optimal design of a monitoring solution for a strawberry crop using LoRaWan technology, it is intended to carry out an implementation in a real crop applying the real requirements of farmers. We hope that the knowledge generated in this work can provide an important contribution to the development of this sector, especially in developing countries such as Ecuador.

3 Methodology

For the design and implementation of the LoRaWAN network and the software solution, Scrum development methodology have been used. Scrum is a framework that reduces the complexity of application development in order to satisfy customer requirements. It has been decided to use this methodology due to its flexibility managing changes and functionality on delivering functional systems in the short term [13].

In the first phase, an identification of the needs of the strawberries farmers was carried out. A field visit was done to learn in detail the current crop management process and the physical variables that is required to be controlled.

Based on the farmers' needs, phase 2 was executed, where the LoRaWAN network was designed taking into account the dimensions of the land and the crop. For this work, we have decided to use the soil moisture as the variable to be measured, since it is one of the most important variables in the strawberry production.

In a third phase, the necessary equipment for the implementation of the LoRaWAN network in strawberry cultivation was chosen, following the design of the previous phase.

In the fourth phase, the web application was developed which will serve as a graphical user interface so that the user can view the data obtained from the implemented LoRaWAN network and the established humidity levels, which will allow farmers to know the appropriate time for watering the crop.

4   Proposed Solution

4.1   Analysis and Design of the LoRaWAN Network

Description of the Case Study. For the actual implementation of the solution, a strawberry crop located in the parish of Tababela belonging to the Metropolitan District of Quito, capital city of Ecuador, was selected. This sector is characterized by having a large influx of farmers who are dedicated to the cultivation of various products such as corn, beans, strawberries, peas, among others.

The strawberry crop belongs to a farmer who has worked all his life planting different products such as corn, beans, and peas. In particular, he has dedicated his whole life to the cultivation of strawberries as the main source of income for his family. As a case study, it was taken a crop that covers approximately 6,900 m^2 with a total of 134 beds and 60,000 strawberry plants. Table 1 shows a summary of the characteristics of the case study.

Description of the Problem. There are several factors that influence the cultivation of strawberries and, on which, depend the good quality of the fruit. One of the most important is the soil moisture making the watering process one of the main activities that the farmer must carry out to obtain good strawberry productivity.

One of the diseases that is produced by high humidity is the gray rot or better known as botrytis. This disease appears as spots on the fruits. And these spots cause buyers of strawberries not to accept them. This disease also infects other strawberries making the problem an important issue for farmers. Therefore, discarding of strawberries is required to prevent its spread, which causes great losses in crops [14].

There are also several diseases or pests that are generated by the lack of humidity, such as the red spider or mites, which are caused by the lack of irrigation in the soil. This plague affects the leaves initially and later the growth of the plant. To counteract the problem, fungicides are usually applied, which generates an increase in expenses for the farmer [14].

Table 1. Characteristics of the case study.

Characteristic	Value
Land area	Approximately 6900 m^2
Type of crop	Strawberries
Plant type	Monterrey
Number of plants	60000
Number of beds	134
Soil type	Sandy

Another effect caused by the poor humidity control, as mentioned by farmers, is the size of the fruits. To obtain good-sized fruits, an adequate irrigation of water is required in the plants. If this does not happen, smaller fruits are obtained generating low incomes for farmers.

For the solution of the aforementioned problems, it is necessary to have adequate humidity control. It is for this reason that the present work seeks to deliver an automatic irrigation control solution based on the IoT technology at low cost solution, so that the farmer can implement it without major inconveniences.

LoRaWAN Network Design. Based on the ground conditions described above and the carried-out literature review, it was decided to use 2 sensors for the LoRaWAN network, one for each irrigation module. The sensors were placed at the depth of the plant roots i.e., 16 cm below the drip tape. In Fig. 1, a diagram of the LoRaWAN network design in strawberry cultivation is presented.

The network consists of two sensors that are located in strategic places in the strawberry crop. These sensors are connected wirelessly to the LoRaWAN network generated by the LPS8 Gateway, which is connected to a The Things Network (TTN) server. On the other hand, the web applications takes the data stored in the TTN server by using the MQTT protocol.

Equipment for the Implementation of the LoRaWAN Network. For the LoRaWAN network, the Dragino brand equipment was chosen because it is open source in both hardware and software, it has a low cost, and because it has extensive documentation.

LoRaWAN Gateway. The selected gateway was the LPS8 Indoor LoRaWAN Gateway. LPS8 is an open source gateway that allows you to create a Wireless LoRa network, which allows data to be sent over extremely long distances [15].

Soil Moisture Sensor. The selected soil moisture sensor was the LSE01 LoRaWAN Soil Moisture & EC Sensor from the Dragino brand (see Fig. 2). LSE01 is a LoRaWAN sensor that measures soil humidity, temperature and conductivity. This sensor uses the Frequency Domain Reflectometry (FDR) method, which measures the dielectric constant of the soil. The device is calibrated for mineral soils that include sandy and clayey soil types [14]. The LSE01 sensor is configured by default as LoRaWAN OTAA Class A, so it has the authentication keys to the server [16].

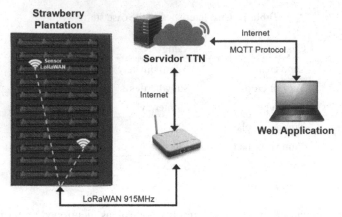

Fig. 1. LoRaWAN Network Design

Fig. 2. LSE01 Soil Moisture Sensor

4.2 Implementation of the LoRaWAN Network

Configuration of the LPS8 Gateway. The LPS8 gateway is configured by default as a WiFi Access Point. To configure the device, it must be connected through a computer either via Wi-Fi or Ethernet Port. In this case, it was connected via WiFi to carry out the configuration. Once in the configuration, the registration to the TTN server was done. For this, the creation of an account in TTN was carried out. Once the gateway was configured to connect to the TTN server, the configuration portal showed the LoRaWAN network connection figure (see Fig. 3).

Configuration of LSE01 Sensors. These devices, as well as the gateway must be registered in the TTN server with their respective keys. After the devices are registered, the sensor must be powered on to join the LoRaWAN network automatically by OTAA

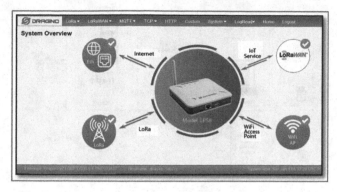

Fig. 3. LPS8 device configuration page after TTN connection.

authentication method. In Fig. 4, you can see the messages sent and received between the server and the sensor at the time of the device joining process.

Fig. 4. Messages exchanged between the LSE01 sensor and the server in the process of joining the network.

4.3 Development of the Web Application

Application Architecture. The web application has been developed in three components. The first component is the application interface with which the user will be able to interact; this component was implemented using HTML, JavaScript and CSS, and the Semantic UI framework was used for the design of the interfaces. The second component is the back-end server in which the Node JS and Express JS frameworks were used. this component allow the connection of the application with the TTN server to send and receive data. Finally, the third component is Firebase, which fulfills the functions of storing the authentication credentials and system data [17, 18].

The system was deployed by using Heroku. This service offers the necessary infrastructure to implement and manage modern web applications based on the use of containers, allowing user access from anywhere in the world [19]. In Fig. 5, the design of the web application architecture is presented.

In sprint 1 of the development stage, the tasks to be implemented i.e., the user registration functionality, were determined. As a result of this sprint, the functional modules shown on Figs. 6 and 7 were obtained.

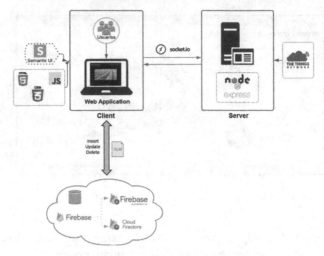

Fig. 5. Web Application Architecture

Fig. 6. Account creation screen.   **Fig. 7.** Login screen

In sprint 1, the application was deployed through a local server using the Node JS and Express JS frameworks. For sprint 2, the crop information management functionality were developed. Through this functionality, the farmer can visualize the soil moisture values of the crop and receive notification of irrigation in the crop. As a result of this sprint, the following screens were obtained (see Fig. 8).

In order to obtain the data from the sensors, the MQTT connection protocol used by the TTN server was used. For the web application development, one of the client libraries of TTN was used i.e. library for Node JS [18]. The data from the server is sent to the client of the web application via the socket.io library and is displayed on the main screen of the application as shown in Fig. 9.

Fig. 8. Crop Record Screen

For the implementation of the irrigation notification requirement in the crop, the optimal levels of soil moisture were established. There are two methods used to measure soil moisture: gravimetric and volumetric [20]. The method used by Dragino sensors is the volumetric one i.e. Volumetric Water Content (VWC). Table 2 shows the soil moisture levels according to [21, 22]. In the event that there are low humidity levels, a message will be displayed at the top of each sensor and the value box in red, as shown in Fig. 9.

Fig. 9. Notification of irrigation in the crop. (Color figure online)

5   Tests

To verify the proper functioning of the developed system, different types of tests were carried out, which are presented below.

Table 2. Soil moisture levels of the case study.

Soil Moisture Levels	Value
Low	Lower than 10%
Optimal	Between 10 and 18%
High	Higher than 18%

5.1 Functionality Tests

In this part, the functionality tests of the application were carried out with people related to the agricultural business and external people. Eight test cases were carried out for the most relevant functionalities of the application.

After having carried out the test cases, a survey was carried out on the 8 people who executed the system tests to verify that the expected results have been met. These people ranged in age from 18 to 50 years.

In most of the survey questions, satisfactory results were obtained. Figure 10 shows the global results obtained in the functionality tests. As can be seen in the figure, 94% of the users were satisfied with the functionality of the developed system.

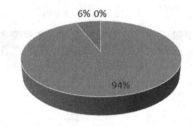

■ Totally Satisfied ■ Parcially Satisfied ■ Non Satisfied

Fig. 10. Result of the Functionality Tests.

5.2 Usability Tests

Usability tests indicate the ease of use of the application from the user's point of view. To carry out these tests, the System Usability Scale (SUS) tool was used, which consists of a 10-question questionnaire with 5 response options with a scale from 1 to 5, where 1 corresponds to totally agree and 5 corresponds to totally disagree [23]. These questions were answered after running each test case. The questions evaluated in this test were the following:

- Do you think you will use the application frequently?
- Do you find this app unnecessarily complex?
- Do you think the app is easy to use?

- Do you think you would need technical support to use the application?
- Do you think that the different functions of the application are well integrated?
- Do you think there are any inconsistencies in the application?
- Do you think most people would learn to use the application quickly?
- Did you find the app too complicated to use?
- Did you feel very safe using the application?
- Did you need to learn several things before you could use the application?

After executing the test, it could be noticed that the question with the lowest score were the number 4 (Do you think you would need technical support to use the application?). The reason of this situation was because of a misunderstanding i.e., most of the testers requested an explanation before running the test cases and they thought it was a technical support. On the other hand, one of the highest scoring questions was the number 2 (Do you find this app unnecessarily complex?). This is because the main objective of the application is to show the data obtained from the sensors to the users in a simple and direct way. Table 3 shows the score obtained by each user according to the SUS scale.

Table 3. Scores obtained according to the SUS scale by each user.

Surveyed Use	SUS Score (over 100)
User 1	82.5
User 2	100
User 3	87.5
User 4	75
User 5	90
User 6	92.5
User 7	87.5
User 8	80
Average	86.87

5.3   Sensor and LoRaWAN Network Test

To test the sensors and LoRaWAN network designed in this work, a coverage test was also carried out. For this purpose, the sensors were placed in the strawberry field that was located at approximately 750 m from the gateway. The gateway was placed inside a house to be connected to the Internet through an Ethernet cable.

The sensors were installed horizontally at the height of the roots of the strawberry plants, approximately 16 cm deep, as shown in Fig. 11.

After installing the sensors, the data collection in the application was verified. Figure 12 shows the real values obtained from the sensors already installed in the strawberry crop.

Through the coverage tests, it was possible to verify one of the main characteristics of the LoRaWAN protocol, which is to achieve communication over long distances.

Fig. 11. Sensor installed on the ground horizontally.

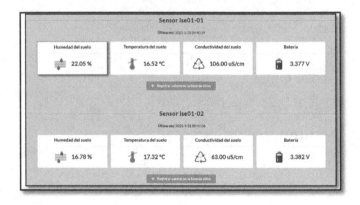

Fig. 12. Reception of Sensors' Data.

6   Conclusions

In the present work, a soil moisture monitoring system was designed and developed using LoRaWAN technology with the aim of improving the production of a strawberry crop.

The in-depth knowledge of the entire strawberry cultivation process from sowing to harvest facilitated obtaining the requirements, as well as a better understanding of the farmers' needs which facilitated the optimal implementation of the project.

Additionally, with the implementation of this project, real values of soil moisture in strawberry cultivation were obtained thanks to the installed sensors. This allowed to improve the cultivation process optimizing time and water resources. It was also possible to avoid some diseases that were caused by excess or insufficient irrigation in the soil.

After having implemented the system, the functionality, usability and coverage tests allowed to know the fulfillment of the objectives set at the beginning of the project. In the tests carried out, a satisfactory result was obtained in terms of the acceptance of the system. In the functionality tests, an acceptance result of 94% was obtained, which means that it was possible to perform most of the functions of the application without problems. In addition, it can be mentioned that the people who tested the application did not need any technical knowledge to achieve the expected results thanks to the friendly user interfaces and the easy flow of the application. On the other hand, usability tests resulted in a score of 86.87 on the SUS scale. This indicates that the application is highly usable and clear based on the user experience. Finally, the coverage tests allowed to know the reach of the LoRaWAN networks. It was found that a single Gateway could cover the entire plantation. This represents a considerable benefit since the sensors can be located in any part of the terrain even kilometers away from the gateway without losing the connection. As a further work, we will try to implement other smart farming solutions using the LoRaWAN technology.

Acknowledgments. The authors would like to thank to Corporación Ecuatoriana para el Desarrollo de la Investigación y Academia - CEDIA for the financial support given to this work through its Fondo Divulga.

References

1. Balaji, S., Nathani, K., Santhakumar, R.: IoT technology, applications and challenges: a contemporary survey. Wirel. Pers. Commun. **108**, 363–388 (2019)
2. Xu, J., Gu, B., Tian, G.: Review of agricultural IoT technology. Artif. Intell. Agric. **6**, 10–22 (2022)
3. Food and Agriculture Organization of the United Nations: The future of food and agriculture. FAO (2018)
4. Entienda la importancia del protocolo LoRaTMen la Agricultura 4.0. https://www.khomp.com/es/protocolo-LoRa-na-agricultura-4-0/. Accesed 06 May 2023
5. Borrero, J. D.: Aplicación de la tecnología LoRaWAN en la agricultura. ResearchGate (2018). https://doi.org/10.13140/RG.2.2.24492.77443
6. Useche, T., Rodriguez, J. C., Estiven, R.: Prototipo de solución IoT con tecnología Lora en monitoreo de cultivos agrícolas. Universidad Distrital Francisco José de Caldas (2018)
7. Universidad de Huelva. https://fundaciondescubre.es/noticias/como-afectan-los-factores-climatologicos-a-la-produccion-de-la-fresa/. Accesed 06 May 2023
8. Cevallos, B., Rubio, S.: Desarrollo de una red IoT con tecnología LoRa para gestión de invernaderos. Universidad Politécnica Salesiana (2021)

9. Enriquez, K., Palacios, M.: Evaluación de tecnologías LPWAN para guiar el planteamiento de soluciones de comunicación de largo alcance para el sector agrícola del Ecuador. Estudio de caso: monitoreo de calidad del suelo, Universidad de las Fuerzas Armadas ESPE (2021)
10. Mañay Chochos, E.: Diseño de una red LoRa de bajo costo para el monitoreo del sector agrícola. Ciencia Latina Revista Científica Multidisciplinar, Ciudad de Mexico (2022)
11. Davcev, D., Mitreski, K., Stefan, T., Nikolovski, V., Koteli, N.: IoT agriculture system based on LoRaWAN. In: Proceedings of 2018 14th IEEE International Workshop on Factory Communication Systems (WFCS), pp. 1–4. IEEE, Imperia, Italy (2018)
12. Citoni, B., Fioranelli, F., Imran, M.A., Abbasi, Q.H.: Internet of Things and LoRaWAN-enabled future smart farming. IEEE Internet Things Mag. 2, 14–19 (2019)
13. Srivastava, A., Bhardwaj, S., Saraswat, S.: SCRUM model for agile methodology. In: Proceedings of the 2017 International Conference on Computing, Communication and Automation (ICCCA), pp. 864–869. IEEE (2017)
14. La Huertina. https://www.lahuertinadetoni.es/plagas-y-enfermedades-de-la-fresa-o-frutilla/. Accesed 06 May 2023
15. Dragino. https://www.dragino.com/downloads/downloads/LoRa_Gateway/LPS8/LPS8_L oRaWAN_Gateway_User_Manual_v1.3.2.pdf. Accessed 06 May 2023
16. Dragino. https://www.dragino.com/downloads/downloads/LoRa_End_Node/LSE01/Test_R eport/LSE01%20V1.0.0%20Power%20Analyze.pdf. Accessed 06 May 2023
17. Moroney, L.: The Definitive Guide to Firebase, 1st edn. Springer, Berkeley (2017)
18. Pramono, L.H., Yana Javista, Y.K.: Firebase authentication cloud service for RESTful API security on employee presence system. In: Proceedings of the 2021 4th International Seminar on Research of Information Technology and Intelligent Systems (ISRITI), pp. 1–6. IEEE, Indonesia (2021)
19. Middleton, N., Schneeman, R.: Heroku: Up and Running, 1st edn. O'Reilly, USA (2014)
20. Chaudhari, P., Tiwari, A.K., Pattewar, S., Shelke, S.N.: Smart infrastructure monitoring using LoRaWAN technology. In: 2021 International Conference on System, Computation, Automation and Networking (ICSCAN), pp. 1–6. IEEE (2021)
21. Landscape Arquitecture Magazine. https://landscapearchitecturemagazine.org/2014/01/21/ soils-the-measure-of-moisture-parked/. Accessed 06 May 2023
22. Observant. https://observant.zendesk.com/hc/en-us/articles/208067926-Monitoring-Soil-Moisture-for-Optimal-Crop-Growth. Accessed 06 May 2023
23. Camille, S., Pham, T., Phillips, R.: Validation of the System Usability Scale (SUS): SUS in the wild. Proc. Hum. Factors Ergon. Soc. Annu. Meet. 57(1), 192–196 (2013)

A Framework for Intelligent Generation of Intrusion Detection Rules Based on Grad-CAM

Xingyu Wang[1,2], Huaifeng Bao[1,2], Wenhao Li[1,2], Haoning Chen[1,2], Wen Wang[1,2(✉)], and Feng Liu[1,2]

[1] Institute of Information Engineering, CAS, Beijing, China
[2] School of Cyber Security, University of CAS, Beijing, China
{wangxingyu,baohuaifeng,liwenhao,chenhaoning,wangwen, liufeng}@iie.ac.cn

Abstract. Intrusion detection systems (IDS) play a critical role in protecting networks from cyber threats. Currently, intrusion detection methods based on artificial intelligen (AI) stand as the mainstream, yet they grapple with the challenges of interpretability and high computational costs. Conversely, rule-based approaches offer ease of comprehension and lower computational overhead, but their development demands extensive expertise. This paper proposes an intelligent framework for generating intrusion detection rules, which integrates the strong representational capabilities of AI detectors while retaining the advantages of rule-based detection. Initially, the framework involves training a TextCNN model for traffic payload classification. The parameters of this model, along with the Gradient-weighted Class Activation Mapping (Grad-CAM) algorithm, are employed to analyze critical fields in captured traffic payloads. Subsequently, a comprehensive list of keywords is obtained through a sensitive words aggregation algorithm, and regular expressions are generated to describe the detection content. These regular expressions undergo fine-tuning to reduce their false positive rate. Furthermore, adhering to the syntax of Suricata rules, they are formulated into intrusion detection rules. The proposed method is evaluated on two publicly available datasets, with experimental results demonstrating commendable detection efficacy for the generated intrusion detection rules.

Keywords: Intrusion detection · grad-cam · rule generation · traffic classification

1 Introduction

The development of computer networks has revolutionized the way people communicate, collaborate, and access information. Concurrently, security issues associated with computer networks have increasingly become a focal point of concern [8]. By scrutinizing network traffic, intrusion detection systems can prevent potential intrusions and ensure the confidentiality, integrity, and availability of the network [1]. In recent years, AI-based intrusion detection methods have proliferated [9]. However, traffic features often fluctuate with changes in the network

L. Franco et al. (Eds.): ICCS 2024, LNCS 14838, pp. 147–161, 2024.
https://doi.org/10.1007/978-3-031-63783-4_12

environment, potentially leading to a high number of false positives [4]. Additionally, the lack of interpretability in artificial intelligence(AI) poses another challenge to the widespread application [10]. This is due to the fact that their output is limited to positive or negative discriminations, and the accuracy of which is difficult to assess effectively. Moreover, neural network models typically incur substantial computational costs. By scrutinizing network traffic, Intrusion detection systems can prevent potential intrusions and ensure the confidentiality, integrity, and availability of the network [1]. In comparison, traditional rule-based methods possess the following advantages:

- Low False Positive Rate: Rule-based intrusion detection systems generally demonstrate lower false positive rates due to their focus on detecting the fundamental payloads of attack behaviors.
- Interpretability: Rule-based intrusion detection systems enable explanations and a better understanding of intrusive behaviors. These rules can be reviewed and validated, making the system's behavior more predictable [6].
- Low Computational Resource Requirement: Rule-based intrusion detection systems demand less computational resources and boast easier deployment when contrasted with machine learning-based counterparts.

These advantages effectively mitigate the limitations of AI. While rule-based intrusion detection possesses the aforementioned merits, it still faces some constraints. Firstly the existing production process of malicious traffic detection rules is inefficient and human resource-consuming. This is because the majority of them are manually extracted by security experts from malicious samples, exploit codes or malicious traffic based on their expertise and experience. Secondly, current manual rule generation methods struggle to promptly generate rules for newly emerging malicious traffic, leading to delayed updates that create opportunities for 1-day attacks. There are existing studies focusing on intelligent intrusion detection rule generation, such as [13,15]. However, existing studies have primarily focused on fields within threat intelligence such as IP and domain. Additionally, challenges arise due to the difficulty of deployment, attributed to the constraints imposed by the syntax of the Suricata rules.

This paper proposes an intelligent intrusion detection rule generation framework. The method begins by training a TextCNN model for traffic payload classification. The Gradient Weighted Class Activation Mapping (Grad-CAM) [18] algorithm is then used to analyze the keywords in the malicious traffic payload from the sample inputs and the well-trained model parameters. A sensitive words aggregation algorithm is devised to obtain a more comprehensive list of keywords. Subsequently, regular expressions are formulated for detection, and through fine-tuning operations, they are further refined to reduce the false positives. Finally, intrusion detection rules are written based on the syntax of Suricata rules [2]. This method formulates rules based on the payload of the attack, specifically targeting the core of the attack behavior. It not only reduces manual dependence but also enhances the interpretability of intrusion detection by focusing on the key aspects of attack payloads.

The main contributions of this paper are as follows:

- We propose an intelligent framework for generating intrusion detection rules based on Grad-CAM. This framework leverages a pre-trained TextCNN model to identify keywords within malicious traffic payloads.
- To achieve more efficient intrusion detection, we introduce a Sensitive Words Aggregation Algorithm aimed at obtaining a more comprehensive list of keywords.
- We have conducted thorough experiments on two publicly available datasets. The results indicate that our method consistently outperforms baselines, demonstrating superior effectiveness in intrusion detection.

The rest of this paper is organized as follows. In Sect. 2, we provide a brief overview of existing research on intrusion detection systems and intelligent rule generation. Section 3 presents a detailed description of the implementation details of the intelligent rule generation method proposed in this paper. In Sect. 4, comprehensive experiments are conducted to validate the effectiveness of the proposed method. Finally, we conclude our work in the last section.

2   Related Work

2.1   Intrusion Detection System

In the realm of network attack detection, intrusion detection plays a pivotal role, and this domain has witnessed substantial research endeavors. Suricata [2], for instance, employs a predefined set of rules to conduct traffic auditing, discerning the presence of attacks therein. With the advancement of machine learning, a plethora of intrusion detection methods empowered by artificial intelligence have emerged. PAYL [22] utilizes the payload field of network traffic as the detection target, employing natural language processing principles to extract features from the payload through 1-gram analysis. This approach achieves a high detection rate with low false positives. ATPAD [17], an end-to-end method, employs an attention mechanism to effectively defend against various payload-based attacks attempting to circumvent detection. Bao et al. [5] introduced a web attack traffic detection method based on graph networks, wherein the payload is represented using graphs, demonstrating superior representation capabilities compared to word vectors. Li et al. [11] utilized a graph matching algorithm to enhance the robustness of traffic identification. The AI-based methods encounter challenges of high computational overhead and low interpretability, while rule-based methods can overcome these drawbacks. However, the generation of rules itself is a complex task. The proposed framework in this study aims to intelligently generate rules, effectively addressing the limitations of rule-based methods.

2.2   Intelligent Intrusion Detection Rule Generation

Due to the efficiency and interpretability advantages of rule-based intrusion detection methods, as well as the inherent difficulty in their manual formulation, there have been research efforts aimed at automating the generation of intrusion detection rules.

Liu et al. [13] proposed a rule generation method based on threat intelligence text, utilizing ResLCNN to extract IOC (Indicators of Compromise) information such as malicious domains from threat intelligence text, subsequently facilitating the creation of detection rules. Basim Mahboob et al. [15] introduced an intrusion detection rule generation method based on a decision tree model, which analyzes the importance of various features and extracts interpretable detection rules from the decision tree. However, the rules generated by this method cannot be directly deployed on common rule-based intrusion detection engines. Li et al. [10] presented a rule extraction method based on an unsupervised detection model. They designed a novel decision tree called CART to approximate the decision boundaries of black-box models, but this method does not inspect malicious traffic payloads. The method proposed in this paper utilizes malicious traffic payloads to generate intrusion detection rules, providing an intuitive depiction of attack behaviors and effectively enhancing the detection accuracy.

3   Methodology

The overall architecture of the rule generation method proposed in this work is illustrated in Fig. 1. Firstly, preprocessing is applied to the payload of malicious traffic. Secondly, the text features of preprocessed characters are vectorized, and a malicious payload classifier is trained based on the TextCNN model. Thirdly, in the Grad-CAM analysis module, traffic undergoes the aforementioned steps, yielding discriminative results. In the case of a positive output, the designed Grad-CAM algorithm is employed to examine the role of each word in the discriminative result, forming a heatmap to describe the contribution of each word to the outcome. Subsequently, through a sensitive words aggregation algorithm, words with lower importance scores that still play a crucial role are included in the list of key words. Finally, referencing this list, regular expressions for detection are formulated, and they are refined into intrusion detection rules, following the syntax of Suricata rules. Subsequent Subsects 3.1–3.5 will provide detailed explanations of the proposed rule generation method in this paper.

3.1   Data Processing

The data preprocessing steps encompass three stages: decoding, generation, and segmentation. To address potential inaccuracies introduced by certain special characters in network traffic payload, such as "?" and "&" in URIs, we decode each payload, including URI, HTML entities, and optional Base64, Unicode, and JavaScript code. This decoding is employed to mitigate the risk of attackers encoding malicious fields. Subsequently, we generalize the decoded payload. Contents like usernames and IP addresses in the payload do not actively contribute to detection; instead, they may increase computational overhead and impact detection effectiveness. The generalization operation enhances feature learning efficiency, preventing excessive false positives in subsequent rule generation. Segmentation involves segmenting the payload into a sequence of words using

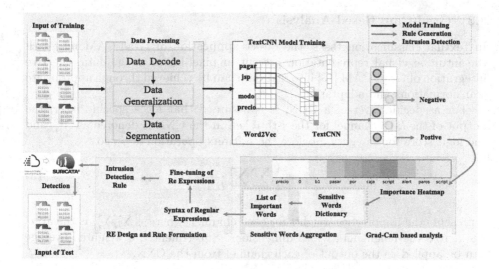

Fig. 1. The system architecture of the proposed method

symbols like "?", "&", and spaces. These three operations prepare the decoded and generalized payload for text feature vectorization.

3.2 TextCNN Model Training

The issue of malicious traffic payload classification fundamentally constitutes a textual classification task. This study employs the TextCNN model [24] for text classification, where TextCNN is composed of convolutional layers and pooling layers, culminating in a fully connected layer for output.

A malicious traffic payload serves as the input, and the text is transformed into word vectors through the application of Word2Vec [7]. Word2Vec is a word embedding technique that involves training a neural network on extensive textual data to learn vector representations for each vocabulary word, capturing the semantic relationships between words. The input text undergoes forward propagation through TextCNN to obtain the model's output:

$$Z = f(\text{Conv}(X) + \text{max\_pooling}(\text{Conv}(X))) \tag{1}$$

where X is the word vector representation of the input text, Conv denotes the convolution operation, f represents the activation function, and Z is the output after convolution and pooling. In the context of TextCNN-based text classification, the loss computation is expressed as:

$$Loss = CrossEntropy(Z, TrueLabels) \tag{2}$$

During the subsequent backpropagation step, the loss is propagated backward through the model to calculate gradients with respect to the input.

3.3 Grad-Cam Based Analysis

In textual classification tasks, the direct application of Grad-CAM may lack the intuitive visual representation present in pixel-based image data, effective integration of Grad-CAM [18] into text data can be achieved through appropriate transformations and adaptations.

For a specific layer of a CNN, it is assumed that for a specific layer, the output of the k-th channel for the j-th token in the CNN is denoted as A_{ij}^k, and the logit score for category c before the softmax operation is y_c, so

$$a_k^c = \frac{1}{T} \sum_i \sum_j \frac{\partial y^c}{\partial A_{ij}^k} \tag{3}$$

represents the importance weights for each channel, where $\frac{1}{T} \sum_i \sum_j$ is the average across the height and width dimensions. Subsequently, a weighted average can be applied to the output of each channel from the CNN.

$$\text{Grad-CAM Map} = activation(\sum_k a_k^c A^k) \tag{4}$$

Finally, based on the equation mentioned above, a heatmap is generated to represent the importance of words towards the classification results by appropriately processing and normalizing the Grad-CAM Map.

Figure 2 serves as an illustrative example of a heat map generated using the proposed methodology outlined above. It is discernible that certain words, highly improbable for usage by regular users, exhibit elevated scores. The heatmap visually represents these words with darker colors, indicative of heightened scores.

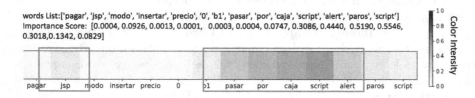

Fig. 2. An example of a heat map generated using the proposed method

3.4 Sensitive Words Aggregation

Observing the payload traffic, it is noticed that the occurrence of certain words, such as "cat", may not be abnormal, as regular users might use this term to refer to the animal. Consequently, its importance score in the aforementioned classification task might not be very high. However, it is also a Linux command used to view file contents. When it appears concurrently with critical system

---

Algorithm 1: Sensitive Words Aggregation Algorithm

Data: *words*, L_{heap}, *indexes*, *dict*
Result: *Res*

1 *Res* ← *indexes*;
2 **for** i ← 1 **to** *length(indexes)* **do**
3      $left = i - 1$, $right = i + 1$;
4   **while** $left > 0$ **do**
5   **if** *words[left]* in *dict* and *left* not in *indexes* **then**
6   *Res.add(left)*;
7            $left = left - 1$;
8   **end**
9   **if** *words[left]* in *dict* and *left* in *indexes* **then**
10           $left = left - 1$;
11   **end**
12   **if** *words[left]* not in *dict* and *left* not in *indexes* **then**
13   break;
14   **end**
15   **end**
16   **while** $right < length(words)$ **do**
17   **if** *words[right]* in *dict* and *right* not in *indexes* **then**
18   *Res.add(right)*;
19           $right = right + 1$;
20   **end**
21   **if** *words[right]* in *dict* and *right* in *indexes* **then**
22           $right = right + 1$;
23   **end**
24   **if** *words[right]* not in *dict* and *right* not in *indexes* **then**
25   break;
26   **end**
27   **end**
28 **end**
29 Return *Res*;

---

files like "/etc/passwd", it can be considered as a potentially malicious command attempting to illicitly access system password information.

In such cases, neglecting its importance may result in overlooking potential attacks. Therefore, after obtaining the heatmap representing the importance of word vectors, this paper employs an attention aggregation algorithm. This algorithm aggregates certain words with lower scores, yet possessing significance, along with closely related words that have higher scores. This aggregated information is utilized for subsequent rule generation. We propose an sensitive words aggregation mechanism. The algorithm's procedure is illustrated as Algorithm 1. The input includes the original payload sequence *words*, arrays L_{heap} containing importance scores for each term generated in the preceding subsection, and an array *indexes* containing the indices of selected crucial terms. Furthermore, a predefined sensitive word dictionary *dict* is supplied. The sensitive word dictionary encompasses various Linux system commands and other words unlikely to be used by ordinary users.

For each index already included in the *indexes* array, maintain an original sliding window of size 0, expanding towards both left and right sides with a step size of 1. If an element is encountered that appears in the sensitive word dictionary but is not present in the *indexes* array, add it to the result array, and continue expanding the sliding window. If an element is encountered that is already in the *indexes* array, increase the window size by 1 without taking any further action. If the window has reached its maximum size or encounters a condition other than the ones mentioned above, the search concludes. The final array of word indices, which is used to generate rules, is the sum of the original *indexes* array and the indices from the aforementioned *Res* array.

3.5 Regular Expression Design and Rule Formulation

In the preceding section, we obtained a sequence of key terms $w1, w2, ...wn$, representing a series of significant words that appear sequentially in a malicious network traffic. In the process of rule creation, this method utilizes the aforementioned key term sequence to describe a scenario based on regular expressions [12], wherein this sequence of key terms appears simultaneously. When all these key terms are present in a test traffic, it is considered indicative of a network attack. The specific regular expression employed is as follows: $/w1.*w2.*....*wn/i$, utilizing a case-insensitive matching to capture the simultaneous occurrence of the specified key terms.

After crafting regular expressions for detection, it is necessary to fine-tune them. The approach involves using these regular expressions to search through negative payloads and observing whether the number of matches is acceptable. If the regex designed for malicious payloads also produces numerous matches, additional terms should be incorporated into the regular expression to ensure it doesn't match negative traffic.

The intrusion detection rules developed in this study must adhere to the syntax constraints of the Suricata engine. Suricata is an open source intrusion detection tool known for its high efficiency and ease of deployment. The intrusion detection rules in Suricata primarily include Rule ID, detection content, and alert content. The detection content is a crucial factor influencing the effectiveness of intrusion detection, as it directly describes the conditions under which an alert will be triggered. Some other crucial keywords for Suricata rules are listed in Table 1.

4 Experiments Evaluation

4.1 Datasets

To validate the detection performance of the proposed model, experiments were conducted using the CSIC2010 [16] and FWAF[1] datasets. CSIC2010 comprises

---

[1] https://github.com/faizann24/Fwaf-Machine-Learning-driven-Web-Application-Firewall.

Table 1. Partial Catalog of Suricata Rule Keywords

Field Name	Meaning	Example
Action	Action executed when rule matching	alert/drop
Protocol	Network protocol	TCP/HTTP
Source IP	IP address of the source	192.168.1.10
Source Port	Port used by the source	54789
Dest. IP	IP address of the destination	any
Dest. Port	Port used by the destination	any
msg	Content of alert triggered	ET SCAN Suspicious Traffic
Signature ID	ID associated with the signature	2019812
Content	Content to be detected	index.php
Pcre	Regularized representation of the content	/[0-9]{6}/

HTTP data traffic generated for e-commerce web applications, consisting of 36,000 normal requests and over 25,000 abnormal requests. The dataset includes various attacks such as SQL injection, buffer overflow, information gathering, file disclosure, CRLF injection, cross-site scripting, server-side includes, parameter tampering, etc. FWAF is a publicly available large-scale malicious request dataset released by Fsecurify, a company dedicated to developing intelligent web firewalls. They combine professional knowledge with heuristic methods to label malicious traffic. The two datasets were divided into training and testing sets in a 8:2 ratio for experimentation. The training set is utilized to train the TextCNN model and generate intrusion detection rules, while the testing set aims to evaluate the effectiveness of the generated rules. It is noteworthy that the classification target of the proposed method in this paper is binary classification.

Additionally, this study captured 100 GB of benign traffic on the laboratory gateway for the purpose of validating whether the generated intrusion detection rules exhibit an acceptable false positive rate and demonstrate efficient detection performance.

4.2   Experimental Environment and Other Configurations

The experiments in this paper were conducted on a 64-bit Windows 10 operating system, with the exception of the suricata-related detection experiments, which were performed on Ubuntu 20.04. All functionality modules were developed and executed in Python 3.9.1. For the vectorization of traffic payloads, a word2vec model was retrained based on a traffic payload vocabulary, and the embedding dimension of word vectors was set to 300. During the training of the TextCNN model, the Adam optimizer was employed with a batch size of 32, a learning rate of 0.001, and 50 training epochs.

4.3 Evaluation Metrics

The result of each classification may either be positive or negative in this binary decision problem. For evaluating the method we proposed, we adopt three widely used metrics in experiments, defined as follows:

$$\text{Accuracy(ACC)} = \frac{TP + TN}{TP + FP + TN + FN} \tag{5}$$

$$\text{False Positive Rate(FPR)} = \frac{FP}{TN + FP} \tag{6}$$

$$\text{False Negative Rate(FNR)} = \frac{FN}{TP + FN} \tag{7}$$

where TP (True Positive) refers to the number of classified as a specific class correctly; FP (False Positive) refers to the number of the misclassified as that class; FN (False Negative) refers to the number of cases that should be classified as that class but misclassified as other classes; TN (True Negative) is the number of cases that predicted as the non-corresponding classes.

4.4 Evaluation Results and Discussion

The experimental evaluation primarily involves the following aspects:

- Accuracy: We evaluate the detection accuracy of the proposed method, particularly examining whether the detection rate diminishes and whether it indeed achieves lower false positive rates.
- Ablation Study: We identify what factors influence the proposed method's effectiveness.
- Computational Consumption: We evaluate whether the rule-based detection method introduced in this paper incurs lower computational overhead compared to AI-based methods.
- Detection Efficiency: Assess whether the proposed method demonstrates higher detection speed during actual deployment testing.

Accuracy. In this study, we compare our proposed method with some feature-based methods and end-to-end methods. Feature-based baselines include HMM-PAYL [3], Logistic Regression (LR) [19], Support Vector Machine (SVM) [19], and Random Forest (RF) [20]. End-to-end methods include TextCNN [23], ATPAD [21], and GraphXSS [14]. For the baselines, we reimplement them with the parameters described in the original papers on the datasets mentioned above (if no relevant experimental results are available). Table 2 demonstrates that our method exhibits relatively good detection performance on both datasets. Generally, our method demonstrates the best performance across most metrics. From the performance metric FNR on CSIC2010, it can be observed that the detection rate of our method has not declined. There is even a slight improvement compared to TextCNN. This is attributed to the precise identification of keywords

Table 2. Comparison of the method in this paper Accuracy (Acc), False Positive Rate(FPR), and False Negative Rate(FNR) with baselines.

Model	CSIC2010			FWAF		
	Acc	FNR	FPR	Acc	FNR	FPR
Handcrafted-features-based Methods						
HMMPAYL	92.63	8.27	7.97	91.26	7.51	8.32
LR	94.68	7.64	4.02	93.57	4.32	3.98
SVM	95.12	5.53	7.04	96.38	4.82	5.21
RF	95.06	3.63	3.27	95.72	3.03	3.56
End-to-end Methods						
GraphXSS	97.21	3.31	2.93	97.44	2.61	2.59
ATPAD	95.86	5.60	0.30	97.93	3.23	3.76
TextCNN	98.48	0.38	2.74	96.67	4.66	3.79
Ours	98.51	2.90	0.08	98.94	3.12	0.57

in the Grad-CAM analysis stage and the synergistic effects of subsequent steps. The FNR slightly decreases compared to TextCNN on the FWAF dataset, due to the greater diversity of data in this dataset. However, this decrease is within an acceptable range. Additionally, it is worth noting that our method exhibits outstanding performance in the FPR performance metric, significantly lower than the baseline methods, showcasing the advantages of rule-based intrusion detection. This suggests that our approach integrates the powerful representational capabilities of an AI detector while retaining the advantages of rule-based detection.

Meanwhile, rules generated based on two datasets were deployed, and false positive tests were conducted using the benign traffic captured as described in Sect. 3.1. The experimental results indicate that, with 100 GB of traffic, only 0 and 3 false positives were respectively generated for the two rule sets. This suggests that when these rules are formally deployed in the future, the likelihood of encountering numerous false positives is low, rendering them practically usable.

Ablation Experiments. We conducted ablation experiments to assess the impact of the sensitive words aggregation algorithm and the fine-tuning of regular expressions on the detection results. In this work, the sensitive words aggregation algorithm aims to obtain a more comprehensive list of key terms, while the fine-tuning of regular expressions is intended to reduce potential false positives. In the ablation experiments, we compared the original experimental results with those where the Sensitive Words Aggregation step or the Regular Expressions Fine-tuning step was omitted. The results are illustrated in Fig. 3. As anticipated, the detection performance deteriorates when either step is omitted. Specifically, the fine-tuning of regular expressions has a more significant impact on FPR, while its effect on FNR is relatively minor, indicating its role in reducing false positives. On the other hand, the Sensitive Words Aggregation Algorithm has a substantial impact on both FNR and FPR, as a more comprehensive list of key terms enables more precise filtering of malicious traffic. The experimental results align with expectations, confirming the beneficial impact of both modules on the detection outcomes.

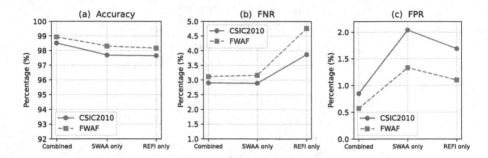

Fig. 3. The comparative results of the ablation experiments for the Sensitive Words Aggregation(SWA) Algorithm and Regular Expressions Fine-tuning (REFI). Each subplot presents the contrasts in ACC, FNR, and FPR between the proposed method and the scenarios where only SWA or only REFI is conducted.

Computational Consumption. To validate whether the computational overhead required by the proposed method is acceptable in practical applications, we recorded the total training durations in comparison to end-to-end baseline methods. As shown in Table 3, the training duration of our method is slightly less than ATPAD but more than TextCNN. This discrepancy is attributed to the additional rule generation process following the training of a TextCNN model. However, the time difference indicates that the rule generation step is relatively time-efficient. Thus, the computational overhead of our method falls within reasonable limits, providing favorable conditions for its practical deployment. Additionally, we recorded the maximum memory consumption during training on the CSIC2010 and FWAf datasets, which were 1581 MB and 932 MB, respectively. Considering the powerful computational capabilities commonly available in contemporary computers, such memory consumption is deemed acceptable.

Detection Efficiency. As shown in Table 3, we also recorded the time required for end-to-end baseline methods and our proposed method to process each traffic flow during the testing phase. It is observed that the rule-based method proposed in this paper has the lowest processing time on both datasets. This is expected, as intrusion detection methods based on Suricata are generally faster than other model-based methods, often implemented in Python. Additionally, it is worth noting that, compared to AI models, rule-based methods offer more straightforward deployment and practical advantages when applied to high-speed traffic systems.

Table 3. Comparison of the method in this paper Acc training time consumption and detection efficiency with end-to-end methods. The **Train** column is the total training time and the **Test** column is the average time to classify a single instance in the test set.

Model	CSIC2010		FWAF	
	Train	Test	Train	Test
TextCNN	36 m 47 s	0.61 ms	38 m 26 s	0.59 ms
ATPAD	52 m 49 s	0.81 ms	46 m 25 s	0.79 ms
Ours	44 m 12 s	0.51 ms	43 m 58 s	0.47 ms

5 Conclusion and Future Work

This paper introduces an intelligent generation framework for intrusion detection rules based on Grad-CAM. The core idea is to assess the importance of each word in attack payloads based on the Grad-CAM algorithm and the parameters of a pre-trained TextCNN model. Furthermore, through operations such as sensitive words aggregation, this keywords list is further refined, ultimately generating intrusion detection rules. Experimental results demonstrate that the proposed method exhibits a commendable generation rate, low computational overhead, and relatively superior detection performance.

In future work, optimization of the proposed method could be considered in the following aspects. Firstly, the method designed in this paper primarily targets traffic using the HTTP protocol. We should explore how to intelligently generate rules in scenarios involving encrypted traffic. Secondly, we plan to explore methods to enhance the generalization capabilities of intrusion detection rules, facilitating their practical application in cybersecurity.

Acknowledgement. This work was supported by the National Key R&D Program of China with No. 2021YFB3101402.

References

1. Ahmad, Z., Shahid Khan, A., Wai Shiang, C., Abdullah, J., Ahmad, F.: Network intrusion detection system: a systematic study of machine learning and deep learning approaches. Trans. Emerg. Telecommun. Technol. **32**(1), e4150 (2021)
2. Albin, E., Rowe, N.C.: A realistic experimental comparison of the suricata and snort intrusion-detection systems. In: 2012 26th International Conference on Advanced Information Networking and Applications Workshops, pp. 122–127. IEEE (2012)
3. Ariu, D., Tronci, R., Giacinto, G.: Hmmpayl: an intrusion detection system based on hidden Markov models. Comput. Secur. **30**(4), 221–241 (2011)
4. Arp, D., et al.: Dos and don'ts of machine learning in computer security. In: 31st USENIX Security Symposium (USENIX Security 22), pp. 3971–3988 (2022)

5. Bao, H., et al.: Payload level graph attention network for web attack traffic detection. In: Mikyska, J., de Mulatier, C., Paszynski, M., Krzhizhanovskaya, V.V., Dongarra, J.J., Sloot, P.M. (eds.) ICCS 2023. LNCS, vol. 14077, pp. 394–407. Springer, Cham (2023). https://doi.org/10.1007/978-3-031-36030-5_32
6. Caswell, B., Beale, J., Baker, A.: Snort intrusion detection and prevention toolkit. Syngress (2007)
7. Di Gennaro, G., Buonanno, A., Palmieri, F.A.: Considerations about learning word2vec. J. Supercomput. 1–16 (2021)
8. Jacobs, A.S., Beltiukov, R., Willinger, W., Ferreira, R.A., Gupta, A., Granville, L.Z.: AI/ML for network security: the emperor has no clothes. In: Proceedings of the 2022 ACM SIGSAC Conference on Computer and Communications Security, pp. 1537–1551 (2022)
9. Li, J., Pan, Z.: Network traffic classification based on deep learning. KSII Trans. Internet Inf. Syst. **14**(11) (2020)
10. Li, R., Li, Q., Zhang, Y., Zhao, D., Jiang, Y., Yang, Y.: Interpreting unsupervised anomaly detection in security via rule extraction. In: Thirty-Seventh Conference on Neural Information Processing Systems (2023)
11. Li, W., Zhang, X.Y., Bao, H., Wang, Q., Li, Z.: Robust network traffic identification with graph matching. Comput. Netw. **218**, 109368 (2022)
12. Li, Y., Krishnamurthy, R., Raghavan, S., Vaithyanathan, S., Jagadish, H.: Regular expression learning for information extraction. In: Proceedings of the 2008 Conference on Empirical Methods in Natural Language Processing, pp. 21–30 (2008)
13. Liu, L., Zhao, Q., Zheng, R., Tian, Z., Sun, S.: An automatically generated intrusion detection rule method based on threat intelligence. Comput. Eng. Des. **43**(1), 1–8 (2022)
14. Liu, Z., Fang, Y., Huang, C., Han, J.: GraphXSS: an efficient XSS payload detection approach based on graph convolutional network. Comput. Secur. **114**, 102597 (2022)
15. Mahbooba, B., Timilsina, M., Sahal, R., Serrano, M.: Explainable artificial intelligence (XAI) to enhance trust management in intrusion detection systems using decision tree model. Complexity **2021**, 1–11 (2021)
16. Nguyen, H.T., Torrano-Gimenez, C., Alvarez, G., Petrović, S., Franke, K.: Application of the generic feature selection measure in detection of web attacks. In: Herrero, Á., Corchado, E. (eds.) CISIS 2011. LNCS, vol. 6694, pp. 25–32. Springer, Heidelberg (2011). https://doi.org/10.1007/978-3-642-21323-6_4
17. Qin, Z.-Q., Ma, X.-K., Wang, Y.-J.: Attentional payload anomaly detector for web applications. In: Cheng, L., Leung, A.C.S., Ozawa, S. (eds.) ICONIP 2018. LNCS, vol. 11304, pp. 588–599. Springer, Cham (2018). https://doi.org/10.1007/978-3-030-04212-7_52
18. Selvaraju, R.R., Cogswell, M., Das, A., Vedantam, R., Parikh, D., Batra, D.: Grad-cam: visual explanations from deep networks via gradient-based localization. In: Proceedings of the IEEE International Conference on Computer Vision, pp. 618–626 (2017)
19. Smitha, R., Hareesha, K.S., Kundapur, P.P.: A machine learning approach for web intrusion detection: MAMLS perspective. In: Wang, J., Reddy, G.R.M., Prasad, V.K., Reddy, V.S. (eds.) Soft Computing and Signal Processing. AISC, vol. 900, pp. 119–133. Springer, Singapore (2019). https://doi.org/10.1007/978-981-13-3600-3_12
20. Tama, B.A., Nkenyereye, L., Islam, S.R., Kwak, K.S.: An enhanced anomaly detection in web traffic using a stack of classifier ensemble. IEEE Access **8**, 24120–24134 (2020)

21. Wang, J., Zhou, Z., Chen, J.: Evaluating CNN and LSTM for web attack detection. In: Proceedings of the 2018 10th International Conference on Machine Learning and Computing, pp. 283–287 (2018)
22. Wang, K., Stolfo, S.J.: Anomalous payload-based network intrusion detection. In: Jonsson, E., Valdes, A., Almgren, M. (eds.) RAID 2004. LNCS, vol. 3224, pp. 203–222. Springer, Heidelberg (2004). https://doi.org/10.1007/978-3-540-30143-1_11
23. Yu, L., et al.: Detecting malicious web requests using an enhanced textCNN. In: 2020 IEEE 44th Annual Computers, Software, and Applications Conference (COMPSAC), pp. 768–777. IEEE (2020)
24. Zhang, T., You, F.: Research on short text classification based on textCNN. In: J. Phys. Conf. Ser. **1757**, 012092 (2021). IOP Publishing

BotRGA: Neighborhood-Aware Twitter Bot Detection with Relational Graph Aggregation

Weiguang Wang[1,2(✉)], Qi Wang[3], Tianning Zang[1,2], Xiaoyu Zhang[1,2], Lu Liu[4], Taorui Yang[5], and Yijing Wang[1]

[1] Institute of Information Engineering, Chinese Academy of Sciences, Beijing, China
{wangweiguang,zangtianning,zhangxiaoyu,wangyijing}@iie.ac.cn
[2] School of Cyber Security, University of Chinese Academy of Sciences, Beijing, China
[3] National Computer Network Emergency Response Technical Team/Coordination Center of China, Beijing, China
[4] China Assets Cybersecurity Technology Co., Ltd., Beijing, China
[5] Department of Accounting, The School of Business, Durham University, Durham, UK
cqlm57@durham.ac.uk

Abstract. With the rapid development of AI-based technology, social bot detection is becoming an increasingly challenging task to combat the spread of misinformation and protect the authenticity of online resources. Existing graph-based social bot detection approaches primarily rely on the topological structure of the Twittersphere but often overlook the diverse influence dynamics across different relationships. Moreover, these methods typically aggregate only direct neighbors based on transitive learning, limiting their effectiveness in capturing the nuanced interactions within evolving social Twittersphere. In this paper, we propose BotRGA, a novel Twitter bot detection framework based on inductive representation learning. Our method begins with extracting the semantic features from Twitter user profiles, descriptions, tweets and constructing a heterogeneous graph, where nodes represent users and edges represent relationships. We then propose a relational graph aggregation method to learn node representations by sampling and aggregating the features from both direct and indirect neighbors. Additionally, we evaluate the importance of different relations and fuse the node's representations across diversified relations with semantic fusion networks. Finally, we classify Twitter users into bots or genuine users and learn model parameters. Extensive experiments conducted on two comprehensive Twitter bot detection benchmarks demonstrate that the superior performance of BotRGA compared to state-of-the-art methods. Additional studies also confirm that the effectiveness of our proposed relational graph aggregation, semantic fusion networks, and strong generalization ability to new and previously unseen user communities.

Keywords: Twitter Bot Detection · Relational Graph Aggregation · Semantic Representation Learning

L. Franco et al. (Eds.): ICCS 2024, LNCS 14838, pp. 162–176, 2024.
https://doi.org/10.1007/978-3-031-63783-4_13

1 Introduction

Along with the rapid development of artificial intelligence and Natural Language Processing (NLP) technology, social network bots have been widely used in various social network platforms, posing great challenges to the authenticity and information security of social networks. These social bots can realistically mimic human social behavior and language habits, and are used by malicious operators to spread disinformation, manipulate public sentiment and political interference. For example, in the past few years, Twitter bots have participated in US presidential election intervention [1], spread false information [2], and promote extremist ideologies [3]. Moreover, with the emergence of ChatGPT, the detection of social bots has become an urgent problem to be solved [4].

Earlier machine learning based Twitter bot detection methods generally utilize feature engineering to extract features from user profiles, and then use traditional machine learning algorithms to classify social robots [5, 6]. However, the heavy reliance on analytical experience and subjective judgment in feature engineering leads to significant limitations for such methods in detecting the sophisticated and diversified social robots. Deep learning based methods use social network users profiles and posting contents as input to the neural network, and identify social bots by building a series of convolutional, recurrent neural network and other deep learning models [7–11]. These methods only consider the user profile and textual information, without utilizing the relations in social networks, making it difficult to achieve effective results in detecting the constantly evolving social bots. With the rapid development of graph neural networks, more and more methods have been proposed for detecting social bots by using graph neural networks for deep analysis of social network structures [12–15]. The above graph-based methods have achieved high recognition accuracy in social bot detection task, but they overlook the different influence weights of diversified relationship types in social networks and only consider the direct relationships between nodes in graph. Moreover, these methods generate node representations with transductive learning, fail to achieve strong generalization in evolving real-world social networks with dynamically added new nodes and relationships.

In light of these challenges, we propose a novel Twitter bot detection framework BotRGA (**Bot** Detection with **R**elational **G**raph **A**ggregation). Specifically, we construct a heterogeneous relational graph to present the Twitter social networks and adopt an inductive learning method to obtain user semantic representations by sampling and aggregating information from one's direct and indirect neighbors in the local neighborhood, then comprehensively integrate the user representations by semantic fusion networks across diversified relationships and conduct bot detection. Our main contributions are summarized as follows:

- We propose to comprehensively leverage the local neighborhood information and diversified relations in heterogeneous relational graphs constructed from real-word Twittersphere, and adopt an inductive method to learn user representation.
- On this basis, we propose BotRGA: a novel social bot detection framework. It is an end-to-end bot detector that uses relational graph aggregation to learn user representation under different relations, then obtain the final node's representations by semantic fusion networks across diversified relations and conduct bot detection.

- We have conducted sufficient experiments to evaluate our proposed BotRGA and compared our method with state-of-the-art baseline methods. Experimental results demonstrate that our proposed method is more efficient and generalized than baseline methods.

2 Related Work

2.1 Graph Neural Network

Graph Neural Network is a deep learning based method for processing graph domain information. The notion of graph neural networks was initially outlined by Gori et al. [16] and further elaborated in Scarselli et al. [17]. These early studies fall into the category of recurrent graph neural networks (RecGNNs), and Li et al. [18] proposed an gated graph sequence neural networks to solve the challenges in previous research. In follow-up works, Kipf et al. [19] further simplify the graph convolutions through a localized first-order approximation and present graph convolutional networks (GCN). Hamilton et al. [20] propose the GraphSAGE framework based on node sampling and features aggregating. It can efficiently generate node embeddings by leveraging neighbor features. Inspired by the attention mechanisms, Velickovic et al. [21] present the graph attention networks (GAT) for node classification. Schlichtkrull et al. [22] proposed Relational Graph Convolutional Networks (RGCN) and apply GCN framework to modeling relational data. In recent years, due to the powerful expressive power of graph structures, GNN is widely used in social network analysis such as node classification, link prediction and graph community detection.

2.2 Twitter Bot Detection

Twitter bots are automated accounts run by software, pose a serious threat to the authenticity and integrity of online platforms. How to effectively detect social bots is a difficult challenge. Existing Twitter bot detection methods mainly fall into three categories: feature-based methods, text-based methods, and graph-based methods.

Feature-based methods generally focused on manually designed features and combined them with traditional machine learning classifiers. These methods conduct feature engineering based on handcrafted user features extracted from user profiles. Yang et al. [6] utilize minimal account metadata and labeled datasets to detect social bots. Davis et al. [8] leverage more than 1,000 user features to evaluate the extent to which a Twitter account exhibits similarity to the known characteristics of social bots. Wu et al. [9] adopt user behavioral sequences and characteristics as features for classifiers to detect bots. However, evolving bots can evade the detection of feature-based approaches by creating deceptive accounts with manipulated metadata and stolen tweets from genuine users [2].

Text-based methods adopt NLP techniques to detect Twitter bots with their historical tweets and user descriptions. Kudugunta et al. [7] proposed a bot detection framework based on contextual LSTM (Long Short-Term Memory) and exploits both user tweet content and account metadata. Wei et al. [10] use word embeddings to encode user historical tweets and adopt a Bi-directional LSTM to distinguish Twitter bots from human accounts. David et al. [29] present a BERT (Bidirectional Encoder Representation from

Transformers) based bot detection model to analyze tweets written by bots and humans. However, text-based methods are easily deceived when advanced bots post stolen tweets and descriptions from genuine users [2].

Graph-based methods utilize the graph structure to represent the various users and diversified relationships in Twitter social networks, and attempt to separate bots and humans based on the graph structure. Ali Alhosseini et al. [14] adopt graph convolutional networks to aggregate the features of user node and conduct bot detection. Feng et al. [13] proposed BotRGCN framework and represent Twitter networks and uses R-GNNs for social bot detection. Lei et al. [15] propose a Twitter Bot detection framework BIC and employs a text-graph interaction module to enable information exchange across modalities in the learning process. The above works used graph convolutional networks to achieve higher detection accuracy than traditional classification methods in robot detection tasks, but both they ignored the different influence weights of diversified relationship types in social node classification. To address the aforementioned issues, Feng et al. [14] construct heterogeneous information networks and propose relational graph transformers to model influence intensity with the attention mechanism and learn node representations to detect social bot, and achieves state-of-the-art performance among graph-based methods. However, this method only considers node's one hop direct relationship in the relationship graph, and the processing of node features is relatively universal, without considering the theme and emotional features of tweet text. Recent years, new advanced AI-based social bots appeared on social platforms from time to time, with the ability to imitate humans and evade detection. The above approaches could not naturally generalize to those new unseen nodes of bots in Twittersphere, because they generate node embeddings by transductive learning and prediction on nodes in a single and fixed graph [20].

3   Methodology

3.1   Overview

Figure 1 presents an overview of our proposed neighborhood-aware and relational graph aggregation Twitter bot detection framework BotRGA. Specifically, we first extract the semantic features encoding from Twitter user profiles, and construct a heterogeneous graph with users as nodes and relationships as edges. We learn node representations by sampling and aggregating its neighbor features under each relationship with our proposed relational graph aggregation. After that, we evaluate the importance of different relations and fuse the node's representations across diversified relations with semantic fusion networks. Finally, we classify Twitter users into bots or genuine users and learn model parameters.

3.2   Feature Encoding

Similar to the Twitter feature extraction method in [13], we separately extracted Twitter user profile metadata and textual embedding, the metadata includes user categorical and numerical metadata and textual embeddings include the semantic representation of user

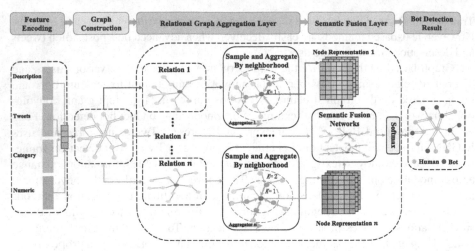

Fig. 1. Overview of our Twitter bot detection with relational graph aggregation framework.

description and historical tweets, and then we use a concatenate function to fuse the four types of features to form the initial encoding of user nodes, which can be defined as:

$$x_{init}^i = CONCAT(r_{des}; r_{tweet}; r_{cat}; r_{num}) \in \mathbb{R}^{D \times 1} \tag{1}$$

where x_{init}^i denotes the initialization encoding of the i-th user node, $r_{des}, r_{tweet}, r_{cat}, r_{num}$ are respectively denotes the representation of description, tweet, category and numerical information extracted from a Tweet user, and D is the user embedding dimension, the detailed encoding strategy is as follows.

Since users of the Twittersphere come from different countries and regions, they usually have different language habits for personalized expressions to share their life experiences and observations on Twittersphere, so the description and tweets in user profiles often contain different languages, a quantity of emoji and named-entity, those textual content imply rich semantics topics and personal sentiments. In order to fully utilize the semantic information form user profiles, different from previous related works using a universal NLP model BERT to handle tweet text, we adopt TweetNLP [23] to learn the semantic embedding on user description and historical tweets, which is a Transformer-based language models and specialized on Twitter social media text. Furthermore, we use TweetNLP to extract topics and sentiments contained in user tweets to form numerical features, which will be introduced in later.

For numerical features, the method of directly using the current number of posts, likes, followers, and following in user profiles as features ignores the factor of user survival time. Therefore, we combine the above values with user survival time as numerical features. In addition, with the support of AI based chatbot technologies such as ChatGPT, it is becoming increasingly difficult to distinguish a tweet is published by human or social bots in terms of syntax and semantics. However, there are still significant differences in posting emotions, topic scope, and other aspects. Therefore, we analyze and extract the topics and sentiments of user tweets, and abstract them into numerical features. The numerical features we extracted are shown in Table 1.

Table 1. Numerical user properties adopted in BotRGA.

Feature Name	Description
Active_days	number of Twitter account active days
Followers_growth_rate	followers_count/active_days
Followings_growth _rate	followings_count/active_days
Tweets_growth_rate	tweets_count/active_days
Status_growth_rate	status_count/active_days
Name_digit_length	number of digits in screen name
Name_upper_length	number of upper case in screen name
Screen name length	number of screen name character
Tweet_neutral_sentiments	number of user tweets with neutral sentiment
Tweet_topics	number of topics in user tweets

3.3 Graph Construction

Relation Graph can be expressed formally as $G(V, E, R)$, where V denotes the user nodes in Twittersphere social graph, E denotes the edges which connecting different user nodes, and R denotes the diversified relationships between user nodes.

We then construct a heterogeneous information network to represent the Twittersphere, which take Twitter users as nodes and take diversified relations types as different edges to connect different user in the social relational graph. We denote the set of relations in the network as R while our framework supports multiple type of relation settings.

In order to better fuse and utilize the four types of feature vectors, we transform the initialization value of the node encoding $x^{(0)}$ with a fully connected layer to serve as initial features in the GNNs, i.e.,

$$x^{(0)} = \sigma (W_0 \cdot x_{init}^i + b_0) \tag{2}$$

where W_0 and b_0 are learnable parameters, σ denotes nonlinearity and we use leaky-relu as σ.

3.4 Relational Graph Aggregation

In order to comprehensively utilize the user node's neighbor features information and its own features to reveal the deep semantic representations under relation $r(r \epsilon R)$, at the same time, to ensure has high generalization and performance in social graph analysis scenarios with large amounts of data and dynamic updates, we propose relational graph aggregation mechanism, a GNN architecture that separately learns embeddings by sampling and aggregating features from a node's local neighborhood on different relationships, formulated as:

$$h_{SN_i^r}^{r(k)} = AGGREGATE_k^r(h_u^{r(k-1)}, \forall u \in SN_i^r) \tag{3}$$

$$h_i^{r(k)} = sigmoid(W_k^r \cdot CONCAT(h_i^{r(k-1)}, h_{SN_i^r}^{r(k)}) + b_k^r) \tag{4}$$

where SN_i^r denotes the sampled i-th node's neighbors set under relation i, $h_u^{r(k-1)}$ denotes the representation of $k-1$ depth neighbor u in sampling set SN_i^r under relation i, $h_{SN_i^r}^{r(k)}$ denotes aggregated representation of the i-th node's neighbors, $h_i^{r(k)}$ denotes the learned representation of i-th node, $AGGREGATE_k^r$ is the aggregation function of k depth under relation i, W_k^r and b_k^r denote learnable parameters.

The size of SN_i^r is set to 25, the depth k is set to 2 by default following GraphSAGE [20]. We use max-pooling aggregator as the aggregation function due to empirical performance. After aggregating *depth-K* neighborhood node information, we obtain a new node embedding $h_i^{r(k)}$.

In order to obtained smooth representation learning results, we adopt the gate mechanism to obtain the representation of node i by:

$$h_i^{r(K)} = tanh(h_{SN_i^r}^{r(k)}) \odot h_i^{r(k)} + x_i^{(0)} \odot (1 - h_i^{r(k)}) \tag{5}$$

where $\odot$ denotes the Hadamard product operation, $x_i^{(0)}$ denotes the initial features of node i in the GNNs, and $h_i^{r(K)}$ is the learned representation of node i for relation r in *depth-K*.

3.5 Semantic Fusion Networks

To aggregate more comprehensive semantic information, the multiple features needed to be revealed by different relation-paths. Moreover, the weights of relationships are different, treating each relationship equally weakens the semantic features which are aggregated by some more important relationships.

In order to address these issues, we propose a novel relation-based attention mechanism to obtain the importance of different relation-paths then utilized to aggregated various semantic information across different relationship to learn the node's semantic representation, defined as:

$$\alpha_d^{r(l)} = \sigma\left(\frac{1}{|V^r|} \sum_{i \in V^r} (q_d^{(l)^T} \cdot \tanh(W_d^{r(l)} \cdot h_i^{r(l)} + b_d^{r(l)}))\right) \tag{6}$$

where $\alpha_d^{r(l)}$ denotes the learned importance of relation r at the d-th attention head, $|V^r|$ denotes the number of nodes under relation r, σ is sigmoid function, $q_d^{(l)^T}$ is semantic attention vector of relation r at the d-th attention head in layer l, $q_d^{(l)^T}$, $W_d^{r(l)}$, $b_d^{r(l)}$ are learned parameters.

We then aggregate node information based on edge relationships with different weights, the formula is as follows:

$$x_i^{(l)} = \frac{1}{D} \sum_{d=1}^{D} [\sum_{r \in R} \alpha_d^{r(l)} \cdot h_i^{r(l)}] \tag{7}$$

where $x_i^{(l)}$ denotes the learned representation of node i aggregated from different relations in layer l, $h_i^{r(l)}$ denotes the results of relational graph transformers and D is the number of attention heads.

3.6 Learning and Optimization

After L layers of GNNs messages passing, we obtain the final node representations $x^{(L)}$, and transform them with an output layer and a softmax layer to get Twitter bot detection result, *i.e.*,

$$\hat{y}_i = softmax(W_O \cdot \sigma(W_L \cdot x_i^{(L)} + b_L) + b_O) \tag{8}$$

where $\hat{y}_i$ is out model's prediction of user i, all W and b are learnable parameters. We then adopt supervised annotations and a regularization term to train out model, formulated as:

$$Loss = - \sum_{i \in Y}[y^i log(\hat{y}_i) + (1 - y^i)log(1 - \hat{y}_i)] + \lambda \sum_{\omega \in \theta} \omega^2 \tag{9}$$

where Y is the annotated user set, y^i is the ground-truth labels, θ denotes all trainable parameters in the model and λ is a hyperparameter.

4 Experiments

4.1 Dataset

In order to verify the effectiveness of our proposed model, the data set needs to have a certain graph structure type. We conducted our experiments on two public data sets with topological relationships, TwiBot-20 [24] and TwiBot-22 [25], which are more representative of the current social network environment.

TwiBot-20 and TwiBot-22 are comprehensive Twitter bot detection benchmarks and provide user follow relationships to support graph-based methods. TwiBot-20 is proposed in 2020 and includes 229,573 Twitter users, 33,488,192 tweets and 455,958 follow relationships. TwiBot-22 is the largest public dataset to date for Twitter bot detection and includes 1,000,000 Twitter users, 88,217,457 tweets and 170,185,937 follow relationships. An overview of the datasets is provided in Table 2.

Table 2. Database Overview.

Dataset	Account	Bot	Human	Tweets	Edges
TwiBot-20	229,573	5,273	6,589	33,488,192	33,716,171
TwiBot-22	1,000,000	139,943	860,057	88,217,457	170,185,937

4.2 Baselines and Experiment Setting

In order to validate the effectiveness of our proposed Twitter bot detection model, we compare our graph-based approach with the following methods:

Yang et al. (2020) [6] use random forest classifier with minimal user metadata and derived features.

Kudugunta et al. (2018) [7] propose to jointly leverage user tweet semantics and user metadata.

Botometer (2016) [8] is a bot detection service that leverages more than 1,000 user features.

Wei et al. (2019) [10] use recurrent neural networks to encode tweets and classify users based on their tweets.

Alhosseini et al. (2019) [14] use graph convolutional networks to learn user representations and conduct bot detection.

BotRGCN (2021d) [13] constructs a heterogeneous graph to represent the Twitter-sphere and adopts relational graph convolutional networks for representation learning and bot detection.

Feng et al. (2022) [14] propose relational graph transformers to model heterogeneous influence between users and use semantic attention networks to aggregate messages across users and relations and conduct heterogeneity-aware Twitter bot detection.

BotBuster (2022) [13] is a social bot detection system that processes user metadata and textual information to enhance cross-platform bot detection.

We use pytorch [26], pytorch lightning [27], torch geometric [28] for an efficient implementation of our proposed Twitter bot detection framework. We conduct all experiments on a server with 2 T V100 GPUs with 32 GB memory, 32 CPU cores, and 300GB CPU memory. To directly and fairly comparing with previous works, we follow the same train, valid and test splits provided in the benchmark.

4.3 Main Results

We then benchmark these bot detection models on TwiBot-20 [24] and TwiBot-22 [25] and present results in Table 3, which demonstrates that:

- BotRGA consistently and significantly outperforms all baseline methods across the two datasets. Specifically, compared with the previously state-of-the-art method proposed by Feng et al. [14], BotRGA achieves 1.2% higher accuracy and 1.1% higher F1-score on TwiBot-20, and also provides a gain of 1.7% F1-score compared with the second best method on TwiBot-22.
- Graph-based methods for Twitter bot detection, such as BotRGA (Ours), BotRGCN [13], and Feng et al. [14], demonstrate higher classification effectiveness compared to traditional non-graph methods like Yang et al. [6] and Kudugunta et al. [7]. This underscores the critical importance of leveraging the topological structure for node classification tasks in social networks.
- We propose the first relation-based and neighborhood-aware bot detection frameworks, which achieves the best performance on a comprehensive benchmark. Our results highlight the necessity of aggregating semantic information from diverse user

relationships, validating the effectiveness of our approach in addressing this challenge. Additionally, our method outperforms existing approaches, emphasizing its potential for advancing Twitter bot detection capabilities.

Table 3. Accuracy and binary F1-score of Twitter bot detection systems on two datasets. Bold indicates the best performance, underline the second best. This table indicates that the results of our method BotRGA is significantly better than the second best baseline.

Model	TwiBot-20		TwiBot-22	
	Accuracy	F1-score	Accuracy	F1-score
Yang et al.	0.8191	0.8546	0.7508	0.3659
Kudugunta et al.	0.8174	0.7515	0.6578	0.5167
Botometer	0.4801	0.6266	0.4990	0.4275
Wei et al.	0.7126	0.7533	0.7020	0.5360
Alhosseini et al.	0.6813	0.7318	0.4772	0.3810
BotRGCN	0.8462	0.8707	0.7887	0.5499
Feng et al.	0.8664	0.8821	0.7650	0.4294
BotBuster	0.7724	0.8118	0.7406	0.5418
BotRGA(Ours)	**0.8783**	**0.9035**	**0.7947**	**0.5671**

4.4 Ablation Study

In order to effectively identify social bot on the Twittersphere, we propose a novel graph-based social bot detection method BotRGA, which comprehensively utilizes and integrates user information and topology in social networks. Especially, we adopt the follower and following relationships between different users as edges to construct a heterogeneous relational graph.

To prove the effectiveness of our graph construction method, we remove different types of edges and obtain results under the ablation setting in Table 4. It is illustrated that the graph with both follower and following edges outperforms any reduced settings. The results also prove the effectiveness of our proposed method to construct heterogeneous relational graph of Twittersphere.

Upon obtaining a Heterogeneity graph, we propose relational graph aggregation to learn node representation by fusion relation-based local neighborhood property information. To prove the effectiveness of our proposed GNN architecture, we conduct ablation study on relational graph aggregation and semantic fusion networks separately, and report the results under different settings in Table 5. It is illustrated that the values of accuracy and F1-score drop significantly when removing our proposed Relational Graph Aggregation and Semantic Fusion Networks independently. Furthermore, when adopting 3 popular GNN models and 4 commonly used fusion algorithms to replace our proposed

Table 4. Ablation studying removing different relationship of our constructed Relational Graph.

Ablation Settings	Accuracy	F1-score
only follower relationship	0.8531	0.8623
only following relationship	0.8587	0.8649
follower + following relationship(homogeneous)	0.8652	0.8726
follower + following relationship(heterogeneous)	**0.8783**	**0.9035**

GNN architecture respectively, the values of accuracy and F1-score have increased but are still lower than our model. The experiment results proved that the relational graph aggregation and semantic fusion networks are all essential parts of our proposed GNN architecture.

To sum up, both our constructed relational graph and our proposed GNN architecture are effectiveness and contribute to our model's outstanding performance.

Table 5. Ablation study of our proposed GNN architecture.

Ablation Settings	Accuracy	F1-score
remove Relational Graph Aggregation	0.8571	0.8691
remove Semantic Fusion Networks	0.8605	0.8758
replace Relational Graph Aggregation with GAT	0.8646	0.8853
replace Relational Graph Aggregation with GCN	0.8625	0.8812
replace Relational Graph Aggregation with RGCN	0.8655	0.8854
Summation as Semantic Fusion Networks	0.8663	0.8862
Mean pooling as Semantic Fusion Networks	0.8651	0.8825
Max pooling as Semantic Fusion Networks	0.8676	0.8863
Min pooling as Semantic Fusion Networks	0.8545	0.8801
full model	**0.8783**	**0.9035**

4.5   Generalization Study

As Twitter bots are constantly evolving in real world [2], the research on effective detection of bots calls for models to better generalize to unseen user accounts. To this end, we evaluate BotRGA's generalization ability to detect bots in unseen user nodes in TwiBot-20 [24] benchmark. Specifically, TwiBot-20 [24] has collected Twitter user accounts which created between 2008 and 2020, so we divide the dataset by year and take the accounts created in 2020 as testing set, the accounts before 2020 as training set and validation set. This division ensures that the user accounts in the testing set are unseen in training. After that, we use BotRGA and baseline models to conduct generalization studies in the above dataset. The result is presented in Fig. 2, the graph-based models BotRGA, Feng et al. [14] and BotRGCN [13] generally outperform the feature-base models. From the result we can see, our proposed BotRGA achieves the highest accuracy of 0.869 and F1-score of 0.8907 among these models, and its accuracy outperforms the second-best model of Feng et al. [14] by 2.7% on unseen accounts. This indicates that incorporated with relational graph aggregation and semantic fusion networks, BotRGA could better generalize to previously unseen user accounts.

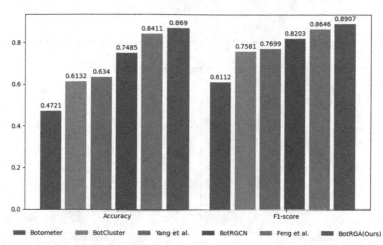

Fig. 2. Generalization study on TwiBot-20 [24] indicates that we proposed BotRGA is better at generalizing to unseen user nodes than baselines approaches.

4.6 Representation Learning Study

Using GNNs for social bot detection is essentially to learn the user's representation to distinguish humans and bots in social networks. In order to verify the effectiveness of our proposed social bot detection model for user representation learning, we present the T-SNE [30] plot of user representation of our method BotRGA and baselines using Twibot-20 [24] dataset in Fig. 3. The result illustrates that the learned representation of our proposed BotRGA and Feng et al. [14] have clearer discrimination between the group of human and bots than the other two baselines on Twittersphere. Compared with Feng's model, the representations learned by our model have higher purity in the clusters of human and bots, it indicates that our proposed method could learn higher quality Twitter user representation to distinguish the group of social bots and human in Twittersphere.

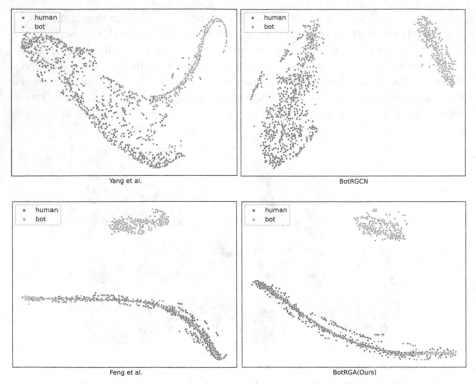

Fig. 3. Plots of Twitter user representations learned with our model and different baselines on TwiBot-20, the result indicates that the learned representations of BotRGA(Ours) have higher discrimination between the group of human and bots than baseline models.

5   Conclusion and Future Work

With the rapid development of AI-based technology, social bot detection has gradually to be an important and challenging task. We propose BotRGA, a novel Twitter bot detection framework with neighborhood-aware and relational graph aggregation to inductively learn the user deep semantic representations in heterogeneous social network. Extensive experiments demonstrate that BotRGA significantly advances the state-of-the-art on two Twitter bot detection benchmarks. Further studies demonstrate the effectiveness of our relation-based features aggregation strategy, and prove the superior generalization ability and representation learning efficiency of BotRGA.

Social network is a dynamically updated temporal network, and the user's survival status and mutual relationships continue to change over time. Especially with the support of large language models, such as ChatGPT, social bots have become more humanoid, how to detect social bots accurately and effectively becomes very difficult. In future work, we plan to experiment with more diversified ways to face the dynamic update scenario of social networks and extend our graph-based detection approach.

Acknowledgments. This work was supported by the Defense Industrial Technology Development Program (Grant JCKY2021906A001), and the National Natural Science Foundation of China (NSFC) (Grant 62376265).

References

1. Deb, A., Luceri, L., Badaway, A., Ferrara, E.: Perils and challenges of social media and election manipulation analysis: the 2018 us midterms. In: Companion Proceedings of the 2019 World Wide Web Conference, pp. 237–247 (2019)
2. Cresci, S.: A decade of social bot detection. Commun. ACM **63**(10), 72–83 (2020)
3. Berger, J.M., Morgan, J.: The ISIS Twitter census: defining and describing the population of ISIS supporters on Twitter (2015)
4. Ferrara, E.: Social bot detection in the age of ChatGPT: challenges and opportunities. In: First Monday (2023)
5. Cresci, S., Di Pietro, R., Petrocchi, M., Spognardi, A., Tesconi, M.: DNA-inspired online behavioral modeling and its application to spambot detection. IEEE Intell. Syst. **31**(5), 58–64 (2016)
6. Yang, K.C., Varol, O., Hui, P.M., Menczer, F.: Scalable and generalizable social bot detection through data selection. Proc. AAAI Conf. Artif. Intell. **34**(01), 1096–1103 (2020)
7. Kudugunta, S., Ferrara, E.: Deep neural networks for bot detection. Inf. Sci. **467**, 312–322 (2018)
8. Davis, C.A., Varol, O., Ferrara, E., Flammini, A., Menczer, F.: Botornot: a system to evaluate social bots. In: Proceedings of the 25th International Conference Companion on World Wide Web, pp. 273–274 (2016)
9. Wu, J., Ye, X., Mou, C.: Botshape: a novel social bots detection approach via behavioral patterns. arXiv preprint arXiv:2303.10214 (2023)
10. Wei, F., Nguyen, U.T.: Twitter bot detection using bidirectional long short-term memory neural networks and word embeddings. In: Proceedings of the 2019 First IEEE International Conference on Trust, Privacy and Security in Intelligent Systems and Applications, pp. 101–109 (2019)
11. Ng, L.H.X., Carley, K.M.: Botbuster: Multi-platform bot detection using a mixture of experts. In: Proceedings of the International AAAI Conference on Web and Social Media, vol. 17 (2023)
12. Ali Alhosseini, S., Bin Tareaf, R., Najafi, P., Meinel, C.: Detect me if you can: Spam bot detection using inductive representation learning. In: Companion Proceedings of the 2019 World Wide Web Conference, pp. 148–153 (2019)
13. Feng, S., Wan, H., Wang, N., Luo, M.: BotRGCN: Twitter bot detection with relational graph convolutional networks. In: Proceedings of the 2021 IEEE/ACM International Conference on Advances in Social Networks Analysis and Mining, pp. 236–239 (2021)
14. Feng, S., Tan, Z., Li, R., Luo, M.: Heterogeneity-aware twitter bot detection with relational graph transformers. Proc. AAAI Conf. Artif. Intell. **36**(4), 3977–3985 (2022)
15. Lei, Z., et al.: BIC: Twitter bot detection with text-graph interaction and semantic consistency. arXiv preprint arXiv:2208.08320 (2022)
16. Gori, M., Monfardini, G., Scarselli, F.: A new model for learning in graph domains. In: Proceedings of the 2005 IEEE International Joint Conference on Neural Networks, vol. 2 (2005)
17. Scarselli, F.: The graph neural network model. IEEE Trans. Neural Netw. **20**(1), 61–80 (2008)
18. Li, Y., Tarlow, D., Brockschmidt, M., Zemel, R.: Gated graph sequence neural networks. arXiv preprint arXiv:1511.05493 (2015)

19. Kipf, T.N., Welling, M.: Semi-supervised classification with graph convolutional networks. arXiv preprint arXiv:1609.02907 (2016)
20. Hamilton, W., Ying, Z., Leskovec, J.: Inductive representation learning on large graphs. Adv. Neural Inf. Process. Syst. **30** (2017)
21. Veličković, P., Cucurull, G., Casanova, A., Romero, A., Lio, P., Bengio, Y.: Graph attention networks. arXiv preprint arXiv:1710.10903 (2017)
22. Schlichtkrull, M., Kipf, T.N., Bloem, P., van den Berg, R., Titov, I., Welling, M.: Modeling relational data with graph convolutional networks. In: Gangemi, A., Navigli, R., Vidal, M.-E., Hitzler, P., Troncy, R., Hollink, L., Tordai, A., Alam, M. (eds.) ESWC 2018. LNCS, vol. 10843, pp. 593–607. Springer, Cham (2018). https://doi.org/10.1007/978-3-319-93417-4_38
23. Camacho-Collados, J., et al.: TweetNLP: cutting-edge natural language processing for social media. arXiv preprint arXiv:2206.14774 (2022)
24. Feng, S., et al.: Twibot-20: a comprehensive twitter bot detection benchmark. In: Proceedings of the 30th ACM International Conference on Information & Knowledge Management (2021)
25. Feng, S., et al.: TwiBot-22: towards graph-based Twitter bot detection. Adv. Neural Inf. Process. Syst. **35** (2022)
26. Paszke, A., et al.: Pytorch: an imperative style, high-performance deep learning library. Adv. Neural Inf. Process. Syst. **32** (2019)
27. Falcon, W.A.: Pytorch lightning. Homepage: https://lightning.ai
28. Fey, M., Lenssen, J.E.: Fast graph representation learning with PyTorch Geometric. arXiv preprint arXiv:1903.02428 (2019)
29. Dukić, D., Keča, D., Stipić, D.: Are you human? Detecting bots on Twitter using BERT. In: International Conference on Data Science and Advanced Analytics (DSAA) (2020)
30. Van der Maaten, L., Hinton, G.: Visualizing data using t-SNE. J. Mach. Learn. Res. **9**(11) (2008)

SOCXAI: Leveraging CNN and SHAP Analysis for Battery SOC Estimation and Anomaly Detection

Amel Hidouri[✉], Slimane Arbaoui[✉], Ahmed Samet, Ali Ayadi,
Tedjani Mesbahi, Romuald Boné, and François de Bertrand de Beuvron

Université de Strasbourg, Institut National des Sciences Appliquées (INSA
Strasbourg), CNRS, ICube Laboratory UMR 7357, Strasbourg 67000, France
{amel.hidouri,slimane.arbaoui,ahmed.samet,ali.ayadi,
tedjani.mesbahi,romuald.bone,francois.beuvron}@insa-strasbourg.fr

Abstract. In the domain of battery energy storage systems for Electric
Vehicles (EVs) applications and beyond, the adoption of machine learning techniques has surfaced as a notable strategy for battery modeling.
Machine learning models are primarily utilized for forecasting the forthcoming state of batteries, with a specific focus on analyzing the State-
of-Charge (SOC). Additionally, these models are employed to assess the
State-of-Health (SOH) and predict the Remaining Useful Life (RUL)
of batteries. Moreover, offering clear explanations for abnormal battery usage behavior is crucial, empowering users with insights needed
for informed decision-making, build trust in the system, and ultimately
enhance overall satisfaction. This paper presents SOCXAI, a novel algorithm designed for precise estimation of batteries's SOC. Our proposed
model utilizes a Convolutional Neural Network (CNN) architecture to
efficiently estimate the twenty five future values of SOC, rather than
a single value. We also incorporate a SHApley Additive exPlanations
(SHAP)-based post-hoc explanation method into our method focusing
on the current feature values for deeper prediction insights. Furthermore,
to detect abnormal battery usage behavior, we employ a 2-dimensional
matrix profile-based approach on the time series of current values and
their corresponding SHAP values. This methodology facilitates the detection of discords, which indicate irregular patterns in the battery usage.
Our extensive empirical evaluation, using diverse real-world benchmarks,
demonstrates our approach effectiveness, showcasing its superiority over
state-of-the-art algorithms.

Keywords: Machine Learning · Battery · Explainability · SOC
estimation · Data Mining

1 Introduction

The integration of technological advancements across industries has significantly
enhanced the accessibility and the generation of industrial time series data, a

© The Author(s), under exclusive license to Springer Nature Switzerland AG 2024
L. Franco et al. (Eds.): ICCS 2024, LNCS 14838, pp. 177–191, 2024.
https://doi.org/10.1007/978-3-031-63783-4_14

trend expected to persist with the emergence of Industry 4.0 [10]. This development has led to the inevitable generation of vast datasets, underscores the need for versatile methods to effectively mine this information. This transition involves the utilization of data mining techniques to extract valuable insights from large datasets, particularly within the industrial sector. A key challenge in data mining revolves around identifying significant sub-sequences within time series data, where "interesting" patterns may include both repetitive and singular occurrences or deviations from the norm. Meanwhile, battery management has become a focal point within this context. As batteries play a critical role in various applications, understanding their behavior through time series analysis is essential.

The viability of Electric vehicles (EVs) is predominantly contingent upon the performance, range, lifetime cost-effectiveness, and safety of their batteries. At present, rechargeable lithium-ion (Li-ion) batteries are the preferred choice for EVs due to their favorable energy density and lifespan. The high energy density of Li-ion batteries allows for more energy storage in a relatively compact size, which is crucial for maximizing the driving range of EVs [11].

Therefore, ensuring the efficiency and safety of these advanced batteries is becoming increasingly crucial. Effectively managing batteries within a system demands detailed modeling to accurately predict their condition, with particular focus on metrics such as State of Charge (SOC) and State of Health (SOH). These metrics offer crucial insights into remaining energy, power delivery capacity, and overall cell life. Nevertheless, assessing residual lithium in batteries is a challenging task, necessitating precise algorithms embedded within Battery Management Systems (BMS). These algorithms, often leveraging mathematical models or Machine Learning (ML) techniques, play a pivotal role in estimating the battery's states, including SOC and SOH levels, using data such as terminal voltage, terminal current, and surface temperature. These measures are useful comprehending the remaining driving range of an EV or in designing a battery that will exhibit optimal performance in real-world conditions. Often, in this work we are interested in the SOC of the battery within a single charge/discharge cycle. The SOC of a battery refers to the current level of energy stored in the battery, expressed as a percentage of its total capacity. It indicates how much charge is remaining in the battery relative to its fully charged state and it can be computed using the following formula (1) [7]:

$$\text{SOC } (\%) = \left(\frac{\text{Ongoing capacity}}{\text{Total capacity}} \right) \times 100 \tag{1}$$

Recently, there has been a noticeable shift towards employing data mining tools to facilitate eXplainable Artificial Intelligence (XAI). This involves utilizing data mining techniques to elucidate and interpret black-box models, thereby improving transparency and comprehensibility in the decision-making processes of these models, particularly in safety-critical applications, especially within industries or vehicles. As a result, the development of XAI techniques has become a priority, aiming to provide insights into AI decision-making processes and make their outputs interpretable to end-users. In this context, SHAP, a model-agnostic approach [14] based on Shapely index, has gained significant popularity in recent years for explaining a wide range of ML models.

In the realm of industrial time series data analysis, examining subsequences for similarities or disimilarities provides valuable insights and explanations regarding the state of a product or process. Commonly utilized terms in literature to describe patterns within sequential data are time series motif and time series discord. Time series motifs predominantly emphasize similarities, while discords concentrate on dissimilarities. Mining time series data, particularly through discord identification, is a topic of extensive research. As a result, the field of time series anomaly detection has witnessed a remarkable surge in interest, with hundreds of algorithms proposed over the last two decades [1,12].

Our contribution is twofold, aiming to address both the prediction task and the requirement for explainability concurrently. More specifically, this paper presents a novel approach leveraging Convolutional Neural Networks (CNN) to estimate the SOC of batteries. Unlike existing methods, the proposed model predicts 25 SOC values rather than a single value, thereby offering more detailed insights into battery behavior for the next minutes. To provide explanations for the SOC predictions, we employ the SHAP model, that analyzes the contributions of current values towards SOC predictions. This is complemented by the application of a two-dimensional anomaly detection model, enabling us to identify the factors influencing SOC predictions.

2 Related Work

Abundant literature has been dedicated to the task of SOC estimation. Indeed, tremendous progress has been made in developing efficient algorithms that can estimate its future state, i.e., SOC. Two branches of works can characterize existing battery models: model-based approaches and data-driven methods. The former consists of the equivalent circuit models (ECMs) which is based on empirical knowledge and experimental data. Batteries are represented by groups of electrical components, such as resistors and capacitors, forming resistor-capacitor networks that are used to monitor the battery's behavior at different time constants associated with the diffusion and charge-transfer processes [8,9,17]. Although, this model is used as main battery models that are widely used in the BMS of EVs for online SOC estimations due to their low computational demands, the accuracy is usually limited to the parameterized range of the model. A further improvement on model-based methods is about development of Physics-Based Models (PBMs) [3,5], with the pseudo two-dimensional (P2D) model being the most notable. The P2D model provides insights into battery internal dynamics. However, managing its equations is complex and demands significant computational resources, making it impractical for real-time applications. Moreover, PBMs often overlook details about material information. The second line of research pertains to Data-Driven Models (DDMs), which have garnered considerable attention for their adaptability, model-free advantages, and the capacity to handle high degrees of nonlinearity. Possessing self-learning capabilities and robust generalization ability, DDMs are particularly effective for estimating SOC within nonlinear systems. Typically, these systems are constructed using various machine learning techniques, including neural networks [2,6,21], support vector machines [22], to predict SOC without the necessity for a prior knowledge.

3   A Glimpse at Times Series

In this section, we introduce the definitions and concepts that we will use throughout this paper.

Definition 1 (Time Series). *A time series $T \in \mathcal{R}^n$ denoted as $T = [t_1, \ldots, t_n]$ is a time-ordered sequence of values.*

Definition 2 (Subsequence). *A subsequence $T_{i,m} \in \mathcal{R}^m$ of T is a continuous subset of values from T of length m, starting from position i. Formally, $T_{i,m} = [t_i, t_{i+1}, \ldots, t_{i+m-1}]$.*

By selecting any subsequence $T_{i,m}$ as a query and computing its distance from all subsequences within the time series T, then sequentially saving the distances in an array, we generate a distance profile.

Definition 3 (Distance Profile). *A distance profile D_i of a time series T is an ordered array of Euclidean distances between the query subsequence $T_{i,m}$ and all subsequences in time series T. Formally, $D_i = [d_{i,1}, d_{i,2}, \ldots, d_{i,n-m+1}]$ where $d_{i,j}$ for $i \geq 1, j \leq n - m + 1$ is the Euclidean distance between $T_{i,m}$ and $T_{j,m}$.*

In the distance profile D_i of query $T_{i,m}$, the ith position represents the distance between the query and itself, resulting in a value of 0. Values preceding and following position i are nearly zero, indicating overlapping subsequences with the query. We focus solely on non-self-matches, disregarding these self-matches.

Definition 4 (Non-self Match). *In a time series T, with a subsequence $T_{p,m}$ of length m beginning at position p and a matching subsequence $T_{q,m}$ starting at q, $T_{p,m}$ is a non-self match to $T_{q,m}$ with distance $d_{p,q}$ if $|p - q| \geq m$.*

Definition 5 (Time Series Discord). *In time series T, with a subsequence $T_{d,m}$ of length m starting at position d is considered a discord of T if the distance between $T_{d,m}$ and its nearest non-self match is the largest among all subsequences. Formally, for every $T_{c,m} \in T$, with the non-self matching sets M_D of $T_{d,m}$, and non-self matching set M_C of $T_{c,m}$, $min(d_d, M_D) > min(d_c, M_C)$.*

The Matrix Profile (MP) [23] is the most used solution to compute discords within time series data.

Definition 6 (Matrix Profile). *The Matrix Profile P of a time series T is a vector that records the z-normalized Euclidean distance between each subsequence and its nearest non-self match. Formally, $P = [min(D_1), min(D_2), \ldots, min(D_{n-m+1})]$ where $D_{1 \leq i \leq n-m+1}$ represents the distance profile of the query subsequence $T_{i,m}$ in time series T.*

Definition 7 (Multidimensional Time Series). *A multidimensional time series* $T \in \mathcal{R}^{n \times d}$ *is a n-sized set of d co-evolving time series. Formally,* $T = [T^1, T^2, \ldots, T^d]$.

When extending the matrix profile to multidimensional time series, we introduce a new structure called the multidimensional matrix profile. This adaptation facilitates the analysis of pattern similarity and dissimilarity across multiple dimensions within the time series data.

Definition 8 (Multidimensional Matrix Profile). *Given a multidimensional time series* $T = [T^1, T^2, \ldots, T^d]$ *with d time series, each of length n, the multidimensional matrix profile is constructed by aggregating the matrix profiles* MP_i *of all d time series. It stores the z-normalized Euclidean distance between each subsequence and its nearest neighbor across all dimensions.*

In this paper, we are interested in multi-dimensional time series that exhibit discords that may be present on a subset of dimensions, we call such anomalies a *K-dimensional anomaly*.

Definition 9 (K-Dimensional Anomaly). *A K-dimensional anomaly appears on at least K of the time series* $T = [T^1, T^2, \ldots, T^d]$. *When k equals the total number of time series, such a k-dimensional anomaly is referred to as a natural anomaly [20].*

Definition 10 (Natural Anomaly). *Given a multidimensional time series* $T = [T^1, T^2, \ldots, T^d]$ *consists of d times series and X a K-dimensional-anomaly in T, X is a natural anomaly if k is equal to the total number of dimensions on which the anomaly is observed.*

Natural anomaly detection is particularly intriguing because simply declaring the presence of an anomaly is not enough. It is more valuable to identify which specific dimensions, such as sensors, are involved, especially when their number is significant. In this work, our focus is on identifying natural anomalies and pinpointing the specific time series associated with them.

The MP technique [23] has emerged as a valuable tool for uncovering various properties of time series data across a wide range of applications, including seismology, medicine, and vocalization analysis. This technique has demonstrated its utility in identifying numerous structural elements within time series datasets, such as repeated behaviors, known as motifs [13], as well as anomalies, referred to as discords [4,16,23], shapelets among others. Indeed, the field of time series discord detection has been gaining increasing interest within the domain of data mining [23].

4 Convolutional Neural Network-Based Model for Battery SOC Estimation

In this section, we present the methodology for constructing a deep learning model to estimate the SOC values of Li-ion cells, starting from the dataset used to train and test the model until the model architecture.

4.1   Dataset

In order to develop a model able to estimate the SOC of real-world driving cycles, the dataset includes 142 cycles from the Massachusetts Institute of Technology (MIT) battery dataset [18] and a further 425 cycles from the National Institute of Applied Sciences (INSA) [7]. From the MIT dataset, we selected exactly two cycles from of the 72 identified charges policies, ensuring a broad representation of charging conditions. In addition, we incorporated the Basytec XCTS system for assessing lithium ferrophosphate (LFP) battery cells, identical to those featured in the MIT dataset. This advanced system enables us to conduct tests employing diverse protocols, including the Worldwide Harmonized Light Vehicle Test Procedure (WLTP) and the Assessment and Reliability of Transport Emission Models and Inventory Systems (ARTEMIS) cycle. The testing regimen comprises two primary phases: charge and discharge as shown in Fig. 1. During the charge phase, we employ the classic CC-CV (constant current-constant voltage) method. This involves applying a constant current to the battery cells, followed by a constant voltage, a process that is crucial for accurately simulating the charging behavior of batteries in practical applications. For the discharge phase, we emulate real-world driving conditions by integrating multiple driving cycles. This phase encompasses regenerative braking, a critical feature that recuperates energy dissipated during vehicle deceleration and braking, effectively recharging the battery cells.

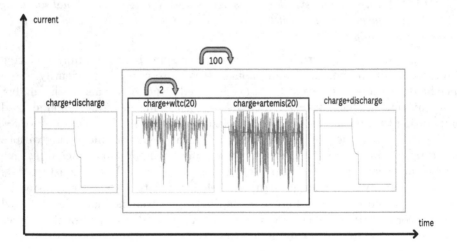

Fig. 1. Structure of the test protocols.

4.2   Data Preprocessing

In the obtained dataset from each cycle, comprising measurements of current, voltage, and temperature recorded during tests, the direct measurement of SOC is not feasible and requires estimation. To initiate a supervisory learning process, for our model, accurately computing the SOC is indispensable. For this purpose, we employ the Coulomb counting technique [19], which involves the cumulative integration of current over time to estimate the SOC. This technique is computed using the following Eq. 2:

$$\text{SOC}(t) = \begin{cases} \text{SOC}(t-1) + \frac{1}{\text{cap}} \sum_{i=1}^{n_c} I_c(t_i)\Delta t_i, & \text{if charging} \\ 1 + \frac{1}{\text{cap}} \sum_{i=1}^{n_d} I_d(t_i)\Delta t_i, & \text{if discharging} \end{cases} \tag{2}$$

where:

- $\text{SOC}(t)$ represents the SOC at time t,
- $\text{SOC}(t-1)$ is the previous SOC value,
- $I_c(t_i)$ and $I_d(t_i)$ denote the charging and discharging currents respectively, recorded at time t_i, with negative current values during the discharge.
- Δt_i signifies the time intervals between consecutive measurements,
- n_c and n_d are the number of measurements taken during the charging and discharging phases respectively.
- cap represents the capacity of the battery.

During the charging phase, the SOC is updated by summing the integrated current over the duration of the charge. Conversely, during the discharging, the initial SOC is subtracted from 1 (assuming a full charge), and the integrated current over the discharge duration is added.

Following the calculation of SOC values, we perform the min-max normalization technique to scale our features within a 0 to 1 range, as presented in Eq. 3. More precisely, this technique was applied first to the training set, which contains 70% of the total cycles including an equal proportion from both the MIT and INSA datasets. Subsequently, we adopted the same minimum and maximum values obtained from the training set to normalize the test set.

$$\text{normalize value} = \frac{\text{data} - \min(\text{data})}{\max(\text{data}) - \min(\text{data})} \tag{3}$$

As depicted in Fig. 2, in our approach, we set the input window size to 100 and the output window size to 25. The choice of these window sizes is strategic; the input window of 100 allows the model to consider a substantial sequence of data points, providing a comprehensive view of the battery's behavior leading up to the current state. This size ensures that the model has enough context to understand the temporal dynamics of SOC changes. The output window of 25, on the other hand, enables the model to predict the SOC for the next 25 time intervals based on the input sequence, offering a detailed forecast that can be invaluable for real-time battery management and planning. More interestingly, in the MIT dataset, the size of 25 corresponds to the SOC values for the next two minutes, whereas in the INSA dataset, it represents the SOC for the following minute.

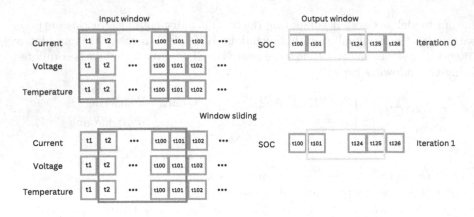

Fig. 2. The sliding window technique.

4.3   Model Architecture

In this work, we propose a data-driven model that leverages the power of CNNs to estimate the SOC for lithium-ion battery cells. CNNs are renowned for their efficacy in processing and analyzing structured grid data, making them ideally suited for interpreting time series data, such as the SOC estimation from battery cycles. Here, we explain the fundamental building blocks of our CNN architecture and their roles within the model:

- *Input layer*: the first layer that receives the raw input data.
- *Convolutional layer*: this layer applies convolutional operations to the input data allowing to capture spatial hierarchies and features d
- *Activation layer*: following convolution, an activation function, commonly the Rectified Linear Unit (ReLU), is applied to introduce non-linearity and enhance the model's ability to make predictions on previously unseen data.
- *Pooling layer*: down-sample the spatial dimensions of the input data, reducing its computational complexity. Max pooling and average pooling are common techniques used in this layer.
- *Fully connected layer*: neurons in this layer are connected to all neurons in the previous layer, resembling a traditional neural network. It helps in learning global patterns and their relationships.
- *Flattening layer*: before entering the fully connected layer, the multi-dimensional data is flattened into a one-dimensional vector. This step prepares the data for the fully connected layers.
- *Output layer*: the final layer outputs the SOC estimation.

In our CNN model, we employ the ReLU activation function[1] in all layers except the final layer, where a Sigmoid activation function[2] is utilized. The architecture of this model, as depicted in Fig. 3, reflects this design choice with the parameters employed for each layer.

---

[1] https://www.tensorflow.org/api_docs/python/tf/keras/activations/relu.
[2] https://www.tensorflow.org/api_docs/python/tf/keras/activations/sigmoid.

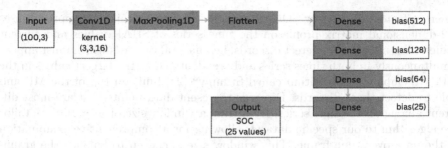

Fig. 3. Architecture of the proposed CNN model.

4.4 Explaining Predictions

Efforts to enhance the interpretability and transparency of deep learning models, particularly complex CNNs for SOC estimation, have led to the integration of XAI techniques. These methods aim to elucidate the decision-making process of models, bridging the gap between advanced computational algorithms and human understanding. Among the various XAI methodologies, SHAP [14] stands out for its comprehensive approach to quantifying the influence of each feature on the model's output. Our idea to explaining predictions involves employing a post-hoc SHAP model applied to the output of our model. This allows us to obtain the contributions of the three features: current (I), voltage (V), and temperature (T) toward the SOC estimation. This analysis is visually represented in Fig. 4, where each feature's impact on the CNN model's predictions is clearly illustrated.

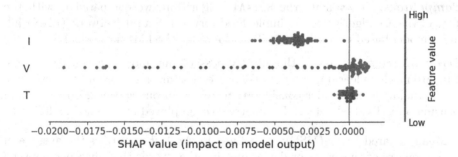

Fig. 4. Feature importance analysis with SHAP.

Upon analysis, we observed that among all the time series provided in the input, only the current exhibits the highest contribution to the predictions. Therefore, we aim to construct a time series using the SHAP values of the current.

Detecting Discords. Detecting discords within time series data is a crucial aspect of interpreting complex patterns, especially when assessing the impact of different factors on battery SOC estimation. To achieve this, our approach uses a 2-dimensional matrix profile on the time series of SHAP values and current values. This allows us to detect discords, i.e., natural discords which must appear simultaneously in both times series and are identified as the largest values in the MP. We employ an algorithm called Stumpy[3] [24] built on top of the MP and able to detect these discords. Discords represent data points that are most different among all the time series. By setting a window size of $m = 100$, we tailor the algorithm to our specific dataset, allowing for a comprehensive examination of the data over time frame. This window size is chosen to balance the granularity of analysis with computational efficiency, ensuring that the algorithm can effectively detect discords without being hindered by excessive detail or data volume.

5   Experimental Evaluation

This section outlines the experimental setup, the comparative analysis with existing state-of-the-art models, and the metrics employed to assess performance.

5.1   Experimental Protocol

Implementation. All experiments were conducted on a machine with an Intel Core i5 12th generation CPU, a NVIDIA GeForce RTX 3070 GPU with 6 GB of VRAM, 32 GB of RAM, and a 512GB SSD. This machine provided the necessary computing power to train and test the models efficiently and effectively.

Competitors. To evaluate the SOCXAI algorithm, we compared it with the state-of-the-art algorithms: a simple Feed-forward Neural Network (FNN) [6], and a model based on Long Short-Term Memory (LSTM) networks [2].

Error Metrics. To assess the effectiveness of our model, we utilized three standard machine learning metrics. These metrics take into account the complete set of window values under consideration, which in our case is 25, rather than focusing on individual values. The metrics we employed are defined as follows:

- Mean Squared Error (MSE): As defined in Eq. 4, it quantifies the average of the squares of the errors or deviations, in other words the difference between the estimator and what is estimated. The MSE is calculated as follows (where n represents the total number of samples):

$$MSE = \frac{1}{n} \sum_{i=1}^{n} (predicted\ value_i - observed\ value_i)^2 \qquad (4)$$

---

[3] https://stumpy.readthedocs.io/en/latest/#.

– Mean Absolute Error (MAE): As defined in Eq. 5, the MAE measures the average absolute difference between the predicted and actual values. This metric offers insight into the magnitude of errors in the model's predictions:

$$MAE = \frac{\sum_{i=1}^{n} |predicted\ value_i - observed\ value_i|}{n} \tag{5}$$

– Root Mean Squared Error (RMSE): As defined in Eq. 6, it is the square root of the MSE and is commonly used to measure the average error between the predicted and actual values in the same units as the original data:

$$RMSE = \sqrt{\frac{\sum_{i=1}^{n} (predicted\ value_i - observed\ value_i)^2}{n}} \tag{6}$$

5.2 Results

SOC Estimation. In Table 1, we conducted a comparative analysis between SOCXAI model and other baseline models, namely FNN and LSTM, using the dataset presented in Sect. 4.1.

Table 1. SOC estimation model performance results in dataset.

Metric	SOCXAI	FNN	LSTM
MAE	0.0143	0.0510	0.0201
MSE	0.0016	0.0065	0.0021
RMSE	0.040	0.0809	0.0464

We evaluated the models based on error metrics including MSE, MAE, and RMSE. The experimental results demonstrate that our model outperforms the baseline models across all error metrics, marking a threefold improvement in accuracy over the FNN model. This performance is attributed to the utilization of CNN layers, which excel in capturing data dependencies and reducing model complexity through by using fewer parameters compared to LSTM layers.

Figure 5 and Fig. 6 depict the SOC estimation values generated by the models alongside the true SOC values for two randomly selected cycles from the MIT and INSA test sets, respectively. The x-axis represents the time, while the y-axis represents the SOC values which range from 0 to 1, representing 0% to 100% charge. These figures clearly demonstrate that the SOCXAI and LSTM models provide estimations closely aligned with the true SOC values, significantly outperforming the FNN model. We note here that the proposed model can be generalized to other types of lithium batteries; one simply needs to retrain it.

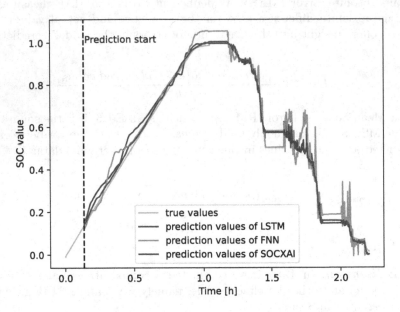

Fig. 5. Comparison of model prediction with LSTM and FNN on INSA driving cycle.

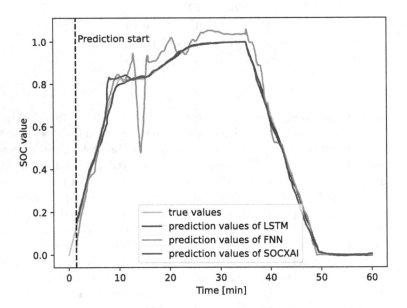

Fig. 6. Comparison of model prediction with LSTM and FNN on MIT driving cycle.

Explaining Predictions. As noted in [15], the SHAP model may sometimes provide imprecise or misleading assessments of relative feature importance, particularly failing to capture inter-feature relationships. Our investigation focuses on this aspect, especially in the context of regression problems related to SOC estimation. Consequently, if the SHAP method highlights significant contributions of current values within identified anomalies, we intend to evaluate the consistency between SHAP assessments and the regions within the current values indicative of anomalies. This approach enhances our understanding of battery behavior and facilitates effective anomaly detection. Our findings confirm that the SHAP model effectively explains the presence of anomalies as shown in Fig. 7 (top), indicating deviations from expected patterns. Moreover, the absence of conserved behavior in the time series underscores the efficacy of SHAP in elucidating abnormal occurrences. Notably, the contributions of SHAP values slightly increase within the subsequences where anomalies are detected, further affirming the model's ability to capture and explain these irregularities. Practically, one explanation for the abnormal behavior, as illustrated after zooming into the discord in Fig. 7 (bottom), of the battery could be attributed to a voltage measurement issue. This issue affects the current values as the driving cycle progresses until it reaches a voltage threshold. However, the rate at which the voltage reaches this threshold can vary, indicating the inconsistent behavior of the current time series values.

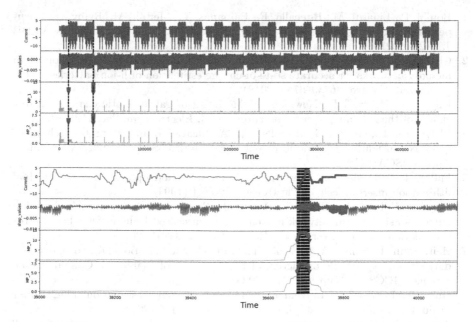

Fig. 7. Illustration of 2-dimensional anomaly detection of current and SHAP values for subsequence m = 100.

6   Conclusion and Perspectives

In this paper, we introduced a novel algorithm named SOCXAI, aimed at estimating the SOC of batteries. This algorithm distinguishes itself by its capability to predict not just a single future SOC value but 25 future values. Its application extends beyond simple constant discharge scenarios to encompass real-world driving cycles and various charging policies. Furthermore, it provides explanations for these predictions using the SHAP model. Additionally, we proposed an anomaly detection method using the concept of natural anomalies, highlighting abnormal battery usage patterns that deviate from expected behavior.

In future work, we aim to enhance the performance of the proposed method by exploring more advanced techniques such as utilizing a sliding window with a dynamically varying size. Furthermore, we intend to expand this model to deal with a new type of battery known as sodium-ion batteries.

Acknowledgment. This research received partial funding from the French National Research Agency (ANR) under the project 'ANR-22-CE92-0007-02'. Additionally, support was provided by the European Union through the Horizon Europe program and the innovation program under 'GAP-101103667'.

References

1. Boniol, P., Linardi, M., Roncallo, F., Palpanas, T., Meftah, M., Remy, E.: Unsupervised and scalable subsequence anomaly detection in large data series. VLDB J. 1–23 (2021)
2. Chemali, E., Kollmeyer, P.J., Preindl, M., Ahmed, R., Emadi, A.: Long short-term memory networks for accurate state-of-charge estimation of li-ion batteries. IEEE Trans. Ind. Electron. 6730–6739 (2018)
3. Doyle, M., Fuller, T.F., Newman, J.: Modeling of galvanostatic charge and discharge of the lithium/polymer/insertion cell. J. Electrochem. Soc. 1526 (1993)
4. El Khansa, H., Gervet, C., Brouillet, A.: Application of matrix profile techniques to detect insightful discords in climate data. Int. J. Soft Comput. Artif. Intell. Appl. (IJSCAI) (2022)
5. Fuller, T.F., Doyle, M., Newman, J.: Simulation and optimization of the dual lithium ion insertion cell. J. Electrochem. Soc. 1 (1994)
6. He, W., Williard, N., Chen, C., Pecht, M.: State of charge estimation for li-ion batteries using neural network modeling and unscented Kalman filter-based error cancellation. Int. J. Electr. Power Energy Syst. 783–791 (2014)
7. Heitzmann, T., Samet, A., Mesbahi, T., Soufi, C., Jorge, I., Boné, R.: Sochap: a new data driven explainable prediction of battery state of charge. In: Computational Science – ICCS 2023, pp. 463–475 (2023)
8. Huria, T., Ludovici, G., Lutzemberger, G.: State of charge estimation of high power lithium iron phosphate cells. J. Power Sources, 92–102 (2014)
9. Johnson, V.: Battery performance models in advisor. J. Power Sources, 321–329 (2002)
10. Kashpruk, N., Piskor-Ignatowicz, C., Baranowski, J.: Time series prediction in industry 4.0: a comprehensive review and prospects for future advancements. Appl. Sci. (2023)

11. Lee, J., Sun, H., Liu, Y., Li, X.: A machine learning framework for remaining useful lifetime prediction of li-ion batteries using diverse neural networks. Energy AI, 100319 (2024)
12. Li, G., Jung, J.J.: Deep learning for anomaly detection in multivariate time series: approaches, applications, and challenges. Inf. Fusion, 93–102 (2023)
13. Linardi, M., Zhu, Y., Palpanas, T., Keogh, E.: Matrix profile x: Valmod-scalable discovery of variable-length motifs in data series. In: Proceedings of the 2018 International Conference on Management of Data, pp. 1053–1066 (2018)
14. Lundberg, S.M., Lee, S.I.: A unified approach to interpreting model predictions. In: Advances in Neural Information Processing Systems (2017)
15. Marques-Silva, J., Huang, X.: Explainability is not a game. arXiv preprint arXiv:2307.07514 (2023)
16. Nakamura, T., Imamura, M., Mercer, R., Keogh, E.: Merlin: parameter-free discovery of arbitrary length anomalies in massive time series archives. In: 2020 IEEE International Conference on Data Mining (ICDM), pp. 1190–1195 (2020)
17. Plett, G.L.: Extended Kalman filtering for battery management systems of lipb-based HEV battery packs: Part 3. State and parameter estimation. J. Power Sources, 277–292 (2004)
18. Severson, K.A., et al.: Data-driven prediction of battery cycle life before capacity degradation. Nat. Energy, 383–391 (2019)
19. Stefanopoulou, A., Kim, Y.: System-level management of rechargeable lithium-ion batteries. Rechargeable Lithium Batteries, 281–302 (2015)
20. Tafazoli, S., Keogh, E.: Matrix profile xxviii: discovering multi-dimensional time series anomalies with k of n anomaly detection. In: Proceedings of the 2023 SIAM International Conference on Data Mining (SDM), pp. 685–693 (2023)
21. Tian, J., Chen, C., Shen, W., Sun, F., Xiong, R.: Deep learning framework for lithium-ion battery state of charge estimation: Recent advances and future perspectives. Energy Storage Mater. 102883 (2023)
22. Yan, Q.: SOC prediction of power battery based on SVM. In: 2020 Chinese Control And Decision Conference (CCDC), pp. 2425–2429 (2020)
23. Yeh, C.C.M., et al.: Matrix profile I: all pairs similarity joins for time series: a unifying view that includes motifs, discords and shapelets. In: 2016 IEEE 16th International Conference on Data Mining (ICDM), pp. 1317–1322 (2016)
24. Zhu, Y., et al.: Matrix profile ii: exploiting a novel algorithm and GPUs to break the one hundred million barrier for time series motifs and joins. In: 2016 IEEE 16th International Conference on Data Mining (ICDM), pp. 739–748 (2016)

Towards Detection of Anomalies in Automated Guided Vehicles Based on Telemetry Data

Paweł Benecki[1]([✉])[ID], Daniel Kostrzewa[1][ID], Marek Drewniak[3][ID], Bohdan Shubyn[1,2][ID], Piotr Grzesik[1][ID], Vaidy Sunderam[4][ID], Boleslaw Pochopien[5][ID], Andrzej Kwiecien[6][ID], Bozena Malysiak-Mrozek[6][ID], and Dariusz Mrozek[1][ID]

[1] Department of Applied Informatics, Silesian University of Technology, Gliwice, Poland
pawel.benecki@polsl.pl
[2] Department of Telecommunications, Lviv Polytechnic National University, Lviv, Ukraine
[3] AIUT Sp. z o.o. (Ltd.), Gliwice, Poland
mdrewniak@aiut.com
[4] Department of Computer Science, Emory University, Atlanta, GA 30322, USA
[5] Department of Graphics, Computer Vision and Digital Systems, Silesian University of Technology, Gliwice, Poland
[6] Department of Distributed Systems and Informatic Devices, Silesian University of Technology, Gliwice, Poland

Abstract. The rapid evolution of smart manufacturing and the pivotal role of Automated Guided Vehicles (AGVs) in enhancing operational efficiency, underscore the necessity for robust anomaly detection mechanisms. This paper presents a comprehensive approach to detecting anomalies based on AGV telemetry data, leveraging the potential of machine learning (ML) algorithms to analyze complex data streams and time series signals. By focusing on the unique challenges posed by real-world AGV environments, we propose a methodology that integrates data collection, preprocessing, and the application of specific AI/ML models to accurately identify deviations from normal operations. Our approach is validated through extensive experiments on datasets featuring anomalies caused by mechanical wear or excessive friction and issues related to tire and wheel damage, employing LSTM and GRU networks, alongside traditional classifiers like K-nearest neighbors and SVM. The results demonstrate the efficacy of our method in forecasting momentary power consumption as an indicator of mechanical anomalies, and in classifying wheel-related issues with high accuracy. This work not only contributes to the enhancement of predictive maintenance strategies but also provides valuable insights for the development of more resilient and efficient AGV systems in smart manufacturing environments.

Keywords: automated guided vehicles · anomaly detection · telemetry anomaly detection · machine learning

L. Franco et al. (Eds.): ICCS 2024, LNCS 14838, pp. 192–207, 2024.
https://doi.org/10.1007/978-3-031-63783-4_15

1   Introduction

The smart manufacturing industry frequently relies on automated component or manufactured products delivery performed by Automated Guided Vehicles (AGVs) [9]. These vehicles autonomously transport goods to or between assembly stations on production lines, await unloading after successful docking, and then return to begin another operational cycle. The uninterrupted production in manufacturing that relies on a fleet of AGVs necessitates continuous monitoring of the vehicles and their characteristics through telemetry, detecting anomalies and, thus, predicting upcoming failures [23]. These tasks would not be possible without observing various signals captured by onboard gauges and sensors. Small IoT devices installed on AGVs may collect data, communicate with other devices and central servers to exchange the data, and perform more sophisticated analyses supporting predictive maintenance tasks [8, 25]. They can gather signal values and sensor readings directly from the PLC controllers onboard the AGV, transmitting them to analytical systems for further analysis.

Deviations from the normal operation of an AGV are typical indicators of wear of its components, adverse environmental impact, or human error or influence [24]. For example, the progressive change in the wheel diameter resulting from its wear has a negative impact on the accuracy of the planned route in the odometry system. The slippery ground may result in inaccurate turns and route errors or incorrect docking. Excessive vehicle load causes increased energy consumption, reduces the use time of AGVs, and increases the risk of wear of their parts. Frequent stops caused by people entering the vehicle's route not only disrupt the entire delivery schedule but also cause parts to wear out faster. All these deviations can be captured by prior observation of various internal signals of the vehicle and their subsequent analysis.

Recent works in anomaly detection predominantly rely on employing Artificial Intelligence (AI) algorithms and analyzing data streams with various Machine Learning (ML) models [11]. However, despite the general inference capabilities of the existing AI/ML algorithms, each detection task requires a separate reasoning model that should be deployed specifically for the problem and particular data preparation. Detecting anomalies in real AGV environments that produce industrial data streams and expose time series signals poses unique challenges beyond simple training and testing various ML models [22].

The paper advances smart manufacturing by introducing ML-based methods for detecting anomalies in AGVs, focusing on mechanical wear and tire or wheel damage. It fills a gap in predictive maintenance, enhancing AGV operational efficiency and reliability. By providing accurate forecasts of power consumption and classifying wheel-related issues, this research not only improves current AGV management but also sets a foundation for future advancements in the field.

The remainder of this paper is organized as follows: Sect. 2 reviews the current literature on anomaly detection in AGVs, highlighting the key methodologies and gaps that this study aims to address. Section 3 delves into the proposed methodology for anomaly detection, including data collection, pre-processing and specific AI/ML models used. Section 4 describes the experimental datasets in detail, covering their collection, characterization and justification for their use. Section 5 presents the results of our anomaly detection experiments, offering insights into the performance and effectiveness of the models. Section 6 reflects the results, discussing their implications for smart manufacturing and AGV operations, and outlines future research directions.

2   Related Works

The growing adoption of AGVs across diverse sectors, particularly in logistics and manufacturing, has sparked substantial research endeavors aimed at improving their operational efficiency and dependability. A key focus of these endeavors is the advancement of sophisticated techniques for anomaly detection based on AGV telemetry data, optimization of energy consumption, and the utilization of ML for predictive analytics on time series data. This section delves into the state-of-the-art methodologies and their implications for AGV technology.

The capability to detect anomalies based on AGV telemetry data is critical for predictive maintenance (PdM) and operational efficiency. The study by Malhotra et al. stands out for its pioneering use of LSTM networks, offering a robust framework for identifying anomalies in time-series data [14]. Complementing this, Hundman et al. explored the use of LSTMs and nonparametric dynamic thresholding for detecting spacecraft anomalies, illustrating the potential of these methods in complex operational contexts akin to AGV environments [12]. The versatility of LSTM models in capturing temporal dependencies makes them particularly suited for AGV telemetry, where anomalies must be detected in real-time to prevent operational disruptions.

Efficient energy management is crucial for sustainable AGV operations. The work by Khan et al. in modeling AGV energy consumption laid the groundwork for integrating predictive analytics into energy management strategies [17]. The application of ML for time series anomaly detection transcends the specific use case of AGVs, offering a wealth of methodologies that can be tailored to this context. Lai et al. presented a comprehensive approach using Recurrent Neural Networks (RNNs) for modeling both long- and short-term temporal patterns while analyzing AGV telemetry data [13]. Additionally, the review by Zhang et al. on deep learning for financial time series provides a solid foundation for adopting similar techniques in the operational analysis of AGVs [1].

Integration of edge computing and IoT is crucial for enhancing real-time data processing capabilities. The study by Shi et al. emphasizes the role of edge computing in facilitating the real-time analysis of AGV data, thereby enabling more immediate and localized decision-making processes [20]. This approach significantly reduces latency in anomaly detection and energy consumption optimization, essential for maintaining continuous and efficient AGV operations. In addition to ML and edge computing, the application of Federated Learning (FL) techniques in monitoring AGV is gaining interest. FL allows for the decentralized processing of data, enabling AGVs to learn from distributed data sources without the need to centralize sensitive information. Our previous studies [21] were focused on the effectiveness of using FL for AGV to improve the forecast of signals in time and the effectiveness of this approach in terms of energy consumption. We also used LSTM networks to forecast momentary energy consumption (MEC), the signal identified as the one reflecting possible anomalies in the AGV operation. However, none of these works were focused on specific anomalies of the AGVs working within their operational environment. This paper extends these efforts toward real anomaly detection.

3   Anomaly Detection Based on AGV Telemetry Data

Anomaly detection in AGVs requires telemetry and exchanging data between various IT systems within a factory. The topology of the system composition is usually complex, and anomalies usually occur rarely. These factors hinder the development of accurate anomaly detection models but can be mitigated by extracting appropriate data, integrating them, and performing appropriate experimental scenarios engaging AGV vehicles.

3.1   Data in Intralogistics Systems

Modern intralogistic systems are composed not only of the hardware layer, which consists of AGVs and other types of transportation robots. IT systems which enable various operability functions are equally important. One of the most crucial is the Transportation Management System class (TMS), which evolved from the fleet management system. Among solving traffic problems, its tasks are related to formulating transportation orders, selecting and dispatching AGVs for specific tasks, managing the logic of transportation flow, scheduling and reporting tasks as well as providing detailed diagnostics of the state and behavior of the fleet. In parallel, the fleet of AGVs needs to communicate and cooperate with industrial environments and infrastructure. Therefore, integration with existing industrial third-party systems, such as Warehouse Management Systems (WMS), Manufacturing Execution Systems (MES), Business Intelligence (BI), or Computerized Maintenance Management Systems (CMMS), is required.

Intralogistics systems operate on many communication layers, like data exchange with field devices, acquisition of traceability, process data, or generating asset management information. Due to this fact, a vast amount and diversity of data can be used for multiple cases, from general process control tasks through optimization of transportation orders based on current utilization and energy consumption of the fleet to support the maintenance, e.g., via the calculation of Overall Equipment Effectiveness (OEE) or utilization in predictive maintenance approaches based on data mining.

3.2   Overview of the Methodology

Anomaly detection in the AGV operation follows the general methodology illustrated in Fig. 1. The approach begins with the stage of data generation from

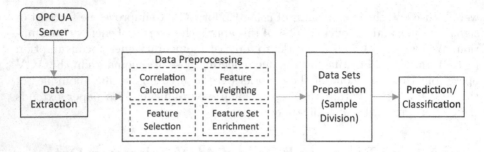

Fig. 1. Overview of the general methodology applied for predicting anomalies in the AGV operation.

AGV onboard devices and systems as well as from the fleet supervisory layer. Signals are acquired either directly from measurement devices such as battery management systems providing momentary currents, voltages, and temperatures of the power system or from process controllers, which generate statuses, diagnostic information, or states of work of AGVs. An important factor is that data generation and collection from multiple endpoints are synchronized in time. This guarantees that, e.g., calculations of MPC are reliable and refer to corresponding statuses and states of work of AGV and its devices. The methodology covers capturing data from intralogistics systems and exposing them through the OPC UA Server, from which they are extracted in the form of a collection of time series signals (a wide time-dependent data stream). Then, we perform a data preprocessing step that may include finding correlations between signals in the data stream, feature selection, and weighting based on the calculated correlation coefficients, and feature set enrichment by deriving other features based on existing ones or based on a wider view of feature's values (e.g., within a time window). This phase is followed by the preparation of data sets, which covers more or less sophisticated strategies for data/sample division. Finally, depending on the anomaly, forecast or classification is performed, which employs dedicated AI/ML algorithms. Data preprocessing steps and data set preparation strategies depend on the specific use case for anomaly detection and the requirements of the forecasting/classification algorithms applied.

3.3   Anomalies Caused by Mechanical Problems

The operability of AGVs leads to the degradation of their components. This can be caused by many factors, but in most cases, they result from the mechanical wear of onboard components. Exposure to negative phenomena like vibrations, strokes, overloading, or working in hazardous conditions shortens the lifetime of traction systems or onboard electronics. As a result, AGVs can require more frequent maintenance, which is costly and excludes them from production. Mechanical wear, particularly in bearings and components experiencing larger friction in vehicles, involves several mechanisms, among them the following [10]:

1. Adhesive wear, occurs when surfaces weld together and tear apart, which is common in bearings.
2. Abrasive wear caused by particles or uneven sliding across a surface.
3. Corrosive wear, chemical or electrochemical reaction with the environment.
4. Fatigue wear due to cyclic loading.

Modeling or examining real worn systems is either hard or ineffective due to the time necessary to observe the actual wear [10, 15, 16]. One of the possibilities to observe similar effects is overloading a vehicle. This can indeed be used to emulate the situation where the wear has already occurred, simulating the reduced performance and altered operational characteristics that worn components would exhibit [6]. This method does not accelerate wear but instead aims to mimic the effects of wear on the vehicle's performance.

3.3.1 Modelling. We assume that a vehicle operating with payloads up until a chosen threshold is in normal operation, and payloads above the threshold are emulating more mechanical wear and friction. Therefore, we trained the forecasting models on a subset of normal data and evaluated on other non-overlapping sequences from normal data and data considered anomalous. The description of the data used in this experiment is provided in Sect. 4.1.

Models used here forecast short-term (a 10-s ahead forecast horizon Δt) MPC using features from the whole acquired telemetry. This is a common approach in such telemetry-related tasks, which allows taking measures when the expected energy consumption differs significantly from the observed values, which often indicate an anomalous event [5, 12, 19]. A concise overview of how the forecasting operates is illustrated in Fig. 2.

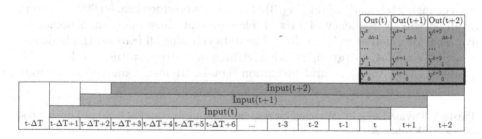

Fig. 2. Forecasting of MPC in time windows of a size ΔT moving over an input time series X. Δt elements ahead are forecasted (forecast horizon). Only the first elements y_0 of forecasts are taken to form an output sequence (elements in a bold frame).

With such an approach, it is possible to evaluate if the model forecasts values correctly during normal operations and if its performance deteriorates with more unusual patterns, which is an expected behavior here. In the context of mechanical wear and vehicles, this method allows for predictive maintenance strategies. By detecting anomalies early, maintenance can be scheduled proactively to address wear and tear before it leads to failure [7].

3.4 Anomalies Caused by Tire and Wheel Damage

One of the mechanical wears that lead to the lowered performance of AGVs is the uncontrolled and unintended change of diameter of traction wheels. In all types of kinematic models, AGVs use odometry for maneuvering. The odometry is calculated from pulses or the frequency measured by encoders installed on traction axes. This requires prior knowledge of wheel diameter, as the odometry transforms the number of pulses or the frequency into the traveled distance. Then, the distance can be used for closed-loop control when providing drive commands to motors. However, if the diameter changes due to its wear or uncontrolled change, e.g., during abrasion or sticking of dust, the odometry is calculated incorrectly. This increases the error of calculated traveled distance in time. As a result, the navigation system generates an increased number of corrections and, through this, increased consumption of energy as well as faster wear of mechanical components. To keep a good quality of navigation, it is important to control the condition of the traction wheels.

4 Datasets

The experiments were conducted using data obtained from an actual industrial CoBotAGV known as Formica. This AGV is a product developed by AIUT Ltd. and smarticized with AI by the Silesian University of Technology [22].

4.1 Test Drives with Changing Payload Weight

Our investigations rely on data acquired in October 2022 based on test drives with changing payload weight (Fig. 3). These tests produced ca. 50,000 time steps acquired with a frequency of 1 Hz. Table 1 presents how many data points are available for different weight values. The data contains 56 features (numeric and boolean) covering energy signals, left/right motor drive statuses, vehicle PLC signals, LED statuses, natural navigation signals, odometry, and safety statuses.

Table 1. Payload weight (Wt) and corresponding sample counts (# Smpl.) during test drives in October 2022

Wt [kg]	# Smpl.	Wt [kg]	# Smpl.	Wt [kg]	# Smpl.	Wt [kg]	# Smpl.	Wt [kg]	# Smpl.
0	276	100	812	200	6336	300	1943	400	756
20	860	120	915	220	888	320	1722	420	1070
40	804	140	739	240	722	340	1727	440	2190
60	807	160	893	260	676	360	1599	480	943
80	860	180	825	280	2071	380	1644	498	17564

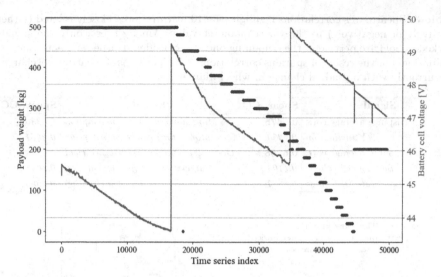

Fig. 3. Payload weight (blue) changes during test drives in October 2022. The test drives also included different levels of battery voltage (red), with two charges in between. (Color figure online)

4.2 Distorted Natural Navigation Data

This data comes from test runs executed in July and August 2023, which included changes in the diameter of one of the wheels passed to the natural navigation subsystem. The value was not changed physically but by software means, however, it resulted in natural navigation corrections anyway through mechanisms described in Sect. 3.4. The data consists of ca. 121,000 time steps acquired with a frequency of 1 Hz. Table 2 presents how many data points are available for different wheel diameters, and Fig. 4 presents a short outline of how natural navigation is distorted with false gradually changed wheel diameter.

Table 2. Wheel diameters (Ø) used during test drives in July and August 2023 with sample counts (# Smpl.) for each diameter.

Ø [mm]	# Smpl.	Ø [mm]	# Smpl.	Ø [mm]	# Smpl.	Ø [mm]	# Smpl.	Ø [mm]	# Smpl.
52.90	54494	54.75	2162	56.55	262	59.51	2449	61.51	2431
53.16	1299	54.97	1797	57.93	5031	59.77	2858	61.83	2385
53.42	1939	55.23	5340	58.19	3019	60.04	1795	62.09	2158
53.69	2195	55.49	1929	58.46	577	60.31	2053		
53.95	2065	55.76	2308	58.72	999	60.57	2035		
54.22	2516	56.03	2319	58.99	2714	60.99	2009		
54.48	2161	56.29	2321	59.23	1505	61.26	2485		

Table 3. Spearman's correlations highlight the 10 features most closely related (either positively or negatively) to the wheel diameter value. Among these, only the signals marked in bold are pertinent; the remaining ones are considered either artifacts (such as inclinations) or the result of spurious correlations, which arise as the battery discharges concurrently with a gradual change in wheel diameter.

Signal	Spear. CC	Signal	Spear. CC
Distance average corr.	**0.6475**	*Momentary current consumption*	*0.3336*
Y inclination	*0.6360*	*Momentary power consumption*	*0.2481*
Difference heading average corr.	**0.5268**	*Odometry: cumulative distance right*	*0.2253*
Battery cell voltage	*0.4103*	*Cumulative energy consumption*	*0.1674*
State Of Charge	*0.3887*	*X inclination*	*0.1593*

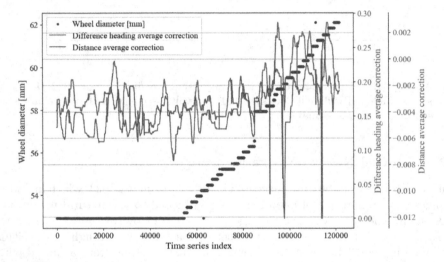

Fig. 4. Wheel diameter (blue) changes and resulting natural navigation correction values (red and green) during test drives in July and August 2023 (Color figure online)

5   Experimental Validation

This section details the methodology and experimental validation results aimed at detecting anomalies in mechanical systems and components due to wear or excessive friction and wheel degradation. Through a series of experiments utilizing recurrent neural network (RNN) models, this research investigates the efficacy of forecasting models on datasets emulating mechanical anomalies by varying payload thresholds. The experiments extend previous work by exploring the impact of model training and validation under conditions of normal and overload test drives, employing momentary power consumption (MPC) as a predictive metric. Additionally, we include the detection of wheel-related anomalies, such as excessive tire wear. Utilizing traditional two-class classifiers, we aim to thoroughly assess and enhance the system's diagnostic capabilities concerning wheel integrity.

5.1   Detecting Potential Mechanical Wear or Excessive Friction

The first series of experiments was focused on finding anomalies caused by mechanical wear or excessive friction. As mentioned in Sect. 3.3 those were emulated by changing payload over the chosen threshold. The input dataset described in Sect. 4.1 was used to train a 2-hidden layers (with 80 units per layer) LSTM (Long Short-Term Memory) and 2-hidden layers (with 80 units per layer) GRU (Gated Recurrent Unit) models forecasting MPC, which proved to be effective in our previous works [2–4]. Similarly, the input window size was set to $\Delta T = 50$ elements and the forecast horizon to $\Delta t = 10$.

The objective of this fragment of experimental validation was to check if it is possible to train and validate a model that gives good forecasting quality on normal data (without excessive load) and the quality of its results deteriorates on data resulting from overload test drives. To achieve that, the dataset described in Sect. 4.1 was split into fragments based on payload weight, each fragment ranging 40 kg of payload. The structure of the resulting division is shown in Table 4. A fixed boundary in payload weight was set first to 200 kg and in a second part to 320 kg. Data below the thresholds was treated as *normal* and above – as *anomalous*.

The models were trained on part of data from *normal* ranges: 0–40, 80–120, 160–200 for the first and 0–40, 80–120, 160–200, and 240–280 for the second part. The rest of data from *normal* was treated as *normal test* set and all data above the threshold was used as *anomalous test* set. The training was executed for 200 epochs with early stopping after 20 epochs without improvement of the loss function.

The forecasting quality was evaluated using the Mean Square Error (MSE) metric computed over the resulting output sequence as shown in Fig. 2. Additionally, input data was preprocessed to 1) contain or not the MPC feature [4] and 2) use or not feature weighting [3]. That, together with two models (LSTM/GRU), gave eight experiments for the two previously mentioned dataset splits. Additionally, to assess whether the resulting forecasting errors were statistically significantly different from forecasts on *normal test* fragments, the Student's t-test was conducted.

The results are presented in Table 5. It can be observed that errors between expected (forecasted) and actual values of MPC are significantly larger than in *normal test* set in most of the ranges in *anomalous test* sets, especially for the experiment with a threshold set to 320 kg. That would allow to employ an error-thresholded anomaly detection [12]. Also, for most of *normal test* payload ranges, error values do not differ significantly from the expected values. However, for cases where this does not stand (e.g., range 280–320 in Table 6), the errors still can be thresholded not to report false positive anomalies. Also, it was confirmed that the weighting of features contributes to better forecasting [4]. Additional processing (e.g., error thresholding [12]) is required after forecasting to achieve the final anomaly detection result. Also, this method allows for near real-time processing, since similarly as in [4], the processing time for the whole test set is ca. 0.1–0.2 s (the machine used is equipped with an AMD Ryzen 5 processor, NVIDIA GeForce GTX 1080 Ti GPU, and 16 GB of installed RAM).

Table 4. Payload weight (Wt) distribution in the dataset in terms of sample counts (# Smpl.).

Wt [kg]	# Smpl.	Wt [kg]	# Smpl.	Wt [kg]	# Smpl.	Wt [kg]	# Smpl.
0–40	1940	160–200	7161	320–360	3326	480–500	17564
40–80	1667	200–240	1610	360–400	2400		
80–120	1727	240–280	2747	400–440	3260		
120–160	1632	280–320	3665	440–480	943		

Table 5. Results of forecasting for models trained on data assuming payloads < 200 kg to be normal. Numbers stand for MSE computed on values inferred on test sequences. Greater > /less than < marks denote values statistically significantly different from the *normal test* set and the direction of inequality. Ranges 40–80 and 120–160 constitute a *normal test* set, thus, they are presented separately.

MPC present	Weighted	Model	MSE - normal test			MSE - anomalous test							
			Whole	40-80	120-160	200-240	240-280	280-320	320-360	360-400	400-440	440-480	480-500
−	−	GRU	0.095	0.091	0.099	0.117	0.113	0.173 >	0.186 >	0.243 >	0.371 >	0.424 >	0.542 >
−	−	LSTM	0.124	0.152	0.096 <	0.156 >	0.171 >	0.218 >	0.212 >	0.302 >	0.498 >	0.418 >	0.511 >
−	+	GRU	0.104	0.100	0.108	0.091	0.093	0.138 >	0.169 >	0.223 >	0.343 >	0.381 >	0.506 >
−	+	LSTM	0.121	0.106	0.136	0.114	0.121	0.153 >	0.195 >	0.245 >	0.373 >	0.372 >	0.522 >
+	−	GRU	0.088	0.089	0.087	0.120 >	0.103	0.140 >	0.203 >	0.229 >	0.379 >	0.509 >	0.518 >
+	−	LSTM	0.139	0.163	0.113	0.167	0.142	0.148	0.211 >	0.250 >	0.458 >	0.509 >	0.536 >
+	+	GRU	0.093	0.083	0.104	0.088	0.175 >	0.156 >	0.202 >	0.175 >	0.335 >	0.317 >	0.439 >
+	+	LSTM	**0.082**	0.079	0.086	0.091	0.089	0.137 >	0.189 >	0.213 >	0.342 >	0.427 >	0.551 >

5.2   Detecting Problems with Wheels

This experimental section fragment focuses on finding anomalies that are caused by excessive tire wear or objects that are accidentally attached to the wheel surface. The test data is described in Sect. 4.2 and relies on distortions in the natural navigation subsystem (NN). As it was described previously, the problems are visible through larger amounts of corrections reported by the NN.

Table 6. Results of forecasting for models trained on data assuming payloads $< 320\,\mathrm{kg}$ to be normal. Numbers stand for MSE computed on values inferred on test sequences. Greater $>$ / less than $<$ marks denote values statistically significantly different from the *normal test* set and the direction of inequality. Ranges 40–80, 120–160, 200–240, and 280–320 constitute the *normal test* set, thus, they are presented separately.

MPC present	Weighted	Model	Whole	40–80	120–160	200–240	280–320	320–360	360–400	400–440	440–480	480–500
					MSE - *normal test*				MSE - *anomalous test*			
–	–	GRU	0.119	0.096 <	0.112	0.112	0.135 >	0.199 >	0.299 >	0.531 >	0.413 >	0.537 >
–	–	LSTM	0.122	0.117	0.102	0.123	0.134	0.213 >	0.264 >	0.463 >	0.413 >	0.497 >
–	+	GRU	0.112	0.110	0.089 <	0.101	0.127 >	0.168 >	0.238 >	0.384 >	0.410 >	0.541 >
–	+	LSTM	0.145	0.161	0.139	0.126	0.149	0.200 >	0.245 >	0.533 >	0.723 >	0.516 >
+	–	GRU	0.114	0.094 <	0.098	0.095 <	0.139 >	0.167 >	0.213 >	0.312 >	0.348 >	0.468 >
+	–	LSTM	0.104	0.084 <	0.109	0.092	0.117	0.192 >	0.220 >	0.329 >	0.332 >	0.459 >
+	+	GRU	0.095	0.076 <	0.096	0.080 <	0.110 >	0.154 >	0.183 >	0.309 >	0.330 >	0.484 >
+	+	LSTM	**0.094**	0.062 <	0.082	0.088	0.115 >	0.158 >	0.200 >	0.337 >	0.357 >	0.468 >

We employed traditional two-class classifiers such as K-nearest neighbors, Naive Bayes, Decision Tree, Random Forest, and Support Vector Machines (SVM). Hyperparameters of the models were not tuned – defaults from SciKit-Learn [18] were used. The key component of classification is to properly select features that are fed to the classifier [26, 27]. Since we have the prior knowledge that anomalies in the dataset are arising with more modified wheel diameter, it was expedient to use the features that are most correlated with the diameter. As shown in Table 3, the most correlated features are corrections from NN together with some other falsely correlated signals. Thus, we decided to use NN correction signals as input for the classifiers, namely *distance average correction* and *difference heading average correction*.

During experiments, it was also planned to check whether feature engineering improves the classification. Thus, two additional moving mean [27] of size 300 were added for both features, and different feature set configurations were passed to the classifiers. The data was labeled as a "normal" class where the diameter of the wheel was not changed and "abnormal"in other cases (ca. 45%/55% of data points respectively). 20% of all data was taken as a training set, and the remaining 80% was used as a test set. Such a setting was enough to train a well-working model and evaluate it on a broader test set. Different combinations of features mentioned above were examined. Results for classifiers are reported in Table 7.

It can be noted that classification works very well using basic classifiers, such as K-nearest neighbors. More sophisticated methods like SVM were not performing well here. Even using single correction features results in 91–94% accuracy, combining them gives almost 99%, and adding the moving average increases the classification result to nearly 1.0. The limitation of such an approach is related

Table 7. Results of classification to normal/abnormal wheel diameter based on natural navigation data. The first four columns denote whether a feature is present in the input data. DHAC – difference heading average correction, DAC – distance average correction, MA stands for moving average. The best results in each column are in bold.

DHAC	DHAC(MA)	DAC	DAC(MA)	K-nn accuracy	K-nn AUC-ROC	Gauss. NB accuracy	Gauss. NB AUC-ROC	Dec. tree accuracy	Dec. tree AUC-ROC	Rand. forest accuracy	Rand. forest AUC-ROC	SVM accuracy	SVM AUC-ROC
+	−	−	−	0.915	0.964	0.592	0.685	0.936	0.935	0.936	0.968	0.672	0.732
−	+	−	−	0.732	0.821	0.590	0.684	0.698	0.694	0.698	0.805	0.669	0.732
−	−	+	−	0.926	0.970	0.712	0.770	0.943	0.946	0.943	0.973	0.713	0.818
−	−	−	+	0.807	0.885	0.711	0.770	0.767	0.764	0.767	0.872	0.712	0.818
+	+	−	−	0.927	0.978	0.632	0.686	0.956	0.956	0.963	0.992	0.661	0.739
+	−	+	−	0.986	0.996	**0.723**	0.770	0.987	0.987	0.989	0.999	0.770	0.863
+	−	−	+	0.980	0.996	0.722	0.771	0.987	0.987	0.990	0.999	0.768	0.864
−	+	+	-	0.976	0.996	0.718	0.770	0.986	0.985	0.990	0.999	0.762	0.866
−	+	−	+	0.981	0.997	0.720	0.771	0.976	0.976	0.982	0.998	0.772	0.867
−	−	+	+	0.944	0.984	0.711	0.771	0.964	0.964	0.972	0.994	0.717	0.819
+	+	+	−	0.997	0.999	0.717	0.756	0.994	0.994	0.998	1.000	0.773	0.862
+	+	−	+	0.998	**0.999**	0.719	0.756	0.994	0.994	0.998	1.000	0.766	0.866
+	−	+	+	0.998	0.999	0.713	**0.784**	0.994	0.994	0.998	1.000	0.774	0.867
−	+	+	+	0.997	0.999	0.714	0.784	0.994	0.994	0.998	1.000	0.772	0.874
+	+	+	+	**0.999**	**0.999**	0.716	0.772	**0.996**	**0.996**	**0.999**	**1.000**	**0.775**	**0.880**

to the possibility of real-time operation. Although the classification requires no history (sequence) of samples, implicitly using the moving average of length L involves a delay of $L/2$ time steps to process the classification. However, if it is possible to stay with ca. 1% point lower accuracy, then real-time processing is feasible (i.e., for the combination $+ - +-$ in Table 7).

6 Discussion and Conclusions

This study demonstrated the application of machine learning techniques for the detection of anomalies based on Automated Guided Vehicle (AGV) telemetry data, a critical aspect of maintaining operational efficiency in smart manufacturing environments. Through the application of LSTM and GRU models, along with traditional machine learning classifiers, we have addressed two significant types of anomalies that can affect AGVs: mechanical wear or excessive friction and tire or wheel damage.

The experiments conducted here showed that the proposed approach can forecast momentary power consumption and thus be used to find potential mechanical issues. Similarly, the classification of wheel-related anomalies achieved very good accuracy, highlighting the effectiveness of our feature selection and engineering approach. These results underscore the potential of machine learning in enhancing predictive maintenance strategies, thereby reducing downtime and improving the reliability of AGVs in industrial settings.

However, the implementation of such systems does not come without challenges. The collection and preprocessing of telemetry data require careful consideration to ensure the quality and relevance of the information being analyzed. Additionally, the dynamic nature of manufacturing environments means that models must be continually updated and refined to adapt to new conditions and anomalies. The limitation of our research is that changes in power consumption can be due to other reasons than mechanical issues. Addressing that topic would need additional thorough research.

In conclusion, this research contributes valuable insights into detecting anomalies in AGV operations, offering an approach to improving the resilience and efficiency of smart manufacturing systems. Future work could focus on expanding the types of anomalies detectable by our models, improving the automation of data preprocessing, and exploring the integration of these models into real-time AGV management systems for immediate anomaly detection and response.

Acknowledgments. The research was supported by the Norway Grants 2014-2021 operated by the National Centre for Research and Development under the project "Automated Guided Vehicles integrated with Collaborative Robots for Smart Industry Perspective" (Project Contract no.: NOR/POL-NOR/CoBotAGV/0027/2019-00), the proquality grant (02/100/RGJ23/0026) of the Rector of the Silesian University of Technology, Gliwice, Poland, and by Statutory Research funds of the Department of Applied Informatics, Silesian University of Technology, Gliwice, Poland (grants no. 02/100/BKM24/TBD, BKM/RAu7/2024, 02/110/BK_24/1028 and 02/100/BK_24/0035).

References

1. Bao, W., Yue, J., Rao, Y.: A deep learning framework for financial time series using stacked autoencoders and long-short term memory. PLoS ONE **12** (2017)
2. Benecki, P., Kostrzewa, D., Grzesik, P., Shubyn, B., Mrozek, D.: Forecasting of energy consumption for anomaly detection in automated guided vehicles: Models and feature selection. In: 2022 IEEE International Conference on Systems, Man, and Cybernetics (SMC), pp. 2073–2079 (2022)
3. Benecki, P., Kostrzewa, D., Grzesik, P., Shubyn, B., Mrozek, D.: Optimizing telemetry signal influence for power consumption prediction. In: Proceedings of Gecco 2024, pp. 51–52. ACM, New York, NY, USA (2023)
4. Benecki, P., et al.: Effective prediction of energy consumption in automated guided vehicles with recurrent and convolutional neural networks. In: 2023 IEEE International Conference on Big Data (BigData), pp. 5024–5030 (2023)
5. Blázquez-García, A., Conde, A., Mori, U., Lozano, J.A.: A review on outlier/anomaly detection in time series data. ACM Comput. Surv. (CSUR) **54**(3), 1–33 (2021)

6. Budinski, K.G.: Guide to Friction, Wear and Erosion Testing. ASTM International West Conshohocken, PA (2007)

7. Carrasco, J., et al.: Anomaly detection in predictive maintenance: a new evaluation framework for temporal unsupervised anomaly detection algorithms. Neurocomputing **462**, 440–452 (2021)

8. Cheng, C.F., Srivastava, G., Lin, J.C.W., Lin, Y.C.: A consensus protocol for unmanned aerial vehicle networks in the presence of Byzantine faults. Comput. Electr. Eng. **99**, 107774 (2022)

9. Cupek, R., et al.: Autonomous guided vehicles for smart industries – the state-of-the-art and research challenges. In: Krzhizhanovskaya, V.V., et al. (eds.) ICCS 2020. LNCS, vol. 12141, pp. 330–343. Springer, Cham (2020). https://doi.org/10.1007/978-3-030-50426-7_25

10. Gillespie, T.: Fundamentals of vehicle dynamics. SAE International (2021)

11. Han, Z., Zhao, J., Leung, H., Ma, K.F., Wang, W.: A review of deep learning models for time series prediction. IEEE Sens. J. **21**(6), 7833–7848 (2019)

12. Hundman, K., Constantinou, V., Laporte, C., Colwell, I., Soderstrom, T.: Detecting spacecraft anomalies using LSTMs and nonparametric dynamic thresholding. In: Proceedings of ACM SIGKDD 2018, pp. 387–395. ACM, New York, NY, USA (2018)

13. Lai, G., Chang, W.C., Yang, Y., Liu, H.: Modeling long- and short-term temporal patterns with deep neural networks (2018)

14. Malhotra, P., Ramakrishnan, A., Anand, G., Vig, L., Agarwal, P., Shroff, G.: LSTM-based encoder-decoder for multi-sensor anomaly detection (2016)

15. Minaker, B.P.: Fundamentals of Vehicle Dynamics and Modelling: A Textbook for Engineers with Illustrations and Examples. Wiley, Hoboken (2019)

16. Nelson, W.B.: Accelerated Testing: Statistical Models, Test Plans, and Data Analysis. Wiley, Hoboken (2009)

17. Niestrój, R., Rogala, T., Skarka, W.: An energy consumption model for designing an agv energy storage system with a PEMFC stack. Energies **13**(13) (2020)

18. Pedregosa, F., et al.: Scikit-learn: machine learning in python. J. Mach. Learn. Res. **12**(Oct), 2825–2830 (2011)

19. Ruff, L., et al.: A unifying review of deep and shallow anomaly detection. Proc. IEEE **109**(5), 756–795 (2021)

20. Shi, W., Cao, J., Zhang, Q., Li, Y., Xu, L.: Edge computing: vision and challenges. IEEE Internet Things J. **3**(5), 637–646 (2016). https://doi.org/10.1109/JIOT.2016.2579198

21. Shubyn, B., et al.: Federated learning for anomaly detection in industrial IoT-enabled production environment supported by autonomous guided vehicles. In: Groen, D., de Mulatier, C., Paszynski, M., Krzhizhanovskaya, V.V., Dongarra, J.J., Sloot, P.M.A. (eds.) ICCS 2022. LNCS, vol. 13353, pp. 409–421. Springer, Cham (2022). https://doi.org/10.1007/978-3-031-08760-8_35

22. Steclik, T., Cupek, R., Drewniak, M.: Automatic grouping of production data in Industry 4.0: the use case of internal logistics systems based on automated guided vehicles. J. Comput. Sci. **62**, 101693 (2022)

23. Wang, B., Huo, D., Kang, Y., Sun, J.: AGV status monitoring and fault diagnosis based on CNN. J. Phys. Conf. Ser. **2281**(1), 012019 (2022)

24. Wang, B., Chen, Y., Liu, D., Peng, X.: An embedded intelligent system for online anomaly detection of unmanned aerial vehicle. J. Intell. Fuzzy Syst. **34**(6), 3535–3545 (2018). https://doi.org/10.3233/JIFS-169532

25. Wang, W., Srivastava, G., Lin, J.C.W., Yang, Y., Alazab, M., Gadekallu, T.R.: Data freshness optimization under CAA in the UAV-aided MECN: a potential game perspective. IEEE Trans. Intell. Transp. Syst. (2022)
26. Witten, I.H., Frank, E.: Data mining: practical machine learning tools and techniques with Java implementations. ACM SIGMOD Rec. **31**(1), 76–77 (2002)
27. Zheng, A., Casari, A.: Feature Engineering for Machine Learning: Principles and Techniques for Data Scientists. O'Reilly Media, Inc., Sebastopol (2018)

Analysis of Marker and SLAM-Based Tracking for Advanced Augmented Reality (AR)-Based Flight Simulation

Onyeka J. Nwobodo[1]([⊠]) [iD], Godlove Suila Kuaban[2] [iD], Tomasz Kukuczka[1] [iD], Kamil Wereszczyński[1] [iD], and Krzysztof Cyran[1] [iD]

[1] Department of Computer Graphics, Vision and Digital Systems, Silesian University of Technology, Akademicka 2A, 44-100 Gliwice, Poland
{onyeka.nwobodo,tomasz.kukuczka,kamil.wereszczynski,
krzysztof.cyran}@polsl.pl
[2] Institute of Theoretical and Applied Informatics, Polish Academy of Sciences, Baltycka 5, 44-100 Gliwice, Poland
gskuaban@iitis.pl

Abstract. Augmented reality (AR)-based flight simulation reshapes how pilots are trained, offering an immersive environment where commercial and fighter pilots can be trained at low cost with minimal use of fuel and safety concerns. This study conducts a pioneering comparative analysis of marker-based tracking and SLAM technologies within the Microsoft HoloLens 2 platform, mainly focusing on their efficacy in landing manoeuvre simulations. Our investigation incorporates an experimental setup where marker-based tracking overlays interactive video tutorials onto a simulated cockpit, enhancing the realism and effectiveness of landing procedures. The experiment demonstrates that marker-based systems ensure high precision within 5 cm and 15 cm from the HoloLens 2 camera, proving indispensable for procedural training that requires exact overlay precision. Conversely, the native SLAM algorithm, while lacking the same level of precision, offers flexibility and adaptability by accurately mapping the cockpit and superimposing virtual information in dynamic, markerless conditions. The study juxtaposes these technologies, revealing a trade-off between precision and adaptability, and suggests an integrative approach to leverage their respective strengths. Our findings provide pivotal insights for developers and training institutions to optimize AR flight simulation training, contributing to advanced, immersive pilot training programs.

Keywords: Augmented Reality · Marker-based Tracking · SLAM-based tracking · Flight simulation · Microsoft HoloLens 2

1 Introduction

Augmented Reality (AR)-based flight simulation is reshaping how pilots are trained, offering an immersive environment where commercial and fighter pilots

L. Franco et al. (Eds.): ICCS 2024, LNCS 14838, pp. 208–222, 2024.
https://doi.org/10.1007/978-3-031-63783-4_16

can be trained at low cost with minimal use of fuel and safety concerns. Also, the continuous quest for enhanced pilot training methodologies is crucial in the aviation industry, where safety and proficiency are paramount. Traditional flight simulation has long been a cornerstone of pilot training, offering a controlled environment to hone essential skills. However, these conventional methods, including Full-Flight Simulators (FFS) and Computer-Based Training (CBT), often lack the dynamic and immersive qualities of real-world flight, potentially impacting the effective transfer of skills to trainee pilots and their overall engagement [1]. This gap underscores the need for innovative training solutions to provide more realistic and engaging training environments.

Augmented Reality (with its ability to overlay digital information onto the real world) emerges as a promising technology to bridge the gap in traditional flight simulators, offering a dynamic and immersive environment to train pilots. By enhancing the realism of flight simulation through AR, training programs can simulate complex flight scenarios in a risk-free environment, fostering improved situational awareness and decision-making skills among trainees [2].

The adoption of AR in flight simulation leverages advanced tracking technologies such as marker-based tracking and Simultaneous Localization and Mapping (SLAM) tracking, each with distinct advantages and limitations regarding accuracy, flexibility, and applicability in diverse training scenarios.

Despite AR technologies' potential benefits, a comprehensive comparison of marker based and SLAM tracking methods in the context of AR flight simulation training effectiveness still needs to be discovered. This study's main contributions are summarised as:

- We aim to fill this gap by introducing a novel comparative analysis of these tracking technologies within the Microsoft HoloLens 2 platform. We specifically focus on the distinct impacts of marker-based and SLAM tracking on AR flight simulation training quality, aiming to develop more immersive and effective training programs.
- We further contribute to the field by carrying out an experiment to track video on markers for a landing tutorial, a critical manoeuvre in pilot training. This experiment explores the practical applications of marker-based tracking in enhancing the realism and effectiveness of landing simulations, thereby addressing an essential aspect of pilot training.
- By systematically evaluating marker-based and SLAM tracking in Hololens 2 through a mixed-methods approach, we address crucial research questions regarding accuracy, user experience, training effectiveness, and cost-effectiveness.

Our findings aim to empower developers and training institutions to select the optimal tracking solution for their specific needs, thereby contributing to the development of more immersive and compelling pilot AR training programs. In light of the limitations of existing training methods and the potential of AR to enhance pilot training, our study not only addresses a significant gap in the literature but also proposes a practical experiment with direct implications for training effectiveness. Through this comprehensive approach, we seek to contribute to the ongoing evolution of pilot training methodologies, ensuring they are both practical and engaging in preparing pilots for the complexities of modern aviation.

2   Related Works

In this section, we present a state-of-the-art overview of marker and SLAM-based tracking technologies for AR systems. AR marker-based tracking uses distinct artificial markers for camera positioning and orientation. Systems employ various shapes and features: a) InterSense utilizes concentric rings, needing at least four for precision [3]. b) QR Codes are famed for fast scanning and high data capacity [4]. c) Visual Code identifies markers via image processing and databases [5]. d) Vuforia's customization feature allows for using any selected image as a marker, enabling personalized marker design for specific applications. Figure 1 depicts these markers.

(a) InterSense   (b) QR Codes   (c) Visual Code   (d) Customised

Fig. 1. Various AR Tracking Markers

Marker-based tracking offers high precision, making it ideal for accurately overlaying virtual elements onto real cockpits [6]. This approach can enhance realism and facilitate procedural training. Ribeiro et al.[7] demonstrate its application in UAV pilot training using printed markers and overlays, creating a cost-effective and accessible approach. Wallace et al. [8] leverage ArUco tags to overlay virtual gauges on a physical instrument panel, allowing immersive training while maintaining natural tactile interaction. However, marker setup and potential occlusion issues might pose challenges in dynamic scenarios [9].

SLAM-based tracking, on the other hand, adapts to dynamic environments but may face accuracy limitations compared to marker-based tracking, particularly during emergency procedures [10]. However, this adaptability eliminates setup requirements and facilitates training in unpredictable scenarios. Sun and Li [11] propose a system that translates user movements and control inputs into augmented visuals, eliminating the need for expensive physical mockups. Wang and Zhou [8] outline a system utilizing real-time data acquisition and intelligent identification of cockpit elements to enhance learning, reduce errors, and improve efficiency. A critical examination of existing studies reveals discrepancies in these technologies' effectiveness and implementation challenges. For instance, while Nwobodo et al. [12] review and evaluate SLAM methods for AR in flight simulators, emphasizing accuracy challenges in confined spaces and computational demands [13], it is crucial to acknowledge the broader context of technological and educational trends. Integrating adaptive learning technologies and gamification elements in pilot training could complement AR technologies, offering a more holistic approach to training that caters to diverse learning styles and enhances engagement and retention [14].

SLAM empowers robots and AR devices to autonomously build maps and track their location within them, eliminating the dependency on external markers or GPS. Leveraging a mix of sensors such as LiDAR, cameras, and Inertial Measurement Units (IMUs), SLAM technologies have advanced to meet specific application needs, significantly enhancing mapping and tracking accuracy [15]. AR devices like the HoloLens 2 utilize advanced SLAM technology, integrating IMUs and depth cameras to construct a detailed environmental model. This facilitates both navigational tasks and interactive user experiences, as depicted in Fig. 4, with features like loop closure and object manipulation for a fully immersive AR experience.

A sophisticated hybrid approach appears promising for future flight simulation training in AR. This method would harness the precise capabilities of marker-based tracking for essential tasks requiring exactitude, such as instrument manipulation, while simultaneously utilizing the vast, dynamic environments afforded by SLAM technology. Such a multifaceted system would marry the best aspects of both technologies, extracting maximum benefit from the evolving hardware and software proficiencies of the HoloLens.

3   Materials and Methods

This section elaborates on the methodology applied in our study to assess the effectiveness of marker-based and SLAM-based tracking technologies in augmented reality flight simulation training using Microsoft HoloLens 2. We present the camera calibration procedure, which is essential for ensuring the accuracy of both tracking methods. We also present the design of markers and establish a mathematical relationship between the detection range and parameters of the Hololens 2 camera.

3.1   Camera Calibration

Practical AR simulation training necessitates accurate superimposition of virtual content, which hinges on robust camera calibration. As illustrated in Fig. 2, calibration ensures the transformation of 3D world coordinates $\mathbf{E}_w$ into 2D image points $\mathbf{V}_c$, facilitating precise alignment of virtual and real elements. Understanding the intrinsic parameters-focal lengths f_x, f_y and the principal point x_0, y_0-and the extrinsic pose parameters $[R|t]$ is crucial for AR systems to track markers and reference points consistently. These parameters, including the marker size S_M and their distance D from the camera, along with the field of view (FOV), are instrumental in defining the tracking range. The image demonstrates this via the convergence of lines at the principal point, underscoring the geometric basis for optimizing detection thresholds for seamless AR integration. Incorporating these calibration parameters is vital for the realistic rendering of virtual content in pilot training modules, ensuring immersive and compelling skill acquisition.

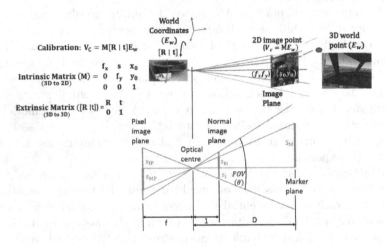

Fig. 2. The figure depicts the AR camera calibration, showing 3D world coordinates $\mathbf{E}_w$ projection onto 2D image points $\mathbf{V}_c$, including the camera's intrinsic matrix M and extrinsic parameters $[R|t]$. It also shows the principal point, FOV lines, marker size S_M, and distance D from the camera. (The lowest part of the figure is adapted from [16])

One of the calibration procedures is to estimate the optimal detection range of the camera. Marker-based tracking efficacy in AR devices like HoloLens 2 hinges on optimal detection range settings. This range, crucial for AR applications like flight simulation in cockpits, influences the required size of the printed, and it is given by [16]

$$D = \frac{S_M}{s_M} \tag{1}$$

where S_M is the printed marker size, and s_M is the normal image plane marker size at a distance $D = 1$ as shown in Fig. 2.

This expression establishes the fundamental relationship that determines the optimal placement of markers to the HMD's camera. This relationship is critical for the HoloLens 2 employed in flight simulation training to ensure that the markers fall within the device's field of view and are of a size that the onboard camera system can reliably detect. From Eq. (1) and Fig. 2, the detection range can be express as

$$D = \frac{S_M \cdot s_I}{s_{IP} \cdot s_{MP}} = \frac{S_M \cdot f}{s_{MP}} \tag{2}$$

where, $s_I = 2 \cdot tan\left(\frac{\theta}{2}\right)$, $s_{IP} = 2f \cdot tan\left(\frac{\theta}{2}\right)$. Also, f is the focal length of the camera, and θ is the field of view angle. Thus, for given parameters of the Hololens (f and s_{MP}), the detection range is proportional to the printed marker size, S_M. The parameter s_{MP} can also be obtained for a given type of Hololens.

Equations (1) and (2) are useful in the estimation of the detection range D and ensuring accurate AR marker tracking in the simulation environment. The practical application of these theoretical principles can be visualized through the marker-tracking process, as shown in Fig. 3.

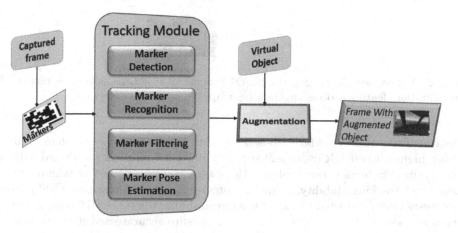

Fig. 3. Flowchart depicting the marker tracking process, from capture to augmentation.

The process begins with capturing a frame containing markers, which are then detected, recognized, and filtered by the tracking module. Following this, the pose of the markers is estimated, allowing virtual objects to be accurately overlaid onto the captured frame, resulting in an augmented reality experience. This sequence of operations is critical for ensuring that AR elements are appropriately aligned with the real world, providing an immersive experience for users, particularly in the demanding context of flight simulation training.

3.2   Experimental Setup

This study explores the profound impact of augmented reality (AR) technologies on flight simulation training, particularly in improving landing manoeuvres and comprehension of cockpit instrumentation using Microsoft HoloLens 2 and Unity software. The research involved developing an advanced AR flight simulation application integrating Unity's AR Foundation and the Mixed Reality Toolkit (MRTK) to blend real-world cockpit settings with interactive virtual flight

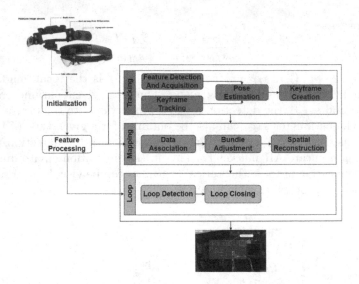

Fig. 4. A flowchart illustrating the SLAM process in HoloLens, including phases of initialization, feature processing, tracking, mapping, and loop detection.

instruments seamlessly. Marker-based tracking technologies were employed, utilizing high-contrast QR codes and specially designed markers positioned within the optimal detection range of the HoloLens 2. This ensured maximum visibility and tracking stability, guided by precise spatial calculations. Calibration processes were conducted to adjust the interpupillary distance (IPD) for individual users, along with sensor optimization, to ensure accurate and stable tracking. The markers were printed in sizes of S_{M1}: 10 cm × 10 cm, S_{M2}: 16.2 cm × 16.2 cm, and S_{M3}: 29 cm × 20.4 cm, respectively.

During the experiment, the users were asked to wear the Head-mounted device (HMD) comfortably and suitably to track the markers in the cockpit. The localization speed, the predicted and actual positions of the markers, and the overlayed objects' positions were recorded as the user moved their head to track each marker in the cockpit. The experiment aimed to position the markers within optimal ranges in Figs. 6 and 7 to improve pilot training. Similarly, SLAM tracking was implemented to dynamically map the cockpit environment, with MRTK configurations meticulously adjusted to enhance environmental understanding and computational efficiency. This led to the projection of highly interactive flight instruments and instructional content directly into the user's field of view, significantly enriching the training environment. The evaluation of the study focused on the effectiveness of these tracking technologies in maintaining overlay accuracy and system responsiveness and providing an immersive, realistic experience. The primary goal was to enhance pilot training efficiency by integrating AR into flight simulation, representing a significant advancement in training methodologies and promising improved learning outcomes through enriched, interactive experiences. Figure 5 depicts the experiment setup for marker and SLAM tracking in unity

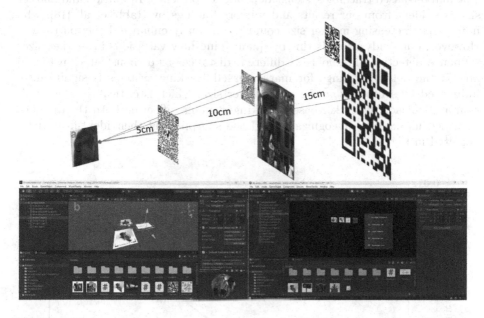

Fig. 5. a) Depicts markers at varying distances for HoloLens 2 tracking calibration. b) Shows a Unity interface setting up a simulated cockpit with marker detection and SLAM configuration for AR development.

4   Result and Discussion

This study conducted a comparative analysis of marker-based tracking and SLAM technologies within the Microsoft HoloLens 2 for advanced flight simulation training. Our systematic approach yielded nuanced insights into the performance and applicability of each system under varying detection ranges. The marker-based system demonstrated high precision within 5 cm and 15 cm from the HoloLens 2 camera in well-lit environments, ensuring reliability for procedural training requiring exact overlay precision. Including interactive video tutorials on landing manoeuvres, we further enhanced trainees' understanding of complex tasks, merging virtual learning with physical interaction.

In their study, Cheng et al. [17] highlighted that marker-based AR provides high positional accuracy, vital for precise overlays in flight simulation, as confirmed by our findings. They noted potential instabilities such as shakiness, which are linked to marker quality and AR SDKs and issues that are mitigated in our controlled simulation settings. Conversely, our results on SLAM reflect Cheng et al.'s observations on the adaptability of markerless AR. Utilizing GPS and gyroscopes offers flexibility without fixed markers, enhancing spatial awareness and cognitive skills in flight training scenarios despite its slightly lower accuracy. The marker-based tracking systems' dependency on sensor resolution and marker size is evident from our results and mirrors findings by Rabbi et al. [18], who noted that increasing marker size could significantly enhance detection ranges. However, our study extends this by quantifying how variations in marker sizes influence detection thresholds at different distances as demonstrated in Figs. 6 and 7, the detection range for marker-based tracking systems is significantly influenced by sensor resolution and marker size, which is critical for the precision and efficacy of marker-based tracking systems, affecting both the distance at which markers are recognizable and the sharpness of their identification as depicted in Fig. 8.

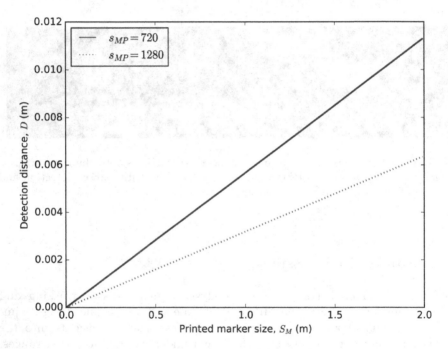

Fig. 6. Variation of minimum detection distance, D with printed marker size S_M for various marker size in the pixel image plane, s_{MP}.

Figures 6 and 7 demonstrate the relationship between the detection distance, D (in meters, m) and the printed marker size, S_M (in meters, m) for various values of the marker sizes in the pixel image plane, s_{MP} (in pixel). The focal length, f, of the device is obtained from the camera specifications, such as image width and height and the field of view angle, θ. The image size in the pixel image plane, s_{IP} and the image size in the normal image plane can then be obtained from the camera specifications and thus, $f = s_{IP}/s_I$. In the plots in Figs. 6 and

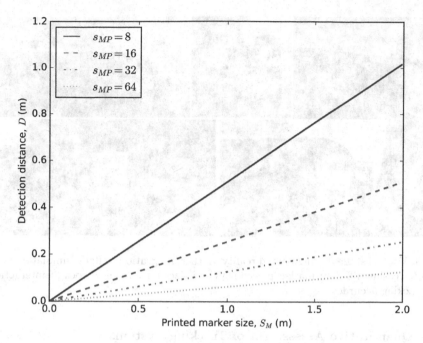

Fig. 7. Variation of maximum detection distance, D with printed marker size S_M for various marker size in the pixel image plane, s_{MP}.

7, we used the nominal values of the focal length and the Field of view angle of the Hololens 2 (e.g., $f = 1.08\,\text{mm}$ and $\theta = 96.1°$). Thus, with given values of S_M, f, θ, and s_{MP} for a given device, the detection distance can be obtained from Eq. (2).

Aligning seamlessly with the physical markers, this intervention significantly enhanced trainees' comprehension of complex tasks while maintaining interaction with the physical cockpit, effectively bridging the gap between virtual learning and practical execution. However, SLAM rose to the challenge in the adaptability arena. Its ability to map and project virtual information onto any surface without relying on pre-placed markers offered an immersive, dynamic experience perfect for honing spatial awareness and broader skill sets, as shown in Fig. 9. Our analysis revealed a sweet spot for each technology. Marker-based training, potentially leading to improved objective flight data, shines in focused skill development. SLAM, on the other hand, fosters real-world, transferable skills through its immersive and adaptable nature. Cost-wise, marker-based systems offer long-term savings for targeted training, whereas SLAM, despite higher initial costs, presents a scalable solution across different scenarios. Regarding user experience, marker-based tracking assures precision task confidence, and SLAM enhances cognitive skills and situational awareness through engagement.

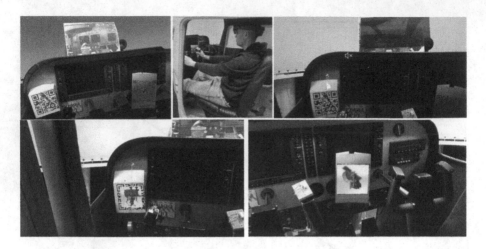

Fig. 8. Various stages of augmented reality marker integration in flight simulator training, showcasing different marker placements and sizes for enhanced cockpit interaction and tracking accuracy.

4.1   Quantitative Assessment of Tracking Systems

The efficacy of the marker-based and SLAM tracking systems was quantitatively assessed through key performance metrics: accuracy, precision, and error rates, as encapsulated in Table 1. The accuracy and precision rates were calculated based on the proportion of correct detections.

Mean Absolute error (MAE) and standard deviation of error (SD) were computed using the following standard formulas:

Mean Absolute Error (MAE):

$$MAE = \frac{1}{n} \sum_{i=1}^{n} |x_i - \hat{x}_i| \tag{3}$$

where n is the number of measurements, x_i is the true position, and $\hat{x}_i$ is the predicted position.

Standard Deviation of Error (SD):

$$SD = \sqrt{\frac{1}{n-1} \sum_{i=1}^{n} (x_i - \bar{x})^2} \tag{4}$$

where $\bar{x}$ is the mean of the observed errors.

In our analysis, as depicted in Fig. 10, the marker-based system reported an accuracy rate of 98.5% and a precision rate of 97.8%, with an MAE of 0.5 cm and an SD of 0.3 cm. The SLAM system had an accuracy rate of 96.2% and a precision rate of 94.5%, with an MAE of 1.2 cm and an SD of 1.0 cm. These show that

Fig. 9. SLAM tracking in AR for procedural training: showcasing interactive checklists and equipment diagrams to boost pilot understanding and interaction.

Table 1. Performance metrics of marker-based and SLAM systems.

Metrics	Marker-Based Systems	SLAM Systems
Accuracy Rate (%)	98.5	96.2
Precision Rate (%)	97.8	94.5
Mean Absolute Error (cm)	0.5	1.2
Standard Deviation of Error (cm)	0.3	1.0
Localization Speed (Milliseconds)	16	30
Scenario Versatility Index (1 scenario)	1	1
Scenario Versatility Index (2 scenarios)	1	2
Scenario Versatility Index (3 scenarios)	1	3
Scenario Versatility Index (4 scenarios)	1	4
Scenario Versatility Index (5 scenarios)	1	5
Cognitive Skills Improvement Factor	4	7

the marker-based system outperforms the SLAM system regarding accuracy and precision, although the SLAM offers greater flexibility in dynamic environments. The derived error metrics underscore the trade-offs inherent to each system: marker-based tracking provides higher reliability in controlled settings. However, SLAM tracking's greater adaptability in changing conditions is evidenced by a broader error distribution and localization speed. While marker-based systems boast rapid localization speeds at 16 milliseconds, indicative of their efficiency in stable environments, SLAM systems demonstrate a notable proficiency with a localization speed of 30 ms. This slightly increased time consumption is offset

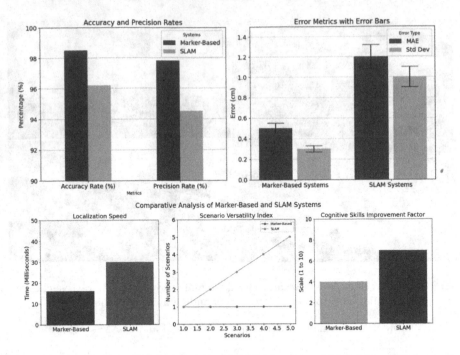

Fig. 10. A comparison of marker-based and SLAM systems highlighting differences in accuracy, precision, error, localization speed, adaptability, and cognitive impact.

by its remarkable adaptability, enabling real-time responsiveness in dynamic scenarios. The versatility index in Fig. 10 displays the scenario versatility index, comparing the adaptability of marker-based systems, which excel at stable tasks such as precise video overlays for landing manoeuvres, to SLAM systems, which scale from essential instrument readings to complex, real-time interactions like equipment checks and adaptive navigation. Marker-based systems held a steady index of 1. In contrast, SLAM systems advanced from 1 to 5, gauging their performance by the users' accuracy in engaging with augmented content and their adaptability to dynamic scenarios.

The cognitive skills enhancement is quantified, showcasing that SLAM systems achieved a prominent boost in cognitive abilities with a score of 7, surpassing the marker-based systems, which scored 4 on a scale of 1 to 10. This was derived from objective data, such as response times and checklist accuracy, alongside users' self-evaluations of situational awareness and adaptability. Through normalization and weighting, these measures established SLAM's significant role in advancing pilot training with realistic, engaging simulations that align with modern aviation's complexities.

5   Conclusion

Our comprehensive analysis provides critical insights into the comparative performance of marker and SLAM-based tracking within Microsoft HoloLens 2 for advanced flight simulation training. The results underscore marker-based systems' superior precision and reliability, which is particularly beneficial for procedural training where exact overlay precision is paramount. Interactive video tutorials further augment these systems, effectively enhancing comprehension of complex tasks by bridging the virtual-physical interface. Conversely, SLAM technology demonstrates remarkable adaptability and immersive capabilities, making it ideal for dynamic environments and broader skill development. Though SLAM may incur higher initial costs, its scalability and flexibility across varied training scenarios offer long-term benefits.

The findings of this study, while focused on flight simulation, resonate across AR education and training sectors. The high fidelity of marker-based systems is indispensable for tasks that demand high precision, particularly in controlled settings where the consistency of virtual overlays directly impacts learning outcomes. Conversely, the adaptability of SLAM shines in scenarios that benefit from less regimented, more versatile interaction. The balance between accuracy, precision, and error variability underscores the advantage of marker-based systems for stability. Nevertheless, it illuminates SLAM's broader educational promise, enhancing cognitive skills and situational awareness through its immersive nature.

For the future direction, the emerging trends in AR flight simulation training suggest a blended approach that merges the precision of marker-based tracking for critical tasks such as instrument operation with the extensive environments enabled by SLAM technology. This dual system could leverage the advancing capabilities of devices like the HoloLens to enhance training efficacy. Integrating biofeedback tools such as EEG and GSR could offer deeper insights into trainees' psychological states, allowing the development of adaptive training that responds to individual learning needs. Imagine AR simulations that adjust complexity based on the user's biofeedback, creating a tailored training experience that optimizes skill acquisition and minimizes stress. Additionally, incorporating deep learning for better marker recognition and environmental mapping could enrich simulation training. Such technological progress, especially in wearable AR, is crucial for the comprehensive application of AR in aviation training. However, incorporating biofeedback raises ethical issues, particularly around data privacy. Strict anonymization protocols and informed consent are essential to maintaining ethical standards and respecting trainee privacy.

Acknowledgments. This publication was supported by the Department of Computer Graphics, Vision and Digital Systems, under statute research project (Rau6, 2024), Silesian University of Technology (Gliwice, Poland).

Conflict of Interest. The authors declare no conflict of interest.

References

1. Chae, S., Yim, S., Han, Y.: Flight simulation on tiled displays with distributed computing scheme, pp. 1–6 (2012)
2. Brown, C., Hicks, J., Rinaudo, C.H., Burch, R.: The use of augmented reality and virtual reality in ergonomic applications for education, aviation, and maintenance. Ergon. Des. **31**(4), 23–31 (2023)
3. Naimark, L., Foxlin, E.: Circular data matrix fiducial system and robust image processing for a wearable vision-inertial self-tracker. In: Proceedings. International Symposium on Mixed and Augmented Reality, pp. 27–36, IEEE (2002)
4. Deineko, Z., Kraievska, N., Lyashenko, V.: QR code as an element of educational activity (2022)
5. Rohs, M., Gfeller, B.: Using camera-equipped mobile phones for interacting with real-world objects (2004)
6. Armbrister, L.P.: Automation in aviation: an advancement or hindrance to aviation safety? Ph.D. thesis, Texas Southern University (2023)
7. Ribeiro, R., et al.: Web AR solution for UAV pilot training and usability testing. Sensors **21**(4), 1456 (2021)
8. Wallace, J.W., Hu, Z., Carroll, D.A.: Augmented reality for immersive and tactile flight simulation. IEEE Aerosp. Electron. Syst. Mag. **35**(12), 6–14 (2020)
9. Araar, O., Mokhtari, I.E., Bengherabi, M.: PDCAT: a framework for fast, robust, and occlusion resilient fiducial marker tracking. J. Real-Time Image Proc. **18**(3), 691–702 (2021)
10. Xing, Z., Zhu, X., Dong, D.: DE-SLAM: slam for highly dynamic environment. J. Field Robot. **39**, 528–542 (2022)
11. Su, P., Luo, S., Huang, X.: Real-time dynamic slam algorithm based on deep learning. IEEE Access **10**, 87754–87766 (2022)
12. Nwobodo, O.J., Wereszczyński, K., Cyran, K.: Slam methods for augmented reality systems for flight simulators. In: Mikyska, J., de Mulatier, C., Paszynski, M., Krzhizhanovskaya, V.V., Dongarra, J.J., Sloot, P.M. (eds.) ICCS 2023. LNCS, vol. 14073, pp. 653–667. Springer, Cham (2023). https://doi.org/10.1007/978-3-031-35995-8_46
13. Skurowski, P., Myszor, D., Paszkuta, M., Moroń, T., Cyran, K.A.: Energy demand in AR applications-a reverse ablation study of the Hololens 2 device. Energies **17**(3), 553 (2024)
14. Noh, D.: The gamification framework of military flight simulator for effective learning and training environment (2020)
15. Kazerouni, I.A., Fitzgerald, L., Dooly, G., Toal, D.: A survey of state-of-the-art on visual slam. Expert Syst. Appl. **205**, 117734 (2022)
16. Nitschke, C.: Marker-based tracking with unmanned aerial vehicles. In: 2014 IEEE International Conference on Robotics and Biomimetics (ROBIO 2014), pp. 1331–1338. IEEE (2014)
17. Cheng, J.C., Chen, K., Chen, W.: Comparison of marker-based AR and markerless AR: a case study on indoor decoration system. In: Lean and Computing in Construction Congress (LC3): Proceedings of the Joint Conference on Computing in Construction (JC3), pp. 483–490 (2017)
18. Rabbi, I., Ullah, S., Javed, M., Zen, K.: Analysis of artoolkit fiducial markers attributes for robust tracking. In: International Conference of Recent Trends in Information and Communication Technologies (IRICT'14). University Technology Malaysia, pp. 281–290. Citeseer (2014)

Automated Prediction of Air Pollution Conditions in Environment Monitoring Systems

Dawid Białka, Małgorzata Zajęcka[iD], Ada Brzoza-Zajęcka[✉][iD],
and Tomasz Pełech-Pilichowski[iD]

AGH University of Science and Technology, Kraków, Poland
{mzajecka,abrzoza,tomek}@agh.edu.pl

Abstract. This paper aims to explore the problem of air pollution forecasting, especially the particulate matter (PM) concentration in the air. Other quantities such as air temperature, atmospheric pressure, and relative humidity are also considered. Moreover, a large part of the discussion in this paper can be extended and applied to a variety of other quantities which are stored and expressed as data series. The goal is to evaluate different time series forecasting models on a selected air pollution data set. The proposed model is compared with other implemented state-of-the-art methods in order to validate whether it could be a reliable pick for air pollution forecasting problem.

Keywords: Air Pollution · Sensor Networks · Severe environmental conditions · Forecasting Models · Time series · Internet of Things · Energy crisis

1 Introduction

Forecasting severe environmental conditions is very useful to protect the health and well-being of citizens. In general, forecasting atmospheric phenomena is rather complicated and requires very complex numerical models. On the other hand environmental pollution is related to human activity, which in some situations may be easier to forecast with the help of modern algorithms, techniques of data analysis, and artificial intelligence methods. In the first section we present the motivation and goal of our research and the related work. The rest of the paper is organized as follows: Sect. 2 presents the proposed solution and in Sect. 3 we discuss the obtained results. Section 4 concludes the paper.

1.1 Motivation and Goal

The increased air pollution leads to many diseases and premature deaths [20, 32]. The smallest particles, like PM2.5, are the most threatening. The situation is

© The Author(s), under exclusive license to Springer Nature Switzerland AG 2024
L. Franco et al. (Eds.): ICCS 2024, LNCS 14838, pp. 223–238, 2024.
https://doi.org/10.1007/978-3-031-63783-4_17

even worse in developing countries, which do not pay much attention to that issue [3].

The environment monitoring stations typically provide only real-time air quality information, which cannot be used to notify people in advance. Therefore, the best way to prevent people from entering areas with high concentration of air pollutants is to forecast the trend of air pollution and alert citizens before the hazardous situation occurs. However, it is challenging to forecast air pollution because it is affected by complex factors, such as air pollution accumulation, meteorology, traffic flow and industrial emissions. It is difficult to obtain sufficient data to model each factor. Therefore, there is a need to develop better models, which yields more accurate results.

Main objective of the research is to explore the problem of air pollution forecasting, especially the particulate matter (PM) concentration in the air. Other quantities such as air temperature, atmospheric pressure, and relative humidity are also considered. The goal is to evaluate different time series forecasting models on a selected air pollution data set. The proposed deep Echo State Network (ESN) model is compared with other implemented state-of-the-art methods in order to validate whether it could be a reliable pick for air pollution forecasting problem. Moreover, a large part of the discussion in this work can be extended and applied even further, to a variety of other quantities which are stored and expressed as data series.

1.2   Literature Review

In case of air pollution data, there are basically three types of time series forecast models that are being used [4]: numerical models, statistical models, and machine learning models.

Numerical Models. Numerical models use a range of equations and mathematical functions to describe and simulate the physical processes that contribute to air pollution. The Atmospheric Dispersion Modelling System (ADMS) [17,26] employs a three-dimensional Gaussian model to estimate the dispersion of pollutants. The California Puff Model (CALPUFF) [23] is a three-dimensional Lagrangian puff dispersion and transport model which allows to simulate the discrete and transform processes of matter emitted from a source. It can be combined with other models to produce better results [1,27]. The Comprehensive Air Quality Model (CMAQ) [4] is a comprehensive air quality modeling system that consists of multiple processors and chemical-transport models.

Statistical Models. Statistical analysis models are a common way of performing time series forecasting [21]. The Autoregressive Integrated Moving Average (ARIMA) [2,16,25] and The Seasonal Autoregressive Integrated Moving Average [18,22] (SARIMA) are models that are widely used for time series forecasting. The VARMA [9,10,33] builds on the principles of VAR [6,8] models by incorporating moving average terms. Forecasting air pollution by applying VARMA model yielded better results that ARIMA model [10]. Prophet [24,28,31] is an open-source time series forecasting model. It is designed to handle the complexities of time series data such as seasonality, trends and holidays.

Machine Learning Models. Machine learning models are more difficult to create and are more complex but they are able to provide a very accurate and repeatable predictions [15]. Multilayer Perceptrons (MLP) [7] were used to construct a high accuracy model for predicting air pollution in London. MLP and LSTM (Long Short-Term Memory) were combined for hourly prediction of PM10 in Lima [5]. Deep spatial-temporal ensemble model coupling many stacked LSTM models were used by Wang et al. [12] for air quality prediction. Alléon, Antoine, et al. [13] created a large-scale air quality forecasting model using the Convolutional LSTM (ConvLSTM) network called PlumeNet. Xu, Xinghan et al. [30] utilized the Echo State Network (ESN) and improved particle swarm optimization (IPSO) to forecast PM2.5 concentrations in Beijing. However, their analysis did not incorporate meteorological data or account for spatial dependencies in their models.

2   The Proposed Solution

In this chapter improvements to the typical approach of time series forecasting are shown along with proposed models architectures. Based on their popularity in related work the selected models are: state-of-the-art models Long-Short Term Memory (LSTM), Multilayer Perceptron (MLP), and the Echo State Network (ESN) model, which this paper aims to evaluate.

A typical approach to time series forecasting is to make models that are trained on historical data obtained from a single source to forecast future values. In the case of air pollution forecasting, that approach can be improved in two ways. First, we propose that air pollution time series forecasting can be enhanced by incorporating weather forecasts as an additional factor in the predictive models. Second, we intend to consider that the air pollution is not limited to a single location but the pollution from one area can impact pollution levels in other areas. For instance, a strong wind can carry emissions from a different station, thereby influencing the air pollution levels at the first station. Therefore, it is also necessary to account for the spatial dependencies and interactions between different locations when forecasting air pollution values.

2.1   Proposed Architecture of the Models

Every model is tuned for each of four scenarios:

1. Base - model is trained only on historical air pollution.
2. Meteorological data - model is trained on historical air pollution and meteorological data.

3. Weather forecast improvement - model is trained using historical air pollution and meteorological data, as well as incorporating additional information from weather forecasts.
4. Spatial dependencies improvement - model undergoes training using a comprehensive data set that includes historical air pollution data, meteorological information, weather forecasts and additionally data from other stations.

Figure 1 illustrates the common workflow for the entire forecasting process, which was used for all of the proposed models to forecast PM2.5 values. The workflow consists of following steps: *Collect* - air pollution and meteorological data along with weather forecast data is collected, based on which the models will be trained; *Data fusion* - air pollution and meteorological data for one station, for which the forecasts will be made is fused with air pollution data from other stations and meteorological data forecast to provide the model with more context; *Data cleaning, transformation and normalization* - the process of preparing the data to be fed into the models. This include operations like removing outliers, interpolating missing data, normalizing the data and changing the data format to one that the model will accept and understand as an input; *Model training, evaluation and tuning* - feeding the training and test data, parameters tuning; *Forecasting based on input data* - validating the model by providing the input data and comparing the obtained forecasts with ground truth values.

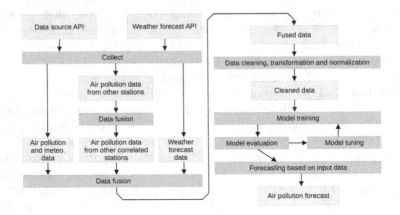

Fig. 1. High level workflow of the whole forecasting process.

ESN architecture which comprises an input node, reservoir node and readout node. In the first two scenarios, we propose a straightforward architecture, which includes a single reservoir node and a readout node. The reservoir node is connected to both the input and readout node. In the following two more complex scenarios, we introduced a deep ESN architecture shown in Fig. 2. It incorporates three interconnected reservoirs, each connected to the input and readout node.

The LSTM architecture which is common for every scenario and consists of five consecutive layers: LSTM layer, dropout layer, LSTM layer, dense layer and reshape layer. Parameters and hyperparameters differ for the scenarios and are explained in the Sect. 2.2 in more depth.

The MLP architecture which is also common for every scenario and comprises of five consecutive layers: dense layer, dropout layer, two dense layers and reshape layer. Parameters and hyperparameters vary across the scenarios.

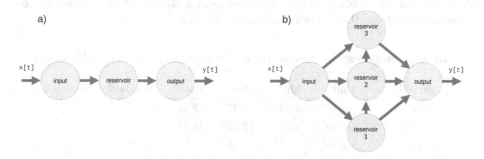

Fig. 2. The proposed deep ESN architecture with one reservoir node (a) and three reservoir nodes (b) and with one readout node.

2.2   A Sample Implementation

This section outlines the practical implementation of the aforementioned methodology visualised in Fig. 1 which authors used on specific data. This resulted in the development of models capable of forecasting future air pollution values.

Data Collection. The AIRLY[1] service was chosen as a source of air pollution data. It is an organization that monitors and gathers air pollution and meteorological data across Europe, including Poland and Krakow city, North America, and South-East Asia. The data collection process utilizes a network of sensors distributed in various locations. These measurements are subsequently transmitted to a central server and stored in a database.

A Python script was developed to collect the data by making requests for the previous 24 h' worth of data from 194 stations in and around Krakow. The data used in this study spans from 11th March 2022 to 16th March 2023, providing a substantial timeframe for analysis. The whole data set contains 1635934 single measurements from all stations.

---

[1] https://airly.org/.

Data Description. Each measurement in the data set comprises the hourly average values of the following variables: date time of the start and end of the measurement $[yyyy - mm - dd\ hh : mm : ss]$, PM1, PM10, PM2.5 $[\mu g/m^3]$, temperature $[^\circ C]$, pressure $[hPa]$, humidity $[\%]$, wind speed $[m/s]$ and wind bearing $[azimuth]$. There is also information about station id, longitude and latitude.

The further analysis primarily focuses on a specific station with ID 1026, located close to the city center of Kraków, Poland. The measurements from that station have no outliers and have a low number of missing values. Total of 8760 samples are available for that station and a statistical description of the measurements from that station are shown in Table 1.

Table 1. Statistical description of data obtained from station with ID 1026.

Variable	Unit	Min	Max	Mean	Standard deviation
PM1	$\mu g/m^3$	0.03	66.27	12.32	8.92
PM10	$\mu g/m^3$	0.31	129.08	24.48	19.01
PM2.5	$\mu g/m^3$	0.15	101.96	18.12	14.02
Temperature	$^\circ C$	−16.35	34.58	10.55	8.49
Pressure	hPa	988.35	1041.88	1015.75	8.66
Humidity	%	28.93	100	72.12	11.83
Wind speed	m/s	0.26	40.71	10.17	6.25
Wind bearing	$azimuth$	0	359.98	193.62	94.43

Data Cleansing. The FBEWMA (forwards-backwards exponential weighted moving average) was employed for outlier detection and removal [11]. The method was appropriately adjusted and consists of the following steps:

- Removing measurements that exceed the maximum or minimum value determined based on domain knowledge.
- Calculating FBEWMA for the whole time interval and determining the threshold value used to remove the measurements.
- Calculating the distance between FBEWMA and measurement value and removing measurements that exceed the determined threshold.

Out of the total 8760 samples, 25 of them were found to be invalid due to null values, and there were 115 missing measurements (gaps) in the data set [19]. To address these gaps and create a continuous set of measurements, rows with all values set to null apart from date and time were inserted into the data set for all missing points.

Data Exploration. No apparent trend was evident for any of the features. The data comes from human activities that disrupt natural processes and therefore we do not check nor assume a normal distribution of samples. However, daily seasonality was visible in all features except for pressure. Additionally, both PM and temperature values exhibit yearly seasonality. One interesting observation can be spotted in Fig. 3. As the temperature and wind speed value rise, the PM values evidently decrease. Those two features are strong candidates for being included in the training set.

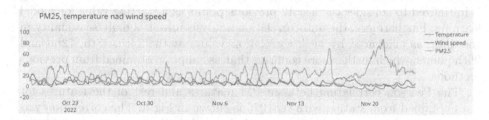

Fig. 3. PM2.5, temperature and wind speed on the same chart for data obtained from station with ID 1026.

Seasonality Analysis. Fast Fourier Transform (FFT) was utilized to extract important frequencies from the PM2.5 time series. The peak frequencies were detected near one year and one day, which corresponds to the information obtained from visual examination during data exploration. Subsequently, seasonal decomposition was conducted using the day frequency.

Stationary Assessment. Two tests were performed to assess stationarity: the ADF test for difference stationarity and the KPSS test for trend stationarity. In order for the time series to be both difference and trend stationary, the null

Table 2. Results of ADF and KPSS tests.

Variable	ADF		KPSS	
	Test statistic	5% critical value	Test statistic	5% critical value
PM1	−7.93	−2.86	2.21	0.463
PM10	−7.86	−2.86	1.95	0.463
PM2.5	−7.71	−2.86	2.48	0.463
Temperature	−3.16	−2.86	7.16	0.463
Pressure	−7.25	−2.86	0.08	0.463
Humidity	−9.81	−2.86	0.54	0.463
Wx	−12.19	−2.86	1.7	0.463
Wy	−8.9	−2.86	0.74	0.463

hypothesis should be rejected in the ADF test and not rejected in the KPSS test. The null hypothesis is rejected at a significance level of less than 5% when the test statistic value is smaller than the critical value. Table 2 shows that the null hypothesis is rejected in the ADF test for all variables, while the null hypothesis is not rejected in the KPSS test for all variables except for pressure.

Correlation Analysis. The autocorrelation analysis of the PM2.5 feature revealed that current points are significantly influenced by the previous 12 points, with a correlation strength exceeding 50%. Based on this information, the model is initialized to consider the last 12 previous points as a starting value for model learning. Furthermore, the autocorrelation analysis unveils a daily seasonality in the data, as evidenced by peak correlation values at the 24th, 48th, 72nd and 96th points, which furthermore confirms that assumption obtained from previous methods.

The Pearson correlation between PM features and rest of the features for data obtained from station with ID 1026 is shown in Fig. 4. The correlation was the strongest for temperature and for wind speed, indicating a strong relationship. The correlation is relatively lower for humidity and pressure, and the wind bearing shows the lowest correlation. The correlation value is almost 1 between all PM features, thus indicating a very significant dependency.

The spatial correlation between PM2.5 for data obtained from station with ID 1026 and PM2.5 for data obtained from other stations decreases linearly as the distance increases.

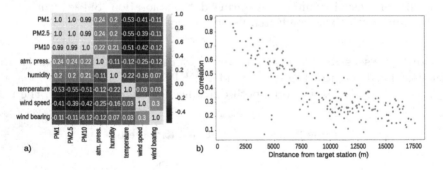

Fig. 4. a) Pearson correlation matrix of features obtained from station ID 1026. b) Pearson correlation of PM2.5 data between station ID 1026 and other stations.

Data Normalization. For both the LSTM and MLP models, the data was normalized by min-max scaling to a range between 0 and 1. The data for ESN was scaled using z-score scaling to improve performance.

Feature Engineering and Selection. The following new features were introduced: x and y wind speed vector components, day and year signal, and weekend flag. Based on the analysis, we chose the features included in the training set for every testing scenarios.

- First scenario - PM2.5, day signal, year signal, weekend flag
- Second scenario - PM2.5, day signal, year signal, weekend flag, temperature, pressure, wind speed x vector component and wind speed y vector component
- Third scenario - PM2.5, day signal, year signal, weekend flag, temperature, pressure, wind speed x vector component and wind speed y vector component. Forecast values of temperature, pressure, wind speed x vector component and wind speed y vector component
- Fourth scenario - PM2.5, day signal, year signal, weekend flag, temperature, pressure, wind speed x vector component, wind speed y vector component, PM2.5 of correlated stations. Forecast values of temperature, pressure, wind speed x vector component and wind speed y vector component. Only the stations with the correlation coefficient greater than 0.6 and with low amount of invalid data were chosen.

Forecasting Models Creation and Optimization. The forecast is made 6 h ahead for PM2.5 value for the target station with ID 1026 utilizing multi-step forecasting approach. The data was split into training, validation and test set in standard 80%: 10%: 10% ratio. The hyperparameters were tuned by utilizing random search and manual tuning. Models are presented with the best achieved configuration. We implemented the models in Python, utilizing the TensorFlow and ReservoirPy libraries.

Forecasting Models Testing and Final Evaluation. In the final step of the process, the models were tested using the data from the test set. This testing phase was conducted for each model and each scenario. The performance of the models was evaluated and compared using two commonly used metrics: Mean Squared Error (MSE) [29] and R-squared score [14].

3 Results

This section is an overview of the final performance results for all models across all testing scenarios.

3.1 First Scenario: Base Scenario

In this scenario, all models are trained on 12 previous hours of PM2.5 value, day signal, year signal and weekend flag to forecast future 6 h of PM2.5 value. For every hour of forecast, the MSE values, R-squared scores, and the percentages of forecasted value points for which the distance from the ground truth is less than 10% of maximum value of the test set are shown in Table 3.

In the base scenario the ESN model demonstrates superior performance compared to other models. It achieves the highest R-squared scores and percentage of points close to ground truth for all forecast hours. It also achieves the lowest MSE for all forecast hours, except for the third hour.

Table 3. MSE values highlighted in red are the smallest for given hour. R-squared values and percentage values highlighted in red are the greatest for given hour.

Model	1 h	2 h	3 h	4 h	5 h	6 h
MSE						
ESN	11.9	34.35	56.34	76.99	98.27	119.13
LSTM	14.18	37.36	62.26	87.9	113.37	136.3
MLP	14.71	34.37	56.05	78.32	100.02	119.02
R-squared						
ESN	0.96	0.88	0.82	0.75	0.67	0.59
LSTM	0.95	0.86	0.77	0.66	0.55	0.45
MLP	0.94	0.87	0.78	0.68	0.59	0.52
Percentage of points						
ESN	97.58	89.4	82.03	75.35	70.62	66.94
LSTM	96.27	88.56	81.09	72.35	66.51	58.46
MLP	95.92	89.15	81.56	75.14	71.06	68.84

3.2 Second Scenario: Meteorological Data Scenario

The models in this scenario forecast 6 future hours of PM2.5 value based on 12 past hours of PM2.5 value, day signal, year signal, weekend flag, temperature, pressure, wind speed x vector component and wind speed y vector component.

Table 4. MSE values highlighted in red are the smallest for given hour. R-squared values and percentage values highlighted in red are the greatest for given hour.

Model	1 h	2 h	3 h	4 h	5 h	6 h
MSE						
ESN	15.42	31.64	46.48	62.33	73.99	83.28
LSTM	20.07	34.37	46.24	59.77	72.06	83.18
MLP	17.53	32.76	49.26	66.24	79.34	92.32
R-squared						
ESN	0.95	0.89	0.85	0.78	0.74	0.70
LSTM	0.93	0.86	0.82	0.77	0.72	0.69
MLP	0.94	0.87	0.8	0.73	0.68	0.63
Percentage of points						
ESN	95.50	89.98	84.36	76.72	73.37	70.63
LSTM	95.57	90.09	86.35	80.23	77.65	73.98
MLP	95.79	90.08	83.43	78.92	73.91	71.81

The MSE values, R-squared scores, and the percentages of forecasted value points for which the distance from the ground truth is less than 10% of maximum value of the test set are shown in Table 4.

For the second scenario the LSTM model yields the best results. It has the highest percentage of forecast points close to the ground truth, except for the first hour. Additionally, the MSE is the lowest for the last four forecast hours when using the LSTM model. Only the R-squared score is greater for all forecast hours for the ESN.

3.3   Third Scenario: Weather Forecast Improvement Scenario

In the third scenario, the models are trained to forecast the future values of PM2.5 for the next 6 h based on the previous training set extended with weather forecast data. Features used for training include the 12 previous hours of PM2.5 values, day signal, year signal, weekend flag value, temperature, pressure, wind speed x vector component and wind speed y vector component as well as the 6 future points of temperature, pressure, wind speed x vector component, and wind speed y vector component.

The MSE values, R-squared scores, and the percentages of forecasted value points sufficiently close to the ground truth are shown in Table 5.

For the third scenario the LSTM model again yields the best results. It has the highest percentage of forecast points for which the distance between the ground truth and forecast value is less than 10% of the maximum for all forecast hours. The MSE is the lowest for the last three forecast hours when using the LSTM model. Only the R-squared score is greater for almost all forecast hours for the ESN model.

Table 5. MSE values of forecasts for the third scenario for every hour. MSE values highlighted in red are the smallest for given hour.

Model	1 h	2 h	3	4 h	5 h	6 h
	MSE					
ESN	16.62	34.23	50.05	60.53	72.79	82.01
LSTM	22.31	34.94	47.69	56.42	65.9	74.23
MLP	16.74	32.91	50.51	63.38	78.89	90.02
	R-squared					
ESN	0.95	0.89	0.84	0.79	0.75	0.71
LSTM	0.92	0.84	0.80	0.78	0.74	0.72
MLP	0.91	0.83	0.76	0.70	0.66	0.60
	Percentage of points					
ESN	95.73	88.74	81.56	79.72	75.34	72.93
LSTM	95.80	90.43	85.41	80.51	77.82	74.09
MLP	93.23	87.04	81.68	79.65	75.73	73.75

3.4   Fourth Scenario: Spatial Dependencies Improvement Scenario

In the last scenario PM2.5 data from other correlated stations was added to the training set. Thus the features used for training include the 12 previous hours of PM2.5 values, PM2.5 values obtained from 11 correlated stations, day signal, year signal, weekend flag, temperature, pressure, wind speed x vector component and wind speed y vector component, as well as the 6 future points of temperature, pressure, wind speed x vector component, and wind speed y vector component.

The MSE values, R-squared scores, and the percentages of forecasted value points sufficiently close to the ground truth are shown in Table 6.

Table 6. MSE values of forecasts for the fourth scenario for every hour. MSE values highlighted in red are the smallest for given hour.

Model	1 h	2 h	3 h	4 h	5 h	6 h
	MSE					
ESN	19.04	36.38	53.46	65.22	77.34	90.22
LSTM	19.96	37.38	49.64	61.29	73.07	78.90
MLP	20.83	41.71	54.95	68.65	83.81	93.60
	R-squared					
ESN	0.95	0.87	0.79	0.72	0.69	0.66
LSTM	0.92	0.86	0.81	0.76	0.70	0.66
MLP	0.92	0.81	0.72	0.66	0.62	0.58
	Percentage of points					
ESN	95.65	87.56	76.95	72.39	67.70	64.63
LSTM	95.12	86.55	82.39	76.42	73.21	70.84
MLP	93.51	86.63	78.06	75.14	74.03	72.62

For this scenario, overall, the LSTM model again yields the best results. It achieves the highest R-squared scores for the last four forecast hours and the highest percentage of forecast points where the distance between the ground truth and forecast value is sufficiently small. Furthermore, the LSTM model exhibits the lowest mean squared error (MSE) for the last four forecast hours.

3.5   Comparison

From the analysis of the Tables 3, 4, 5 and 6, several observations can be made. For the last three forecast hours the third scenario stands out with the LSTM model achieving the lowest MSE and the highest percentage of forecast points within the desired range. On the other hand, the ESN model performs the best in terms of R-squared scores for the same hours.

However, for the first three hours, the metrics are better for the first three scenarios. This suggests that incorporating weather forecast data into the training set improves the model's ability to forecast long-term values (last three hours), but it may have a negative impact on forecasting the first three hours.

Another interesting finding is that for none of the forecast hours or metrics the fourth scenario performs the best. This could be due to the simplicity of the models in handling both temporal and spatial dependencies, as well as the large number of features in this scenario.

Overall, these results highlight the importance of considering the specific requirements and characteristics of the forecasting task when selecting and designing the appropriate model and training set.

3.6   Forecasting Visualization

The visualizations of forecasts of PM2.5 values for the third scenario three hours in the future compared to ground truth values from validation set are presented in Fig. 5 for visual overview purposes. Every point on the graph is a single forecast of PM2.5 value for three hours in the future based on 12 points of input data (each point consisting of the features of third scenario described in Sect. 2.2) compared to the PM2.5 ground truth value.

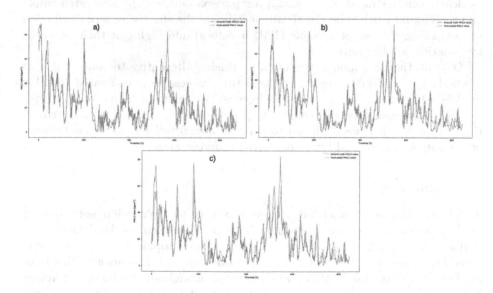

Fig. 5. Forecasts of PM2.5 values for three hours in the future obtained from a) ESN, b) LSTM, c) MLP model.

4 Summary and Future Work

4.1 Designed Models and Methods

The ESN model performed very well in terms of speed, outpacing both the LSTM and MLP models while maintaining relatively satisfactory performance. The ESN model was almost 60 times faster in fitting the data compared to the LSTM model and 5 times faster than MLP model for the third and most favorable scenario. However, when considering the forecast evaluation metrics, the LSTM model achieved the best results, while the MLP model exhibited the poorest performance.

Training the models solely on historical PM2.5 data yielded satisfactory results for the short-term forecasts. However, for long-term forecasts, the performance deteriorated significantly. By incorporating additional meteorological data in the second scenario the long-term forecasts improved.

Furthermore, the inclusion of weather forecast data further enhanced the long-term forecasting capabilities. However, incorporating spatial dependencies actually led to a decrease in the performance of all models. This suggests that the designed models might be too simplistic to effectively capture and utilize spatial and temporal features in the data simultaneously.

The obtained results from the models can be regarded as highly satisfactory, as clearly demonstrated by the visual comparison between the forecasted values and the ground truth. The visualizations validate the effectiveness of the models in forecasting the target variable (PM2.5 values) and highlight their potential for practical applications.

Overall, the ESN model presents a promising alternative to state-of-the-art methods by significantly reducing the fitting process time compared to other models, while achieving a comparable forecast performance. The results are specific to a particular geographical region, period of time, and measurement conditions. Therefore it is very difficult to perform a qualitative comparison between our results and presented by authors from different areas.

4.2 Improvements

To further enhance the modeling process, it is advisable to develop more complex models capable of effectively capturing both temporal and spatial dependencies within the data. One approach is to design a more sophisticated models with more layers and longer training time. Another approach to achieve this is by implementing ensemble learning, where multiple models are combined to leverage their individual strengths. In this context, one model can be specifically designed to capture temporal dependencies, while another model can focus on spatial dependencies.

By integrating these two models through ensemble learning techniques, such as model averaging or stacking, it becomes possible to leverage the collective intelligence of both models and potentially achieve superior performance. This approach takes advantage of the complementary nature of temporal and spatial information, allowing for a more comprehensive understanding of the underlying patterns and dynamics in the data.

In addition to model complexity, allocating more computational resources can enable more extensive model tuning. This involves exploring a wider range of hyperparameter configurations and optimizing the model settings to achieve the best possible performance.

In future work, there are several areas to explore for enhancing the models and their applications. The presented approach can be extended and evaluated for other quantities apart from PM2.5, such as NO, NO_2, SO_2 or O_3. Another aspect is to investigate alternative methods for selecting stations, such as employing Granger causality analysis. This might help identify influential stations better than Pearson Correlation. Additionally, expanding the dataset by collecting real-time weather data alongside air pollution and meteorological data, thus making the training data set larger can potentially lead to better forecasts.

Acknowledgements. The research presented in this paper was partially financed by the funds of Polish Ministry of Education and Science assigned to AGH University.

References

1. Abdul-Wahab, S., Sappurd, A., Al-Damkhi, A.: Application of California Puff (CALPUFF) model: a case study for Oman. Clean Technol. Environ. Policy **13**, 177–189 (2011)
2. Abhilash, M.S.K., Thakur, A., Gupta, D., Sreevidya, B.: Time series analysis of air pollution in Bengaluru using ARIMA model. In: Perez, G.M., Tiwari, S., Trivedi, M.C., Mishra, K.K. (eds.) Ambient Communications and Computer Systems. AISC, vol. 696, pp. 413–426. Springer, Singapore (2018). https://doi.org/10.1007/978-981-10-7386-1_36
3. Amegah, A., Jaakkola, J.: Household air pollution and the sustainable development goals. Bull. World Health Organ. **94**, 215 (2016)
4. Bai, L., Wang, J., Ma, X., Lu, H.: Air pollution forecasts: an overview. Int. J. Environ. Res. Public Health **15**, 780 (2018)
5. Cordova, C.H., Portocarrero, M.N.L., Salas, R., Torres, R., Rodrigues, P.C., López-Gonzales, J.L.: Air quality assessment and pollution forecasting using artificial neural networks in metropolitan Lima-Peru. Sci. Rep. **11**(1), 24232 (2021)
6. Davis, R.A., Zang, P., Zheng, T.: Sparse vector autoregressive modeling. J. Comput. Graph. Stat. **25**, 1077–1096 (2016)
7. Fionn, M.: Multilayer perceptrons for classification and regression. Neurocomputing **2**, 183–197 (1991)
8. Gholamzadeh, F., Bourbour, S.: Air pollution forecasting for Tehran city using Vector Auto Regression. In: 6th Iranian Conference on Signal Processing and Intelligent Systems (ICSPIS), December 2020
9. Guo, H., Liu, X., Sun, Z.: Multivariate time series prediction using a hybridization of VARMA models and Bayesian networks. J. Appl. Stat. **43**, 2897–2909 (2016)
10. Hajar, H., Benjamin, H.: Multivariate time series modelling for urban air quality. Urban Climate **37**, 100834 (2021)
11. Hodge, V., Austin, J.: A survey of outlier detection methodologies. Artif. Intell. Rev. **22**, 85–126 (2004)

12. Junshan, W., Guojie, S.: A deep spatial-temporal ensemble model for air quality prediction. Neurocomputing **314**, 198–206 (2018)
13. Junshan, W., Guojie, S.: PlumeNet: large-scale air quality forecasting using a convolutional LSTM network arXiv:2006.09204 (2020). Accessed 26 Mar 2023
14. Lewis-Beck, M.S., Skalaban, A.: The R-squared: some straight talk. Polit. Anal. **2**, 153–171 (1990)
15. Masini, R.P., Medeiros, M.C., Mendes, E.F.: Machine learning advances for time series forecasting. J. Econ. Surv. **37**, 76–111 (2023)
16. Mondal, P., Shit, L., Goswami, S.: Study of effectiveness of time series modeling (Arima) in forecasting stock prices. Int. J. Comput. Sci. Eng. Appl. **4**, 13 (2014)
17. Nasstrom, J., Sugiyama, G., Leone, J., Ermak, D.: A real-time atmospheric dispersion modeling system. Technical report, Lawrence Livermore National Lab.(LLNL), Livermore, CA (United States) (1999)
18. Peng, C., Aichen, N., Duanyang, L., Wei, J., Bin, M.: Time series forecasting of temperatures using SARIMA: an example from Nanjing. IOP Conf. Ser. Mater. Sci. Eng. **394**, 052024 (2018)
19. Pratama, I., Permanasari, A.E., Ardiyanto, I., Indrayani, R.: A review of missing values handling methods on time-series data. In: 2016 International Conference on Information Technology Systems and Innovation (ICITSI) (2016)
20. Radim, S., et al.: Health impact of air pollution to children. Int. J. Hyg. Environ. Health **216**, 533–540 (2013)
21. Richard, W., Marcus, O.: Judgemental and statistical time series forecasting: a review of the literature. Int. J. Forecast. **12**, 91–118 (1996)
22. Samal, K.K.R., Babu, K.S., Das, S.K., Acharaya, A.: Time series based air pollution forecasting using SARIMA and prophet model, pp. 80–85 (2019)
23. Scire, J.S., Strimaitis, D.G., Yamartino, R.J., et al.: A user's guide for the Calpuff dispersion model. Earth Tech. Inc. **521**, 1–521 (2000)
24. Setianingrum, A.H., Anggraini, N., Ikram, M.F.D.: Prophet model performance analysis for Jakarta air quality forecasting, pp. 1–7 (2022)
25. Shumway, R.H., Stoffer, D.S.: ARIMA models. Time series analysis and its applications (2017)
26. Snoun, H., Kanfoudi, H., Christoudias, T., Chahed, J.: One-way coupling the Weather Research and Forecasting Model with ADMS for fine-scale air pollution assessment. Colloque scientifique National, February 2017
27. Tartakovsky, D., Broday, D.M., Stern, E.: Evaluation of AERMOD and CALPUFF for predicting ambient concentrations of total suspended particulate matter (TSP) emissions from a quarry in complex terrain. Environ. Pollut. **179**, 138–145 (2013)
28. Taylor, S.J., Letham, B.: Forecasting at scale. Am. Stat. **72**(1), 37–45 (2018)
29. Wallach, D., Goffinet, B.: Mean squared error of prediction as a criterion for evaluating and comparing system models. Ecol. Model. **44**, 299–306 (1989)
30. Xu, X., Ren, W.: Application of a hybrid model based on echo state network and improved particle swarm optimization in PM2.5 concentration forecasting: a case study of Beijing, China. Sustainability **11**, 3096 (2019)
31. Ye, Z.: Air pollutants prediction in Shenzhen based on ARIMA and Prophet method. E3S Web of Conferences, January 2019
32. Yin, F., Jinhua, C., Jun, S., Han, S.: Spatial effects of air pollution on public health in China. Environ. Resource Econ. **73**, 229–250 (2018)
33. Yunus, K., Chen, P., Thiringer, T.: Modelling spatially and temporally correlated wind speed time series over a large geographical area using VARMA. IET Renew. Power Gener. **11**, 132–142 (2017)

$\mu Chaos$: Moving Chaos Engineering to IoT Devices

Wojciech Kalka[1]([⊠]) [iD] and Tomasz Szydlo[1,2]([⊠]) [iD]

[1] Institute of Computer Science, AGH University of Krakow, Al. Mickiewicza 30,
30-059 Krakow, Poland
{wkalka,tomasz.szydlo}@agh.edu.pl
[2] Newcastle University, Newcastle Upon Tyne NE1 7RU, UK
tomasz.szydlo@newcastle.ac.uk

Abstract. The concept of the Internet of Things (IoT) has been widely used in many applications. IoT devices can be exposed to various external factors, such as network congestion, signal interference, and limited network bandwidth. This paper proposes an open-source $\mu Chaos$ software tool for the ZephyrOS real-time operating system for embedded devices. The proposed tool intends to inject failures into device's applications in a controlled manner to improve their error-handling algorithms. The proposed novel framework fills the gap in the chaos engineering tools for edge devices in the cloud-edge continuum. In the paper, we also discuss the typical failures of IoT devices and the potential use cases of the solution.

Keywords: IoT · Chaos Engineering · fault injection · Zephyr-OS

1 Introduction

According to predictions, the number of IoT devices is expected to cross 29 billion in 2030. This is an increase of about 300% compared to 2019 [1]. Nowadays, IoT devices become ubiquitous in many areas of life so they can be exposed to various external factors. The spectrum of their operation covers from smart home applications, through transport or industrial areas, to extreme conditions, such as sensors working at fire sites to track the safety of firefighters [8]. Therefore, ensuring the work continuity of IoT devices is a significant challenge.

Nowadays, more and more companies are paying attention to the proper testing of products before they are launched into production. It is crucial to provide consumers with an appropriate quality and resilience of services [19]. This has given rise to the concept of *Chaos Engineering*, which Netflix introduced [3] in 2010. Before that, engineers were forced to introduce constant improvements and add new functionalities to handle noticed problems. The systems' complexity caused that there was not possible to predict all possible failures during the design time [7]. This led to the creation of Chaos Monkey[1] a tool for the unpredictable termination of system parts, including virtual machines and containers,

---

[1] Netflix, https://netflix.github.io/chaosmonkey/.

L. Franco et al. (Eds.): ICCS 2024, LNCS 14838, pp. 239–254, 2024.
https://doi.org/10.1007/978-3-031-63783-4_18

to test whether the system is fault-tolerant. However, chaos engineering could be deployed not only in IoT but also in e.g. blockchain technologies [20].

In the IoT domain, the failure resiliency problem is more complex due to its multilayered architecture. The outages and other disturbances could appear in many places along the sensor data processing paths. The IoT systems generally comprise three layers - perception, edge, and the cloud. Each layer has its weak spots that can cause failures. In this paper, we will focus on the perception layer where IoT devices are prone to errors stemming from various factors. Since the most commonly used type of communication is wireless connectivity, network congestion, signal interference, and limited bandwidth can disrupt communication between devices and cause delays or failures in data transmission. Modern microcontrollers allow for more complex operations, enabling TinyML algorithms, which are susceptible to sensor data failures. These examples and many others conclude that chaos engineering has open challenges in IoT devices.

Several chaos engineering tools exist for cloud services, such as Chaos Twin [4], Chaos Monkey, CHAOSORCA [13], Chaos Recommendation Tool [18], or Chaos Mesh for Kubernetes. However, there are few known solutions for IoT end devices, but they mainly work in static software testing, e.g., Chaos Duck [5]. Other tools were also created for faults injection, but they are focused on the network layer using customized MQTT broker commonly used in IoT [6]. This paper proposes an open-source $\mu Chaos$ software component for the real-time operating system ZephyrOS for embedded devices supported by organizations like Google, Intel, and Nordic Semiconductor. By enabling $\mu Chaos$ component in the OS, a user can inject several types of faults, including the ones related to the sensors, hardware, network, and application. The $\mu Chaos$ exposes a flexible API that can be easily interacted with via serial port or console, enabling interaction during the tests. In the paper, we will discuss types of faults in IoT devices and show how the proposed novel tools and the software library work in the selected use cases.

The paper is organised as follows. Section 2 presents related work, and Sect. 3 introduces fault taxonomy in IoT and faults in IoT end devices. The next section describes the concept and the details of the designed tool. Section 5 presents the use case and experiments, while the last section gives conclusions.

1.1   Challenges

Performing chaos engineering techniques for IoT devices requires facing the following challenges: (i) identification of the common faults related to the IoT devices, (ii) introduction of the failure scenarios mimicking real failures, (iii) seamless integration of the chaos engineering tools with the operating systems for embedded devices.

1.2 Contribution

This paper proposes a novel chaos engineering tool for IoT devices encompassing several types of failures and failure scenarios. The following are the paper's key contributions: (i) analysis of the common failures targeting IoT devices and the mimicking scenarios, (ii) implementation of the μChaos tool for Zephyr OS, (iii) evaluation of the solution in typical IoT use cases.

2 Related Work

Due to the complexity resulting from the multi-layer architecture of IoT systems, it is challenging to have a holistic tool to test them through the full architecture stack. Therefore, different testing methods depend on which part or layer of the system is analysed. Properly chosen testing strategy and selected tools make the debugging and development process more efficient. Finally, the chaos engineering tools should allow for finding the bugs early in development, saving time and reducing project costs. This section presents a selected tool related to chaos engineering, listed in Table 1.

Chaos Monkey[2] developed by Netflix, aims to terminate in a controllable manner production instances of one or more virtual machines. These machines can run some microservices, e.g. video transcoding or streaming. It is possible to configure the scheduler to trigger failures at times when they can be better monitored. Another tool, *Gremlin* [2] is provided as a SaaS (Software as a Service) platform[3]. It allows testing cloud-based applications and infrastructure. The tool works with the most popular platforms like AWS or Azure. It has an agent that should be installed on a user container or virtual machine. Gremlin provides different possibilities for failure introduction, including computational resources, network, and system states by killing processes.

Chaos Mesh[4] is an open-source tool for cloud-based solutions, widely used with Kubernetes. It supports many types of faults, i.e. network latency, packet loss, HTTP communication latency, CPU race, system time changes, and platform faults like AWS node restart. It supports losing packets, pressure on the CPU, increasing physical disk load, filling disks, and killing or stopping processes. *Chaos Twin* [4] is a simulation tool created to work with cloud-based systems. It allows the creation of the digital twin of a system under test and then evaluates its performance on a business level by finding the most optimized architecture. Digital twin is a concept which occurs in more research [17]. The tool focuses on three types of errors, including partial or complete data centre failures, incorrect operation of virtual machines, and delays in communication between data centres. Simulations are run until the specified number of iterations is reached.

---

[2] Netflix, https://github.com/Netflix/chaosmonkey.
[3] Gremlin, https://github.com/gremlin/chaos-engineering-tools.
[4] The Linux Foundation, https://chaos-mesh.org/.

Chaos Duck[5] is a tool for testing IoT devices that emulates fault injection attacks. Injecting faults via hardware is difficult and expensive, so *Chaos Duck* uses a tested program binary file. It disassembles a chosen file and collects information about the code, e.g. branch instructions, static variables, and address space. *Chaos Duck* produces many faulted binaries and executes each by collecting statistics about the impact of each fault. The tool supports x86 and ARM architectures. The following error types are available - *bit flip (FLP)*, *byte zeroing*, and adding *conditional, and unconditional branches*.

Custom MQTT Broker[6] is a tool created for testing systems that use communication via MQTT. The authors have chosen a broker as a crucial element in the distributed systems, independent from factors such as complexity, and focused on the network layer as a place to inject failures. Messages are gathered and modified before sending to the clients. Each rule consists of the *topic* and an array of filters named *operators*. Operators transform the messages and pass them to the next one as an additional parameter. The tool has four operators - *map, randomDelay, message buffering*, and *randomDrop*.

The discussed tools do not focus on the internal peripherals of the IoT devices, which are essential from a data processing perspective. They are fairly directed to cloud layer of the IoT systems. Therefore, the proposed in this work μ*Chaos* tool is designed to fit into the embedded operating system, acting as the intermediate layer between the sensors and the application.

Table 1. Chaos Engineering testing tools for IoT systems.

Tool name	Layer	CPU	Memory	Peripheral Device	Virtual Machines	Network	Sensor	Application
Chaos Monkey	Could	○	○	○	●	○	○	○
Gremlin	Cloud	●	●	●	○	●	○	●
Chaos Mesh	Cloud	●	●	○	○	●	○	●
Chaos Twin	Cloud	○	○	○	●	●	○	○
Chaos Duck	End device	○	○	○	○	○	○	●
Custom MQTT Broker	Edge	○	○	○	○	●	●	○
μChaos	IoT	●	●	○	○	○	●	●

● - Supported
○ - Not Supported

2.1   IoT Faults Taxonomy

Due to the complexity of IoT systems, errors can occur at different layers - cloud (e.g. server failures), edge (e.g. package loss, incorrect data processing) and end devices (e.g. sensor damage). This makes it challenging to create a very detailed IoT failure taxonomy. However, referring to works describing faults in WSN (Wireless Sensors Networks) [12], combined with the information regarding IoT faults [10], it is possible to introduce a general classification, divided into different categories. Figure 1 shows the taxonomy of IoT faults proposed in this paper and

---

[5] Igor Zavalyshyn, https://github.com/zavalyshyn/chaosduck.
[6] SIGNEXT, https://github.com/SIGNEXT/instrumentable-aedes.

divided into four categories of faults affiliation, bahaviour, time, system layer, and location.

From the system's operational perspective, the main distinctive factor is the type of failure, which might be either hard or soft. A soft fault is when a device can respond and communicate, but the system knows it is damaged, e.g. it sends invalid data. On the contrary, a hard error is when the component or the device is not detectable or completely broken. Low battery and no power are good examples [15].

Another issue is related to the time domain during which that failure might appear. This category covers permanent, transient, or intermittent failures. Long-term faults are usually permanent. An example of such an error may be the previously mentioned power supply problem or a damaged communication module. Intermittent faults may be caused by changes in the operating environment, such as sensors covered by, e.g. worms, dust, plants, shadows, or objects placed by humans. Intermittent faults last longer than transient faults. The interval between their appearance might vary and be unpredictable [11].

The next group of faults is related to the system layers. End devices are very susceptible to errors, which may come from the outside environment they are working in or might result from their invalid operation. Each IoT device can be described from two perspectives - hardware and software. The hardware is mainly CPU, communication module, power module and all kinds of sensors. Any of these items may be damaged or contain manufacturing errors. The most common software fault is a problem with memory management. This can lead to memory leaks and device malfunction. Other common issues are sensor calibration, data processing, and packet forming. Then, the edge layer might be susceptible to failures such as insufficient network bandwidth, resulting in packet loss or incorrect processing/filtering of received data. Finally, in the cloud layer, the failure might appear with service configurations, database failures, errors in the application or the wrong selection of algorithms for processing the results.

The last group considered in the taxonomy is location-based faults. These can be internal system faults, such as connectivity or infrastructure problems. Defects from the external environment are important, as they significantly impact end devices. Changes such as temperature spikes can interfere with measurements, and electromagnetic interference or discharges can damage the device.

2.2 Operating Systems for IoT

There are many operating systems for embedded systems on the market. Most are open-source, and choosing the right system depends on the project's requirements. Some big tech companies have invested in developing Real-time Operating Systems (RTOS), for example, Amazon or Microsoft. One of the most popular is FreeRTOS maintained mainly by Amazon. It provides the RTOS API, with no support for other functionality such as GPIO or serial drivers, which are left to the developer to create.

The Zephyr OS is governed by the Linux Foundation. Its architecture is similar to Linux, e.g. the kernel is configured similarly. It is intended for complex

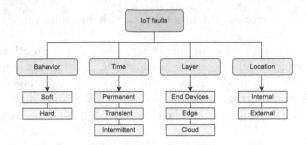

Fig. 1. IoT faults taxonomy.

and high-level embedded applications. It has ready-to-use libraries and supports various hardware target processors and development boards. It is more portable than other operating systems but could be more complicated for less experienced users. It has more than 450 officially supported boards and a lot of ready-to-use examples of system features.

Mbed OS is an RTOS for IoT solutions based on Cortex-M boards. It provides an abstraction layer for C/C++ applications that ensures portability between platforms supporting this operating system. There are all necessary operating system components e.g. Semaphores, Queues, Threads, Mutexes and Scheduler for switching between application activities.

Since 2019, Microsoft has supported the development of Azure RTOS ThreadX. One of the most important features is an advanced multitasking solution named *preemption-threshold scheduling*. It allows a sub-microsecond context switching and was a topic of academic research [16]. Manufacturers emphasize that performance is the feature that distinguishes ThreadX from other RTOSs. One of the solutions is the *Picocernel* design, where services don't have many layers, so additional overhead in function call is removed. Moreover, interrupt handling is also optimized because only scratch registers are processed.

3   Chaos Tool Concept

This section presents the concept of chaos engineering tool and its working principles. We will analyze and model faults introduced by the tool and discuss its integration with the resource-constrained Zephyr operating system.

3.1   IoT Faults Modeling

In this paper, we will mainly focus on IoT end device faults. For this group of devices, we are dividing them based on three criteria - hardware, software, and data. IoT devices are often specialized to work in specific environments and are most likely optimized for low energy consumption. Based on [14], we have collected selected faults in the Table 2, Table 3, and Table 4 along with their descriptions.

The first group of failures is associated with data processing. Data failures might result in the inability to access or utilize it effectively. These failures can occur due to various reasons, including technical issues, human error, software bugs, cybersecurity breaches, natural disasters, and hardware malfunctions. The selected failures are presented in Table 2.

The significant challenge is the proper handling of software failures, as it can lead to the dysfunction of the entire device or its most important parts, e.g. communication with the rest of the system. The selected failures are presented in Table 4. Dealing with these failures in microcontroller-based devices with limited memory and computing resources is demanding. Every software failure, which forces the service staff to go and recheck the device state or program, generates additional costs that should be avoided.

The last group of failures covers hardware ones. Depending on the conditions, it is critical to properly secure the device to protect it from external conditions and possible mechanical damage because it can manifest itself in many ways and be challenging to detect. Selected failures are summarized in Table 3.

Table 2. Data faults

Fault	Classification	Description	Inject
Outlier	Soft, Intermittent	Single, unexpected value that comes a lot over normal measurements. It could be caused by some hardware problems like unstable sensor connection or external factors e.g. electromagnetic pulse.	Pseudo-random value, chosen by the user and added to the original measurement, occurs with pseudo-random frequency.
Spike	Soft, Transient	Set of data with a value different than expected, often as a result of supply or connection failures.	The peak in read sensor data, which rises and slopes symmetrically in a number of samples set by the user.
Offset/ Rota- tion	Soft, Transient	Some value constantly added to the output. If the sensor position was changed due to some case or mounting system damage, it may produce different values.	Sample is increased every time by a constant percentage part of measurement, chosen by the user.
Stuck at value	Soft, Permanent	Sensor readings stay the same for a period of time. The reason may be a sensor malfunction.	Value returned by a sensor is constant and taken from the first sample after command execution.
Noise	Soft, Permanent	An additional, approximately constant value changing the read results may be related to a change in the environment in which the sensor operates, e.g. covering by another object, dirt, or change in ambience temperature.	Similar as in 'Outlier' however a sample is increased by value in every measurement.

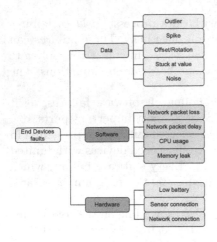

Fig. 2. IoT end devices faults.

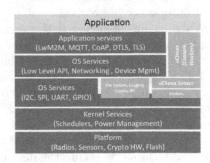

Fig. 3. ZephyrOS high-level architecture with *μChaos* components marked orange.

Table 3. Hardware faults

Fault	Classification	Descripiton	Inject
Low battery	Hard, Permanent	The amount of energy drawn from the battery prevents proper operation	Device battery measured voltage value is reduced by the value and with timestamp selected by the user
Sensor connection	Soft, Transient	IoT devices usually have sensor connected. Every mechanical connection has limits to its strength. Prolonged exposure to external conditions may cause some damages. There are also many other situations that are not related to harsh environmental conditions	Zephyr has a special value to signal, that the microcontroller is unable to communicate with the sensor. This value is returned with a pseudo-random range set by the user

3.2 μChaos Tool Design

Among the open-source Real-time Operating Systems (RTOS) solutions available on the market and discussed in Sect. 2.2, we have chosen Zephyr OS. The most important aspect is that it supports sensors, memory management and protocols used in IoT and provides modularity. Noteworthy is also the fact that it has a significant and growing community size. Another aspect was to ensure that μChaos did not take up too much RAM and ROM resources due to the fact that it is a library dedicated to embedded devices.

As shown in Fig. 3 the tool comprises two components - *μChaos Sensor* and *μChaos Console*. The first one, *μChaos Sensor*, is designed to manipulate the measurement values. The second one, *μChaos Console*, is implemented as a separate thread working in the background and manages the sensor intermediate layer and introduces failures. Its functionality is exposed via UART

Table 4. Software faults

Fault	Classification	Descripiton	Inject
CPU usage	Hard, Transient	When too many tasks in the operating system work concurrently and there are no more available resources to do other operations. There may also be a bug in the software in tasks switching and one or more still occupy the CPU	Set of threads to possibly add to the application. Their number with all necessary parameters like priority, stack size, and area has to be determined by the user at program compilation time. At the application beginning all defined threads are ready to start and the user can freely start them and then suspend and resume
Memory leak	Hard, Intermittent	One of the common mistakes, especially when it comes to IoT devices, as developers have very little available memory and computing resources. It is quite easy to come over the size of the array with e.g. measurements	Blocks of dynamically allocated memory with the size chosen by the user. Each block has a name to handle and a pointer to allocated memory. Blocks could be created at any time the application is running

(Universal Asynchronous Receiver-Transmitter) and thus can be controlled by external physical components but also is exposed to Zephyr Console where failures might be introduced via interactive command line console. The tool is provided as an open-source library, available on GitHub[7].

3.3 Data Faults Injection

Data faults are the most common group of failures available in *μChaos Sensor* component of the library. *μChaos Sensor* has been designed as a wrapper on the sensor subsystem provided by Zephyr OS. After activating μChaos, each time the Sensor's methods are called, it redirects to the library function. It uses low-level functions to obtain the data, and then before passing the values to

---

[7] https://github.com/wkalka/uChaos.

the user application code, failures are introduced, as discussed in the previous section. Every type of fault has its own method to manipulate the data or driver behaviour properly. Thanks to this, the applications' operations on the sensors remain unchanged.

One of the main goals of this project was to design the library coherently with the sensor drivers provided by the Zephyr OS. Therefore, the failure types are associated with the types of sensors and not the particular devices. To use μChaos in a project, the user has to initialize its components at the beginning. For the sensors part, e.g. there is a parameter that indicates the number of used sensors in the application. All settings could be changed in the configuration file, e.g. which types of failures will be used in the application.

3.4   Hardware Faults Injection

Hardware faults group is implemented by μ*Chaos Battery* component. This part of μ*Chaos* library is based on Zephyr OS Analog-to-Digital Converter (ADC) driver. Similarly to μ*Chaos Sensor* it uses low-level system functions to obtain the tested voltage. After every `adc_raw_to_millivolts_dt` ADC driver function call, which returns the measured value expressed in millivolts, the context is redirected to `uChaosBattery_RawToMillivoltsDt`. Inside this method, the data is manipulated to simulate malfunctions in a device's power supply. Voltage drops may occur for a certain number of measurements or continuously until the minimum permissible battery voltage is reached.

3.5   Software Faults Injection

In μChaos, a software faults are realised by μ*Chaos CPU*, where a CPU is loaded by additional operations e.g. variable iteration or mathematical calculations, which don't have any impact on the tested application. The aim is to capture a free CPU computational resources, to keep a CPU as long as possible in non-idle state. This could lead to limiting some functions of threads critical for application like communication with a network, monitoring power supply and measuring some crucial metrics. This part of μChaos library doesn't redirect behaviour of custom system functions. User has to define a thread with selected priority and name to recognize the thread via μ*Chaos Console*. After the application begins running all μChaos threads are in *prestart* state and then the user can start them or suspend them via certain commands e.g. *load_ add LoadThread*.

4   Use Case and Evaluation

In the evaluation, we consider a typical IoT device used to monitor industrial machines. In the target machine, we monitor its temperature and vibrations using a temperature sensor and an accelerometer. The IoT device is battery-powered. In the evaluation, we are focusing only on the perception layer. Therefore, we are analysing only the raw values provided by the sensors. Experiments realise different scenarios, each representing a distinct type of data errors shown in Fig. 2.

We have prepared experiments with two types of sensors (Fig. 4) - accelerometer ADXL345 and temperature sensor DPS310. Sensors were placed on a separate, small development board with connectors[8],[9]

The application was run on nRF52 DK, a development kit from Nordic Semiconductor with nRF52832 Soc (System on Chip)[10] It supports wireless communication like Bluetooth LE, Bluetooth mesh ANT and NFC.

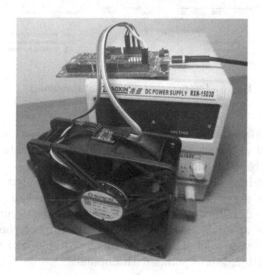

Fig. 4. Testbed setup.

4.1 Temperature

Scenario for the temperature sensor contains producing anomaly. During the experiment, the sensor was subjected to a heat source after the first 10 samples were taken, and then a further 50 samples were captured. The heat source was then removed, and a final 150 measurements were taken to bring the sensor back to near its initial temperature. However, this took about 100 measurements due to the inertia of the sensor.

In the next experiment iteration, fault injection began after the first 5 samples (indicated by a black vertical line), before the heat source approach, and ended before the last 10 samples of the experiment. Regarding data anomaly, outliers

---

[8] OKYSTAR, ADXL345, https://www.okystar.com/product-item/adxl345-digital-3-axis-gravity-acceleration-sensor-oky3247/.

[9] Seeed Studio, DPS310, https://wiki.seeedstudio.com/Grove-High-Precision-Barometric-Pressure-Sensor-DPS310/.

[10] Nordic Semiconductor, nRF52 DK, https://www.nordicsemi.com/Products/Development-hardware/nrf52-dk.

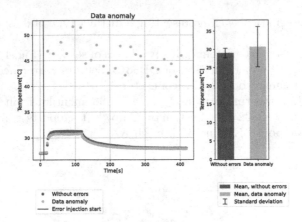

Fig. 5. Temperature sensor data anomaly.

appeared in the data every 6–10 measurements with a level between 50% and 70% of the original value. Separated points in Fig. 5 show data anomalies.

The temperature value chart is accompanied by a comparison of the average value and standard deviation across the entire measurement range. These result validate the experiment's assumption, that the mean is slightly higher than the clear signal, indicating that some changes have occurred in the data. This suggests that anomalies do not significantly impact the overall average value. However, they do contribute to an increased standard deviation due to the presence of scattered data points with anomalies.

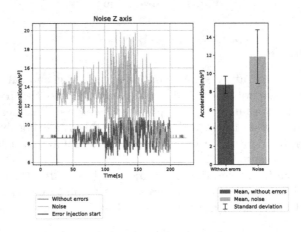

Fig. 6. Accelerometer Z axis noise.

4.2   Acceleration

In this case, we had a different scenario, where the noise error was injected. The acceleration was measured every 500 ms and displayed on the device console. The sensor was mounted on top of a computer fan, while the fan speed was voltage-controlled by the laboratory power supply. Measurements were made at four fan speeds, including idle. Scenario consists of 450 samples. After every 100 samples, the speed was increased, and the last 50 measurements were at idle speed again. In case containing faults injection, it started after the first 50 samples of idle speed and finished after 50 measurements of the highest fan speed. The fault was applied to each accelerometer axis.

The black vertical line points to the moment of faults injection start. The axis Z showed the direction where the vibrations had the greatest impact including also the gravitational acceleration. The range of the added noise value was between 40% and 70% of the original values, as shown in Fig. 6. In analyzing the provided information, it is evident that the graph displaying the noisy data is noticeably shifted from the original graph. This observation is further supported by an approximate increase of 40%–45% in the average value. Additionally, the high standard deviation value indicates a substantial absolute value of noise, aligning with the assumptions made for the experiment, ranging from 40%–70%.

4.3   Battery Measurement

As the majority of IoT devices are battery-powered, monitoring of energy source level is essential. Below some voltage value, typically 1.7 V–1.8 V, most micro-controllers cannot work properly. Information about the low battery state could prevent the device from being useless by enabling power-save modes. Remaining in the deep discharge state is also destructive for the battery itself. For this purpose, we have created *μChaos Battery*. This component functionality allows the simulation of an unexpected power consumption increase as a rapidly decreased battery voltage level. A sudden, increased energy consumption was simulated by reducing the read battery voltage value by 10 mV, every 2 measurements, until the minimum voltage value was obtained, which for the type of battery used was approximately 1.8 V.

4.4   CPU Usage

To analyze the behaviour of the embedded firmware facing the high CPU util-isation in the tool, we have implemented a separate thread which, at regular intervals, performs a simple variable increment operation for approximately 300 milliseconds. Subsequently, the thread waits for the next operation for 1 s, effectively increasing the CPU load. The user can initiate and resume specific threads through the *μChaos Console* in *μChaos CPU* and has the flexibility to define the tasks of the load functions and the threads utilized. As subsequent threads loading the system were launched, the time between subsequent temperature measurements increased from 2 s to 16 s. The threads were then gradually disabled to decrease again the interval between measurements.

4.5 Memory Consumption

Table 5 shows the size of each μChaos component. In accordance with the assumptions outlined in Sect. 3.2, one of the key design considerations for out library was its imperative to occupy minimal space in both ROM and RAM, given the constrained resources typically available in embedded systems. As it could be seen, due to the need to operate on strings and handle many commands, *μChaos Console* takes up most of the RAM and ROM. Following the tests conducted on the nRG52 DK board, it can be inferred that μChaos aligns with the initial assumptions. The proportion of RAM it utilizes amounts to about 2% of the total, while in terms of ROM, it accounts for below 1%.

Table 5. μChaos memory size

Component	ROM [kB]	RAM [kB]
μChaos Sensor	1,21	0,11
μChaos Battery	0,3	0,04
μChaos Console	1,58	1,32
μChaos CPU	0,27	0,05

5 Summary

The paper discusses the novel *μChaos* tool for Zephyr OS based embedded IoT devices. It enables chaos engineering to be applied directly to low-power battery-operated devices. In the paper we have also discussed the common types and sources of the failures of the devices. Finally, we have presented the scenarios showing the usage of the developed tool in sensor-based applications for IoT devices.

The proposed tool can be augmented by the edge and cloud chaos engineering tools composing the holistic failure injection solution for the IoT continuum. Further improvements should also concern other data source modalities, including audio data from the microphones for keyword spotting applications and predictive maintenance and video sources for tiny object detection and image recognition. Another possible direction of library improvement is exploring faults and disturbances in communication between devices in wireless networks like Wi-Fi, Matter, and Bluetooth. A diversity in types of networks and protocols creates interesting opportunities.

Acknowledgments. The research presented in this paper received funding from Polish Ministry of Science and Education assigned to AGH University of Krakow.

Disclosure of Interests. The authors have no competing interests to declare that are relevant to the content of this article.

References

1. Vailshery, L.S.: Number of Internet of Things (IoT) connected devices worldwide from 2019 to 2021, with forecasts from 2022 to 2030, 22 November 2022. https://www.statista.com/statistics/1183457/iot-connected-devices-worldwide/. Accessed 20 Mar 2023

2. Butow, T.: Chaos Engineering: the history, principles, and practice, May 2021. https://www.gremlin.com/community/tutorials/chaos-engineering-the-history-principles-and-practice. Accessed 20 Mar 2023

3. Sharieh, S., Ferworn, A.: Securing APIs and chaos engineering. In: 2021 IEEE Conference on Communications and Network Security (CNS), Tempe, AZ, USA, pp. 290–294 (2021). https://doi.org/10.1109/CNS53000.2021.9705049

4. Poltronieri, F., Tortonesi, M., Stefanelli, C.: ChaosTwin: a chaos engineering and digital twin approach for the design of resilient IT services. In: 2021 17th International Conference on Network and Service Management (CNSM), Izmir, Turkey, pp. 234–238 (2021). https://doi.org/10.23919/CNSM52442.2021.9615519

5. Zavalyshyn, I., Given-Wilson, T., Legay, A., Sadre, R., Rivière, E.: Chaos duck: a tool for automatic IoT software fault-tolerance analysis. In: 2021 40th International Symposium on Reliable Distributed Systems (SRDS), Chicago, IL, USA, pp. 46–55 (2021). https://doi.org/10.1109/SRDS53918.2021.00014

6. Duarte, M., Dias, J.P., Ferreira, H.S., Restivo, A.: Evaluation of IoT self-healing mechanisms using fault-injection in message brokers. In: 2022 IEEE/ACM 4th International Workshop on Software Engineering Research and Practices for the IoT (SERP4IoT), Pittsburgh, PA, USA, pp. 9–16 (2022). https://doi.org/10.1145/3528227.3528567

7. Basiri, A., et al.: Chaos engineering. IEEE Softw. **33**(3), 35–41 (2016). https://doi.org/10.1109/MS.2016.60

8. Ahmed, A., et al.: Self-extinguishing triboelectric nanogenerators. Nano Energy **59**, 336–345 (2019). https://doi.org/10.1016/j.nanoen.2019.02.026. ISSN 2211-2855

9. Priyadarshini, I., Bhola, B., Kumar, R., So-In, C.: Novel cloud architecture for Internet of Space Things (IoST). IEEE Access **10**, 15118–15134 (2022). https://doi.org/10.1109/ACCESS.2022.3144137

10. Anandayuvaraj, D., Davis, J.C.: Reflecting on recurring failures in IoT development. In: Proceedings of the 37th IEEE/ACM International Conference on Automated Software Engineering (ASE 2022), pp. 1–5. Association for Computing Machinery, New York (2023). https://doi.org/10.1145/3551349.3559545. Article 185

11. Moridi, E., Haghparast, M., Hosseinzadeh, M., Jassbi, S.J.: Fault management frameworks in wireless sensor networks: a survey. Comput. Commun. **155**, 205–226 (2020). https://doi.org/10.1016/j.comcom.2020.03.011. ISSN 0140-3664

12. Adday, G.H., Subramaniam, S.K., Zukarnain, Z.A., Samian, N.: Fault tolerance structures in wireless sensor networks (WSNs): survey, classification, and future directions. Sensors **22**(16), 604 (2022). https://doi.org/10.3390/s22166041

13. Simonsson, J., Zhang, L., Morin, B., Baudry, B., Monperrus, M.: Observability and chaos engineering on system calls for containerized applications in Docker. Futur. Gener. Comput. Syst. **122**, 117–129 (2021). https://doi.org/10.1016/j.future.2021.04.001. ISSN 0167-739X

14. Ni, K., et al.: Sensor network data fault types. ACM Trans. Sen. Netw. **5**(3) (2009). https://doi.org/10.1145/1525856.1525863. Article 25, 29 pages

15. Predictive power consumption adaptation for future generation embedded devices powered by energy harvesting sources. Microprocess. Microsyst. **39**(4), 250–258 (2015). https://doi.org/10.1016/j.micpro.2015.05.001
16. Wang, Y., Saksena, M.: Scheduling fixed-priority tasks with preemption threshold. In: Proceedings Sixth International Conference on Real-Time Computing Systems and Applications, RTCSA 1999 (Cat. No.PR00306), Hong Kong, China, pp. 328–335 (1999). https://doi.org/10.1109/RTCSA.1999.811269
17. Fogli, M., Giannelli, C., Poltronieri, F., Stefanelli, C., Tortonesi, M.: Chaos engineering for resilience assessment of digital twins. IEEE Trans. Ind. Inform. **20**(2), 1134–1143 (2024). https://doi.org/10.1109/TII.2023.3264101
18. Verma, M., et al.: A chaos recommendation tool for reliability testing in large-scale cloud-native systems. In: 2024 16th International Conference on COMmunication Systems & NETworkS (COMSNETS), Bengaluru, India, pp. 270–272 (2024). https://doi.org/10.1109/COMSNETS59351.2024.10427311
19. Oussane, S., Benkaouha, H., Djouama, A.: Fault tolerance in The IoT: a taxonomy based on techniques. In: 2023 Third International Conference on Theoretical and Applicative Aspects of Computer Science (ICTAACS), Skikda, Algeria, pp. 1–8 (2023). https://doi.org/10.1109/CTAACS60400.2023.10449571
20. Zhang, L., Ron, J., Baudry, B., Monperrus, M.: Chaos engineering of ethereum blockchain clients. Distrib. Ledger Technol. **2**(3) (2023). https://doi.org/10.1145/3611649. Article 22, 18 pages

Enhancing Lifetime Coverage in Wireless Sensor Networks: A Learning Automata Approach

Jakub Gąsior[✉]

Department of Mathematics and Natural Sciences, Cardinal Stefan Wyszyński
University, Warsaw, Poland
j.gasior@uksw.edu.pl

Abstract. This paper focuses on enhancing the lifespan of the Wireless
Sensor Network (WSN) by integrating a distributed Learning Automaton into its operation. The proposed framework seeks to determine an
optimized activity schedule that extends the network's lifespan while
ensuring that the monitoring of designated target areas meets predefined
coverage requirements. The proposed algorithm harnesses the advantages
of localized algorithms, including leveraging limited knowledge of neighboring nodes, fostering self-organization, and effectively prolonging the
network's longevity while maintaining the required coverage ratio in the
target field.

Keywords: Wireless sensor networks · Self-organization · Learning
automata · Maximum lifetime coverage problem

1 Introduction

Wireless Sensor Network (WSN) is a distributed network comprised of small,
battery-powered devices, referred to as sensors, capable of sensing and collecting data from their surrounding environment. These sensors communicate with
one another wirelessly using radio frequency waves and collaborate to perform
specific tasks, such as monitoring environmental parameters like temperature,
humidity, or air quality. Data is then collected and sent for further processing
via a specialized sink node.

The sensors in WSNs are typically low-power and have limited computing
capabilities, making energy efficiency a crucial aspect of their design. Lifetime
optimization in WSNs refers to maximizing the duration of operation or the
network lifetime of a WSN. The lifetime of a WSN is defined as the time elapsed
between the deployment of the network and the time when the first node in the
network runs out of energy.

This work will focus on the power management aspect of maximizing the
lifetime of WSNs. Usually, a collective of sensors that oversee specific areas
often display redundancy, where multiple sensors can cover the same monitored
targets, and the forms of redundancy can vary. The optimal utilization of this
redundancy within WSNs and determining the potential scheduling sequence

L. Franco et al. (Eds.): ICCS 2024, LNCS 14838, pp. 255–262, 2024.
https://doi.org/10.1007/978-3-031-63783-4_19

of sensors is essential in prolonging the network's lifetime. Effectively resolving this coverage issue can lead to the indirect maximization of the WSN lifetime. Therefore, implementing scheduling schemes that alternately regulate the active and sleep states of the sensor nodes, also known as node wake-up scheduling protocols, is a viable technique for enhancing the network's lifetime.

We present a framework for building sensors' activity schedule that addresses the abovementioned challenges, building on the capabilities introduced in prior publications [1,3]. The main contribution of this paper lies in developing and implementing a versatile, LA agent-based model supporting the optimal activity schedule of individual sensors, aiming to extend the network's autonomous lifetime.

The rest of this paper is organized as follows. We introduce the theoretical background of the problem and review related literature in Sect. 2. Section 3 describes our proposed optimization approach. We present the findings of our experiments in Sect. 4. The last section concludes the paper.

2   Theoretical Background

We consider a WSN comprising N sensors $S = \{s_1, s_2, ..., s_N\}$ randomly deployed over a two-dimensional rectangular area of $x \times y [m^2]$. The area contains M targets $T = \{t_1, t_2, ..., t_M\}$ (also called Points of Interest (POI)) that are uniformly distributed with a step of g. All sensors are assumed to have the same sensing range R_s^i and battery capacity b_i.

Each sensor can operate in one of two modes: an *active* mode when the battery is turned on, a unit of energy is consumed, and the POIs within its sensing range are monitored; and a *sleep* mode when the battery is turned off, and the POIs within its sensing range are not monitored.

The i-th sensor's mode during the j-th time interval is denoted by α_i^j, where $\alpha_i^j \in 0, 1$. A value of α_i^j equal to 1 indicates that the i-th sensor is in active mode during the j-th time interval, and 0 indicates that it is in sleep mode. Assuming that battery activation and deactivation occur at discrete time intervals, a quality of service (QoS) measure can be used to evaluate the performance of the WSN. The network coverage can be determined by the ratio of the number of POIs monitored by active sensors to the total number of POIs as follows:

$$COV(t_j) = \frac{|M|_{obs_j}}{|M|}. \tag{1}$$

At any given time, this ratio should not fall below a predetermined value of q ($0 < q \leq 1$). While maintaining complete coverage of the area is desirable, achieving a high coverage ratio (80–90 %) may be more relevant in some cases. The lifetime of a WSN (further denoted as LF(q)) can be defined as the number of k consecutive time intervals t_j in the schedule during which the coverage of the target area is within a range of δ from a specified coverage ratio q, as follows:

$$LF(q) = \max\{k | (\forall j)\, j < k, \quad abs(COV(t_j) - q) \leq \delta\}. \tag{2}$$

Multiple sensors can simultaneously detect the same point in the target area, enhancing data quality or reliability. However, this redundancy can also result in wasted energy [11]. In this study, we approach the point coverage problem in wireless sensor networks as a scheduling problem called the Maximum Lifetime Coverage Problem (MLCP). We aim to extend the network's lifetime by minimizing energy consumption and reducing the number of redundant sensors operating during each time interval.

2.1 Learning Automata

A learning automaton is a self-operating mechanism that responds to a sequence of instructions in a certain way to achieve a particular goal. The automaton either responds to a predetermined set of rules or adapts to the environmental dynamics in which it operates [8].

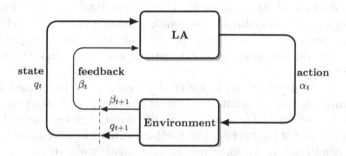

Fig. 1. A feedback loop of learning automata.

The learning process involving the Learning Automata (LA) and a random environment is presented in Fig. 1. Whenever an automaton generates an action α_t, the environment sends a response β_t either penalizing or rewarding the automaton with a specific probability c_i.

Generally, LA can be categorized as a fixed structure LA or a variable structure LA. This paper considers variable structure LA, where the action probability vector is not fixed, and the action probabilities are updated after each iteration. Thus, through interactions with the environment, LAs may adjust their action-selection probabilities by a positive reinforcement (i.e., Reward, Eq. (3)):

$$p_i(t+1) = p_i(t) + a(1 - p_i(t)) \qquad\qquad j = i \qquad\qquad (3)$$
$$p_j(t+1) = (1 - a)p_j(t) \qquad\qquad \forall j, j \neq i$$

or a negative reinforcement (i.e., Penalty, Eq. (4)):

$$p_i(t+1) = (1 - b)p_i(t) \qquad\qquad j = i \qquad\qquad (4)$$
$$p_j(t+1) = \frac{b}{r-1} + (1 - b)p_j(t) \qquad\qquad \forall j, j \neq i$$

Values $p_i(t)$ and $p_j(t)$ are the probabilities of actions α_i and α_j at time t, r is the number of actions, while a and b are the reward and the penalty parameters, respectively. We employ a learning algorithm called *Linear Reward-Penalty* (L_{R-P}) with $a = b$ in our work [8].

2.2   Related Work

In recent years, there has been a growing interest in developing distributed algorithms to address these challenges through reinforcement learning and automata models. [6] introduced an effective scheduling technique named LAML, leveraging learning automata. In this method, each node is equipped with a learning automaton, facilitating the selection of its appropriate state (active or asleep) at any given time.

Subsequently, this research was expanded in [7], where attention was directed toward addressing partial coverage challenges. This scenario involves continuous monitoring of a limited area of interest. The authors introduced the PCLA algorithm to deploy sleep scheduling strategies, demonstrating its effectiveness in selecting sensors efficiently to meet imposed constraints and ensuring favorable performance metrics, including time complexity, working-node ratio, scalability, and WSN lifetime.

In a recent development, [5] introduced an energy-efficient scheduling algorithm utilizing learning automata to address the target coverage problem. This approach allows sensor nodes to autonomously select their operational state. To validate the efficacy of their scheduling method, comprehensive simulations were conducted, comparing its performance against existing algorithms.

In another study, [4] presented a novel on-demand coverage-based self-deployment algorithm tailored for significant data perception in mobile sensing networks. The authors first extend the cellular automata model to accommodate the characteristics of mobile sensing nodes, resulting in a new mobile cellular automata model adept at characterizing spatial-temporal node evolution. Subsequently, leveraging learning automata theory and historical node movement data, they proposed a new mobile cellular learning automata model. This model empowered nodes to intelligently and adaptively determine optimal movement directions with minimal energy consumption.

In their study, [2] utilized a Learning Automata-based model as a routing mechanism in wireless sensor networks, aiming for enhanced energy efficiency and reliable data delivery. The approach aims to calculate the selection probability of the next node in a routing path based on various factors such as node score, link quality, and previous selection probability. Furthermore, they proposed an energy-efficient and reliable routing mechanism by combining learning automata with the A-star search algorithm.

Another contribution by [10] introduced a scheduling technique named Pursuit-LA. Each sensor node in the network was equipped with an LA agent to autonomously determine its operational state to achieve comprehensive target coverage at minimal energy cost.

Lastly, [9] proposed a continuous learning automata-based approach for optimizing sensor angles in Distributed Sensor Networks (DSNs). The method involved continuously adapting sensing angles using LA models. Comparative analysis against a conventional automata-based approach demonstrated the efficacy of the proposed algorithm.

3   Automata-Based Approach to the WSN Lifetime Optimization

In this section, we introduce our proposed methodology. Every sensor node s_i is linked with an automaton LA_i in the setup phase. This automaton randomly chooses one of two available actions (0 - *sleep* or 1 - *active*) and disseminates this decision to n_i immediate neighbors (sensors sharing the same subset of Points of Interest). By the end of this phase, all sensor nodes in the network are informed about their neighboring nodes and the targets under surveillance.

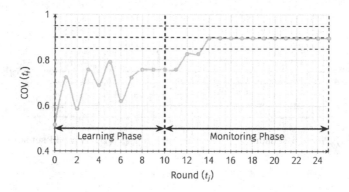

Fig. 2. Illustration of the LA-based WSN scheduling scheme consisting of a learning phase and an operation phase.

Upon completion of this phase, we advance to the learning phase. Here, each LA_i updates its action probability vector based on its chosen action and the actions of its immediate neighbors. Notably, an agent A_i can receive a reward (Eq. (3)) even if it remains inactive ($\alpha_i = 0$), thereby conserving its battery. This is feasible if neighboring sensors collectively cover shared Points of Interest (POIs), provided the number of uncovered POIs remains below a specified threshold value determined by the coverage parameter q. Conversely, if the coverage parameter is not met, the automaton will incur a penalty (Eq. (4)).

On the other hand, if an agent A_i opts to expend its battery energy to cover POIs, it may face penalties if neighboring sensors already cover the same subset of POIs. This introduces a trade-off between achieving the required coverage and preserving battery power. Based on this interaction, each node selects optimal actions based on the acquired information, determining whether to remain active or become idle during the subsequent operation phase. Upon depletion of battery power by a group of sensors, the network must reorganize to restore the required level of coverage. This process is visualized in Fig. 2.

4   Experimental Study

In this section, we aim to evaluate the effectiveness of the proposed algorithm through multiple computer simulations. To accomplish this, we will employ a fixed sensor network, where sensor nodes are randomly positioned within a $1000 : m \times 1000 : m$ area alongside a static deployment of 400 targets. The sensing range of sensors was set at a value of $R_s^i = 175$. The required coverage target was $q = 0.8$ with $\delta = 0.1$. The number of nodes will vary in the range $S = \{16, 36, 49, 100\}$ sensors. The simulations were conducted using Matlab's custom Wireless Sensor Network (WSN) simulator. The results were averaged over ten runs to ensure robustness. Through this evaluation, we seek insights into the algorithm's performance across diverse conditions.

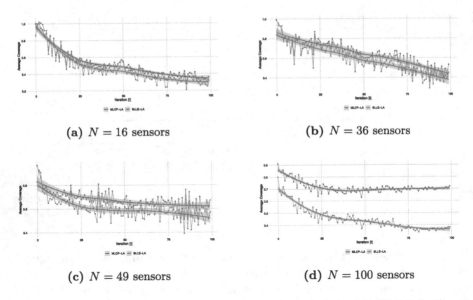

(a) $N = 16$ sensors

(b) $N = 36$ sensors

(c) $N = 49$ sensors

(d) $N = 100$ sensors

Fig. 3. Averaged results of the network coverage acquired by MLCP-LA (red) and SLLE-LA (blue) algorithms for variable number of sensors deployed over the target area with $T = 400$ POIs. (Color figure online)

We begin with a comparison of the proposed approach (denoted as MLCP-LA) with our previous work (further denoted as SLLE-LA) presented in [1,3]. The old solution employed a synchronized local leader election game model to replace a global optimization problem with a problem of searching for Nash equilibrium (NE) by a team of players participating in a non-cooperative game.

The average coverage ratio changes in successive iterations of the algorithm are presented in Fig. 3. The figure visualizes four sample runs of the two algorithms (MLCP-LA in red, SLLE-LA in blue) for WSNs comprised of $N = 16$ (Fig. 3(a)), $N = 36$ (Fig. 3(b)), $N = 49$ (Fig. 3(c)) and $N = 100$ (Fig. 3(d)) sensors, respectively.

In the case of the smaller networks ($N = 16$ and $N = 36$ sensors), the performance of both algorithms is similar, with a slight advantage to the proposed, fully decentralized LA-based solution. However, results become more varied with the rising complexity of the scheduling problem for WSN comprised of $N = 49$ and $N = 100$ sensors. Compared to previous experiments, we can observe more considerable differences in performance between analyzed scheduling solutions, especially in the case of an extensive network comprised of $N = 100$ sensors.

This behavior is mainly consistent with our expectations and the nature of both solutions. The solution presented in this work (MLCP-LA) is a fully decentralized algorithm enabling sensors to manage their sleep/activity cycles based on specific coverage goals. This algorithm has the advantages of being localized, utilizing limited knowledge of neighboring sensors, and self-reorganizing to preserve the required coverage ratio and prolong the WSN's lifetime.

While our other work (SLLE-LA) also employs the LA model as its primary learning loop, it requires further negotiation between players to achieve NE in each local neighborhood. While it does not present a problem for smaller networks, it is clear that SLLE-LA needs to be more scalable to compete in more extensive networks. Therefore, it is more advantageous for sensor networks when nodes learn what actions to take rather than follow a predefined schedule.

5   Conclusion

In this paper, we proposed an algorithm based on the concept of LA to solve MLCP in WSN. This algorithm has the advantages of being localized, utilizing limited knowledge of neighboring sensors, and self-reorganizing to preserve the required coverage ratio and prolong the WSN's lifetime.

Our early research findings demonstrate that the LA agents can achieve an effective solution in a completely decentralized fashion, minimizing battery expenditure and ultimately prolonging the lifetime of the WSNs. Compared to our older works, empirical data suggests that aligning the agents' objectives with the system goal is critical in achieving global efficiency in decentralized learning. Allowing each agent to pursue its objectives selfishly may result in a suboptimal solution. In contrast, we achieved global efficiency by requiring each agent to consider a small group of surrounding agents.

Future work will include studying additional reinforcement learning functions to find better solutions to the studied problem. For example, *Linear Reward-Epsilon-Penalty* when $b << a$ and *Linear Reward-Inaction* when $b = 0$. An additional study of the relation between the experimental parameters (density of sensors and targets, regularity of their distribution, variable sensing range of nodes, and battery levels) and the achieved coverage and lifetime results will follow.

References

1. Gąsior, J., Seredyński, F.: A learning automata-based approach to lifetime optimization in wireless sensor networks. In: Rutkowski, L., Scherer, R., Korytkowski, M., Pedrycz, W., Tadeusiewicz, R., Zurada, J.M. (eds.) ICAISC 2021. LNCS (LNAI), vol. 12854, pp. 371–380. Springer, Cham (2021). https://doi.org/10.1007/978-3-030-87986-0_33
2. Gudla, S., Kuda, N.R.: Learning automata based energy efficient and reliable data delivery routing mechanism in wireless sensor networks. J. King Saud Univ. Comput. Inf. Sci. **34**(8, Part B), 5759–5765 (2022)
3. Gąsior, J., Seredyński, F., Hoffmann, R.: Towards self-organizing sensor networks: game-theoretic ε-learning automata-based approach. In: Cellular Automata - 13th International Conference on Cellular Automata for Research and Industry, ACRI 2018, Como, Italy, 17–21 September 2018, Proceedings, pp. 125–136 (2018)
4. Lin, Y., Wang, X., Hao, F., Wang, L., Zhang, L., Zhao, R.: An on-demand coverage based self-deployment algorithm for big data perception in mobile sensing networks. Futur. Gener. Comput. Syst. **82**, 220–234 (2018)
5. Manju, Chand, S., Kumar, B.: Target coverage heuristic based on learning automata in wireless sensor networks. IET Wirel. Sensor Syst. **8**(3), 109–115 (2018)
6. Mostafaei, H., Esnaashari, M., Meybodi, M.R.: A coverage monitoring algorithm based on learning automata for wireless sensor networks (2014)
7. Mostafaei, H., Montieri, A., Persico, V., Pescapé, A.: A sleep scheduling approach based on learning automata for WSN partial coverage. J. Netw. Comput. Appl. **80**, 67–78 (2017)
8. Narendra, K.S., Thathachar, M.A.L.: Learning Learning Automata: An Introduction. Prentice-Hall Inc, Upper Saddle River (1989)
9. Qarehkhani, A., Golsorkhtabaramiri, M., Mohamadi, H., Yadollahzadeh-Tabari, M.: Solving the target coverage problem in multilevel wireless networks capable of adjusting the sensing angle using continuous learning automata. IET Commun. **16**(2), 151–163 (2022)
10. Upreti, R., Rauniyar, A., Kunwar, J., Haugerud, H., Engelstad, P., Yazidi, A.: Adaptive pursuit learning for energy-efficient target coverage in wireless sensor networks. Concurr. Comput. Pract. Exp. **34**(7), e5975 (2022)
11. Yetgin, H., Cheung, K.T.K., El-Hajjar, M., Hanzo, L.H.: A survey of network lifetime maximization techniques in wireless sensor networks. IEEE Commun. Surv. Tutor. **19**(2), 828–854 (2017)

Solving Problems with Uncertainties

A Rational Logit Dynamic for Decision-Making Under Uncertainty: Well-Posedness, Vanishing-Noise Limit, and Numerical Approximation

Hidekazu Yoshioka$^{1(\boxtimes)}$![ORCID], Motoh Tsujimura^2 ![ORCID], and Yumi Yoshioka^3 ![ORCID]

^{1} Japan Advanced Institute of Science and Technology, 1-1 Asahidai, Nomi 923-1292, Japan
yoshih@jaist.ac.jp
^{2} Doshisha University, Karasuma-Higashi-iru, Imadegawa-dori, Kamigyo-ku,
Kyoto 602-8580, Japan
mtsujimu@mail.doshisha.ac.jp
^{3} Gifu University, Yanagido 1-1, Gifu 501-1193, Japan
yoshioka.yumi.k6@f.gifu-u.ac.jp

Abstract. The classical logit dynamic on a continuous action space for decision-making under uncertainty is generalized to the dynamic where the exponential function for the softmax part has been replaced by a rational one that includes the former as a special case. We call the new dynamic as the rational logit dynamic. The use of the rational logit function implies that the uncertainties have a longer tail than that assumed in the classical one. We show that the rational logit dynamic admits a unique measure-valued solution and the solution can be approximated using a finite difference discretization. We also show that the vanishing-noise limit of the rational logit dynamic exists and is different from the best-response one, demonstrating that influences of the uncertainty tail persist in the rational logit dynamic. We finally apply the rational logit dynamic to a unique fishing competition data that has been recently acquired by the authors.

Keywords: Rational Logit Function · Limit Evolution Equation · Computational Modeling and Application

1 Introduction

1.1 Research Background

Modeling social dynamics has been a central research topic to better understand and manage human interactions in the contemporary world. Social dynamics can be modelled through the evolutionary game theory where interactions among (infinitely) many agents are described through an evolution equation. Analysis of this equation is therefore a central issue in the theory. Both trajectory and its equilibrium of a reasonable solution to the evolution equation are important to analyze transitions of the social state of interest. Such examples include but are not limited to the best-response dynamic [1],

© The Author(s), under exclusive license to Springer Nature Switzerland AG 2024
L. Franco et al. (Eds.): ICCS 2024, LNCS 14838, pp. 265–279, 2024.
https://doi.org/10.1007/978-3-031-63783-4_20

replicator dynamic [2], projection dynamic [3], gradient dynamic [4], and logit dynamic [5]. Evolutionary systems generalizing the above-mentioned dynamics have also been proposed along with their well-posedness and stability results [6–8].

We focus on the logit dynamic [5] in a continuous action space as it serves as the simplest evolution equation for decision-making of individuals in an uncertain environment. The logit dynamic reads the following evolution equation governing a time-dependent probability measure μ on a compact set Ω (the equation will be defined more rigorously in Sect. 2):

$$\frac{d\mu}{dt} = \frac{\exp(\eta^{-1}U(x;\mu))}{\int_\Omega \exp(\eta^{-1}U(y;\mu))dy} - \mu \text{ for time } t > 0 \tag{1}$$

subject to a prescribed initial condition. Here, U is a bounded utility depending on the social state μ as a probability measure on the collection of individuals' actions parameterized in Ω, and $\eta > 0$ is a noise intensity. Phenomenologically, a larger η implies a larger uncertainty that individuals face in the given environment during the decision-making. The noise regularizes the dynamic through the logit function. An equilibrium of (1) is interpreted as a maximizing μ of the utility U under uncertainty.

The logit dynamic (1) is an evolution equation nonlocal in space. The first term in the right-hand side of (1) uses the logit function, which is here the exponential function so that the term serves as a softmax function. A stationary state of the dynamic (1) under the vanishing-noise limit $\eta \to +0$, if it exists, approximates a Nash equilibrium of the maximization problem of the utility U with respect to probability measures μ [5]. The dynamic for a nonzero η serves as a model of dynamic decision-making in an uncertain environment, which is of a separate interest.

It has recently been found that the use of a different logit function in the dynamic (1), such as a polynomial function, instead of the exponential function substantially affects the equilibrium profile of the dynamic. Lahkar et al. [9] numerically demonstrated that different logit functions lead to different equilibria for small η. Subsequently, Yoshioka [10] theoretically showed along with examples that the growth speed of the logit function controls the equilibrium; a different growth speed may give a different equilibrium. Their analysis clarified influences of the logit functions on the equilibria, whereas the analysis of time-dependent solutions is still scarce.

1.2 Objectives and Contributions

The objectives as well as contributions of this study are formulation, analysis, discretization, and computational application of a generalized logit dynamic: an evolution equation using a wider class of logit function than the classical one (1). As a logit function, we use the κ-exponential function with $\kappa \in [0, 1]$, which is a positive, increasing, and strictly convex function [11] (Fig. 1):

$$e_\kappa(z) = \left(\kappa z + \sqrt{\kappa^2 z^2 + 1}\right)^{\frac{1}{\kappa}}, \ z \in \mathbb{R} \tag{2}$$

with the convention $e_0(z) = \exp(z)$. The κ-exponential function is therefore a generalization of the exponential one. An elementary calculation yields

$$\frac{de_\kappa(z)}{dz} = \frac{1}{\sqrt{\kappa^2 z^2 + 1}} e_\kappa(z), \ z \in \mathbb{R} \tag{3}$$

showing that the derivative of e_κ is bounded in each compact set of $\mathbb{R}$.

We demonstrate that the resulting logit dynamic, which we call the rational logit dynamic because it uses a rational(-like) function (2) in the softmax part, admits a unique measure-valued solution. Moreover, we explicitly obtain the evolution equation corresponding to the vanishing-noise limit $\eta \to +0$ along with a convergence result, with which we can track solution trajectories with and without uncertainties. Rational logit functions also recently found their applications in accelerating the convergence of machine learning algorithms [12, 13] (Fig. 1).

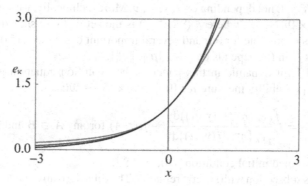

Fig. 1. Profiles of κ-exponential functions: κ equals 0 (black), 0.5 (blue), and 1 (red) (Color figure online).

We also propose a provably-convergence finite difference method for the rational logit dynamic. Finally, we apply the generalized logit dynamic to a unique fishing competition data that has been recently acquired by the authors, so that this study covers an engineering application.

The rest of this paper is organized as follows. Section 2 formulates the rational logit dynamic, performs its mathematical analysis, and presents its finite difference discretization. Section 3 shows a computational application of the rational logit dynamic. Section 4 concludes this study and its future perspectives. Appendix contains a technical proof and the data used in our application.

2 Formulation and Analysis

2.1 Rational Logit Dynamic

We consider a decision-making problem of individuals parameterized by the closed unit interval $\Omega = [0, 1]$. Borel algebra on Ω is denoted as $\mathfrak{B}$. The set of probability measures on Ω is denoted as $\mathfrak{M}$. The variational norm $\|\cdot\|$ for $\mu \in \mathfrak{M}$ is given by [5]

$$\|\mu\| = \sup_g \left| \int_\Omega g(y)\mu(\mathrm{d}y) \right|, \quad g : \Omega \to [-1, 1] \text{ is measurable.} \tag{4}$$

The difference $\|\mu - \nu\|$ for $\mu, \nu \in \mathfrak{M}$ is defined as in (4) where μ is formally replaced by $\mu - \nu$. The utility $U : \Omega \times \mathfrak{M} \to \mathbb{R}$ is bounded and Lipschitz continuous:

$$|U(x; \mu) - U(y; \nu)| \leq K(|x - y| + \|\mu - \nu\|) \text{ for any } x, y \in \Omega \text{ and } \mu, \nu \in \mathfrak{M} \tag{5}$$

with a constant $K > 0$ not depending on x, y, μ, ν. More technically, we should consider a utility $U : \Omega \times \mathfrak{M} \to \mathbb{R}$, where $\mathfrak{F}$ ($\mathfrak{M} \subset \mathfrak{F}$) is the set of finite signed measures on Ω. However, this is possible for our and several important cases (see (15)) by assuming that the constant K in (5) depends on $\max\{\|\mu\|, \|\nu\|\}$.

The rational logit dynamic in this paper is the evolution equation governing the time-dependent probability measure $\mu : [0, +\infty) \times \mathfrak{B} \to \mathfrak{M}$:

$$\frac{\mathrm{d}\mu(t, A)}{\mathrm{d}t} = \frac{\int_A e_\kappa\left(\eta^{-1}U(y; \mu)\right)\mathrm{d}y}{\int_\Omega e_\kappa\left(\eta^{-1}U(y; \mu)\right)\mathrm{d}y} - \mu(t, A) \text{ for any } A \in \mathfrak{B} \text{ and } t > 0 \tag{6}$$

subject to a prescribed initial condition $\mu(0, \cdot) \in \mathfrak{M}$.

We close this subsection with several remarks. The time derivative in (6) is understood in the strong sense with the variational norm (e.g., Mendoza-Palacios and [2]). The Lipschitz condition (5) is a typical assumption for logit dynamics [9, 10], and is actually not so restrictive as the constant K is allowed to be arbitrary large if it is bounded. The κ-exponential function seems to be less famous than the q-exponential one often used in the Tsallis formalism [14]:

$$e_{(q)}(z) = (1 + (1 - q)z)^{\frac{1}{1-q}}, \; z \in \mathbb{R} \text{ if } 1 + (1 - q)z > 0 \tag{7}$$

with a parameter $q > 0$. The q-exponential function reduces to the exponential one at $q = 1$ like the κ-exponential one. The q-exponential function is explicitly connected to heavy-tailed noises [10] particularly the generalized extreme value one [15], and therefore admits a firm background in the context of logit dynamics; however, a serious disadvantage of the q-exponential function is that it is not defined over $\mathbb{R}$. The κ-exponential function completely avoids this theoretical issue. As shown later, what is important in the rational logit dynamic (6) is the growth speed of the logit function, namely the function $e_\kappa\left(\eta^{-1}(\cdot)\right)$, for a large argument: $e_\kappa(z) \sim O\left(z^{1/\kappa}\right)$ for a sufficiently large $z > 0$. In this view, a similar result will be expected to follow when using the q-exponential function with the choice $q = 1 - \kappa \in (0, 1)$. The dynamic (6) still governs a maximizer of the utility U under uncertainty where the noise has a heavier tail than that in the classical one (1), such that the heaviness increases as κ does.

2.2 Well-Posedness and Stability

The unique existence of solutions to the rational logit dynamic follows by a direct application of Proof of Theorem 3.4 in Larkar and Riedel [5]. Here, a solution to a (rational) logit dynamic is a time-dependent probability measure $\mu(t, \cdot) \in \mathfrak{M}$ for $t \in [0, +\infty)$ that is continuously differentiable with respect to $t > 0$. The only difference between their and our logit dynamics are the logit functions. To apply their result, it suffices to show the following technical inequality for any $x \in \Omega$ and $\mu, \nu \in \mathfrak{M}$:

$$\left| \frac{e_\kappa \left(\eta^{-1} U(x; \mu) \right)}{\int_\Omega e_\kappa \left(\eta^{-1} U(y; \mu) \right) dy} - \frac{e_\kappa \left(\eta^{-1} U(x; \nu) \right)}{\int_\Omega e_\kappa \left(\eta^{-1} U(y; \nu) \right) dy} \right| \leq L \| \mu - \nu \|. \tag{8}$$

Here, $L > 0$ is a constant independent from $x \in \Omega$ and $\mu, \nu \in \mathfrak{M}$. The inequality (8) follows from the Lipschitz continuity U assumed in (5) and its boundedness, the Lipschitz continuity of the κ-exponential function in each compact set in $\mathbb{R}$ that follows from (3), strict positivity of the κ-exponential function in each compact set in $\mathbb{R}$ that follows from its definition, and the fact that a composition of functions that are Lipschitz continuous is again a Lipschitz continuous function. With the help of these continuity conditions, we can use the generalized Picard–Lindelöf theorem (e.g., Theorem A.3 in Lahkar et al. [9]). Consequently, we obtain Proposition 1 below.

Proposition 1
The rational logit dynamic (6) given a prescribed initial condition $\mu(0, \cdot) \in \mathfrak{M}$ admits a unique solution $\mu(t, \cdot) \in \mathfrak{M}$ for $t \in [0, +\infty)$.

Having proven the unique existence of solutions to the rational logit dynamic, we analyze its vanishing-noise limit $\eta \to +0$. With the notations analogous to (6), the limit equation, the dynamic obtained under this limit, is inferred by formally taking the limit $\eta \to +0$ in the logit function considering (2):

$$\frac{d\mu(t, A)}{dt} = \frac{\int_A \max\{U(y; \mu), 0\}^{1/\kappa} dy}{\int_\Omega \max\{U(y; \mu), 0\}^{1/\kappa} dy} - \mu(t, A) \quad \text{for any } A \in \mathfrak{B} \text{ and } t > 0 \tag{9}$$

with the initial condition being unchanged. The following proposition is the main theoretical contribution of this study, which states some stability of the rational logit dynamic (6) with respect to $\eta > 0$ as well as its convergence to the limit one (9).

Proposition 2
Assume that $U(x; \mu) \geq \underline{U}$ with a constant $\underline{U} > 0$ for any $x \in \Omega$ and $\mu \in \mathfrak{M}$. Fix constants $\eta_0 > 0$ and $T > 0$. Solutions to (6) with $\eta \in (0, \eta_0]$ and (9) with the same prescribed initial condition are denoted as μ_η and μ_0, respectively. Then, it follows that.

$$\lim_{\eta \to +0} \sup_{0 \leq t \leq T} \| \mu_0(t, \cdot) - \mu_\eta(t, \cdot) \| = 0. \tag{10}$$

See, **Appendix** for the proof of Proposition 2. Note that the constants $\eta_0, T > 0$ can be taken to be arbitrary large if they are bounded. Proposition 2 therefore shows that the limit Eq. (9) indeed serves as a vanishing-noise limit of (6). The positivity assumption of U was imposed for a technical reason, but it may be removed in some case as computationally demonstrated in Sect. 3. We can also obtain the parameter continuity result below:

$$\lim_{\eta_2 \to \eta_1} \sup_{0 \le t \le T} \left\| \mu_{\eta_1}(t, \cdot) - \mu_{\eta_2}(t, \cdot) \right\| = 0 \tag{11}$$

with $\eta \in (0, \eta_0]$ under the assumption of Proposition 2. To show this, it suffices to see the triangle inequality

$$\left\| \mu_{\eta_1}(t, \cdot) - \mu_{\eta_2}(t, \cdot) \right\| \le \left\| \mu_{\eta_1}(t, \cdot) - \mu_0(t, \cdot) \right\| + \left\| \mu_0(t, \cdot) - \mu_{\eta_2}(t, \cdot) \right\|. \tag{12}$$

2.3 Numerical Discretization

We present a finite difference discretization of the rational logit dynamic (6). Set the resolution $N \in \mathbb{N}$ with $N \ge 2$. The domain Ω is uniformly discretized into N cells

$$\Omega_i = [(i-1)/N, i/N) \ (i = 1, 2, ..., N-1) \text{ and } \Omega_N = [1 - 1/N, 1]. \tag{13}$$

The midpoint of Ω_i is denoted as $x_{i-1/2}$. The dynamic (6) with the choice $A = \Omega_i$ is discretized as follows:

$$\frac{d\hat{\mu}(t, \Omega_i)}{dt} = \frac{N^{-1} e_\kappa \left(\eta^{-1} U \left(x_{i-1/2}; \hat{\mu} \right) \right)}{\sum_{j=1}^{N} N^{-1} e_\kappa \left(\eta^{-1} U \left(x_{j-1/2}; \hat{\mu} \right) \right)} - \hat{\mu}(t, \Omega_i) \text{ for } i = 1, 2, ..., N \text{ and } t > 0. \tag{14}$$

We used the fact that the length of each cell is N^{-1} and $\hat{\mu}(t, \Omega_i)$ is the discretization of $\mu(t, \Omega_i)$. The initial condition $\mu(0, \cdot)$ will also be discretized in each cell.

To complete the spatial discretization (14), we specify the form of the utility U. For that purpose, we assume the following widely-used form [2, 9, 10]:

$$U(x; \mu) = \int_\Omega f(x, y) \mu(dy) \text{ for and } x \in \Omega \text{ and } \mu \in \mathfrak{M} \tag{15}$$

with a function $f : \Omega \times \Omega \to \mathbb{R}$ that is continuous on the compact set $\Omega \times \Omega$, and hence bounded as well as uniformly continuous on $\Omega \times \Omega$. We apply the discretization

$$U(x; \mu) = \int_\Omega f(x, y) \mu(dy) \to \sum_{k=1}^{N} f \left(x_{j-1/2}, x_{k-1/2} \right) \hat{\mu}(t, \Omega_k) = U \left(x_{j-1/2}; \hat{\mu} \right) \tag{16}$$

in (14). The full discretization of the Eq. (14) is finally obtained by applying the common first-order forward Euler method that evaluates the right-hand side of (14) explicitly. The limit Eq. (9) can also be discretized using the same finite difference method by directly utilizing (15)–(16). By using each $\hat{\mu}(t, \Omega_i)$, the space semi-discretized solution to the rational logit dynamic is potentially defined as the time-dependent measure μ_N satisfying the equality:

$$\mu_N(t, A) = \int_A N \sum_{i=1}^{N} \mathbb{I}_{\Omega_i}(y)\hat{\mu}(t, \Omega_i)dy \text{ for any } A \in \mathfrak{B} \text{ and } t \in [0, +\infty) \qquad (17)$$

where $\mathbb{I}_S(\cdot)$ is the indicator function of set S such that $\mathbb{I}_S(y) = 1$ if $y \in S$ and is 0 otherwise. This μ_N is checked to be a probability measure at each $t \in [0, +\infty)$ by substituting $A = \Omega$ in (17) considering the identity $\int_{\Omega_i} N dy = 1$.

We state that the solution to the semi-discretized dynamic (14) exists and converges to a solution to the original one. The proof of the proposition below is omitted here because it is based on Proof of Proposition 3.1 of Lahkar et al. [16], where it suffices to use the Lipschitz continuity like that for Proposition 1.

Proposition 3
Assume $\kappa(0, 1]$ and (15) with $f : \Omega \times \Omega \to \mathbb{R}$ that is continuous on $\Omega \times \Omega$. Solutions to (6) and (14) with initial conditions $\mu(0, \cdot)$, $\mu_N(0, \cdot)$ are denoted as μ and μ_N, respectively. Also assume that $\lim_{N \to +\infty} \|\mu_N(0, \cdot) - \mu(0, \cdot)\| = 0$. Then, it follows that.

$$\lim_{N \to +\infty} \|\mu_N(t, \cdot) - \mu(t, \cdot)\| = 0 \text{ at each } t > 0 \qquad (18)$$

Note that a similar result applies to the limit equation. Hence, the assumption $\eta > 0$ can be formally replaced by $\eta \geq 0$ in Proposition 3. Finally, this proposition still holds true if a sufficiently regular term satisfying the Lipschitz continuity of the form (5) is added to the right-hand side of (15) (see, (19)–(20) in the next section).

3 Application

We apply the rational logit dynamic (6) and the finite difference method to an engineering problem related to inland fisheries. The authors have been studying environmental and ecological dynamics of the Hii River, San-in area, Japan since 2015. This river is a major habitat of the migratory fish *Plecoglossus altivelis altivelis*, which is one of the most famous inland fishery resources in Japan (e.g., Murase and Iguchi [17]). Toami (i.e., casting net) is a major fishing method for catching the fish *P. altivelis*.

The Hii River Fishery Cooperative (HRFC) authorizing inland fisheries in the Hii River is holding the Toami competition in each year. Participants of the competition are resisted in pairs. The total number of pairs in each competition was 10 to 16 in recent years. The registered pairs compete the total catch of the fish *P. altivelis* by the Toami angling in the two hours (typically 10:00 a.m. to 12:00 a.m.) in a fixed day. All participants must use casting nets under a common regulation. Top few pairs (about three pairs in each year) are awarded by the HRFC. Table 2 in **Appendix** summaries the number of catches of the fish *P. altivelis* in each pair in each Toami competition.

The total number of pairs and their catches are different among different competitions. We therefore normalize the data of the fish catches in each competition by the division by the maximum (e.g., 82 in 2023 and 43 in 2016). The fish catches in each competition then belong to $\Omega = [0, 1]$, to which our formulation applies. In this view, the fish catch can be understood as the normalized harvesting effort. The data used in our application is an ensemble of the normalized data in each year in Table 2.

We model the Toami competition by the following utility

$$U(x; \mu) = \int_\Omega \left\{ -ax^2 + b|x - y|^c \right\} \mu(\mathrm{d}y) + d \max\left\{ \alpha - \int_x^1 \mu(\mathrm{d}y), 0 \right\} \quad (19)$$

with $a, b, c, d \geq 0$ and $\alpha \in (0, 1)$. Each term in the right-hand side of (19) is explained as follows. The first term represents the harvesting cost that is assumed to be quadratic with respect to the effort. The quadratic assumption of the cost is typical in the control and optimization problems. The second term represents the individual difference modelled in a way that this term is larger when the fish catches between different pairs are more different, and hence is expected to induce a competition. The third term represents the awarding scheme by HRFC for the Toami competition that the $100\alpha\%$ upper tier is awarded. This term is continuous in x if μ has a density. The utility (19) models a competition with an award mechanism. Rigorously, to apply our mathematical framework to (19), the last term should be regularized. We use the following continuous regularization of the step function:

$$\int_x^1 \mu(\mathrm{d}y) = \int_0^1 \mathbb{I}_{(x,1]}(y) \mu(\mathrm{d}y) \rightarrow \int_0^1 \max\left\{ 0, \min\left\{ 1, \frac{y - x + \varepsilon}{\varepsilon} \right\} \right\} \mu(\mathrm{d}y) \text{ with } \varepsilon > 0 \quad (20)$$

Indeed, we have the Lipschitz continuity

$$\left| \int_0^1 \max\left\{ 0, \min\left\{ 1, \frac{y - x + \varepsilon}{\varepsilon} \right\} \right\} (\mu(\mathrm{d}y) - v(\mathrm{d}y)) \right| \leq \|\mu - v\| + \varepsilon^{-1}|x - z| \quad (21)$$

for any $x, z \in \Omega$ and $\mu, v \in \mathfrak{M}$. In the sequel, we choose $\varepsilon = N^{-1}$.

We need to identify the model parameters in (19) as well as those in the rational logit dynamic. We set $d = 1$ without significant loss of generality. Further, we set $c = 1$ to simplify the problem. We choose $\alpha = 0.2$ considering the award scheme employed by the HRFC. The remaining parameters a, b, η, κ are determined by a trial-and-error approach so that the sum of the squares of the relative errors of the average and standard deviation of x is minimized between the empirical and computational results. The computational resolution to discretize the rational logit dynamic and its limit equation is $N = 500$ with the time step $\Delta t = 0.001$. The initial condition is the uniform distribution in Ω. The solution to the discretized dynamic is judged to be stationary when the computed probability density function (PDF) given by $N\hat{\mu}$ satisfies.

$$\max_{1\le i\le N} |N\hat{\mu}(t_{k+1}, \Omega_i) - N\hat{\mu}(t_k, \Omega_i)| \le \delta\left(= 10^{-11}\right) \text{ at some } k \ge 0. \qquad (22)$$

The stationary solution is then given by $\hat{\mu}\left(t_{k+1}, \Omega_{(\cdot)}\right)$ with the smallest k satisfying the condition (22). We assume the stationary state against the empirical data.

The identified parameter values are $a = 0.27$, $b = 0.23$, $\eta = 0.01$, $\kappa = 1.0$. The empirical average and standard deviation of the normalized harvesting effort x are 0.32471 and 0.30352, respectively; the theoretical average and standard deviation using the fitted model are 0.32471 and 0.30377, respectively. The relative errors of the average and standard deviation between the empirical and theoretical ones are 0.0000011 and 0.00025, respectively, that are sufficiently small. Figure 2 compares the empirical and theoretical PDFs of the effort, suggesting their reasonable agreement. Particularly, the model captures the bimodal nature of the empirical data.

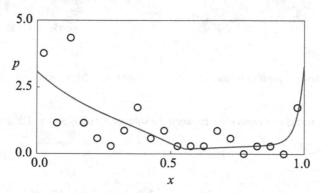

Fig. 2. Comparison of the empirical PDF (black) and theoretical one (blue). (Color figure online)

Having identified the model from the data, we conduct a sensitivity analysis of the rational logit dynamic. We focus on the two parameters characterizing the dynamics, which are the noise intensity η and shape parameter κ.

Firstly, we analyze η. Figure 3 shows the computed profiles of the PDF p for different values of η including the extreme case $\eta = 0$, where the other parameters remain the same with those identified from the data. As indicated in Fig. 3, increasing η flattens the profile of p, which is in accordance with the fact that η is the intensity of noises blurring the information available for anglers. The bimodal nature of p is preserved during increasing η. The profiles between the cases $\eta = 0$ and $\eta = 0.01$ are visually almost the same with each other. We then evaluate the convergence speed of the numerical solutions with $\eta > 0$ to that of $\eta = 0$ using the maximum norm at time $t = 1$, $t = 10$, and $t = 100$ as shown in Table 1. The convergence speed is slightly better than the first order in η, not violating the theory (32).

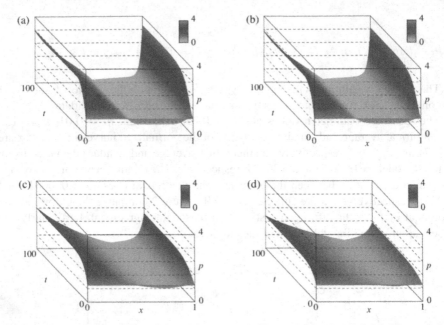

Fig. 3. The transient profiles of the PDF: (a) $\eta = 0$ (b) $\eta = 0.01$, (c) $\eta = 0.05$, (d) $\eta = 0.1$.

Table 1. The maximum norm error between the solutions with $\eta > 0$ and that with $\eta = 0$.

		η			
		0.0001	0.001	0.01	0.1
$t = 1$	Error	8.33E−05	3.74E-03	1.15E−01	1.03E+00
	Convergence rate	1.65.E+00	1.49.E+00	9.50.E−01	
$t = 10$	Error	2.44E−05	2.40E−03	1.73E−01	1.94E+00
	Convergence rate	1.99.E+00	1.86.E+00	1.05.E+00	

Secondly, we analyze influences of the shape parameter κ on solutions to the rational logit dynamic. Figure 4 compares stationary PDFs of the logit dynamic for different values of κ, where the other parameters remain the same with those identified from the data. The maximum values of p in the right-most cell are 14.60 for $\kappa = 0.1$ and 16.01 for $\kappa = 0$, which are placed above the figure panel. Decreasing κ from the identified value 1 toward 0 sharpens the profile of stationary p, while maintaining the flat part that exists around $0.6 \leq x \leq 0.9$. The stationary PDFs are entirely positive for both large and small κ, and hence the support of p is identical to Ω. Studying a more complex problem with a support expansion/contraction will be an interesting problem.

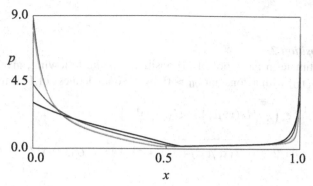

Fig. 4. Comparison of the stationary PDFs: $\kappa = 0$ (red), $\kappa = 0.1$ (green), $\kappa = 0.5$ (blue), $\kappa = 1$ (black). (Color figure online)

4 Conclusion

We proposed a rational logit dynamic where a rational function based on the κ-exponential function replaced the exponential function in the soft max part. It was proven that the rational logit dynamic is well-posed and has a parameter continuity with respect to the noise intensity that may vanish. We pointed out that the vanishing-noise limit of the rational logit dynamic depends on the shape parameter κ of the logit function, while such a phenomenon is not encountered in the classical logit dynamic, thereby the rational logit dynamic should be understood as a qualitatively different model from the classical one. We finally demonstrated a unique application of the rational logit dynamic to the fishing competition data.

The proposed logit function can be generalized because our analysis implied that what determines behavior of the rational logit dynamic under a small-noise limit is not the whole profile of the logit function but its asymptotic for a large argument. We will generalize the proposed mathematical framework based on this conjecture. The proposed logit function grows slower than the exponential one, while the case where it grows faster than the exponential one, like $\exp(x^2)$, will also be worth investigating. Examining the logit dynamics against real problems are of importance to deeper study their applicability and limitations. In particular, the regularity condition assumed the utility, such as the Lipschitz continuity with respect to the action and probability measure, should also be investigated in future so that the logit dynamic can be formulated under a weaker assumption. Establishment of a provably convergent as well as more efficient numerical method to discretize the logit dynamics will be a critical issue as well. Exploring the connection of the logit dynamics to mean filed games [18] has been under investigations by the authors.

Acknowledgments. The authors would like to express their gratitude towards the officers and members of HRFC for their supports on our field surveys. This study was supported by JSPS grants No. 22K14441 and No. 22H02456.

Disclosure of Interests. The authors have no competing interests to disclose.

Appendix

Proof of Proposition 2.

The key estimate in the proof of Proposition 2 is the following; given $\kappa \in (0, 1]$, $\mu \in \mathfrak{M}$, $\eta \in (0, \eta_0]$ with a constant $\eta_0 > 0$, by the boundedness of U it follows that

$$
\left| (\eta/(2\kappa))^{1/\kappa} e_\kappa \left(\eta^{-1} U(x; \mu) \right) - U(x; \mu)^{1/\kappa} \right|
$$

$$
= \left| \left(U(x; \mu)/2 + \sqrt{U(x; \mu)^2/4 + \eta^2 \kappa^{-2}/4} \right)^{\frac{1}{\kappa}} - U(x; \mu)^{1/\kappa} \right| \leq C\eta \tag{23}
$$

where $C > 0$ **is a constant independent from** η. This independence is crucial in our context, particularly when considering the limit $\eta \to +0$ (see (27)–(28) below).

Now, for any $t \in (0, T]$ and $A \in \mathfrak{B}$, it follows that

$$
\frac{d}{dt} \left(\mu_0(t, A) - \mu_\eta(t, A) \right)
$$

$$
= \frac{\int_A U(y; \mu_0)^{1/\kappa} dy}{\int_\Omega U(y; \mu_0)^{1/\kappa} dy} - \frac{(\eta/(2\kappa))^{1/\kappa} \int_A e_\kappa \left(\eta^{-1} U(y; \mu_\eta) \right) dy}{(\eta/(2\kappa))^{1/\kappa} \int_\Omega e_\kappa \left(\eta^{-1} U(y; \mu_\eta) \right) dy} - \left(\mu_0(t, A) - \mu_\eta(t, A) \right) \tag{24}
$$

We have the following estimate with a constant $C_1 > 0$ independent from η, μ_0, μ_η:

$$
\int_\Omega U(y; \mu_0)^{1/\kappa} dy (\eta/(2\kappa))^{1/\kappa} \int_\Omega e_\kappa \left(\eta^{-1} U(y; \mu_\eta) \right) dy
$$

$$
\geq \int_\Omega U(y; \mu_0)^{1/\kappa} dy \int_\Omega U(y; \mu_\eta)^{1/\kappa} dy > C_1 > 0 \tag{25}
$$

We also have the following estimate at each $y \in \Omega$:

$$
(\eta/(2\kappa))^{1/\kappa} \int_\Omega e_\kappa \left(\eta^{-1} U(x; \mu_\eta) \right) dx U(y; \mu_0)^{1/\kappa} - \int_\Omega U(x; \mu_0)^{1/\kappa} dx (\eta/(2\kappa))^{1/\kappa} e_\kappa \left(\eta^{-1} U(y; \mu_\eta) \right)
$$

$$
= (\eta/(2\kappa))^{1/\kappa} \int_\Omega e_\kappa \left(\eta^{-1} U(x; \mu_\eta) \right) dx U(y; \mu_0)^{1/\kappa} - \int_\Omega U(x; \mu_0)^{1/\kappa} dx U(y; \mu_0)^{1/\kappa}
$$

$$
+ \int_\Omega U(x; \mu_0)^{1/\kappa} dx U(y; \mu_0) - \int_\Omega U(x; \mu_0)^{1/\kappa} dx (\eta/(2\kappa))^{1/\kappa} e_\kappa \left(\eta^{-1} U(y; \mu_\eta) \right)
$$

$$
= \int_\Omega \left((\eta/(2\kappa))^{1/\kappa} e_\kappa \left(\eta^{-1} U(x; \mu_\eta) \right) - U(x; \mu_0)^{1/\kappa} \right) dx U(y; \mu_0)^{1/\kappa}
$$

$$
+ \int_\Omega U(x; \mu_0)^{1/\kappa} dx \left\{ U(y; \mu_\eta) - (\eta/(2\kappa))^{1/\kappa} e_\kappa \left(\eta^{-1} U(y; \mu_\eta) \right) \right\}
$$

$$
\leq C_2 \left\{ \begin{array}{l} \int_\Omega \left| (\eta/(2\kappa))^{1/\kappa} e_\kappa \left(\eta^{-1} U(x; \mu_\eta) \right) - U(x; \mu_0)^{1/\kappa} \right| dx \\ + \left| U(y; \mu_\eta) - (\eta/(2\kappa))^{1/\kappa} e_\kappa \left(\eta^{-1} U(y; \mu_\eta) \right) \right| \end{array} \right.
$$

$$
\leq C_2 \left\{ \begin{array}{l} (\eta/(2\kappa))^{1/\kappa} \int_\Omega \left| e_\kappa \left(\eta^{-1} U(x; \mu_\eta) \right) - e_\kappa \left(\eta^{-1} U(x; \mu_0) \right) \right| dx \\ + \int_\Omega \left| (\eta/(2\kappa))^{1/\kappa} e_\kappa \left(\eta^{-1} U(x; \mu_0) \right) - U(x; \mu_0)^{1/\kappa} \right| dx \\ + \left| U(y; \mu_\eta) - (\eta/(2\kappa))^{1/\kappa} e_\kappa \left(\eta^{-1} U(y; \mu_\eta) \right) \right| \end{array} \right. \tag{26}
$$

with a constant $C_2 > 0$ independent from μ_0, μ_η. By (23) and $U(x; \mu_\eta) > 0$, in (26) we obtain

$$\left| U(y; \mu_\eta) - (\eta/(2\kappa))^{1/\kappa} e_\kappa\left(\eta^{-1} U(y; \mu_\eta)\right) \right| \le C\eta, \tag{27}$$

$$\int_\Omega \left| (\eta/(2\kappa))^{1/\kappa} e_\kappa\left(\eta^{-1} U(x; \mu_0)\right) - U(x; \mu_0)^{1/\kappa} \right| dx \le C\eta, \tag{28}$$

and

$$(\eta/(2\kappa))^{1/\kappa} \int_\Omega \left| e_\kappa\left(\eta^{-1} U(x; \mu_\eta)\right) - e_\kappa\left(\eta^{-1} U(x; \mu_0)\right) \right| dx \le C_3(\eta_0) \| \mu_\eta - \mu_0 \| \tag{29}$$

with a constant $C_3(\eta_0) > 0$ independent from η, μ_0, μ_η. By (24)–(29), we obtain

$$\frac{d}{dt}\left(\mu_0(t, A) - \mu_\eta(t, A)\right) \le C_4(\eta_0)\left(\eta + \| \mu_\eta(t, \cdot) - \mu_0(t, \cdot) \|\right) \tag{30}$$

with a constant $C_4(\eta_0) > 0$ independent from η, μ_0, μ_η. Hence, by integrating (30) for $(0, t)$, and taking the variational norm yields

$$\| \mu_\eta(t, \cdot) - \mu_0(t, \cdot) \| \le 2 \int_0^t C_4(\eta_0)\left(\eta + \| \mu_\eta(s, \cdot) - \mu_0(s, \cdot) \|\right) ds. \tag{31}$$

Applying a classical Gronwall lemma to (31) yields

$$\| \mu_\eta(t, \cdot) - \mu_0(t, \cdot) \| \le C_5(T, \eta_0)\eta \exp(C_5(\eta_0)T) \tag{32}$$

with a constant $C_5(T, \eta_0) > 0$ depending on η_0, T but not on μ_0, μ_η. The conclusion (10) directly follows from (32).

$\square$

Collected Data

The number of catches of the fish *P. altivelis* in each pair in each Toami competition is summarized in the ascending order in Table 2. We consider that this kind of fish catch data is useful because it can be utilized not only for our study but also for other studies by other researchers. Members of each pair were anonymized. The Toami competition has been basically held by HRFC in each summer. It was not held in 2020, 2021, 2022 due to the outbreak of the coronavirus disease 2019. The data before 2015 may exist but was not available for us.

Table 2. The number of catches of the fish *P. altivelis* in each group in each year.

2016 Aug 7	2017 Aug 6	2018 Aug 5	2019 Aug 4	2023 Jul 30
1	0	2	0	0
1	5	6	1	0
2	5	6	1	3
2	8	8	2	9
4	9	17	4	9
5	16	17	4	9
6	17	21	4	10
6	21	23	5	12
8	22	24	6	12
10	25	53	16	13
14	28		25	24
16	41		28	30
21	42		29	31
32			35	56
36			41	82
43				

References

1. Swenson, B., Murray, R., Kar, S.: On best-response dynamics in potential games. SIAM J. Control. Optim. **56**(4), 2734–2767 (2018)
2. Mendoza-Palacios, S., Hernández-Lerma, O.: The replicator dynamics for games in metric spaces: finite approximations. In: Ramsey, D. M., Renault, J. (eds.) Advances in dynamic games: games of conflict, evolutionary games, economic games, and games involving common interest, pp. 163–186. Birkhäuser, Cham. (2020)
3. Harper, M., Fryer, D.: Lyapunov functions for time-scale dynamics on Riemannian geometries of the simplex. Dyn. Games Appl. **5**, 318–333 (2015)
4. Friedman, D., Ostrov, D.N.: Evolutionary dynamics over continuous action spaces for population games that arise from symmetric two-player games. J. Econ. Theory **148**(2), 743–777 (2013)
5. Lahkar, R., Riedel, F.: The logit dynamic for games with continuous strategy sets. Games Econom. Behav. **91**, 268–282 (2015)
6. Cheung, M.W.: Pairwise comparison dynamics for games with continuous strategy space. J. Econ. Theory **153**, 344–375 (2014)
7. Harper, M.: Escort evolutionary game theory. Phys. D **240**(18), 1411–1415 (2011)
8. Zusai, D.: Evolutionary dynamics in heterogeneous populations: a general framework for an arbitrary type distribution. Internat. J. Game Theory **52**, 1215–1260 (2023)
9. Lahkar, R., Mukherjee, S., Roy, S.: Generalized perturbed best response dynamics with a continuum of strategies. J. Econ. Theory **200**, 10539 (2022)

10. Yoshioka, H.: Generalized logit dynamics based on rational logit functions. Dyn. Games Appl. In press (2024)
11. Kaniadakis, G.: New power-law tailed distributions emerging in κ-statistics. Europhys. Lett. **133**(1), 10002 (2021)
12. Mei, J., Xiao, C., Dai, B., Li, L., Szepesvári, C., Schuurmans, D.: Escaping the gravitational pull of softmax. Adv. Neural. Inf. Process. Syst. **33**, 21130–21140 (2020)
13. Li, G., Wei, Y., Chi, Y., Chen, Y.: Softmax policy gradient methods can take exponential time to converge. Math. Program. **201**, 707–802 (2023)
14. Abe, S.: Stability of Tsallis entropy and instabilities of Rényi and normalized Tsallis entropies: a basis for q-exponential distributions. Phys. Rev. E **66**(4), 046134 (2002)
15. Nakayama, S., Chikaraishi, M.: A unified closed-form expression of logit and weibit and its application to a transportation network equilibrium assignment. Transport. Res. Procedia **7**, 59–74 (2015)
16. Lahkar, R., Mukherjee, S., Roy, S.: The logit dynamic in supermodular games with a continuum of strategies: a deterministic approximation approach. Games Econom. Behav. **139**, 133–160 (2023)
17. Murase, I., Iguchi, K.I.: High growth performance in the early ontogeny of an amphidromous fish, Ayu *Plecoglossus altivelis altivelis*, promoted survival during a disastrous river spate. Fish. Manage. Ecol. **29**(3), 224–232 (2022)
18. Barker, M., Degond, P., Wolfram, M.T.: Comparing the best-reply strategy and mean-field games: the stationary case. Eur. J. Appl. Math. **33**(1), 79–110 (2022)

Fragmented Image Classification Using Local and Global Neural Networks: Investigating the Impact of the Quantity of Artificial Objects on Model Performance

Kwabena Frimpong Marfo[1], Małgorzata Przybyła-Kasperek[1]([✉]),
and Piotr Sulikowski[2]

[1] Institute of Computer Science, University of Silesia in Katowice,
Będzińska 39, 41-200 Sosnowiec, Poland
{kmarfo,malgorzata.przybyla-kasperek}@us.edu.pl
[2] Faculty of Computer Science and Information Technology, West Pomeranian
University of Technology in Szczecin, ul. Żołnierska 49, 71-210 Szczecin, Poland
piotr.sulikowski@zut.edu.pl

Abstract. This paper addresses the challenge of classifying objects based on fragmented data, particularly when dealing with characteristics extracted from images captured from various angles. The complexity increases when dealing with fragmented images that may partially overlap. The paper introduces a classification model utilizing neural networks, specifically multilayer perceptron (MLP) networks. The key concept involves generating local models based on local tables comprising characteristics extracted from fragmented images. Since the local tables may have different sets of attributes due to varying perspectives, missing attributes in the tables are imputed by introducing artificial objects. The local models, now with identical structures are created and the aggregation of these models into a global model is carried out using weighted averages. The model's efficacy is evaluated against existing literature methods using various metrics, demonstrating superior performance in terms of F-measure and balanced accuracy. Notably, the paper investigates the impact of the number of generated artificial objects on classification quality, revealing that a higher number generally improves results.

Keywords: Fragmented Image Classification · Neural Networks · Artificial Objects · Characteristics Generated From Images

1 Introduction

Every so often in computer vision and object recognition tasks, the goal is not necessarily to create a virtual representation of an object, but to assign it to a certain class, based on its characteristics. Such a situation occurs in instances such as recognizing the architectural style of a building, the type of vehicle or

L. Franco et al. (Eds.): ICCS 2024, LNCS 14838, pp. 280–294, 2024.
https://doi.org/10.1007/978-3-031-63783-4_21

type of land based on a satellite image. The situation becomes more complicated when not a single image represents an entire object, but rather many fragmented images that may partially overlap. Here, one can think about a set of cameras that perceive an image of an object from different angles. Fragmented images can partially overlap – the cameras can observe (in some part) the same fragment of the object. In this paper, an assumption of not having fragmented images as such, but rather characteristics that have been extracted from these images is made. These object characteristics are stored in decision tables, also known as local tables and may contain common attributes and objects. By common objects in tables, we mean a situation where characteristics extracted from images of the same object are stored in different tables. There may be inconsistencies among tables in that an image in question may have been distorted in some way, resulting in a completely different value on conditional attributes or even decision attribute for the same object.

The paper proposes a classification model based on such fragmented data. The main idea of the model is to use neural networks (specifically MLP networks) to generate local models based on each local table. These models are then aggregated into a global model using trained weights from the local models. However, aggregation of MLP networks is not possible to realize without the local models having the same structure. This constraint can be satisfied by ensuring the presence of the same conditional attributes in all local tables, which, of course is not originally fulfilled due to different cameras observing different parts of an object and, consequently, different set of characteristics being stored in local tables. So, to achieve homogeneity in local tables, it is necessary to modify them before generating local models. This is done by generating artificial objects – supplemented with values for missing attributes. After such modification, local models with identical structures are built from local tables. Local models are then aggregated into a global model. Finally, the global model is refined using a small set of objects.

Figure 1 shows the stages of building the global model. In the first and second steps, characteristics of objects are extracted from fragmented images and local decision tables are created with different sets of conditional attributes – sets of attributes are not necessarily disjoint. It should be noted that this part is not addressed directly in this paper, since the data used retains characteristics of objects extracted from images and were obtained from the repository. Then, in order to unify the local tables, values are imputed for missing attributes in all local tables. For an original object, more than one artificial object can be created, making the cardinality objects in local tables dynamic. The fourth step describes building local neural networks, and the fifth aggregates these local networks into a global model – weights from local models are used for this purpose. Finally, this global model is trained using a small set of objects.

The main contribution of this paper is a proposal of a classification model based on characteristics obtained from fragmented images. Comparing the classification quality of the proposed model with known methods from the literature, it was shown that the proposed model, on average, generates better F-measure

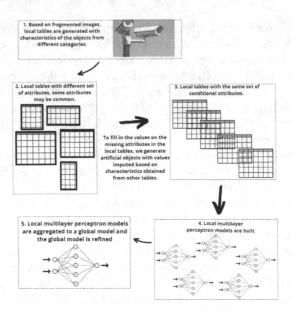

Fig. 1. Global model generation stages.

and balanced accuracy results. An important result of the paper is to examine the effect of the number of artificial objects generated on classification quality. Here, usually a higher number of objects improves the quality of classification. The paper also provides some guidance for which data the proposed model is suitable.

An effective and excellent model for image recognition is convolution neural networks [15]. There are many applications of image-based object recognition, among which are handwriting recognition [13], X-ray analysis [5], plant disease identification [3], facial emotion recognition [1]. However, it should be admitted that there are not many papers dedicated to the subject of processing fragmented images. Papers on fragmented images are usually concerned with completing the object that are presented in a fragmentary way on the image – examples are cracks in a roadway [14] or fragments of a plant leaf [2], coloring and fragmentation of image where the objects are located [9]. Identifying fragmented images is an infrequently addressed area of study. This challenge often involves working with sets of images captured from diverse perspectives, where each image is obtained by a distinct camera viewing the object from different angles [11]. When dealing with fragmented image data, recognizing the depicted object becomes more challenging. In a related paper [7], the proposed methodology involves assembling fragmented portions of photos to reconstruct a complete image, aligning and merging components to form a cohesive whole. It is important to note that this approach assumes non-overlapping fragmented data for effective implementation. The first research on classification based on dispersed and fragmented images was presented in the paper [10]. In this paper we present an approach that builds a common model which given the characteristics of an object, can recognize the class from which it comes from.

In Sect. 2, the proposed classification model is described. Section 3 addresses the data sets that were used and presents the conducted experiments and discussion on obtained results. Section 4 is on conclusions.

2 Model and Methods

There is an assumption that features are extracted from fragmented images and stored in a tabular form. More formally, some characteristics of images are available in a dispersed form, that is, in the form of a set of local tables. A set of decision tables $D_i = (U_i, A_i, d)$, $i \in \{1, \dots, n\}$ is available, where U_i is the universe comprising objects – images; A_i is a set of attributes that describe the image; d is a decision attribute – object shown in the image; $A = \bigcup_{i=1}^{n} A_i$ is the union of attributes present in all local tables. Objects and attributes in local tables can be different but some may be common.

The aim is to generate local neural network models (MLP models) based on each local table. To construct a global model, the structure of such local models must be identical, and this can only be achieved if all local tables have the same sets of attributes. Each table D_i is modified so that the full set of attributes A is included. This is done by generating new objects with completed values on the missing attributes. Suppose the object $\bar{x} \in U_i$ has a decision value v, $d(\bar{x}) = v$, $v \in V^d$, where V^d is the set of values of the decision attribute d and $b \in A \setminus A_i$. For each decision table $D_j, i \neq j$ for which $b \in A_j$ the following values are computed: $MIN_{j,v}^{b} = min_{x \in U_j, d(x)=v} \, b(x)$, $MAX_{j,v}^{b} = max_{x \in U_j, d(x)=v} \, b(x)$, $AVG_{j,v}^{b} = avg_{x \in U_j, d(x)=v} \, b(x)$, $MED_{j,v}^{b} = median_{x \in U_j, d(x)=v} \, b(x)$. In this way, individual values assigned to attribute b for each local table are derived. The final value, which completes the object $\bar{x}$ in table D_i is determined by applying one of four statistical measures (minimum, maximum, mean, or median) to the local values obtained in the previous step. Consequently, there are sixteen potential combinations with one chosen randomly for determining the value of attribute b. As an illustration, consider a scenario where the maximum is selected for calculating local values, and the median is chosen for the aggregate value. In this case, the determination of the value for attribute b is as follows: $b(\bar{x}) = med_{D_j:b \in A_j} MAX_{j,v}^{b}$. This method is repeated for each attribute that does not belong to the set A_i but occurs in other local tables. By the procedure described above, one can generate several artificial objects based on an original object. A parameter k is used to determine the number of artificial objects generated based on a single original object. This expanded approach has been tested, and the corresponding results are detailed in the experimental section of the paper. In this way, a set of modified local tables $\bar{D}_i = (\bar{U}_i, A, d)$, $i \in \{1, \dots, n\}$ with equal sets of attributes is obtained.

The local tables $\bar{D}_i$ are used in subsequent steps for training MLP neural networks. The input layer is defined as the cardinality of A. The number of neurons in the output layer corresponds to the number of decision classes, where each neuron determines the probability of the test object belonging to a specific decision class. In the experimental section, the consideration is given to one or two hidden layers. The number of neurons in the hidden layer is determined proportionally to the number of neurons in the input layer, exploring different proportions ranging from 0.25 to 5 times the number of input layer neurons. In the case of two hidden layers, all combinations of neuron numbers are explored, with the first layer being chosen from the set $\{0.25 \times I, 0.5 \times I, 0.75 \times I, 1 \times I, 1.5 \times I, 1.75 \times I, 2 \times I, 2.5 \times I, 2.75 \times I, 3 \times I, 3.5 \times I, 3.75 \times I, 4 \times I, 4.5 \times I, 4.75 \times I, 5 \times I\}$, and the second layer chosen from the set $\{1 \times I, 2 \times I, 3 \times I, 4 \times I, 5 \times I\}$ where I is the number of neurons in the input layer. The ReLU (Rectified Linear Unit) function is employed as the activation function for the hidden layer. For the output layer, the softmax activation function is utilized. The neural network is trained using the back-propagation method, specifically employing a gradient descent method with an adaptive step size. The model employs the categorical cross-entropy loss function in conjunction with the Adam optimizer for optimal performance.

Since all the local models created have the same structure, the global model is created by a weighted sum of the trained weights from local models. Prior to this aggregation, weights from each local model are adjusted by the formula: $\omega_i = ln(\frac{1-e_i}{e_i})$, where e_i is the classification error of the i-th local model on the training set $\bar{U}_i$. In summary, the global model is created as follows: initially, the network's structure is specified, then its weights assigned based on the weighted sum of weights from the local models. The final stage involves retraining the global network. For this step, training objects must possess values of all attributes A. This can be a certain set of examples/objects that an expert will describe and classify by capturing all the characteristics of an object at once. In the experiments conducted in this paper, such a validation set derived from the test set.

3   Data Sets and Results

3.1   Data and Measures

The proposed system was tested on three data sets from the UC Irvine Machine Learning Repository [6,8,12]. Vehicle Silhouettes – aims to classify vehicle silhouettes into one of four types, considering characteristics extracted from images taken from various angles: eighteen quantitative attributes, four decision classes, 846 objects (592 training, 254 test set). Landsat Satellite – involves classifying earth types in satellite images based on multispectral pixel values in a 3×3 neighborhood: thirty-six quantitative attributes, six decision classes, 6435 objects (4435 training, 1000 test set). Dry Bean – focuses on classifying types of beans using characteristics extracted from high-resolution images subjected to segmentation and feature extraction stages: seventeen quantitative attributes, seven decision classes, 13611 objects (9527 training, 4084 test set).

The data pre-processing involved random dispersion into 3, 5, 7, 9, and 11 local tables. Each local table included a reduced set of attributes with all objects from the original table. The data sets exhibited imbalance, with varying object counts across decision classes in both training and test sets. Two variants were considered for each data set: experiments on dispersed imbalanced data and on balanced data modified using the Synthetic Minority Over-sampling Technique (SMOTE) method [4].

The quality of classification was evaluated based on the test set with the following accuracy measures. Classification accuracy measure (acc) – a fraction of the total number of objects in the test set that were classified correctly; Recall – an assessment of the classifier's ability to correctly recognize a given class; Precision (Prec.) – a measure of how often the classifier does not make a mistake when classifying an object to a given class, F-measure (F-m.) – an assessment of the classifier's ability to keeping accuracies balanced. F-measure $= 2 \cdot \frac{\text{Precision} \cdot \text{Recall}}{\text{Precision} + \text{Recall}}$; Balanced accuracy – an average value of Recall for all decision classes. Balanced accuracy ($bacc$) ensures that the performance assessment considers the classification accuracy of all classes equally.

The effect of the number of artificial objects created based on the original object from the local table on the quality of classification was tested. The following numbers of artificial objects used were examined $\{1, 2, 3, 4, 5\}$. During the experiments, different structures of local networks with one or two hidden layers were also tested. Moreover, different number of neurons in hidden layers were studied. The following values were tested: for the first hidden layer $\{0.25, 0.5, 0.75, 1, 1.5, 1.75, 2, 2.5, 2.75, 3, 3.5, 3.75, 4, 4.5, 4.75, 5\} \times$ the number of neurons in the input layer; for the second hidden layer $\{1, 2, 3, 4, 5\} \times$ the number of neurons in the input layer. A validation set was obtained by dividing the original test set randomly but in a stratified manner into two equal parts. First, one part is used as the validation set (for re-training process) and the second part is used to assess the quality of classification. Then the roles reverse as the second part acts as the validation set. Finally, both results are averaged. Each experiment is repeated three times; in the following tables, all results given are the average of these three runs.

3.2   Results Analysis

Due to space limit, results obtained for all parameters are not shown (however, they will be made available upon request sent to the authors). Tables 1, 2 and 3 show the best (in terms of classification accuracy) results obtained. The tables also show in bold the best result for each of the considered data sets.

The proposed approach was compared with three other approaches. The first approach (MLP ensemble) uses a homogeneous ensemble of MLP networks. The final decision was determined by soft voting since networks generated from the local tables could be not aggregated due to their different structures. The second approach uses an ensemble of classifiers (KNN, DT, NB). This ensemble of classifiers method consists of creating three base classifiers: $k-$nearest neighbors, decision tree and naive bayes classifier based on each local table. The parameter $k = 3$ and the Gini index as a splitting criterion when building decision trees were used. The final decision of the ensemble was also made by soft voting.

Table 1. Results of Prec., Recall, F-m., *bacc* and *acc* for the global neural network.

Data set	No. tables	No. artif. obj.	One hidden layer					Two hidden layer				
			Prec.	Recall	F-m.	*bacc*	*acc*	Prec.	Recall	F-m.	*bacc*	*acc*
Vehicle imbalanced	3	1	0.707	0.627	0.599	0.627	0.627	0.718	0.713	0.696	0.685	0.713
		2	0.671	0.681	0.656	0.655	0.681	0.703	0.715	0.701	0.68	0.715
		3	0.665	0.665	0.665	0.665	0.665	0.711	0.713	0.705	0.684	0.713
		4	0.685	0.68	0.665	0.651	0.68	0.738	0.74	0.737	0.721	**0.74**
		5	0.69	0.685	0.669	0.656	0.685	0.726	0.739	0.723	0.705	0.739
	5	1	0.685	0.673	0.658	0.646	0.673	0.713	0.715	0.707	0.692	0.715
		2	0.682	0.689	0.664	0.653	0.689	0.727	0.734	0.724	0.714	**0.734**
		3	0.669	0.671	0.645	0.639	0.671	0.711	0.72	0.714	0.694	0.72
		4	0.647	0.672	0.641	0.631	0.672	0.741	0.732	0.731	0.714	0.732
		5	0.713	0.689	0.655	0.65	0.689	0.695	0.702	0.693	0.67	0.702
	7	1	0.653	0.682	0.65	0.645	0.682	0.735	0.735	0.727	0.708	**0.735**
		2	0.701	0.697	0.674	0.669	0.697	0.728	0.717	0.711	0.691	0.717
		3	0.703	0.696	0.672	0.669	0.696	0.732	0.724	0.722	0.704	0.724
		4	0.701	0.706	0.688	0.682	0.706	0.725	0.73	0.72	0.701	0.73
		5	0.698	0.707	0.689	0.687	0.707	0.711	0.72	0.713	0.691	0.72
	9	1	0.705	0.697	0.689	0.674	0.697	0.746	0.752	0.739	0.719	**0.752**
		2	0.669	0.678	0.662	0.656	0.678	0.728	0.74	0.727	0.706	0.74
		3	0.706	0.717	0.699	0.688	0.717	0.713	0.723	0.708	0.687	0.723
		4	0.683	0.686	0.661	0.663	0.686	0.737	0.744	0.735	0.718	0.744
		5	0.712	0.703	0.681	0.671	0.703	0.733	0.734	0.723	0.704	0.734
	11	1	0.675	0.694	0.672	0.666	0.694	0.74	0.743	0.732	0.717	0.743
		2	0.709	0.714	0.698	0.69	0.714	0.752	0.756	0.744	0.726	**0.756**
		3	0.673	0.688	0.671	0.659	0.688	0.726	0.739	0.718	0.703	0.739
		4	0.658	0.664	0.641	0.634	0.664	0.75	0.744	0.736	0.724	0.744
		5	0.712	0.694	0.671	0.665	0.694	0.73	0.735	0.726	0.711	0.735
Vehicle balanced	3	1	0.69	0.703	0.678	0.678	0.703	0.722	0.726	0.713	0.702	0.726
		2	0.714	0.685	0.664	0.67	0.685	0.713	0.731	0.717	0.699	0.731
		3	0.687	0.677	0.669	0.654	0.677	0.746	0.743	0.734	0.712	0.743
		4	0.693	0.697	0.684	0.676	0.697	0.751	0.761	0.741	0.728	0.761
		5	0.694	0.703	0.683	0.665	0.703	0.756	0.762	0.753	0.736	**0.762**
	5	1	0.699	0.71	0.689	0.674	0.71	0.748	0.748	0.73	0.725	0.748
		2	0.704	0.72	0.698	0.682	0.72	0.728	0.714	0.705	0.694	0.714
		3	0.692	0.717	0.696	0.684	0.717	0.734	0.734	0.721	0.701	0.734
		4	0.733	0.714	0.671	0.685	0.714	0.749	0.749	0.742	0.725	**0.749**
		5	0.686	0.703	0.682	0.675	0.703	0.723	0.739	0.717	0.702	0.739
	7	1	0.715	0.723	0.708	0.692	0.723	0.751	0.759	0.749	0.732	0.759
		2	0.715	0.726	0.708	0.691	0.726	0.755	0.753	0.75	0.731	0.753
		3	0.724	0.735	0.72	0.7	0.735	0.755	0.765	0.753	0.732	0.765
		4	0.713	0.724	0.684	0.679	0.724	0.767	0.756	0.751	0.734	0.756
		5	0.732	0.724	0.711	0.7	0.724	0.769	0.77	0.758	0.741	**0.77**
	9	1	0.74	0.743	0.732	0.715	0.743	0.744	0.745	0.743	0.721	0.745
		2	0.704	0.724	0.703	0.689	0.724	0.718	0.72	0.71	0.692	0.72
		3	0.707	0.711	0.695	0.681	0.711	0.743	0.741	0.739	0.718	0.741
		4	0.699	0.714	0.701	0.682	0.714	0.744	0.757	0.743	0.723	0.757
		5	0.728	0.722	0.704	0.693	0.722	0.757	0.76	0.747	0.732	**0.76**
	11	1	0.706	0.714	0.701	0.681	0.714	0.768	0.768	0.762	0.746	**0.768**
		2	0.709	0.713	0.704	0.68	0.713	0.702	0.722	0.694	0.684	0.722
		3	0.687	0.705	0.689	0.667	0.705	0.75	0.747	0.742	0.727	0.747
		4	0.71	0.71	0.705	0.682	0.71	0.723	0.73	0.721	0.698	0.73
		5	0.726	0.738	0.719	0.703	0.738	0.749	0.755	0.745	0.726	0.755

Both approaches are implemented in the Python programming language using implementations available in the sklearn library.

The results obtained for the proposed approach and the two approaches discussed above are given in Table 4. The best obtained F-measure, balanced accuracy and accuracy values for each data set and dispersed version are presented

Table 2. Results of Prec., Recall, F-m., *bacc* and *acc* for the global neural network.

Data set	No. tables	No. artif. obj.	One hidden layer					Two hidden layer				
			Prec.	Recall	F-m.	*bacc*	*acc*	Prec.	Recall	F-m.	*bacc*	*acc*
Satellite imba- lanced	3	1	0.8	0.795	0.783	0.762	0.795	0.806	0.818	0.803	0.775	0.818
		2	0.798	0.796	0.778	0.759	0.796	0.82	0.822	0.818	0.798	0.822
		3	0.796	0.792	0.78	0.759	0.792	0.822	0.829	0.822	0.8	**0.829**
		4	0.782	0.799	0.778	0.758	0.799	0.821	0.828	0.822	0.798	0.828
		5	0.795	0.799	0.786	0.763	0.799	0.819	0.822	0.815	0.792	0.822
	5	1	0.806	0.81	0.798	0.773	0.81	0.818	0.824	0.818	0.795	0.824
		2	0.806	0.805	0.787	0.763	0.805	0.827	0.829	0.826	0.803	0.829
		3	0.789	0.801	0.781	0.761	0.801	0.831	0.833	0.829	0.803	0.833
		4	0.791	0.802	0.785	0.763	0.802	0.836	0.841	0.834	0.809	**0.841**
		5	0.793	0.795	0.779	0.757	0.795	0.832	0.839	0.831	0.803	0.839
	7	1	0.799	0.805	0.792	0.77	0.805	0.817	0.821	0.814	0.792	0.821
		2	0.814	0.809	0.798	0.776	0.809	0.831	0.838	0.831	0.807	0.838
		3	0.811	0.804	0.786	0.767	0.804	0.831	0.84	0.832	0.81	**0.84**
		4	0.8	0.804	0.791	0.768	0.804	0.826	0.833	0.824	0.793	0.833
		5	0.802	0.793	0.777	0.759	0.793	0.837	0.839	0.834	0.808	0.839
	9	1	0.8	0.808	0.799	0.775	0.808	0.831	0.832	0.825	0.803	0.832
		2	0.813	0.81	0.791	0.769	0.81	0.825	0.831	0.821	0.797	0.831
		3	0.806	0.813	0.8	0.778	0.813	0.835	0.839	0.827	0.8	**0.839**
		4	0.806	0.808	0.79	0.766	0.808	0.831	0.836	0.829	0.804	0.836
		5	0.818	0.804	0.781	0.759	0.804	0.834	0.839	0.834	0.807	**0.839**
	11	1	0.808	0.813	0.802	0.777	0.813	0.833	0.837	0.832	0.807	0.837
		2	0.815	0.808	0.792	0.77	0.808	0.832	0.839	0.832	0.809	**0.839**
		3	0.817	0.812	0.793	0.769	0.812	0.825	0.833	0.823	0.796	0.833
		4	0.805	0.81	0.796	0.772	0.81	0.825	0.835	0.825	0.798	0.835
		5	0.8	0.808	0,791	0.769	0.808	0.832	0.835	0.827	0.801	0.835
Satellite ba- lanced	3	1	0.734	0.78	0.749	0.712	0.78	0.799	0.806	0.791	0.758	0.806
		2	0.753	0.77	0.747	0.721	0.77	0.802	0.803	0.797	0.768	0.803
		3	0.787	0.773	0.765	0.745	0.773	0.793	0.8	0.792	0.762	0.8
		4	0.773	0.772	0.758	0.734	0.772	0.805	0.809	0.802	0.776	**0.809**
		5	0.784	0.783	0.774	0.748	0.783	0.807	0.808	0.799	0.772	0.808
	5	1	0.771	0.776	0.766	0.732	0.776	0.812	0.811	0.803	0.776	0.811
		2	0.766	0.777	0.755	0.723	0.777	0.805	0.808	0.798	0.765	0.808
		3	0.791	0.791	0.784	0.757	0.791	0.796	0.803	0.789	0.759	0.803
		4	0.801	0.79	0.776	0.747	0.79	0.798	0.803	0.796	0.769	0.803
		5	0.794	0.791	0.78	0.755	0.791	0.807	0.814	0.806	0.775	**0.814**
	7	1	0.796	0.79	0.781	0.753	0.79	0.804	0.807	0.793	0.763	0.807
		2	0.773	0.791	0.768	0.739	0.791	0.808	0.814	0.803	0.774	**0.814**
		3	0.771	0.785	0.768	0.738	0.785	0.812	0.814	0.81	0.784	**0.814**
		4	0.811	0.791	0.771	0.749	0.791	0.807	0.813	0.806	0.778	0.813
		5	0.789	0.78	0.772	0.746	0.78	0.809	0.814	0.807	0.778	**0.814**
	9	1	0.778	0.788	0.764	0.734	0.788	0.818	0.822	0.813	0.786	**0.822**
		2	0.786	0.786	0.775	0.745	0.786	0.808	0.81	0.8	0.77	0.81
		3	0.782	0.78	0.764	0.737	0.78	0.805	0.815	0.803	0.773	0.815
		4	0.787	0.787	0.771	0.742	0.787	0.812	0.819	0.807	0.775	0.819
		5	0.781	0.792	0.771	0.742	0.792	0.803	0.81	0.801	0.769	0.81
	11	1	0.79	0.798	0.785	0.756	0.798	0.812	0.815	0.809	0.783	**0.815**
		2	0.778	0.791	0.773	0.739	0.791	0.803	0.81	0.803	0.773	0.81
		3	0.801	0.795	0.773	0.745	0.795	0.795	0.805	0.795	0.758	0.805
		4	0.79	0.788	0.775	0.745	0.788	0.807	0.812	0.804	0.779	0.812
		5	0.782	0.791	0.774	0.744	0.791	0.797	0.812	0.8	0.77	0.812

in bold in the table. These three measures were chosen for analysis as F-measure and balanced accuracy best illustrate the model's overall ability to correctly identify all decision classes and balance between precision and recall. The accuracy measure was also compared, but it is less significant in general, as it can lead to incorrect conclusions in the case of imbalanced data. As can be seen in the

Table 3. Results of Prec., Recall, F-m., *bacc* and *acc* for the global neural network.

Data set	No. tables	No. artif. obj.	One hidden layer					Two hidden layer				
			Prec.	Recall	F-m.	*bacc*	*acc*	Prec.	Recall	F-m.	*bacc*	*acc*
Dry Bean imbalanced		1	0.913	0.912	0.912	0.923	0.912	0.921	0.92	0.92	0.931	0.92
		2	0.918	0.917	0.917	0.927	0.917	0.921	0.921	0.921	0.931	**0.921**
	3	3	0.912	0.911	0.911	0.921	0.911	0.92	0.919	0.919	0.93	0.919
		4	0.914	0.913	0.913	0.922	0.913	0.92	0.92	0.92	0.93	0.92
		5	0.915	0.914	0.914	0.925	0.914	0.919	0.919	0.919	0.93	0.919
		1	0.912	0.912	0.912	0.922	0.912	0.918	0.918	0.918	0.928	0.918
		2	0.916	0.915	0.915	0.925	0.915	0.92	0.92	0.919	0.93	**0.92**
	5	3	0.914	0.913	0.913	0.923	0.913	0.918	0.917	0.917	0.927	0.917
		4	0.914	0.914	0.914	0.924	0.914	0.918	0.918	0.917	0.927	0.918
		5	0.915	0.915	0.915	0.925	0.915	0.918	0.918	0.918	0.928	0.918
		1	0.914	0.914	0.913	0.923	0.914	0.917	0.917	0.917	0.928	0.917
		2	0.914	0.914	0.914	0.924	0.914	0.919	0.919	0.919	0.93	**0.919**
	7	3	0.914	0.913	0.913	0.923	0.913	0.915	0.915	0.915	0.925	0.915
		4	0.913	0.913	0.913	0.923	0.913	0.916	0.916	0.916	0.927	0.916
		5	0.913	0.913	0.913	0.922	0.913	0.916	0.915	0.915	0.924	0.915
		1	0.915	0.914	0.914	0.924	0.914	0.916	0.915	0.915	0.925	0.915
		2	0.914	0.913	0.913	0.923	0.913	0.918	0.918	0.918	0.929	**0.918**
	9	3	0.914	0.913	0.913	0.923	0.913	0.915	0.915	0.915	0.924	0.915
		4	0.913	0.912	0.912	0.922	0.912	0.914	0.913	0.913	0.922	0.913
		5	0.913	0.912	0.912	0.921	0.912	0.915	0.915	0.915	0.924	0.915
		1	0.914	0.913	0.913	0.922	0.913	0.915	0.914	0.914	0.924	0.914
		2	0.914	0.914	0.914	0.924	0.914	0.918	0.918	0.918	0.928	**0.918**
	11	3	0.912	0.911	0.911	0.92	0.911	0.909	0.908	0.908	0.916	0.908
		4	0.913	0.912	0.912	0.922	0.912	0.912	0.912	0.912	0.92	0.912
		5	0.913	0.913	0.913	0.922	0.913	0.91	0.91	0.91	0.918	0.91
Dry Bean balanced		1	0.913	0.912	0.912	0.922	0.912	0.92	0.919	0.919	0.93	0.919
		2	0.916	0.915	0.915	0.927	0.915	0.924	0.923	0.923	0.935	**0.923**
	3	3	0.912	0.911	0.911	0.921	0.911	0.92	0.92	0.919	0.931	0.92
		4	0.916	0.916	0.916	0.926	0.916	0.92	0.919	0.919	0.929	0.919
		5	0.914	0.914	0.914	0.923	0.914	0.92	0.92	0.919	0.929	0.92
		1	0.916	0.916	0.916	0.926	0.916	0.918	0.917	0.917	0.927	0.917
		2	0.917	0.916	0.916	0.928	0.916	0.921	0.921	0.921	0.932	**0.921**
	5	3	0.914	0.913	0.913	0.923	0.913	0.92	0.919	0.919	0.929	0.919
		4	0.914	0.913	0.913	0.923	0.913	0.918	0.918	0.918	0.929	0.918
		5	0.918	0.917	0.917	0.927	0.917	0.919	0.919	0.919	0.928	0.919
		1	0.913	0.912	0.911	0.922	0.912	0.919	0.919	0.919	0.93	**0.919**
		2	0.914	0.914	0.914	0.923	0.914	0.918	0.918	0.918	0.928	0.918
	7	3	0.914	0.913	0.913	0.922	0.913	0.918	0.918	0.917	0.927	0.918
		4	0.913	0.912	0.912	0.921	0.912	0.919	0.918	0.918	0.928	0.918
		5	0.914	0.913	0.913	0.922	0.913	0.918	0.918	0.917	0.927	0.918
		1	0.913	0.913	0.913	0.922	0.913	0.917	0.917	0.916	0.926	0.917
		2	0.908	0.907	0.907	0.915	0.907	0.919	0.919	0.919	0.929	**0.919**
	9	3	0.912	0.912	0.912	0.921	0.912	0.916	0.916	0.916	0.925	0.916
		4	0.914	0.913	0.913	0.923	0.913	0.917	0.917	0.917	0.926	0.917
		5	0.913	0.912	0.912	0.921	0.912	0.918	0.917	0.917	0.926	0.917
		1	0.912	0.912	0.912	0.922	0.912	0.917	0.917	0.917	0.926	0.917
		2	0.91	0.909	0.909	0.918	0.909	0.916	0.915	0.914	0.925	0.915
	11	3	0.913	0.912	0.912	0.921	0.912	0.92	0.92	0.919	0.929	**0.92**
		4	0.912	0.911	0.911	0.92	0.911	0.917	0.916	0.916	0.926	0.916
		5	0.913	0.913	0.913	0.922	0.913	0.914	0.914	0.913	0.922	0.914

vast majority of cases, the proposed approach gives better results for all three compared measures. However, in the case of the Satellite data set, the proposed approach does not perform well. This is due to the data having the greatest variation in attributes present in local tables(very few overlapping attributes). To conclude, when a camera points at an object and generates attributes the majority of are not present in any other local tables, then the proposed approach does not perform better than the classifier ensemble approach. However, in other cases, when the cameras are more densely arranged, overlapping in terms of attributes then the proposed approach definitely performs better. It should also be noted that the number of objects in the training set does not affect the quality of classification generated by the proposed approach, i.e. for both small and large training sets the proposed approach gives good results.

Now, the results for all three measures generated by the analyzed approaches will be compared. To prove that the obtained differences in F-measure values are significant, the Friedman test was performed. Three dependent samples of 30 observations was used, with the test confirming that there is a statistically significant difference in the F-measure obtained for the three approaches considered, $\chi^2(29,2) = 10.034, p = 0.007$. Additionally, comparative box-plot for the F-measure with three methods was created (Fig. 2). As can be observed, on average, the values of the F-measure for the proposed approach are the largest. The post-hoc Dunn Bonferroni test was also performed which confirmed a significant difference in average F-measure values between the three approaches. The results (significant were presented in bold) can be found in Table 5. In the end, it can be said that the proposed approach improves the quality of classification compared to approaches known from the literature in terms of the F-measure.

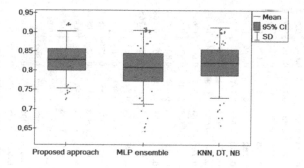

Fig. 2. Comparison of F-measure obtained for approaches: the proposed approach with global model; homogeneous ensemble with MLP networks (MLP ensemble) and the ensemble of classifiers: $k-$nearest neighbors, decision tree and naive bayes classifier (KNN, DT, NB).

Next the Friedman test was performed in order to show that the obtained differences in balanced accuracy values are significant. For balanced accuracy, the Friedman statistics was $7.983, p = 0.018$ indicating that there is a statistically significant difference in the balanced accuracy obtained for the three approaches considered in the paper. Additionally, comparative box-plot for the balanced accuracy was created (Fig. 3). As can be seen also here, the average of balanced accuracy is the highest for the proposed approach. The post-hoc Dunn Bonferroni test confirmed a significant difference in balanced accuracy values between one pair: the proposed approach & MLP ensemble with $p = 0.035$. So it can be

Table 4. Results of Prec., Recall, F-m., *bacc* and *acc* for the global neural network; homogeneous ensemble with MLP networks (MLP ensemble) and the ensemble of classifiers (KNN, DT, NB).

Data set	No. tables	The proposed approach					MLP ensemble					KNN, DT, NB				
		Prec.	Recall	F-m.	bacc	acc	Prec.	Recall	F-m.	bacc	acc	Prec.	Recall	F-m.	bacc	acc
Vehicle imba-lanced	3	0.738	0.74	**0.737**	**0.721**	**0.74**	0.725	0.74	0.728	0.715	0.74	0.701	0.709	0.694	0.686	0.709
	5	0.727	0.734	**0.724**	0.714	**0.734**	0.761	0.724	0.695	**0.722**	0.724	0.718	0.709	0.696	0.69	0.709
	7	0.735	0.735	**0.727**	0.708	0.735	0.73	0.74	0.722	**0.709**	**0.74**	0.69	0.693	0.677	0.67	0.693
	9	0.746	0.752	**0.739**	**0.719**	**0.752**	0.655	0.685	0.652	0.658	0.685	0.708	0.717	0.699	0.694	0.717
	11	0.752	0.756	**0.744**	**0.726**	**0.756**	0.67	0.685	0.676	0.66	0.685	0.705	0.685	0.677	0.672	0.685
Vehicle ba-lanced	3	0.756	0.762	**0.753**	**0.736**	**0.762**	0.742	0.756	0.746	0.729	0.756	0.733	0.732	0.716	0.705	0.732
	5	0.749	0.749	**0.742**	**0.725**	**0.749**	0.582	0.717	0.642	0.661	0.717	0.726	0.728	0.711	0.698	0.728
	7	0.769	0.77	**0.758**	**0.741**	**0.77**	0.711	0.728	0.708	0.699	0.728	0.748	0.752	0.733	0.721	0.752
	9	0.757	0.76	**0.747**	**0.732**	**0.76**	0.66	0.689	0.654	0.649	0.689	0.718	0.728	0.712	0.7	0.728
	11	0.768	0.768	**0.762**	**0.746**	**0.768**	0.683	0.685	0.662	0.656	0.685	0.667	0.677	0.657	0.646	0.677
Satellite imba-lanced	3	0.822	0.829	0.822	0.8	0.829	0.838	0.849	0.826	0.8	0.849	0.868	0.87	**0.868**	**0.848**	**0.87**
	5	0.836	0.841	0.834	0.809	0.841	0.843	0.842	0.804	0.783	0.842	0.863	0.864	**0.86**	**0.835**	**0.864**
	7	0.831	0.84	0.832	0.81	0.84	0.758	0.835	0.79	0.768	0.835	0.855	0.857	**0.852**	**0.823**	**0.857**
	9	0.834	0.839	0.834	0.807	0.839	0.75	0.829	0.783	0.758	0.829	0.856	0.858	**0.851**	**0.82**	**0.858**
	11	0.832	0.839	0.832	0.809	0.839	0.754	0.828	0.783	0.757	0.828	0.851	0.854	**0.844**	**0.811**	**0.854**
Satellite ba-lanced	3	0.805	0.809	0.802	0.776	0.809	0.865	0.868	0.864	0.839	0.868	0.879	0.872	**0.874**	**0.859**	**0.872**
	5	0.807	0.814	0.806	0.775	0.814	0.875	0.88	**0.876**	0.851	**0.88**	0.877	0.871	0.873	**0.856**	0.871
	7	0.812	0.814	0.81	0.784	0.814	0.873	0.869	0.87	0.859	0.869	0.881	0.878	**0.878**	**0.861**	**0.878**
	9	0.818	0.822	0.813	0.786	0.822	0.858	0.86	0.858	0.838	0.86	0.874	0.871	**0.871**	**0.851**	**0.871**
	11	0.812	0.815	0.809	0.783	0.815	0.873	0.87	**0.87**	**0.856**	**0.87**	0.871	0.87	0.87	0.848	0.87
Dry Bean imba-lanced	3	0.921	0.921	**0.921**	**0.931**	**0.921**	0.911	0.91	0.91	0.92	0.91	0.908	0.906	0.906	0.916	0.906
	5	0.92	0.92	**0.919**	**0.93**	**0.92**	0.903	0.903	0.902	0.908	0.903	0.904	0.902	0.901	0.908	0.902
	7	0.919	0.919	**0.919**	**0.93**	**0.919**	0.902	0.901	0.9	0.906	0.901	0.901	0.899	0.899	0.905	0.899
	9	0.918	0.918	**0.918**	**0.929**	**0.918**	0.898	0.896	0.895	0.9	0.896	0.897	0.894	0.893	0.897	0.894
	11	0.918	0.918	**0.918**	**0.928**	**0.918**	0.902	0.901	0.901	0.907	0.901	0.902	0.9	0.9	0.905	0.9
Dry Bean ba-lanced	3	0.924	0.923	**0.923**	**0.935**	**0.923**	0.911	0.911	0.911	0.926	0.911	0.91	0.909	0.909	0.919	0.909
	5	0.921	0.921	**0.921**	**0.932**	**0.921**	0.907	0.907	0.907	0.919	0.907	0.9	0.899	0.898	0.907	0.899
	7	0.919	0.919	**0.919**	**0.93**	**0.919**	0.906	0.905	0.905	0.917	0.905	0.901	0.9	0.899	0.909	0.9
	9	0.919	0.919	**0.919**	**0.929**	**0.919**	0.903	0.902	0.902	0.913	0.902	0.899	0.898	0.898	0.907	0.898
	11	0.92	0.92	**0.919**	**0.929**	**0.92**	0.905	0.904	0.904	0.916	0.904	0.903	0.903	0.902	0.912	0.903

Table 5. p-values for the post-hoc Dunn Bonferroni test for F-measure

	Proposed approach	MLP ensemble	KNN, DT, NB
Proposed approach	–	**0.011**	**0.035**
MLP ensemble	**0.011**	–	1
KNN, DT, NB	**0.035**	1	–

concluded that both the proposed approach and the ensemble of classifiers KNN, DT, NB get the best balanced accuracy results.

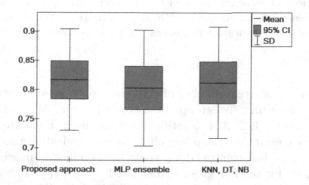

Fig. 3. Comparison of balanced accuracy obtained for approaches: the proposed approach with global model; homogeneous ensemble with MLP networks (MLP ensemble) and the ensemble of classifiers: k−nearest neighbors, decision tree and naive bayes classifier (KNN, DT, NB).

As we know, accuracy values can be deceptive and often do not take minority classes into account, nonetheless, comparative analysis was carried out for this measure. For accuracy, the Friedman statistics was $6.889, p = 0.032$ indicating a reject of the null hypothesis, but as can be seen in Fig. 4, the average accuracy values are similar. Also, the post-hoc Dunn Bonferroni test did not confirm a significant difference between any pair of approaches. Thus, as a conclusion, it can be confirmed that the proposed approach on average improves values of F-measure and balanced accuracy. Of course, comparing the results obtained for each data set separately (Table 4), it can be seen that for some data sets this improvement is significant, while for others the proposed approach does not improve the quality of classification. The situation in which the proposed approach does not do well is when we have a very large number of conditional/descriptive attributes and a relatively small number of local tables/cameras.

Next, the impact of the number of artificial objects used in the proposed approach on the quality of classification is analyzed. The comparison, as before, was be made using three measures: F-measure, balanced accuracy and accuracy.

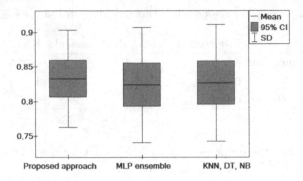

Fig. 4. Comparison of accuracy obtained for approaches: the proposed approach with global model; homogeneous ensemble with MLP networks (MLP ensemble) and the ensemble of classifiers: k−nearest neighbors, decision tree and naive bayes classifier (KNN, DT, NB).

To test whether the different number of artificial objects used generated a significant difference in results, five groups were created 1AO, 2AO, 3AO, 4AO, 5AO – results obtained for 1, 2, 3, 4, 5 artificial objects used. Each group contained 60 observations (results obtained for all data sets and all versions of dispersion). The Friedman test confirmed a statistically significant difference in the F-measure obtained for the five groups considered, $\chi^2(59, 4) = 10.830, p = 0.029$. The Wilcoxon each-pair test confirmed the significant differences between the average F-measure values for the following pairs: 1AO & 4AO with $p = 0.01$, 1AO & 5AO with $p = 0.001$, 3AO & 4AO with $p = 0.005$, 3AO & 5AO with $p = 0.0002$. Additionally, a comparative graph for the F-measure with different number of artificial objects used was created (Fig. 5). It can be seen that the results obtained for using 4 and 5 artificial objects are better than those obtained with fewer artificial objects.

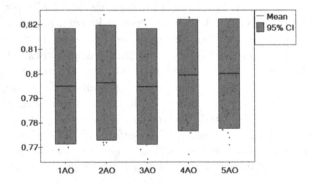

Fig. 5. Comparison of F-measure accuracy obtained for the proposed approach with 1, 2, 3, 4, 5 artificial objects generated (1AO, 2AO, 3AO, 4AO, 5AO).

Similar analyses were performed for balanced accuracy and accuracy. Friedman's test confirmed a significant difference for accuracy with statistics $15.501, p = 0.004$. The Wilcoxon each-pair test confirmed the significant differences between the average accuracy values for the following pairs: 1AO & 4AO with $p = 0.044$, 1AO & 5AO with $p = 0.002$, 2AO & 5AO with $p = 0.009$, 3AO & 4AO with $p = 0.002$, 3AO & 5AO with $p = 0.0001$. Also, the comparative graph for accuracy values (Fig. 6) proves that for 5 and 4 artificial objects the generated results are better. For the balanced accuracy, the Friedman test does not confirm a significant difference in the mean value for different numbers of artificial objects used. Nonetheless, it can be concluded that larger numbers of artificial objects used to build the global model improve its quality. Thus, the proposed method of generating artificial objects with missing values in local tables has a positive effect on the model accuracy, and a larger number of artificial objects increases the quality of the model.

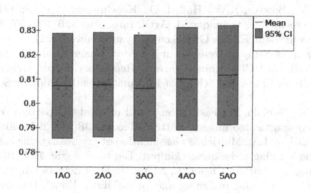

Fig. 6. Comparison of accuracy obtained for the proposed approach with 1, 2, 3, 4, 5 artificial objects generated (1AO, 2AO, 3AO, 4AO, 5AO).

4 Conclusion

In this paper, a situation in which local decision tables containing partial characteristics of objects from fragmented images was considered. Then, based on the local tables, tables with artificial objects were generated by filling the missing values of characteristics. Finally, local models were built based on local tables, which were finally aggregated into a global model.

In conclusion, it is important to note that while the proposed approach consistently enhances classification quality on average, its efficacy may vary across different data sets. The method excels particularly in scenarios with a substantial number of attributes and a relatively small number of local tables/cameras.

In summary, our study establishes the superiority of the proposed approach in terms of F-measure and balanced accuracy, showcasing its potential to elevate classification performance. The positive correlation between the number of artificial objects and model quality reinforces the practical applicability of our method. These findings contribute valuable knowledge to the field of classification based on fragmented images, offering a promising avenue for further research and application in real-world scenarios.

References

1. Canal, F.Z., et al.: A survey on facial emotion recognition techniques: a state-of-the-art literature review. Inf. Sci. **582**, 593–617 (2022)
2. Chaki, J., Dey, N., Moraru, L., Shi, F.: Fragmented plant leaf recognition: bag-of-features, fuzzy-color and edge-texture histogram descriptors with multi-layer perceptron. Optik **181**, 639–650 (2019)
3. Chen, J., Chen, J., Zhang, D., Sun, Y., Nanehkaran, Y.A.: Using deep transfer learning for image-based plant disease identification. Comput. Electron. Agric. **173**, 105393 (2020)
4. Chawla, N.V., Bowyer, K.W., Hall, L.O., Kegelmeyer, W.P.: SMOTE: synthetic minority over-sampling technique. J. Artif. Intell. Res. **16**, 321–357 (2002)
5. Çalli, E., Sogancioglu, E., van Ginneken, B., van Leeuwen, K.G., Murphy, K.: Deep learning for chest X-ray analysis: a survey. Med. Image Anal. **72**, 102125 (2021)
6. Dua, D., Graff, C.: UCI Machine Learning Repository. http://archive.ics.uci.edu/ml. University of California, School of Information and Computer Science, Irvine, CA (2019)
7. Fornasier, M., Toniolo, D.: Fast, robust and efficient 2D pattern recognition for re-assembling fragmented images. Pattern Recogn. **38**(11), 2074–2087 (2005)
8. Koklu, M., Ozkan, I.A.: Multiclass classification of dry beans using computer vision and machine learning techniques. Comput. Electron. Agric. **174**, 105507 (2020)
9. Lin, G., Tang, Y., Zou, X., Cheng, J., Xiong, J.: Fruit detection in natural environment using partial shape matching and probabilistic Hough transform. Precision Agric. **21**, 160–177 (2020)
10. Marfo, K.F., Przybyła-Kasperek, M.: Radial basis function neural network with a centers training stage for prediction based on dispersed image data. In: Mikyška, J., de Mulatier, C., Paszynski, M., Krzhizhanovskaya, V.V., Dongarra, J.J., Sloot, P.M. (eds.) Computational Science – ICCS 2023. ICCS 2023. LNCS, vol. 10476, pp. 89–103. Springer, Cham (2023). https://doi.org/10.1007/978-3-031-36027-5_7
11. Shelepin, Y.E., Chikhman, V.N., Foreman, N.: Analysis of the studies of the perception of fragmented images: global description and perception using local features. Neurosci. Behav. Physiol. **39**(6), 569–580 (2009)
12. Siebert, J.P.: Vehicle Recognition Using Rule Based Methods, Turing Institute Research Memorandum TIRM-87-0.18, March 1987
13. Vashist, P.C., Pandey, A., Tripathi, A.: A comparative study of handwriting recognition techniques. In: 2020 International Conference on Computation, Automation and Knowledge Management (ICCAKM), pp. 456–461. IEEE, January 2020
14. Wu, L., Mokhtari, S., Nazef, A., Nam, B., Yun, H.B.: Improvement of crack-detection accuracy using a novel crack defragmentation technique in image-based road assessment. J. Comput. Civ. Eng. **30**(1), 04014118 (2016)
15. Xin, M., Wang, Y.: Research on image classification model based on deep convolution neural network. EURASIP J. Image Video Process. **2019**, 1–11 (2019)

A Cross-Domain Perspective to Clustering with Uncertainty

Salvatore F. Pileggi[✉]

Faculty of Engineering and IT, School of Computer Science,
University of Technology Sydney, Ultimo, Australia
SalvatoreFlavio.Pileggi@uts.edu.au

Abstract. Clustering in presence of uncertainty may be considered, at the same time, to be a pressing need and a challenge to effectively address many real-world problems. This concise literature review aims to identify and discuss the associated body of knowledge according to a cross-domain perspective. A semi-systematic methodology has allowed the selection of 68 papers, with a priority on the most recent contributions. The analysis has re-marked the relevance of the topic and has made explicit a trend to domain-specific solutions over generic-purpose approaches. On one side, this trend enables a more specific set of solutions within specific communities; on the other side, the resulting distributed approach is not always well-integrated in the mainstream and may generate a further fragmentation of the body of knowledge, mostly because of some lack of abstraction in the definition of specific problems. While these gaps are largely understandable within the research community, a lack of implementations to provide ready-to-use resources is overall critical, looking at a more and more computational and data intensive world.

Keywords: Clustering · Uncertainty Modelling · Uncertainty Management · Unsupervised Learning · Data Analysis · Data Mining

1 Introduction

Empirical observations show an increasing quantity of data with a degree of uncertainty [14]. Indeed, real-world data naturally tends to present uncertainty due to different factors including, among others, human or instrumental errors [81], randomness, imprecision, vagueness and partial ignorance [17].

In general terms, the theoretical impact of data uncertainty, as well as the risk associated with ignoring it (e.g. [19,48]), are well-known issues within the scientific community. Always in general, it strongly suggests, wherever possible, a proper and explicit uncertainty model to effectively support representation, measuring and consequent analysis. From a more practical perspective, more and more studies present a specific focus in a variety of application domains, such as, among the very many, budget impact analysis [59], organizational environments [45] and hydrological data [56]. Such a critical modelling is intrinsically

L. Franco et al. (Eds.): ICCS 2024, LNCS 14838, pp. 295–308, 2024.
https://doi.org/10.1007/978-3-031-63783-4_22

challenging and may require a domain specific-approach, such as for Big Data [29] [79], Visualization [41] and Deep Learning [1].

On the other side, clustering techniques [86] group data points into different clusters based on their similarity. These techniques have been extensively used in a general scientific context and traditional approaches keep evolving as a response to an environment characterised by evolving needs [82]. For instance, clustering is a common class of unsupervised learning [73], often adapted to achieve concrete goals in the different application domains (e.g. [7]), as well as formal classification [8], ontological modelling [49,62] and rule mining [76] commonly rely on clustering techniques.

Intuitively, clustering in a context of uncertainty, or even just potential uncertainty, proposes additional significant challenges on both (i) modelling similarity between uncertain objects and (ii) developing effective and efficient computational methods accordingly [39].

Alternative approaches to deal with uncertainty can be used for different reasons in different contexts. A classification of these techniques is not trivial. For instance, in [36] the authors have identified two main broad categories that aim, respectively, to complement and to generalise probabilistic representations. The former family addresses non-probabilistic uncertainty (typically imprecision, vagueness or gradedness), while the latter targets effective modelling of partial ignorance. More holistically, looking at extensions of traditional methods, three main categories have been summarised in [39]: partitioning clustering, density-based clustering, and possible world approaches. The resulting extended solutions integrate the original semantics with uncertainty modelling.

This papers aims to holistically review the most recent advances in the field of clustering in presence of uncertainty. The focus is on a cross-domain perspective resulting from a contextual analysis that considers the most relevant computational trends and related applications.

Related Work. This work can be framed in the very broad context of uncertain data algorithms and applications [5]. A valuable review specifically on clustering has been provided in 2017 [17]. The focus of such a work is on uncertainty management and associated theoretical formalisms. Other concise contributions aimed at summarising the body of knowledge are relatively old (e.g. [36] and [14]), given the strong and constant advances in the computational world. This paper proposes an additional contribution to the body of knowledge in the field by addressing recent advances minimising the overlapping with existing reviews. Additionally, the provided analysis is performed according to a simple analysis framework, which also include the application domain. It allows to distinguish between generic-purpose and application-specific solutions.

Structure of the Paper. The introductory part of the paper is concluded by a description of the adopted methodology (Sect. 2). The core part of the paper includes two different sections that aims, respectively, to overview the most relevant contributions in literature according to a cross-domain perspective (Sect. 3) and to discuss results looking at major gaps and challenges (Sect. 4). Finally, a conclusions section provides an overview of the work.

2   Methodology and Approach

In order to generate a tangible contribution to the body of knowledge and avoid, in the limit of the possible, overlapping and lack of deepening, this literature review has been conducted by combining typical systematic processes with non-systematic practices.

The mainstream process assumes, as usual, paper retrieval from relevant databases. In this specific case, queries have been performed by simply combining two main keywords, namely *Clustering* and *Uncertainty*.

Selection Criteria and Saturation. The selection of the papers to include in the review has been performed by applying a critical analysis aimed at the identification of the most relevant contributions in the field. Although no pre-defined objective filter (e.g. on time-range) has been applied, the most recent papers have been systematically included in order to highlight the most recent advances in context and to maximise the value provided. The relatively soft selection criteria enabled the retrieval of an important number of papers. However, the selection process has been much more focused in fact. Indeed, after a number of iterations, a feeling of saturation naturally emerged as contributions started to present consolidations of existing concepts rather than novel solutions. This additional non-systematic element has been a determinant to facilitate de facto conciseness at a relevant scale.

Analysis Framework and Limitations. The analysis has been conducted according to two major dimensions: *domain* and *approach*. The former dimension aims primarily to distinguish between generic-purpose and domain specific solutions, while the latter wants to facilitate an overview of major techniques. The presentation of the review (Sect. 3) has been structured looking at the domain. Indeed, the classification of the different approaches is intrinsically more fragmented and not always explicit. In general terms, the classification followed the claims by authors and the original analysis. Non-systematic practices may have introduced biases. It applies mostly to selection criteria. Additionally, because of the high number of existing works distributed in a variety of domains, it is hard to assess the exhaustiveness of the review.

3   A Cross-Domain Analysis

This section has a descriptive focus as it provides an overview of the contributions included in this study.

We deal separately with solutions that present a completely generic focus (referred to as "generic-purpose" and reported in Table 1) and that have been designed within a specific application domain ("domain-specific", Table 2). This generic classification naturally introduces a cross-domain analysis. However, there are not always well defined boundaries as certain applications as identified in the context of this work may present a certain degree of genericness.

Table 1. Generic-purpose selected contributions.

Title/Ref.	Year	Approach
Cloud-Cluster: An uncertainty clustering algorithm based on cloud model [54]	2023	Method
Outlier-robust multi-view clustering for uncertain data [69]	2021	Multi-view Clustering
Multi-view spectral clustering for uncertain objects [68]	2021	Multi-view Clustering
Modeling uncertain data using Monte Carlo integration method for clustering [67]	2019	Monte-Carlo
Optimal clustering under uncertainty [15]	2018	Method
Locally weighted ensemble clustering [33]	2017	Ensemble Clustering
Self-adapted mixture distance measure for clustering uncertain data [51]	2017	Method
Novel density-based and hierarchical density-based clustering algorithms for uncertain data [88]	2017	Hierarchical Clustering
An information-theoretic approach to hierarchical clustering of uncertain data [24]	2017	Hierarchical Clustering
Active Clustering with Model-Based Uncertainty Reduction [84]	2016	Active Clustering
Robust ensemble clustering using probability trajectories [32]	2015	Ensemble Clustering
Large margin clustering on uncertain data by considering probability distribution similarity [85]	2015	PD Similarity
Representative clustering of uncertain data [90]	2014	Framework
Active spectral clustering via iterative uncertainty reduction [80]	2012	Active Clustering
Minimizing the variance of cluster mixture models for clustering uncertain objects [22]	2012	Method
Clustering uncertain data based on probability distribution similarity [39]	2011	PD Similarity
Clustering uncertain data using voronoi diagrams and r-tree index [44]	2010	Method
Subspace clustering for uncertain data [25]	2010	Sub-space clustering
Exceeding expectations and clustering uncertain data [20]	2009	Optimization
Clustering Uncertain Data with Possible Worlds [77]	2009	Method
Clustering Uncertain Data Via K-Medoids [21]	2008	Method
A hierarchical algorithm for clustering uncertain data via an information-theoretic approach [23]	2008	Hierarchical Clustering
Clustering Uncertain Data Using Voronoi Diagrams [43]	2008	Voronoi diagrams
Uncertain fuzzy clustering: Insights and recommendations [65]	2007	Fuzzy Logic
Uncertain fuzzy clustering: Interval type-2 fuzzy approach to c-means [38]	2007	Fuzzy Logic
Density-based clustering of uncertain data [47]	2005	Fuzzy Logic
Hierarchical density-based clustering of uncertain data [46]	2005	Fuzzy Logic

Generic-Purpose Solutions. Among the generic-purpose works, there are two clearly identifiable sub-sets of solutions adopting, respectively, mixed (or not uniquely classifiable) methods [15, 21, 22, 44, 51, 54, 77] and Fuzzy Logic [38, 46, 47, 65]. Smaller classes of solutions adopt Hierarchical Clustering [23, 24, 88], Probability Distribution Similarity [39, 85], Ensemble Clustering [32, 33], Multiview Clustering [68, 69] and Active Clustering [80, 84]. Other methods focus on framework-based solutions [90], Voronoi diagrams [43], Monte-Carlo [67], optimization strategies [20] and Sub-space Clustering [25].

Domain-Specific Solutions. Mixed methods [3, 4, 9, 16, 26, 30, 31, 34, 35, 40, 50, 53, 55, 57, 60, 61, 70, 72, 75], as well as Fuzzy Logic [11, 18, 64, 71], Hierarchical Clustering [74, 78], Optimization [13, 28, 37, 87], and framework-based approaches [27, 66] play a significant role also in a context of domain-specific applications. Other contributions include the adaptation of traditional techniques [12], quality assessment [83], Possible Worlds [52], Distributed Clustering [89], Bayesian Modeling [10], Three-way Clustering [2], review for assessment purpose within a specific domain [42], Rough Set Theory [58], Active Learning [63] and Stochastic Models [6].

Table 2. Domain-specific selected contributions.

Title/Ref.	Year	Approach	Domain
Stochastic economic dispatch of wind power under uncertainty using clustering-based extreme scenarios [6]	2024	Stochastic Model	Energy
Uncertainty clustering internal validity assessment using Fréchet distance for unsupervised learning [64]	2022	Fuzzy Logic	Machine Learning
A three-stage automated modal identification framework for bridge parameters based on frequency uncertainty and density clustering [30]	2022	Method	Engineering
Clustering uncertain graphs using ant colony optimization (ACO) [37]	2022	Optimization	Graphs
Uncertainty-Aware Clustering for Unsupervised Domain Adaptive Object Re-Identification [78]	2022	Hierarchical Clustering	Machine Learning
Decision-based scenario clustering for decision-making under uncertainty [31]	2022	Method	Decision Making
Active domain adaptation via clustering uncertainty-weighted embeddings [63]	2021	Active Learning	Machine Learning
UAC: An Uncertainty-Aware Face Clustering Algorithm [16]	2021	Method	Face Recognition
Uncertainty assessment in reservoir performance prediction using a two-stage clustering approach: Proof of concept and field application [26]	2021	Method	Petroleum Science
Handling uncertainty in financial decision making: a clustering estimation of distribution algorithm with simplified simulation [70]	2020	Method	Decision Making
Ride-sharing under travel time uncertainty: Robust optimization and clustering approaches [50]	2020	Method	Transportation
Deep semantic clustering by partition confidence maximisation [35]	2020	Method	Machine Learning
Uncertainty mode selection in categorical clustering using the rough set theory [58]	2020	Rough Set	Categorical Data
Clustering of electrical load patterns and time periods using uncertainty-based multi-level amplitude thresholding [11]	2020	Fuzzy Logic	Energy
Efficient Assessment of Reservoir Uncertainty Using Distance-Based Clustering: A Review [42]	2019	Review	Petroleum Science
Big-data clustering with interval type-2 fuzzy uncertainty modeling in gene expression datasets [71]	2019	Fuzzy Logic	Genetics
Clustering mining blocks in presence of geological uncertainty [75]	2019	Method	Geology
Efficient and effective algorithms for clustering uncertain graphs [28]	2019	Optimization	Graphs
A three-way clustering approach for handling missing data using GTRS [2]	2018	Three-way Clustering	Missing Data
Clustering uncertain graphs [9]	2017	Method	Graphs
Novel adaptive multi-clustering algorithm-based optimal ESS sizing in ship power system considering uncertainty [87]	2017	Optimization	Energy
Multiple clustering views from multiple uncertain experts [10]	2017	Bayesian Model	Collaborative Environments
Uncertain data clustering in distributed peer-to-peer networks [89]	2017	Distributed Clustering	P2P Network

continued

Table 2. continued

Title/Ref.	Year	Approach	Domain
Efficient clustering of large uncertain graphs using neighborhood information [27]	2017	Framework	Graphs
Clustering based unit commitment with wind power uncertainty [72]	2016	Method	Energy
A framework for clustering uncertain data [66]	2015	Framework	Visualization
Efficient clustering of uncertain data streams [40]	2014	Method	Data Stream
Uncertain data clustering-based distance estimation in wireless sensor networks [55]	2014	Method	Wireless Sensor Network
Weighted graph clustering with non-uniform uncertainties [13]	2014	Optimization	Graphs
Clustering large data with uncertainty [18]	2013	Fuzzy Logic	Large Data
Reliable clustering on uncertain graphs [52]	2012	Possible Worlds	Graphs
Clustering uncertain trajectories [61]	2011	Method	Location Data
Hue-stream: Evolution-based clustering technique for heterogeneous data streams with uncertainty [57]	2011	Method	Data Stream
An algorithm for clustering heterogeneous data streams with uncertainty [34]	2010	Method	Data Stream
On high dimensional projected clustering of uncertain data streams [3]	2009	Method	Data Stream
Clustering trajectories of moving objects in an uncertain world [60]	2009	Method	Location Data
A Framework for Clustering Uncertain Data Streams [4]	2008	Method	Data Stream
Conceptual clustering categorical data with uncertainty [83]	2007	Quality Assessment	Categorical Data
Including probe-level uncertainty in model-based gene expression clustering [53]	2007	Method	Genetics
Pvclust: an r package for assessing the uncertainty in hierarchical clustering [74]	2006	Hierarchical Clustering	Genetics
Uncertain Data Mining: An Example in Clustering Location Data [12]	2006	UK-Means	Location Data

4   Discussion

In order to exhaustively discuss the review, the section is structured in two different subsections to address an overview of the results (Sect. 4.1) and a critical analysis of the major gaps emerged (Sect. 4.2).

4.1   Overview

In quantitative terms, the majority (60%) of the 68 papers selected within the time range 2005–2024 presents a domain-specific focus. As shown in Fig. 1, such a trend becomes more consistent and somehow predominant from 2018 onward. More holistically, the study confirms a substantial research interest in the topic throughout the observation period.

The analysis conducted in this study based on soft-classification allows to overview the application domain (Fig. 2a). Looking at the 41 domain-specific

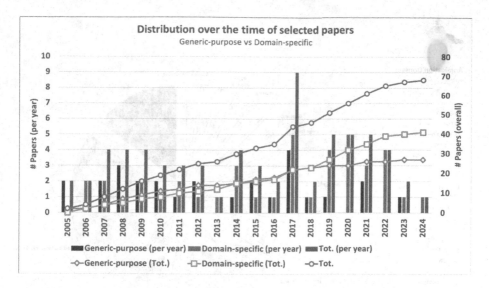

Fig. 1. Distribution over the time of the selected contributions.

papers, as expected, generic application fields, such as *Graphs*, *Data Stream* and *Machine learning*, are quantitatively more relevant, both with large domains (e.g. *Energy*, *Genetics* and *Location Data*). At a more fine-grained level, the review has identified a diverse spectrum reflecting a generic need for clustering in presence of uncertainty.

A more technical perspective is summarised in Fig. 2b. A consistent part (38%) of the considered papers proposes a mixed-method approach, which is generically referred to as *method* in the adopted analysis framework. *Fuzzy Logic*, *Optimization* and *Hierarchical Clustering* are the most popular approaches. In addition, to note a focus on defining analysis *frameworks*, on *Multi-view Clustering*, *Probability Distribution Similarity*, *Ensemble* and *Active Clustering*.

4.2 Major Gaps and Challenges

From a critical perspective, the analysis conducted has allowed the identification of a number of gaps other than the originally reported in the different contributions that are summarised in Table 3.

The review has reiterated the practical relevance of clustering in presence of uncertainty. In such a context, ready-to-use resources in the computational world are crucial and a determinant to consolidate and properly transfer innovation into practice (*G1*).

The cross-domain focus has highlighted and put emphasis on applications to solve real-world problems. The relationship between generic-purpose and domain-specific solutions not always clear (*G2*). The fine-grained application-specific approach makes re-using complex and costly (*G3*). That is because of a lack of abstraction in the formulation of domain-specific problems (*G4*) with a

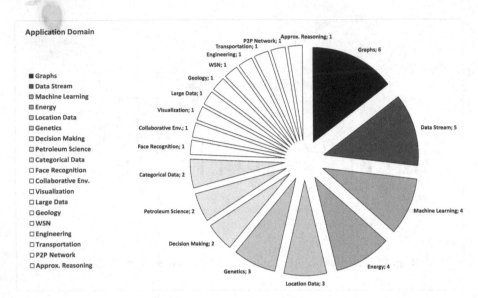

(a) Application Domain.

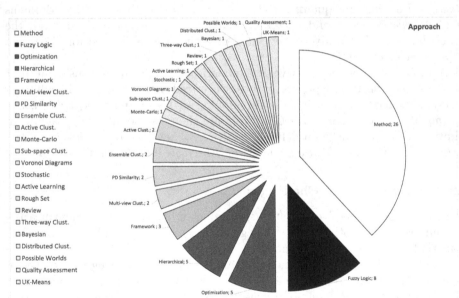

(b) Approaches characterizing the selected solutions.

Fig. 2. Analysis overview.

Table 3. Main gaps.

	Gap
G1	Lack of implementations to provide ready-to-use computational resources
G2	Relationship between generic-purpose and domain-specific solutions not always clear
G3	Fine-grained application-specific approach that doesn't facilitate re-use in a different context
G4	Lack of abstraction in domain-specific approaches
G5	Despite a well-identified research field, solutions are not always discussed in context looking at the exiting body of knowledge

consequent difficulty to generalize solutions or re-using existing ones in a different context.

More in general, despite a well-identified research field, solutions are not always discussed in context looking at the exiting body of knowledge (G5).

5 Conclusions

Given the popularity of clustering techniques within the modern computational world and the intrinsic need to deal with uncertainty in the different application domains, this concise literature review aimed at a cross-domain analysis of the most recent solutions in the field.

Such analysis has firstly re-marked the relevance of the topic and the consequent related research activity. A trend to domain-specific solutions over generic-purpose approaches progressively emerged and became more consistent in the last few years. On one side, this trend enables a more specific set of solutions within specific communities; on the other side, the resulting distributed approach is not always well-integrated in the mainstream and may generate a further fragmentation of the body of knowledge, mostly because of some lack of abstraction in the definition of specific problems.

While these gaps are largely understandable within the research community, a lack of implementations to provide ready-to-use resources is overall critical, looking at a more and more computational and data intensive world.

Acknowledgements. I would like to thank the conference organisers, the workshop chair as well as the anonymous reviewers, who have provided valuable constructive feedback.

References

1. Abdar, M., et al.: A review of uncertainty quantification in deep learning: techniques, applications and challenges. Inf. Fusion **76**, 243–297 (2021)
2. Afridi, M.K., Azam, N., Yao, J., Alanazi, E.: A three-way clustering approach for handling missing data using GTRS. Int. J. Approximate Reasoning **98**, 11–24 (2018)
3. Aggarwal, C.C.: On high dimensional projected clustering of uncertain data streams. In: 2009 IEEE 25th International Conference on Data Engineering, pp. 1152–1154. IEEE (2009)
4. Aggarwal, C.C., Philip, S.Y.: A framework for clustering uncertain data streams. In: 2008 IEEE 24th International Conference on Data Engineering, pp. 150–159. IEEE (2008)
5. Aggarwal, C.C., Philip, S.Y.: A survey of uncertain data algorithms and applications. IEEE Trans. Knowl. Data Eng. **21**(5), 609–623 (2008)
6. Bhavsar, S., Pitchumani, R., Maack, J., Satkauskas, I., Reynolds, M., Jones, W.: Stochastic economic dispatch of wind power under uncertainty using clustering-based extreme scenarios. Electr. Power Syst. Res. **229**, 110158 (2024)
7. Caron, M., Bojanowski, P., Joulin, A., Douze, M.: Deep clustering for unsupervised learning of visual features. In: Proceedings of the European Conference on Computer Vision (ECCV), pp. 132–149 (2018)
8. Castellanos, A., Cigarrán, J., García-Serrano, A.: Formal concept analysis for topic detection: a clustering quality experimental analysis. Inf. Syst. **66**, 24–42 (2017)
9. Ceccarello, M., Fantozzi, C., Pietracaprina, A., Pucci, G., Vandin, F.: Clustering uncertain graphs. Proc. VLDB Endow. **11**(4), 472–484 (2017)
10. Chang, Y., Chen, J., Cho, M.H., Castaldi, P.J., Silverman, E.K., Dy, J.G.: Multiple clustering views from multiple uncertain experts. In: International Conference on Machine Learning, pp. 674–683. PMLR (2017)
11. Charwand, M., Gitizadeh, M., Siano, P., Chicco, G., Moshavash, Z.: Clustering of electrical load patterns and time periods using uncertainty-based multi-level amplitude thresholding. Int. J. Electr. Power Energy Syst. **117**, 105624 (2020)
12. Chau, M., Cheng, R., Kao, B., Ng, J.: Uncertain data mining: an example in clustering location data. In: Ng, W.-K., Kitsuregawa, M., Li, J., Chang, K. (eds.) PAKDD 2006. LNCS (LNAI), vol. 3918, pp. 199–204. Springer, Heidelberg (2006). https://doi.org/10.1007/11731139_24
13. Chen, Y., Lim, S.H., Xu, H.: Weighted graph clustering with non-uniform uncertainties. In: International Conference on Machine Learning, pp. 1566–1574. PMLR (2014)
14. Cormode, G., McGregor, A.: Approximation algorithms for clustering uncertain data. In: Proceedings of the Twenty-Seventh ACM SIGMOD-SIGACT-SIGART Symposium on Principles of Database Systems, pp. 191–200 (2008)
15. Dalton, L.A., Benalcázar, M.E., Dougherty, E.R.: Optimal clustering under uncertainty. PLoS ONE **13**(10), e0204627 (2018)
16. Debnath, B., Coviello, G., Yang, Y., Chakradhar, S.: UAC: an uncertainty-aware face clustering algorithm. In: Proceedings of the IEEE/CVF International Conference on Computer Vision, pp. 3487–3495 (2021)
17. D'Urso, P.: Informational paradigm, management of uncertainty and theoretical formalisms in the clustering framework: a review. Inf. Sci. **400**, 30–62 (2017)
18. Ghosh, S., Mitra, S.: Clustering large data with uncertainty. Appl. Soft Comput. **13**(4), 1639–1645 (2013)

19. Griffin, S.C., Claxton, K.P., Palmer, S.J., Sculpher, M.J.: Dangerous omissions: the consequences of ignoring decision uncertainty. Health Econ. **20**(2), 212–224 (2011)
20. Guha, S., Munagala, K.: Exceeding expectations and clustering uncertain data. In: Proceedings of the Twenty-Eighth ACM SIGMOD-SIGACT-SIGART Symposium on Principles of Database Systems, pp. 269–278 (2009)
21. Gullo, F., Ponti, G., Tagarelli, A.: Clustering uncertain data via K-medoids. In: Greco, S., Lukasiewicz, T. (eds.) SUM 2008. LNCS (LNAI), vol. 5291, pp. 229–242. Springer, Heidelberg (2008). https://doi.org/10.1007/978-3-540-87993-0_19
22. Gullo, F., Ponti, G., Tagarelli, A.: Minimizing the variance of cluster mixture models for clustering uncertain objects. Stat. Anal. Data Min. ASA Data Sci. J. **6**(2), 116–135 (2013)
23. Gullo, F., Ponti, G., Tagarelli, A., Greco, S.: A hierarchical algorithm for clustering uncertain data via an information-theoretic approach. In: 2008 Eighth IEEE International Conference on Data Mining, pp. 821–826. IEEE (2008)
24. Gullo, F., Ponti, G., Tagarelli, A., Greco, S.: An information-theoretic approach to hierarchical clustering of uncertain data. Inf. Sci. **402**, 199–215 (2017)
25. Günnemann, S., Kremer, H., Seidl, T.: Subspace clustering for uncertain data. In: Proceedings of the 2010 SIAM International Conference on Data Mining, pp. 385–396. SIAM (2010)
26. Haddadpour, H., Niri, M.E.: Uncertainty assessment in reservoir performance prediction using a two-stage clustering approach: proof of concept and field application. J. Petrol. Sci. Eng. **204**, 108765 (2021)
27. Halim, Z., Waqas, M., Baig, A.R., Rashid, A.: Efficient clustering of large uncertain graphs using neighborhood information. Int. J. Approximate Reasoning **90**, 274–291 (2017)
28. Han, K., et al.: Efficient and effective algorithms for clustering uncertain graphs. Proc. VLDB Endow. **12**(6), 667–680 (2019)
29. Hariri, R.H., Fredericks, E.M., Bowers, K.M.: Uncertainty in big data analytics: survey, opportunities, and challenges. J. Big Data **6**(1), 1–16 (2019)
30. He, Y., Yang, J.P., Li, Y.F.: A three-stage automated modal identification framework for bridge parameters based on frequency uncertainty and density clustering. Eng. Struct. **255**, 113891 (2022)
31. Hewitt, M., Ortmann, J., Rei, W.: Decision-based scenario clustering for decision-making under uncertainty. Ann. Oper. Res. **315**(2), 747–771 (2022)
32. Huang, D., Lai, J.H., Wang, C.D.: Robust ensemble clustering using probability trajectories. IEEE Trans. Knowl. Data Eng. **28**(5), 1312–1326 (2015)
33. Huang, D., Wang, C.D., Lai, J.H.: Locally weighted ensemble clustering. IEEE Trans. Cybern. **48**(5), 1460–1473 (2017)
34. Huang, G.Y., Liang, D.P., Hu, C.Z., Ren, J.D.: An algorithm for clustering heterogeneous data streams with uncertainty. In: 2010 International Conference on Machine Learning and Cybernetics, vol. 4, pp. 2059–2064. IEEE (2010)
35. Huang, J., Gong, S., Zhu, X.: Deep semantic clustering by partition confidence maximisation. In: Proceedings of the IEEE/CVF Conference on Computer Vision and Pattern Recognition, pp. 8849–8858 (2020)
36. Hüllermeier, E.: Uncertainty in clustering and classification. In: Deshpande, A., Hunter, A. (eds.) SUM 2010. LNCS (LNAI), vol. 6379, pp. 16–19. Springer, Heidelberg (2010). https://doi.org/10.1007/978-3-642-15951-0_6
37. Hussain, S.F., Butt, I.A., Hanif, M., Anwar, S.: Clustering uncertain graphs using ant colony optimization (ACO). Neural Comput. Appl. **34**(14), 11721–11738 (2022)
38. Hwang, C., Rhee, F.C.H.: Uncertain fuzzy clustering: interval type-2 fuzzy approach to c-means. IEEE Trans. Fuzzy Syst. **15**(1), 107–120 (2007)

39. Jiang, B., Pei, J., Tao, Y., Lin, X.: Clustering uncertain data based on probability distribution similarity. IEEE Trans. Knowl. Data Eng. **25**(4), 751–763 (2011)

40. Jin, C., Yu, J.X., Zhou, A., Cao, F.: Efficient clustering of uncertain data streams. Knowl. Inf. Syst. **40**, 509–539 (2014)

41. Kamal, A., et al.: Recent advances and challenges in uncertainty visualization: a survey. J. Visualization **24**(5), 861–890 (2021)

42. Kang, B., Kim, S., Jung, H., Choe, J., Lee, K.: Efficient assessment of reservoir uncertainty using distance-based clustering: a review. Energies **12**(10), 1859 (2019)

43. Kao, B., Lee, S.D., Cheung, D.W., Ho, W.S., Chan, K.: Clustering uncertain data using Voronoi diagrams. In: 2008 Eighth IEEE International Conference on Data Mining, pp. 333–342. IEEE (2008)

44. Kao, B., Lee, S.D., Lee, F.K., Cheung, D.W., Ho, W.S.: Clustering uncertain data using Voronoi diagrams and R-tree index. IEEE Trans. Knowl. Data Eng. **22**(9), 1219–1233 (2010)

45. Karimi, J., Somers, T.M., Gupta, Y.P.: Impact of environmental uncertainty and task characteristics on user satisfaction with data. Inf. Syst. Res. **15**(2), 175–193 (2004)

46. Kriegel, H.P., Pfeifle, M.: Hierarchical density-based clustering of uncertain data. In: Fifth IEEE International Conference on Data Mining (ICDM 2005), pp. 4–pp. IEEE (2005)

47. Kriegel, H.P., Pfeifle, M.: Density-based clustering of uncertain data. In: Proceedings of the Eleventh ACM SIGKDD International Conference on Knowledge Discovery in Data Mining, pp. 672–677 (2005)

48. Kuczenski, B.: False confidence: are we ignoring significant sources of uncertainty? Int. J. Life Cycle Assess. **24**, 1760–1764 (2019)

49. Lee, C.S., Kao, Y.F., Kuo, Y.H., Wang, M.H.: Automated ontology construction for unstructured text documents. Data Knowl. Eng. **60**(3), 547–566 (2007)

50. Li, Y., Chung, S.H.: Ride-sharing under travel time uncertainty: robust optimization and clustering approaches. Comput. Ind. Eng. **149**, 106601 (2020)

51. Liu, H., Zhang, X., Zhang, X., Cui, Y.: Self-adapted mixture distance measure for clustering uncertain data. Knowl.-Based Syst. **126**, 33–47 (2017)

52. Liu, L., Jin, R., Aggarwal, C., Shen, Y.: Reliable clustering on uncertain graphs. In: 2012 IEEE 12th International Conference on Data Mining, pp. 459–468. IEEE (2012)

53. Liu, X., Lin, K.K., Andersen, B., Rattray, M.: Including probe-level uncertainty in model-based gene expression clustering. BMC Bioinform. **8**(1), 1–19 (2007)

54. Liu, Y., Liu, Z., Li, S., Guo, Y., Liu, Q., Wang, G.: Cloud-cluster: an uncertainty clustering algorithm based on cloud model. Knowl.-Based Syst. **263**, 110261 (2023)

55. Luo, Q., Peng, Y., Peng, X., Saddik, A.E.: Uncertain data clustering-based distance estimation in wireless sensor networks. Sensors **14**(4), 6584–6605 (2014)

56. McMillan, H.K., Westerberg, I.K., Krueger, T.: Hydrological data uncertainty and its implications. Wiley Interdiscip. Rev. Water **5**(6), e1319 (2018)

57. Meesuksabai, W., Kangkachit, T., Waiyamai, K.: HUE-stream: evolution-based clustering technique for heterogeneous data streams with uncertainty. In: Tang, J., King, I., Chen, L., Wang, J. (eds.) Advanced Data Mining and Applications: 7th International Conference, ADMA 2011, Beijing, China, 17–19 December 2011, Proceedings, Part II 7, pp. 27–40. Springer, Cham (2011). https://doi.org/10.1007/978-3-642-25856-5_3

58. Naouali, S., Salem, S.B., Chtourou, Z.: Uncertainty mode selection in categorical clustering using the rough set theory. Expert Syst. Appl. **158**, 113555 (2020)

59. Nuijten, M., Mittendorf, T., Persson, U.: Practical issues in handling data input and uncertainty in a budget impact analysis. Eur. J. Health Econ. **12**, 231–241 (2011)

60. Pelekis, N., Kopanakis, I., Kotsifakos, E., Frentzos, E., Theodoridis, Y.: Clustering trajectories of moving objects in an uncertain world. In: 2009 Ninth IEEE International Conference on Data Mining, pp. 417–427. IEEE (2009)

61. Pelekis, N., Kopanakis, I., Kotsifakos, E.E., Frentzos, E., Theodoridis, Y.: Clustering uncertain trajectories. Knowl. Inf. Syst. **28**, 117–147 (2011)

62. Pileggi, S.F.: Ontological modelling and social networks: from expert validation to consolidated domains. In: Mikyška, J., de Mulatier, C., Paszynski, M., Krzhizhanovskaya, V.V., Dongarra, J.J., Sloot, P.M. (eds.) Computational Science – ICCS 2023. ICCS 2023. LNCS, vol. 14077, pp. 672–687. Springer, Cham (2023). https://doi.org/10.1007/978-3-031-36030-5_53

63. Prabhu, V., Chandrasekaran, A., Saenko, K., Hoffman, J.: Active domain adaptation via clustering uncertainty-weighted embeddings. In: Proceedings of the IEEE/CVF International Conference on Computer Vision, pp. 8505–8514 (2021)

64. Rendon, N., Giraldo, J.H., Bouwmans, T., Rodríguez-Buritica, S., Ramirez, E., Isaza, C.: Uncertainty clustering internal validity assessment using Fréchet distance for unsupervised learning. Eng. Appl. Artif. Intell. **124**, 106635 (2023)

65. Rhee, F.C.H.: Uncertain fuzzy clustering: insights and recommendations. IEEE Comput. Intell. Mag. **1**(2), 44–56 (2007)

66. Schubert, E., Koos, A., Emrich, T., Züfle, A., Schmid, K.A., Zimek, A.: A framework for clustering uncertain data. Proc. VLDB Endow. **8**(12), 1976–1979 (2015)

67. Sharma, K.K., Seal, A.: Modeling uncertain data using Monte Carlo integration method for clustering. Expert Syst. Appl. **137**, 100–116 (2019)

68. Sharma, K.K., Seal, A.: Multi-view spectral clustering for uncertain objects. Inf. Sci. **547**, 723–745 (2021)

69. Sharma, K.K., Seal, A.: Outlier-robust multi-view clustering for uncertain data. Knowl.-Based Syst. **211**, 106567 (2021)

70. Shi, W., Chen, W.N., Gu, T., Jin, H., Zhang, J.: Handling uncertainty in financial decision making: a clustering estimation of distribution algorithm with simplified simulation. IEEE Trans. Emerg. Topics Comput. Intell. **5**(1), 42–56 (2020)

71. Shukla, A.K., Muhuri, P.K.: Big-data clustering with interval type-2 fuzzy uncertainty modeling in gene expression datasets. Eng. Appl. Artif. Intell. **77**, 268–282 (2019)

72. Shukla, A., Singh, S.: Clustering based unit commitment with wind power uncertainty. Energy Convers. Manage. **111**, 89–102 (2016)

73. Sinaga, K.P., Yang, M.S.: Unsupervised K-means clustering algorithm. IEEE Access **8**, 80716–80727 (2020)

74. Suzuki, R., Shimodaira, H.: Pvclust: an R package for assessing the uncertainty in hierarchical clustering. Bioinformatics **22**(12), 1540–1542 (2006)

75. Tabesh, M., Askari-Nasab, H.: Clustering mining blocks in presence of geological uncertainty. Min. Technol. **128**, 162–176 (2019)

76. Tew, C., Giraud-Carrier, C., Tanner, K., Burton, S.: Behavior-based clustering and analysis of interestingness measures for association rule mining. Data Min. Knowl. Disc. **28**, 1004–1045 (2014)

77. Volk, P.B., Rosenthal, F., Hahmann, M., Habich, D., Lehner, W.: Clustering uncertain data with possible worlds. In: 2009 IEEE 25th International Conference on Data Engineering, pp. 1625–1632. IEEE (2009)

78. Wang, P., Ding, C., Tan, W., Gong, M., Jia, K., Tao, D.: Uncertainty-aware clustering for unsupervised domain adaptive object re-identification. IEEE Trans. Multimedia (2022)
79. Wang, X., He, Y.: Learning from uncertainty for big data: future analytical challenges and strategies. IEEE Syst. Man Cybern. Mag. **2**(2), 26–31 (2016)
80. Wauthier, F.L., Jojic, N., Jordan, M.I.: Active spectral clustering via iterative uncertainty reduction. In: Proceedings of the 18th ACM SIGKDD International Conference on Knowledge Discovery and Data Mining, pp. 1339–1347 (2012)
81. Weng, C.H., Chen, Y.L.: Mining fuzzy association rules from uncertain data. Knowl. Inf. Syst. **23**, 129–152 (2010)
82. Wierzchoń, S.T., Kłopotek, M.A.: Modern Algorithms of Cluster Analysis, vol. 34. Springer, Cham (2018). https://doi.org/10.1007/978-3-319-69308-8
83. Xia, Y., Xi, B.: Conceptual clustering categorical data with uncertainty. In: 19th IEEE International Conference on Tools with Artificial Intelligence (ICTAI 2007), vol. 1, pp. 329–336. IEEE (2007)
84. Xiong, C., Johnson, D.M., Corso, J.J.: Active clustering with model-based uncertainty reduction. IEEE Trans. Pattern Anal. Mach. Intell. **39**(1), 5–17 (2016)
85. Xu, L., Hu, Q., Hung, E., Chen, B., Tan, X., Liao, C.: Large margin clustering on uncertain data by considering probability distribution similarity. Neurocomputing **158**, 81–89 (2015)
86. Xu, R., Wunsch, D.: Survey of clustering algorithms. IEEE Trans. Neural Networks **16**(3), 645–678 (2005)
87. Yao, C., Chen, M., Hong, Y.Y.: Novel adaptive multi-clustering algorithm-based optimal ESS sizing in ship power system considering uncertainty. IEEE Trans. Power Syst. **33**(1), 307–316 (2017)
88. Zhang, X., Liu, H., Zhang, X.: Novel density-based and hierarchical density-based clustering algorithms for uncertain data. Neural Netw. **93**, 240–255 (2017)
89. Zhou, J., Chen, L., Chen, C.P., Wang, Y., Li, H.X.: Uncertain data clustering in distributed peer-to-peer networks. IEEE Trans. Neural Netw. Learn. Syst. **29**(6), 2392–2406 (2017)
90. Züfle, A., Emrich, T., Schmid, K.A., Mamoulis, N., Zimek, A., Renz, M.: Representative clustering of uncertain data. In: Proceedings of the 20th ACM SIGKDD International Conference on Knowledge Discovery and Data Mining, pp. 243–252 (2014)

Direct Solver Aiming at Elimination of Systematic Errors in 3D Stellar Positions

Konstantin Ryabinin$^{(\boxtimes)}$ ⓘ, Gerasimos Sarras ⓘ, Wolfgang Löffler ⓘ, and Michael Biermann ⓘ

Astronomisches Rechen-Institut, Center for Astronomy of Heidelberg University, Mönchhofstr. 12-14, 69120 Heidelberg, Germany
{konstantin.riabinin,gerasimos.sarras}@uni-heidelberg.de,
{loeffler,biermann}@ari.uni-heidelberg.de

Abstract. The determination of three-dimensional positions and velocities of stars based on the observations collected by a space telescope suffers from the uncertainty of random as well as systematic errors. The systematic errors are introduced by imperfections of the telescope's optics and detectors as well as in the pointing accuracy of the satellite. The fine art of astrometry consists of heuristically finding the best possible calibration model that will account for and remove these systematic errors. Since this is a process based on trial and error, appropriate software is needed that is efficient enough to solve the system of astrometric equations and reveal the astrometric parameters of stars for the given calibration model within a reasonable time. In this work, we propose a novel architecture and corresponding prototype of a direct solver optimized for running on supercomputers. The main advantages expected of this direct method over an iterative one are the numerical robustness, accuracy of the method, and the explicit calculation of the variance-covariance matrix for the estimation of the accuracy and correlation of the unknown parameters. This solver is supposed to handle astrometric systems with billions of equations within several hours. To reach the desired performance, state-of-the-art libraries for parallel computing are used along with the hand-crafted subroutines optimized for hybrid parallelism model and advanced vector extensions of modern CPUs. The developed solver is tested using the mock science data related to the Japan Astrometry Satellite Mission for INfrared Exploration (JASMINE).

Keywords: Astrometry · Data Fitting · Least Squares Method · Eigenproblem · Pseudo-Inverse Matrix · High-Performance Computing · Model Errors

1 Introduction

The Japan Astrometry Satellite Mission for INfrared Exploration (JASMINE) is a proposal by the Institute of Space and Astronautical Science (ISAS) for a near-infrared space telescope mission by the Japan Aerospace Exploration Agency

© The Author(s), under exclusive license to Springer Nature Switzerland AG 2024
L. Franco et al. (Eds.): ICCS 2024, LNCS 14838, pp. 309–323, 2024.
https://doi.org/10.1007/978-3-031-63783-4_23

(JAXA). With its three years mission length JASMINE aims at two complementary science goals: firstly, an astrometric survey of the three-dimensional positions and motions of stars around the centre of our Milky Way, and secondly, a survey aiming at discovering Earth-like exo-planets in the habitable zone around cool red dwarf stars, i.e. stars smaller and cooler than our Sun.

The three-dimensional positions of the stars in the Galactic Centre region are being directly determined by geometric methods, i.e. the change of the aspect angle (parallax angle) of the target star as seen in spring and autumn for opposite locations of the Earth's orbit around the Sun. Since there exist no coordinate markers in space, these aspect angles can only be measured as relative angles between the set of target stars. The mathematical problem of the astrometric data reduction is thus the very same as the geodetic reduction of measured field angles into three-dimensional geographical positions of landscape marks on the surface of Earth. The main difference between geodesy and modern space-based astrometry are the micro-arcsecond accuracy requirements on the calibration of the optical and electronic imaging properties and spatial orientation of the measuring instrument. In space these can only be determined by measuring the very same aspect angles between the different stars.

At the core of the "fine art of astrometry" lies thus finding the best formulation of such a calibration model as well as the subsequent determination of its parameters with high accuracy. As this is an iterative process based on trial and error, there is a need for a software that can solve a given astrometric problem in short time with high accuracy and precision.

The fundamental problem of space-based astrometry is, however, that the number of stars being observed and the number of angles being measured is always extremely large compared to the data storage size and computing power available at the time. In the past the astrometric problem at hand was thus simplified mathematically and the exact solution was approximated iteratively.

The main motivations to choose a direct approach over an iterative one, are the following. From an algorithmic perspective, the direct method terminates in a finite number of steps, thus avoiding arbitrary stopping criteria introduced by an iterative scheme [3]. From an astrometric perspective, only the direct methods can provide estimates of the variances of the unknown parameters and of the residuals based on the covariance matrix which is explicitly calculated and stored in memory for further use. Once these estimates are available, then the scientific investigation of the statistical properties of the solution can start with the aim to identify systematic errors, remove hidden model biases/incompleteness and calibrate out further non-random uncertainties of the model to reach the desired accuracy of the astrometric mission.

This paper is a first step towards the cluster-based implementation of the method and algorithm for a direct non-iterative solution to the JASMINE astrometric problem of determining the three-dimensional positions and velocities of stars around the centre of our Milky Way.

2   The Astrometric Problem

The method for solving geodetic and astrometric problems, the so-called adjustment via mediating observations has been developed in the year 1794 by Carl Friedrich Gauß (eventually published in 1823, [9]) and a few years later in 1805 independently by Adrien-Marie Legendre [12]. In this type of problem the *unknowns* $\boldsymbol{p} = (p_1, \ldots p_\iota, \ldots p_I)^\mathsf{T}$ are not directly accessible to observation, but each of the *observations* $\boldsymbol{o} = (o_1, \ldots o_\ell, \ldots o_L)^\mathsf{T}$ can, in principle, be expressed as a problem-dependent function of these unknowns

$$\boldsymbol{o} = f(\boldsymbol{p}) \tag{1}$$

$$\text{with} \quad o_\ell = f_\ell(p_1, \ldots p_\iota, \ldots p_I). \tag{2}$$

In general the function f will be non-linear. If an estimate $\hat{\boldsymbol{p}}$ for the unknowns $\boldsymbol{p}$ exists, the problem can, however, be linearised around $\hat{\boldsymbol{p}}$ using a Taylor expansion and then neglecting all terms of second order and higher.

In the full multi-dimensional case this kind of linearisation yields

$$
\begin{pmatrix} o_1 \\ \vdots \\ o_\ell \\ \vdots \\ o_L \end{pmatrix}
\simeq
\begin{pmatrix} \hat{o}_1 \\ \vdots \\ \hat{o}_\ell \\ \vdots \\ \hat{o}_L \end{pmatrix}
+
\underbrace{\begin{pmatrix}
\frac{\partial}{\partial p_1} f_1 & \cdots & \frac{\partial}{\partial p_\iota} f_1 & \cdots & \frac{\partial}{\partial p_I} f_1 \\
\vdots & \ddots & \vdots & \ddots & \vdots \\
\frac{\partial}{\partial p_\ell} f_\ell & \cdots & \frac{\partial}{\partial p_\iota} f_\ell & \cdots & \frac{\partial}{\partial p_I} f_\ell \\
\vdots & \ddots & \vdots & \ddots & \vdots \\
\frac{\partial}{\partial p_L} f_L & \cdots & \frac{\partial}{\partial p_\iota} f_L & \cdots & \frac{\partial}{\partial p_I} f_L
\end{pmatrix}}_{\hat{p}}
\begin{pmatrix} p_1 - \hat{p}_1 \\ \vdots \\ p_\iota - \hat{p}_\iota \\ \vdots \\ p_I - \hat{p}_I \end{pmatrix}
\tag{3}
$$

$$\boldsymbol{o} \quad \simeq \quad \hat{\boldsymbol{o}} \quad + \qquad\qquad \boldsymbol{\mathcal{D}} \qquad\qquad\qquad \Delta\hat{\boldsymbol{p}}, \tag{4}$$

where $\boldsymbol{c} = \hat{\boldsymbol{o}} = F(\hat{\boldsymbol{p}})$ are the estimates for the observations $\boldsymbol{o}$ *calculated* from the estimate parameters $\hat{\boldsymbol{p}}$. $\Delta\hat{\boldsymbol{p}}$ are the updates of the estimate unknowns $\hat{\boldsymbol{p}}$. The matrix $\boldsymbol{\mathcal{D}}$ holding the Jacobian derivatives of the functions f_ℓ with respect to the unknowns $\boldsymbol{p}$ is called the *design matrix* of the problem.

Subtracting $\hat{\boldsymbol{o}} = \boldsymbol{c}$ from Eq. (4) we arrive at the linear *observation equation*

$$\boldsymbol{o} - \boldsymbol{c} = \boldsymbol{\mathcal{D}} \, \Delta\hat{\boldsymbol{p}}. \tag{5}$$

The basic interpretation of this observation equation is that the knowledge and understanding of the problem including an initial estimate $\hat{\boldsymbol{p}}$ for the unknowns $\boldsymbol{p}$ is being used to compute in a first step an estimate prediction $\boldsymbol{c}$ where and when the observations $\boldsymbol{o}$ will be made. The inevitable difference between the estimated prediction $\boldsymbol{c}$ and the actual observation $\boldsymbol{o}$ is then used in a second step to compute the update $\Delta\hat{\boldsymbol{p}}$ to the estimate $\hat{\boldsymbol{p}}$ that would be needed to account for that difference between estimated prediction and actual observation, or

$$\boldsymbol{o} - \boldsymbol{c} - \boldsymbol{\mathcal{D}}\Delta\hat{\boldsymbol{p}} \stackrel{!}{=} 0. \tag{6}$$

3   Methods

3.1   The Astrometric Least-Squares Solution

Given the system of linear observation equations

$$l = \mathcal{D}x \qquad (7)$$

with the *unknowns* $x = \Delta\hat{p}$ and differences $l = o - c$ between *observations* and *calculated predictions*, a unique solution exists if the number of unknowns I is equal to the number of observations L and if all observations are linearly independent, i.e. if each observation adds unique information to the system. Then the matrix $\mathcal{D}$ is square, has no zero eigenvalues, has a non-zero determinant and is therefore invertible. The solution is then given by

$$x = \mathcal{D}^{-1}l. \qquad (8)$$

In an astrometric problem we have, however, many more observations than we have unknowns. This problem is redundant to a degree $L - I$. Due to the errors inherent to each observation, the observation equations do, however, not only provide redundant information to the problem but also inconsistencies. There exists no longer an exact solution to the problem.

But the observations can be combined in such a way that an approximate solution can be computed which fits the problem in question to some degree. Following the approach published by Legendre [12] the best approximate solution is given by

$$\hat{x} = (\mathcal{D}^\mathsf{T}\mathcal{D})^{-1}\mathcal{D}^\mathsf{T}l = \mathcal{N}^{-1}\mathcal{D}^\mathsf{T}l, \qquad (9)$$

which minimises the square sum of the residuals, i.e. the quadratic sum of the differences between the individual observations and the computed solution.

If the number of independent, i.e. absolute observations is greater or equal to the number of unknowns, then the matrix $\mathcal{D}$ has full rank and the normal matrix $\mathcal{N} = \mathcal{D}^\mathsf{T}\mathcal{D}$ is regular and positive definite.

If the observations are, however, not independent but, for example, relative to each other, as is the case in astrometry, then the matrix $\mathcal{D}$ is rank deficient. This means that the normal matrix $\mathcal{N}$ is singular and no unique inverse exists. The solution to the problem is no longer unique (with respect to rotation and spin). This can be fixed by adding either constraints to the equations such that the constrained normal matrix becomes regular, or by computing a pseudo-inverse matrix to get one arbitrary astrometric solution and then de-rotating and de-spinning that particular solution in a post-processing step to get the one physical solution.

3.2   The Hipparcos, Gaia and JASMINE Astrometric Solutions

As mentioned already in the introduction, the fundamental problem of space-based astrometry is the sheer amount of unknowns (ten billions in the case of the ESA Gaia astrometry mission) to solve for using an even larger amount of angle measurements (more than two trillion measurements in the case of Gaia), which hitherto prevented any attempt at solving the astrometric problem directly without cutting it into smaller pieces first and then iterating for the full solution.

In the case of the ESA Hipparcos astrometry mission [6] the global astrometric problem with $3 \cdot 10^6$ unknowns was first split into many independent and much smaller astrometric solutions along rings on the celestial sphere made up by the observations during 4.5 revolutions of the satellite around its slowly precessing spin axis. These astrometric solutions computed for each of these rings were then combined into one spherical solution in a second step by simply rotating the rings for a best fit. From this spherical solution, updated spatial orientations of the individual rings were derived and these two steps then iterated for convergence. In addition, the astrometric problem for each individual ring was reduced in size by forward eliminating the attitude unknowns (describing the spatial orientation of the instrument) from the normal equations before the remaining normal matrix was inverted.

In the case of the ESA Gaia astrometry mission [8], the Astrometric Global Iterative Solution (AGIS) [13] is employing a block-iterative scheme, in which the normal matrix for the $8.9 \cdot 10^9$ unknowns is blocked along the astrometric unknowns, the attitude unknowns describing the orientation of the instrument and the calibration unknowns of the instruments, and in which the off-diagonal blocks are all set to zero. This results in a normal matrix in which the three non-zero diagonal blocks are block- or band-diagonal themselves and thus trivially or at least easily inverted. Each of these three main diagonal blocks is inverted and solved for separately by assuming a prior solution for the other two blocks. This solution of one block is serving as prior solution for the other two. The solving of these three blocks is iterated until convergence is achieved.

In contrast to this, the Gaia One Day Astrometric Solution (ODAS) [14] comprised the much smaller but still quite challenging astrometric problem with $5 \cdot 10^5$ unknowns spanning only one ring containing the observations made during about 4.5 revolutions of the satellite around its own slowly precessing spin axis. Here the size of the problem was reduced by only forward-eliminating the astrometric unknowns. The remaining reduced normal matrix was then directly inverted using a singular value decomposition.

The aim of the current work is to find out whether such a direct astrometric solution based on forward elimination and then direct inversion using singular value decomposition can be applied to astrometric problems much larger than the ODAS problem, i.e. to the full JASMINE astrometric problem solving for up to $2.5 \cdot 10^7$ unknowns covering the full 18 months of astrometric mission time. Thus was born the idea of the Astronomisches Rechen-Institut (ARI) JASMINE Astrometric Solution (AJAS).

3.3   The ARI JASMINE Astrometric Solution

The observations in AJAS are two-dimensional. The primary data of these observations are the 2D coordinates (κ, μ) of the corresponding centroid of the light spot (trace of a star detected by a telescope) expressed in the coordinate system of the telescope detector, and the guessed identifier of a source. The pixel coordinates (κ, μ) are determined via segmentation of the exposure into spots and calculation of their centroids. Source identifiers are determined via a so-called cross-match procedure, in which the catalog of known stars is used as *a priori* data. Both procedures are parts of raw data processing and lay beyond the scope of this work.

A system of astrometric equations is built based on some reference frame. In AJAS, we have chosen the telescope's field of view reference frame (FoVRS), because it simplifies the subsequent calculations. Being two-dimensional, each observation contributes twice to the system of astrometric equations spawning two equations for perpendicular directions (η, ζ) of the FoVRS. The unknowns in each equation are the set of calibration (nuisance) and source parameters. The coefficients are the derivatives of field angles η and ζ with respect to the corresponding parameters.

The design matrix of the AJAS astrometric problem is $\mathcal{D} = (\mathcal{C}_0 \, \mathcal{C}_1 \, \mathcal{S})$, where $\mathcal{C}_0$ is a lower-order calibration part encoding the attitude of the satellite telescope, $\mathcal{C}_1$ is a higher-order calibration part taking care of the above-mentioned imperfections of the telescope optics, and $\mathcal{S}$ is a part related to source parameters. In the end, only the unknowns related to $\mathcal{S}$ have a scientific meaning, still, all the unknowns should be determined to ensure the error calculus.

The vector side l of the system is built based on the difference between observed and calculated η and ζ coordinates of the observations:

$$l = \begin{pmatrix} \eta_\ell^{\mathrm{obs}} - \eta_\ell^{\mathrm{calc}} \\ \zeta_\ell^{\mathrm{obs}} - \zeta_\ell^{\mathrm{calc}} \end{pmatrix}_{\ell=\overline{1,L}} = \begin{pmatrix} \eta^{\mathrm{obs}}(\kappa_\ell, \mu_\ell) - \eta^{\mathrm{calc}}(t_\ell, \lambda_\ell) \\ \zeta^{\mathrm{obs}}(\kappa_\ell, \mu_\ell) - \zeta^{\mathrm{calc}}(t_\ell, \lambda_\ell) \end{pmatrix}_{\ell=\overline{1,L}},$$

where L is the total number of observations, t_ℓ is the timestamp of observation ℓ, and λ_ℓ is the source assigned to the observation ℓ.

For the very simple case with a single detector, 4 exposures, 40 observations, 10 sources, 2 source parameters per source, 6 nuisance parameters per observation in the lower-order calibration part, and 6 nuisance parameters per observation in the higher-order calibration part, the design matrix $\mathcal{D}$ looks like in Fig. 1a. For the real data, $\mathcal{D}$ will be way bigger, but the overall structure will be retained.

For a 3-year JASMINE mission, 4 detectors and $\sim 10^6$ exposures are planned. Around $1.15 \cdot 10^5$ sources are expected to be observed with a total number of observations to be around $8.25 \cdot 10^9$ observations, and 5 astrometric parameters per source should be determined (2 celestial coordinates, 2 proper motion components, and parallax). For this amount of input data, the number of rows in $\mathcal{D}$ will be $1.65 \cdot 10^{10}$ (twice as many, as observations, because of two coordinates η and ζ). The number of columns in $\mathcal{C}_0$ will be $2.4 \cdot 10^7$ (count of detectors $\times$ count of exposures $\times$ 2 coordinates $\times$ count of lower-order calibration parameters). The exact number of columns in $\mathcal{C}_1$ depends on the particular higher-order calibration model but is expected to be no more than 100. The number of columns in $\mathcal{S}$ will be $5.75 \cdot 10^5$ (count of sources $\times$ count of source parameters).

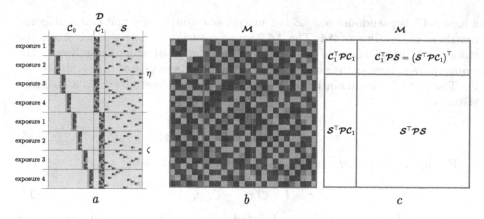

Fig. 1. Structure of the matrix $\mathcal{D}$ (a), structure of the matrix $\mathcal{M}$ (b), and main blocks of the matrix $\mathcal{M}$ (c), gray indicates zero values, green-to-blue indicates non-zero values

Altogether, this gives a size of $\mathcal{D}$ to be in the order of $10^{10} \times 10^7$. Despite sparsity, this matrix is far too big to be treated directly. Instead, another system is to be solved, which is based on the reduced normal matrix $\mathcal{M}$ derived from the normal matrix $\mathcal{N}$:

$$\mathcal{N}x = \mathcal{D}^\mathsf{T}l, \quad \mathcal{N} = \mathcal{D}^\mathsf{T}\mathcal{D} = \begin{pmatrix} \mathcal{C}_0^\mathsf{T}\mathcal{C}_0 & \mathcal{C}_0^\mathsf{T}\mathcal{O} \\ \mathcal{O}^\mathsf{T}\mathcal{C}_0 & \mathcal{O}^\mathsf{T}\mathcal{O} \end{pmatrix}, \quad x = \begin{pmatrix} x_{\mathcal{C}_0} \\ x_{\mathcal{O}} \end{pmatrix}, \quad \mathcal{O} = (\mathcal{C}_1 \; \mathcal{S}).$$

The matrix $\mathcal{M}$ is built using a forward elimination of the $\mathcal{C}_0^\mathsf{T}\mathcal{C}_0$ component of $\mathcal{N}$:

$$\begin{pmatrix} \mathcal{I} & \mathcal{V}\mathcal{O} \\ 0 & \mathcal{M} \end{pmatrix} \begin{pmatrix} x_{\mathcal{C}_0} \\ x_{\mathcal{O}} \end{pmatrix} = \begin{pmatrix} \mathcal{V}l \\ b_{\mathcal{O}} \end{pmatrix},$$

where

$$\mathcal{V} = \left(\mathcal{C}_0^\mathsf{T}\mathcal{C}_0\right)^{-1} \mathcal{C}_0^\mathsf{T}, \quad \mathcal{M} = \mathcal{O}^\mathsf{T}\mathcal{P}\mathcal{O}, \quad b_{\mathcal{O}} = \mathcal{O}^\mathsf{T}\mathcal{P}l, \quad \mathcal{P} = \mathcal{I} - \mathcal{C}_0\mathcal{V}, \quad (10)$$

and $\mathcal{I}$ is the identity matrix. Then, the system to be solved is:

$$\mathcal{M}x_{\mathcal{O}} = b_{\mathcal{O}}. \tag{11}$$

The structure of the matrix $\mathcal{M}$ is shown in Fig. 1b and Fig. 1c. It is symmetric, singular, positive semi-definite. Its size is driven by the sum of the column numbers of $\mathcal{C}_1$ and $\mathcal{S}$ and is expected to be within $10^6 \times 10^6$ for the JASMINE mission.

From (11), the unknowns related to $\mathcal{C}_1$ and $\mathcal{S}$ can be calculated as:

$$x_{\mathcal{O}} = \mathcal{M}^+ b_{\mathcal{O}} = \mathcal{Z}\mathcal{E}^{-1}\mathcal{Z}^\mathsf{T} b_{\mathcal{O}}, \tag{12}$$

where $\mathcal{M}^+$ is pseudo-inverse, $\mathcal{Z}$ is a matrix of eigenvectors, and $\mathcal{E}$ is a diagonal matrix of eigenvalues of $\mathcal{M}$. The $\mathcal{M}^+$ is the covariance matrix which stores the standard error estimates of the higher calibration parameters a s well as of the source parameters and is needed for the statistical analysis of the residuals.

The rest of the unknowns (related to $\mathcal{C}_0$) can be determined via a backsubstitution:

$$x_{\mathcal{C}_0} = \left(\mathcal{C}_0^\mathsf{T}\mathcal{C}_0\right)^{-1} \mathcal{C}_0^\mathsf{T} (l - \mathcal{O}x_\mathcal{O}).$$ (13)

Finally, for the error calculus, the residuals are calculated as

$$r = l - \mathcal{O}x_\mathcal{O} - \mathcal{C}_0 x_{\mathcal{C}_0},$$ (14)

where they should follow a normal distribution based on the assumption that the observational errors are uncorrelated to each other and follow also a normal distribution.

The scientific goal of the JASMINE mission is the determination of the 3D angular positions of the stars on the celestial sphere with a target accuracy of down to $10\,\mu$as. The accuracy of the instrument calibration must therefore be better than this. And that means that we have to calibrate the geometric locations of the photosensitive elements in the focal plane to an accuracy of at least $20\,$nm or 100 silicon atom diameters.

This kind of calibration accuracy cannot be achieved in the laboratory on-ground before launch. Launching the instrument into space will change its geometry in an uncertain way at a far greater scale. The instrument calibration must therefore be based on the very same observational data that determine the stellar positions. This can be achieved by setting up a calibration model and treating its parameters as unknowns of the overall problem. Any error or uncertainty in the formulation of the calibration model will show up as systematic effect in the distribution of the residuals defined in Eq. 14 and their standard errors. These systematics need to be accounted for in the formulation of an improved calibration model and a new solution based on this improved model should be computed to reduce this model uncertainty.

Solving the overall astrometric problem requires therefore a speedy and accurate method with which to compute the solution for a whole series of increasingly more accurate calibration models. The actual and detailed formulation of these calibration models is, however, based on mining the residual data and their standard errors for correlations and systematics.

3.4   Approaches to Build a Direct Astrometric Solver

The direct astrometric solver is defined by Eqs. (11–14). Each formula allows for parallel calculations, therefore, the direct solver can reach correspondingly high performance. The most computationally intensive part is the calculation of the pseudo-inverse matrix $\mathcal{M}^+$ (Eq. (12)), and specifically, the calculation of eigenvalues e and eigenvectors $\mathcal{Z}$ of the matrix $\mathcal{M}$. This is because $\mathcal{M}$ is nearly a dense matrix, while the rest of operands is rather sparse. So, the general architecture of parallelism for the direct solver should be chosen in a way that ensures the best performance of the eigenproblem solution. The rest of the operations should be implemented in the same architecture to minimize the overhead of moving the intermediate data. In this regard, our first experiments were aimed at finding the most optimal hardware and software configuration to solve an eigenproblem of a real symmetric matrix of size $10^6 \times 10^6$ (as this is the upper estimation for the matrix $\mathcal{M}$ in the JASMINE mission).

A state-of-the-art library to solve the eigenproblem for dense matrices is EigenExa [16]. It is written in modern Fortran and implements a hybrid parallelism model based on MPI and OpenMP standards. Compared to ScaLAPACK library [2], which is a *de facto* standard for distributed parallel computations involving linear algebra subroutines, EigenExa has the following distinctive features [16]:

1. The traditional way to solve the eigenproblem in ScaLAPACK is using the Householder transformation [18] of the input symmetric real matrix $\mathcal{M}$ to tridiagonal form $\mathcal{T}$, then performing the Divide-and-Conquer (DC) algorithm [4] or the more modern and efficient algorithm of Multiple Relatively Robust Representations (MRRR) [5] to find the eigenvalues and eigenvectors of $\mathcal{T}$, and then transforming the eigenvectors back to the initial matrix $\mathcal{M}$ (while eigenvalues of $\mathcal{T}$ match eigenvalues of $\mathcal{M}$ since Householder transformation preserves the spectrum). Instead, EigenExa transforms the matrix $\mathcal{M}$ to a pentadiagonal form $\mathcal{P}$, then performs a modified version of the DC algorithm to reveal eigenvalues and eigenvectors and transforms the eigenvectors back to matrix $\mathcal{M}$. This approach reduces the number of numerical operations involved in the calculation [7].
2. EigenExa uses fine-tuned symmetric multiprocessing parallelism inside the distributed processes allowing for a better ratio of computation operations to service operations.
3. EigenExa implements a set of data arrangement optimizations, which improve the CPU cache usage (including reducing the cache thrashing).

These features significantly increase the overall processing speed. According to our experiments on small-scale matrices (order of $10^4 \times 10^4$) and small process grid (4 nodes with 2 cores each), the average speedup of EigenExa compared to ScaLAPACK (with MRRR) is about a factor of 10. For big-scale matrices, as reported by Sakurai et al. [16] and Imamura et al. [11], EigenExa is capable of solving the eigenproblem for $10^6 \times 10^6$ matrix in less than 1 h using 82944 nodes with 8 cores each of the K supercomputer or 4096 nodes with 48 cores each of the Fugaku supercomputer, whereby the scaling is close to linear. This brings us to the conclusion, that the cluster-based architecture with a hybrid model of parallelism is the way to implement a direct astrometric solver to meet our performance requirements.

4   Proposed Architecture of the Direct Astrometric Solver

We adopt the model of hybrid parallel computations within AJAS.

AJAS is a software pipeline written in C++ and optimized for CPU clusters. Its schema is shown in Fig. 2.

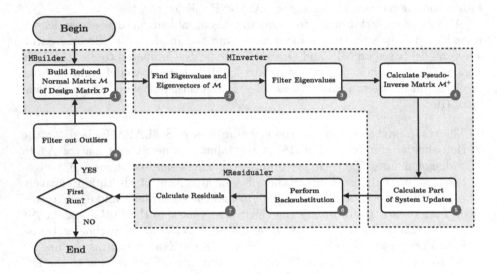

Fig. 2. Direct astrometric solver pipeline

AJAS consists of 3 modules (`MBuilder`, `MInverter`, and `MResidualer`) executed sequentially, while each one performs calculations using a hybrid parallelism. The reason for grouping the pipeline steps into these modules is the semantics of operations. Currently, the outlier filtering step (8) is not yet implemented, but very probably it will be attached to `MResidualer` in the future.

The communication between processes adheres to the Message Passing Interface (MPI) standard. Each process is supposed to run on an individual cluster node, whereby Π available nodes are logically organized as a square-shaped $\left\lfloor \sqrt{\Pi} \right\rfloor \times \left\lfloor \sqrt{\Pi} \right\rfloor$ grid. Each process utilizes multi-threading to organize a fine-grain parallelism. The workload is balanced dynamically based on the actual dimensions of the problem.

4.1   Matrix Building Module

The `MBuilder` module (comprising step (1) in Fig. 2) is responsible for building the reduced normal matrix $\mathcal{M}$ and vector $b_{\mathcal{O}}$ of the astrometric equations system. It computes the following formulas derived from (10):

$$\mathcal{M} = \sum_{n=1}^{N} \sum_{\tau=1}^{\Upsilon} \mathcal{O}_{n\tau}^{\mathsf{T}} \mathcal{P}_{n\tau} \mathcal{O}_{n\tau}, \quad (15) \qquad \boldsymbol{b}_{\mathcal{O}} = \sum_{n=1}^{N} \sum_{\tau=1}^{\Upsilon} \mathcal{O}_{n\tau}^{\mathsf{T}} \mathcal{P}_{n\tau} l_{n\tau}, \quad (16)$$

where $\mathcal{O}_{n\tau} = (\mathcal{C}_{1,n\tau} \ \boldsymbol{\mathcal{S}}_{n\tau})$, $\mathcal{P}_{n\tau} = \boldsymbol{I} - \mathcal{C}_{0,n\tau} \left(\mathcal{C}_{0,n\tau}^{\mathsf{T}} \mathcal{C}_{0,n\tau} \right)^{-1} \mathcal{C}_{0,n\tau}^{\mathsf{T}}$, N is the number of telescope detectors, Υ is the number of exposures made during the mission.

Each process of `MBuilder` composes a block of the matrix $\mathcal{M}$ and a block of the vector $\boldsymbol{b}_{\mathcal{O}}$ by computing $N\Upsilon/\Pi$ "layers" (double-sum internals from (15) and (16)) in a multi-threaded way. The elements of these layers, which belong to the matrix and vector blocks of a particular process, are summed up locally; the rest is transmitted for summation to other processes. The result blocks of $\mathcal{M}$ are then processed by the `MInverter` module. Upon calculation, the intermediate data are stored in binary files for subsequent use by the `MResidualer` module.

To save memory and numerical operations, the layers are compressed exploiting the fact that their components are rather sparse. The zero elements are predicted analytically and excluded from both storage and calculations. To tackle this, a customized code for matrix operations (matrix-matrix and matrix-vector multiplication, matrix addition, etc.) has been implemented. The stored and summed/transmitted layer components with their minimal compression ratios and memory footprint fractions are listed in Table 1. The actual values depend on the calibration model, which will be adjusted during the real mission. On one hand, the more complicated the calibration model is the more unknowns are introduced to the system of astrometric equations but on the other, they will also contain more zero entries. So, the compression ratio will grow.

The communication and data access schemes of `MBuilder` are a potential bottleneck that should be investigated and optimized further.

Table 1. Approximate values for data components compression and memory footprints calculated for the expected JASMINE data

	Element	Compression ratio	Fraction of total memory footprint
Stored on disk	$\mathcal{C}_{0,n\tau}$	2	$1.16 \cdot 10^{-4}$
	$\mathcal{C}_{1,n\tau}$	8	$6.95 \cdot 10^{-4}$
	$\boldsymbol{\mathcal{S}}_{n\tau}$	$1.15 \cdot 10^{5}$	$1.93 \cdot 10^{-4}$
	$l_{n\tau}$	1	$3.86 \cdot 10^{-5}$
	$\left(\mathcal{C}_0^{\mathsf{T}} \mathcal{C}_0 \right)^{-1}$	2	$6.73 \cdot 10^{-7}$
Stored in RAM; transmitted and summed up	$\mathcal{C}_{1,n\tau}^{\mathsf{T}} \mathcal{P}_{n\tau} \mathcal{C}_{1,n\tau}$	61	$1.28 \cdot 10^{-5}$
	$\boldsymbol{\mathcal{S}}_{n\tau}^{\mathsf{T}} \mathcal{P}_{n\tau} \mathcal{C}_{1,n\tau}$	892	$3.47 \cdot 10^{-3}$
	$\boldsymbol{\mathcal{S}}_{n\tau}^{\mathsf{T}} \mathcal{P}_{n\tau} \boldsymbol{\mathcal{S}}_{n\tau}$	$1.24 \cdot 10^{4}$	$9.95 \cdot 10^{-1}$
	$\mathcal{C}_{1,n\tau}^{\mathsf{T}} \mathcal{P}_{n\tau} l_{n\tau}$	4	$1.35 \cdot 10^{-6}$
	$\boldsymbol{\mathcal{S}}_{n\tau}^{\mathsf{T}} \mathcal{P}_{n\tau} l_{n\tau}$	111	$1.93 \cdot 10^{-4}$

4.2   Matrix Inversion Module

The `MInverter` module inverts the previously created matrix $\mathcal{M}$ and calculates the part of the astrometric system's solution implementing Eq. (12). As mentioned above, $\mathcal{M}$ is singular, so it is inverted by a singular value decomposition [1], whereby the diagonalization is performed using the eigenvectors and reciprocal eigenvalues.

The eigenproblem is efficiently solved by the EigenExa library (Fig. 2, step 2). It requires $\mathcal{M}$ to be distributed over the square-shaped $\lfloor\sqrt{\Pi}\rfloor \times \lfloor\sqrt{\Pi}\rfloor$ process grid using the block-cyclic distribution pattern [15], whereby despite the matrix symmetry, all the elements of $\mathcal{M}$ should be presented for the sake of optimal processing. Moreover, to minimize the probability of cache thrashing, blocks are padded in memory with some empty rows and columns (the exact number of which is calculated by EigenExa).

When EigenExa finishes, each cluster node has a corresponding block of the eigenvector's set $\mathcal{Z}$ stored in RAM using the same block-cyclic layout and paddings as the blocks of the input matrix $\mathcal{M}$. Along with it, each node has a copy of the full set of eigenvalues e.

Following the pipeline shown in Fig. 2, the next step is the eigenvalues filtering (step 3). It is very lightweight and performed by each node individually. The filtering zeroes out the expected zero eigenvalues that are numerically close to but not exactly equal to zero.

Next, the pseudo-inverse is computed (step 4). For the sake of efficiency, first, each node scales each i-th column of the $\mathcal{Z}$ block with $\sqrt{1/e_i}$, if $e_i > 0$, and 0, if $e_i = 0$. As long as $\mathcal{M}$ is positive semi-definite, all the eigenvalues are non-negative. This allows us to reformulate Eq. (12) as $\mathcal{M}^+ = \mathcal{Z}_e\mathcal{Z}_e^{\mathsf{T}}$, where $\mathcal{Z}_e$ is scaled $\mathcal{Z}$.

This, in turn, allows the use of the ScaLAPACK `PDSYRK` function that efficiently multiplies a real matrix by its transpose. For this operation, the block-cyclic layout of $\mathcal{Z}$ is reused, so no data is moved or copied in memory. The blocks of multiplication results overwrite the blocks of matrix $\mathcal{M}$, so no additional memory is allocated.

Step 5 is the multiplication of the $\mathcal{M}^+$ matrix by the vector $b_{\mathcal{O}}$, which has been distributed over the nodes in a block-cyclic way by the `MBuilder` module. The ScaLAPACK `PDSYMV` function is used to perform the matrix-vector multiplication. The result vector $x_{\mathcal{O}}$ is gathered on the first node and stored in a file.

4.3   Residuals Calculation Module

The `MResidualer` module adopts Eqs. (13) and (14) (corresponding to the steps (6) and (7) in Fig. 2) split into blocks by detectors and exposures:

$$x_{\mathcal{C}_0,n\tau} = \left(\mathcal{C}_{0,n\tau}^{\mathsf{T}}\mathcal{C}_{0,n\tau}\right)^{-1}\mathcal{C}_{0,n\tau}^{\mathsf{T}}\left(l_{n\tau} - \mathcal{O}_{n\tau}x_{\mathcal{O},n\tau}\right), \tag{17}$$

$$r_{n\tau} = l_{n\tau} - \mathcal{O}_{n\tau}x_{\mathcal{O},n\tau} - \mathcal{C}_{0,n\tau}x_{\mathcal{C}_0,n\tau}. \tag{18}$$

It reuses the components stored by `MBuilder`. The calculations are embarrassingly parallel (having no dependencies and no inter-process communication) and distributed over Π available cluster nodes so that each one has to process $N\Upsilon/\Pi$ blocks.

5 Discussion

To preliminarily validate the AJAS software, we created a set of unit tests for all its modules. For this, we recreated the AJAS pipeline in pure sequential code without any optimizations (using just trivial matrix-matrix and matrix-vector operations) in Python using the NumPy library [10]. This pipeline "twin" can handle very small-scale problems only but allows us to ensure that the AJAS pipeline's numerical part works correctly on all the steps.

To test the accuracy of AJAS, we compared the solution of different small-scale systems found by AJAS, by singular value decomposition [1] of reduced normal matrix $\mathcal{M}$ implemented in Python (using NumPy) and by QR decomposition [17] of the design matrix $\mathcal{D}$ implemented in Julia. The comparison showed the equality of those solutions up to the machine epsilon.

To preliminary test the efficiency of the AJAS implementation, we run it on 2×2 process grid using 4 nodes of the Baden-Württemberg cluster bwUni-Cluster 2.0, which is available for academic use for the universities of Baden Württemberg (Germany). Each node has 40 Intel® Xeon Gold 6230 2.1 GHz CPU cores and 96 Gb RAM. To allow for basic profiling, we also developed a simple generator for mock science data in Julia language that produces artificial observations. The needed amount of calculations in AJAS is mainly driven by the size of the $\mathcal{M}$ matrix, so we generated the data in the amount needed to build matrices of size $m \times m$, where $m = 10^4$; $2 \cdot 10^4$; $3 \cdot 10^4$; $4 \cdot 10^4$.

The approximate time dependencies of the AJAS pipeline steps revealed from the profiling are the following ($m \approx A\Lambda$, A is the number of astrometric parameters, Λ is the number of sources observed, Υ is the number of exposures made, L is the total number of observations made, N is the number of detectors in the telescope):

1 Building of the matrix $\mathcal{M}$: $t_1 = t_1(N\Upsilon, L, m^2) \approx 1.7 \cdot t_2$.
2 Finding eigenvalues and eigenvectors of $\mathcal{M}$: $t_2 = t_2(m^2)$.
3 Filtering eigenvalues: $t_3 = t_3(m) \approx 0$.
4 Calculating pseudo-inverse matrix $\mathcal{M}^{+}$: $t_4 = t_4(m^2) \approx 0.4 \cdot t_2$.
5 Calculating part of system updates: $t_5 = t_5(m) \approx 10^{-3} \cdot t_2$.
6 Performing backsubstitution: $t_6 = t_6(N\Upsilon, L) \ll t_2$.
7 Calculating residuals: $t_7 = t_7(N\Upsilon, L) \ll t_2$.

As seen in the list above, the timing of steps (1)–(5) depends on the size of $\mathcal{M}$ (while the timing of (1) depends mainly on the number of detectors, exposures, and observations). In terms of the amount of computations, step (2) is the heaviest one. However step (1) takes the longest execution time because of the intensive disk and RAM access, as well as inter-process communication. Step

(3) is negligibly fast because it implies just a single scan of eigenvalues zeroing out the values smaller than a given threshold. The timing of steps (6) and (7) depends on the number of detectors, exposures, and observations, but is much smaller than the timing of step (2).

To run a bigger problem that will be close to the real JASMINE size, a larger process grid on the cluster is required. 3 years of JASMINE operation will result in the system of approx. $1.65 \cdot 10^{10}$ astrometric equations with $2.5 \cdot 10^7$ unknowns and the matrix $\mathcal{M}$ will be around $6 \cdot 10^5 \times 6 \cdot 10^5$, which requires 10.5 Tb of RAM and 7 Tb of disk storage to run the pipeline.

6 Conclusion

The paper delineates the novel approach to the development of the direct astrometric solver aimed at revealing stellar properties from the observations made by a space telescope. AJAS, the software implementation of the solver, is optimized for execution on CPU clusters. The computational core is based on the EigenExa, ScaLAPACK, Intel®oneMKL, and Intel®MPI libraries, which ensure high performance in computations and communication between the distributed cluster nodes. Preliminary testing on realistic mock science data and a small-scale process grid on the cluster demonstrate the viability of the proposed approach and architecture. The high performance and numerical accuracy of the solver are crucial to allow for multiple runs in a reasonable time, which is needed to fine-tune the calibration model parameters in order to mitigate the model's uncertainty. Thereby, the correctness of the astrometric solution will be improved by a proper accounting for the stellar properties, orientation of the space telescope, and imperfections of its optics.

The foremost step for future work is the detailed testing of the developed software on the larger process grids on the clusters to reveal and eliminate bottlenecks in computation and communication routines as well as to confirm the predictions of the time needed to handle the astrometric problems of realistic size.

Acknowledgments. This work was financially supported by the German Aerospace Agency (Deutsches Zentrum für Luft- und Raumfahrt e.V., DLR) through grant 50OD2201. The authors acknowledge support by the state of Baden-Württemberg through bwHPC. The authors also thank the National Astronomical Observatory of Japan (NAOJ) fruitful discussions for the mission specifics.

References

1. Ben-Israel, A., Greville, T.N.E.: Generalized Inverses. Theory and Applications. Springer, New York (2003). https://doi.org/10.1007/b97366
2. Blackford, L.S., et al.: ScaLAPACK Users' Guide. Society for Industrial and Applied Mathematics, Philadelphia (1997)

3. Bombrun, A., Lindegren, L., Holl, B., Jordan, S.: Complexity of the Gaia astrometric least-squares problem and the (non-)feasibility of a direct solution method. Astron. Astrophys. **516**, A77 (2010). https://doi.org/10.1051/0004-6361/200913503

4. Cuppen, J.J.M.: A divide and conquer method for the symmetric tridiagonal eigenproblem. Numer. Math. **36**(2), 177–195 (1980). https://doi.org/10.1007/BF01396757

5. Dhillon, I.S., Parlett, B.N., Vömel, C.: The design and implementation of the MRRR algorithm. ACM Trans. Math. Softw. **32**(4), 533–560 (2006). https://doi.org/10.1145/1186785.1186788

6. ESA: The Hipparcos and Tycho Catalogues. ESA SP-1200 (1997). https://www.cosmos.esa.int/web/hipparcos/catalogues

7. Fukaya, T., Imamura, T.: Performance evaluation of the eigen exa eigensolver on oakleaf-FX: tridiagonalization versus pentadiagonalization. In: 2015 IEEE International Parallel and Distributed Processing Symposium Workshop, pp. 960–969 (2015). https://doi.org/10.1109/IPDPSW.2015.128

8. Gaia Collaboration, et al.: The Gaia mission. Astron. Astrophys. **595**, A1 (2016). https://doi.org/10.1051/0004-6361/201629272

9. Gauß, C.F.: Theoria Combinationis Observationum Erroribus Minimis Obnoxiae. In: Commentationes Societatis Regiae Scientiarum Gottingensis recentiores – Classis Physicae, vol. 5. Henrich Dieterich, Gttingen (1823). https://archive.org/details/theoriacombinat00gausgoog

10. Harris, C.R., et al.: Array programming with NumPy. Nature **585**(7825), 357–362 (2020). https://doi.org/10.1038/s41586-020-2649-2

11. Imamura, T., Terao, T., Ina, T., Hirota, Y., Ozaki, K., Uchino, Y.: Performance benchmark of the latest EigenExa on Fugaku (2022). https://sighpc.ipsj.or.jp/HPCAsia2022/poster/108_poster.pdf

12. Legendre, A.M.: Nouvelles Méthodes Pour La Détermination des Orbites des Comètes. Firmin Didot, Paris (1805). https://archive.org/details/61Legendre

13. Lindegren, L., Lammers, U., Hobbs, D., O'Mullane, W., Bastian, U., Hernàndez, J.: The astrometric core solution for the Gaia mission. Overview of models, algorithms, and software implementation. Astron. Astrophys. **538**, A78 (2012).https://doi.org/10.1051/0004-6361/201117905, http://cdsads.u-strasbg.fr/abs/2012A%26A...538A..78L

14. Löffler, W., et al.: The one-day astrometric solution for the Gaia mission. Astron. Astrophys. (in preparation)

15. Ostrouchov, S.: Block Cyclic Data Distribution (1995). https://www.netlib.org/utk/papers/factor/node3.html

16. Sakurai, T., Futamura, Y., Imakura, A., Imamura, T.: Scalable eigen-analysis engine for large-scale eigenvalue problems. In: Sato, M. (ed.) Advanced Software Technologies for Post-Peta Scale Computing, pp. 37–57. Springer, Singapore (2019). https://doi.org/10.1007/978-981-13-1924-2_3

17. Trefethen, L.N., Bau, D.: Numerical Linear Algebra. Society for Industrial and Applied Mathematics, Philadelphia (1997)

18. Wilkinson, J.H.: Householder's method for symmetric matrices. Numerische Mathematik **4**(1), 354–361 (1962). https://doi.org/10.1007/BF01386332

On Estimation of Numerical Solution in Prager&Synge Sense

A. K. Alekseev[1] and A. E. Bondarev[2(✉)]

[1] RSC Energia, Korolev, Russia
[2] Keldysh Institute of Applied Mathematics RAS, Moscow, Russia
bond@keldysh.ru

Abstract. The numerical solution in sense of Prager&Synge is defined as a hypersphere containing a true solution of a system of partial differentiation equations (PDE). In the original variant Prager&Synge method is based on special orthogonal properties of PDE and may be applied only to several equations. Herein, the Prager&Synge solution (center and radius of the hypersphere) is estimated using the ensemble of numerical solutions obtained by independent algorithms. This approach is not problem dependent and may be applied to arbitrary system of PDE. Several options for computation of the Prager&Synge solution are considered. The first one is based on the search for the orthogonal truncation errors and their transformation. The second is based on the orthogonalization of approximation errors obtained using the defect correction method. It applies the superposition of numerical solutions. The third option uses the width of the ensemble of numerical solutions. The numerical tests for the two dimensional inviscid flows are presented that demonstrate the acceptable effectivity of the approximation error estimates based on the solution in the Prager&Synge sense.

Keywords: Prager&Synge method · a posteriori error estimation · ensemble of numerical solutions

1 Introduction

We discuss a notion of numerical solution in the sense of Prager&Synge (some hypersphere containing the true solution) from the viewpoint of the approximation error estimation that is the necessary component of the verification of numerical solutions. The verification of the numerical solution is required by the modern standards [1, 2] and is based on the *a priori* and *a posteriori* error estimates. The Prager&Synge method [3–5] is historically the first approach to *a posteriori* error estimation, unfortunately, highly underestimated. Below we consider an universal version of this method, using notations that follows. $A_h u_h = \rho_h$ is a discrete approximation of a system of partial differential equations, $u_h \in R^M$ denotes the numerical solution (grid function), $\tilde{u}_h \in R^M$ is the projection of a true solution onto the considered grid, $\Delta \tilde{u}_h = u_h - \tilde{u}_h$ is the approximation error, Δu_h is an estimate of the approximation error.

L. Franco et al. (Eds.): ICCS 2024, LNCS 14838, pp. 324–336, 2024.
https://doi.org/10.1007/978-3-031-63783-4_24

A priori error estimates have the form $\| \Delta \tilde{u}_h \| \leq Ch^n$ (n is the order of approximation, h is a grid step). These estimates describe the properties of the algorithm and do not depend on the specific solution. Unfortunately, the unknown constant C restricts the practical applications of this approach.

A posteriori error estimate has the form $\| \Delta \tilde{u}_h \| \leq I(u_h)$ defined by a computable error indicator $I(u_h)$, which depends on the concrete numerical solution u_h and has no unknown constants. The main domain of *a posteriori* error estimation is related with the finite-element analysis [3–8]. For the problems governed by the equations of the hyperbolic or mixed type (typical CFD problems), the progress in the *a posteriori* error estimation is limited. The standards [1, 2] recommend the Richardson extrapolation (RE) [9–11] for the approximation error estimation. However, for the compressible flows, the application of standard RE is impossible due to the absence of the global convergence order [12]. The generalized Richardson extrapolation (GRE) [10, 11] provides an estimate for the spatial distribution of the convergence order. Unfortunately, GRE demonstrates the high computational burden since it requires at least four consequent refinement of the mesh.

Thus, the computationally inexpensive *a posteriori* error estimators are of current interest in the CFD domain. By this reason we consider, herein, the famous Prager&Synge [3–5] method, which seems to be highly underestimated both from the viewpoint of general idea and from the viewpoint of applicability domain. The new options are considered that realize the solution in the sense of Prager&Synge (a hypersphere containing the projection of the true solution on computational grid).

As the first option, we construct an artificial approximation error using the estimate of the truncation errors performed by forward postprocessor and the estimate of the approximation error by an adjoint postprocessor.

As the second option, we construct an artificial solution by the superposition of numerical solutions (obtained by independent algorithms) that provides the approximation error, orthogonal to the error of certain analyzed solution.

Both these options are intrusive and are implemented by rather complex algorithms.

As the third option, we apply the method based on the width of the ensemble of solutions [14], which may be considered as some nonintrusive approximation of the version of the Prager@Synge method, based on the superposition of solutions.

The two-dimensional compressible flows, governed by the Euler equations, are used in the numerical tests in order to compare the above mentioned algorithms. The error estimates by modified Prager&Synge are compared with the "true" discretization errors obtained by subtracting the analytical solutions from the numerical solutions.

2 Prager&Synge Method

The "hypercircle method" by Prager@Synge [3–5] is the oldest algorithm for *a posteriori* error estimation. At present, the term "hypercircle" is used in distant science domains, so, in order to avoid confusion, we mark this method as the "Prager&Synge method". We provide herein some presentation of the original Prager&Synge method with illustrations and citations.

Initially, the Prager&Synge method was applied for the solution of the Poisson equation

$$\nabla^2 u = \rho, \ u_\Gamma = f \tag{1}$$

Two additional linear subspaces of functions are used:

- the subspace of gradients $\partial v/\partial x_i$ ($v \in H^1$) with boundary conditions $v_\Gamma = f$,
- the subspace of functions $q \in H^1$, such that $div(q) = \rho$.

The following orthogonality condition is valid for these functions ($\tilde{u}$ is the exact solutions of the Poisson equation)

$$((\nabla v - \nabla \tilde{u}), (q - \nabla \tilde{u})) = 0. \tag{2}$$

Usually v is assumed to be a numerical solution u_h. The scalar products and norms correspond to L_2 and its finite-dimensional analogue.

The orthogonality relation (Eq. 2) engenders the inequality for the approximation error, since the leg of a right triangle is less the hypotenuse

$$\|\nabla(\tilde{u} - u_h)\| \leq \|\nabla u_h - q\|. \tag{3}$$

So, the Prage@Synge method provides *a posteriori* error estimation from purely geometrical ideas without unknown constants. Since we are unable to compete with authors of the method in vividness, we provide the citation from [5]:

"Imagine a flat piece of country. Imagine in it two strait narrow roads which intersect at right angles. Imagine two blind men set down by helicopter, one on each road. Provide them with a long measuring tape, stretching from the one man to the other man. Let the tape be embossed, so they can read it, even though blind. Without moving, let them tighten the tape and read it. Suppose it reads 100 yards apart.

We are to ask each of them how far he is from the cross roads.

These men, though blind, are very intelligent, and they know, that the hypotenuse of a right-angled triangle is greater than either the other two sides. So each of the men says that his distance from the cross-roads is less than (or possible equal to) 100 yards".

Figure 1 provides the illustration from [5] for this "Two blind men" problem.

Unfortunately, the classic variant of the Prager&Synge method has significant restrictions:

it may be applied to rather narrow domain of equations (Poisson equation, biharmonic equation), which does not include such equations as Euler on Navier-Stokes,

it provides the norm of the error of the solution gradient (Eq. 3) and is unable to estimate the norm $\|\tilde{u} - u_h\|$ of the solution error, which is of the primal interest in practice.

Below we try to demonstrate that the method by Prager&Synge may be significantly modified to avoid these restrictions.

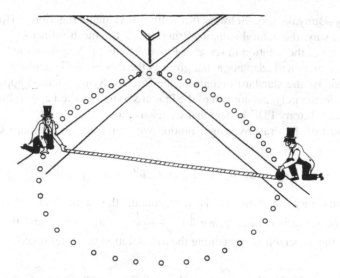

Fig. 1. "Two blind men and their hypercircle [5]"

3   The Numerical Solution in Prager&Synge Sense

The main (and underestimated) idea by Prager&Synge consists in the original concept of the numerical solution. We designate it as the numerical solution in the Prager&Synge sense.

The numerical solution, defined by the standard way, is considered to be an element of the sequence of solutions u_{h_m} at refining grids (grid step $h_m \to 0$ at $m \to \infty$) that is assumed to converge to the exact solution $u_{h_m} \to \tilde{u}_{h_m}$. The mesh adaptation and refinement are the key elements enabling the success of this approach.

In contrast, Synge stated ([4], p. 97): *"In general, **a limiting process is not used, and we do not actually find the solution**…. But although we do not find it, we learn something about its position, namely, that it is located on a certain hypercircle in function space"*.

In the finite-dimensional case the hyprecircle by Prager&Synge is equivalent to the hypersphere. So, the numerical solution in sense of Prager&Synge may be defined as a hypersphere with the center

$$C_h = u_h \tag{4}$$

and a radius R_h

$$\|\tilde{u}_h - C_h\| \le R_h \tag{5}$$

This hypersphere contains the true solution.

Equations (4, 5) demonstrated that the numerical solution by Prager&Synge is naturally related with the *a posteriori* error estimation. In the contrast to the standard approach to approximation, a convergence of numerical solution is not obligatory. The mesh refinement is not mandatory also and should be performed only if the magnitude of R_h is not acceptable from the viewpoint of practical needs (that may be checked using Cauchy–Bunyakovsky–Schwarz inequality for valuable functionals (for example, [14])). In this way, the natural stopping criterion for the mesh refinement termination may be stated. So, the solution in sense of Prager&Synge is free from the "tyranny" of the mesh refinement and adaptation that governs the modern CFD applications.

Unfortunately, the standard domain of the Prager&Synge method applicability is very narrow. Fortunately, the domain of the Prager&Synge method applicability may be extended to an arbitrary PDE system that we try to show.

In the spirit of the Prager&Synge solution we search for an auxiliary solution $u_\perp$ such that

$$(u_h^{(i)} - \tilde{u}_h, u_\perp - \tilde{u}_h)_\Omega = (\Delta \tilde{u}_h^{(i)}, \Delta \tilde{u}_\perp)_\Omega = 0. \tag{6}$$

If the auxiliary solution $u_\perp$ is determined, the numerical solution in the Prager&Synge sense is defined, since $\left\| u_h^{(i)} - \tilde{u}_h \right\| \leq \left\| u_h^{(i)} - u_\perp \right\|$ and the centre and the radius of the hypersphere containing the true solution are determined.

4   The Methods for Approximation of Prager&Synge Solution

We consider several algorithms for calculation of the Prager&Synge numerical solutions. In general, all of them are based on the usage of the ensemble of numerical solutions obtained by independent algorithms. The main differences between these algorithms concern their computational complexity.

4.1   Truncation Error Based Estimation of Prager&Synge Solution

Truncation errors $\delta u_h^{(i)}$ may be estimated either by the differential approximation [18] or by the special postprocessor [19] acting on the numerical solution. Numerical tests [14, 17] demonstrate that the truncation errors $\delta u_h^{(i)}$ on the ensemble of independent solutions are close to orthogonal. The approximation error $\Delta u_h^{(i)}$ may be estimated using the truncation error by solving a special problem for disturbances (see for instance [20])

$$\Delta u_h^{(i)} = A_h^{-1} \delta u_h^{(i)}. \tag{7}$$

However, it is more interesting to express the orthogonality condition (Eq. 6) in terms of nonintrusively computable truncation errors. Let's select some vector $\theta \perp \delta u_h^{(i)}$ and transform it using forward and adjoint operators:

$$\delta u_\perp = A_h A_h^* \theta. \tag{8}$$

Expression (8) may be inverted to obtain

$$\theta = A_h^{-1*} A_h^{-1} \delta u_\perp. \tag{9}$$

The orthogonality condition $\theta \perp \delta u_h^{(i)}$ is equivalent to relation (6) since

$$(\delta u_h^{(i)}, \theta)_\Omega = (\delta u_h^{(i)}, A_h^{-1*} A_h^{-1} \delta u_\perp)_\Omega = (A_h^{-1} \delta u^{(i)}, A_h^{-1} \delta u_\perp)_\Omega = (\Delta u_h^{(i)}, \Delta u_\perp)_\Omega = 0. \tag{10}$$

Finally

$$\Delta u_\perp = A_h^{-1} \delta u_\perp = A_h^* \theta. \tag{11}$$

Equation (11) demonstrates that $\Delta u_\perp$ may be computed nonintrusively using the adjoint postprocessor that implements the solution of adjoint problem (similarly to [16, 19]).

The search for $\theta \perp \delta u_h^{(i)}$ may be conducted as follows:

1. By choice of arbitrary $\theta \perp \delta u_h^{(i)}$ (under condition $\|\theta\| = \left\|\delta u_h^{(i)}\right\|$).
2. By selection of θ that is equal to the truncation error of an additional numerical solution

$$\theta = \delta u_h^{(k)} \tag{12}$$

The latter option is caused by the numerical tests [17] that demonstrate the truncation errors, corresponding to independent algorithms, to be close to orthogonal.

If $\delta u_\perp = A_h A_h^* \theta$ is available, we may compute the auxiliary solution by the defect correction approach

$$u_\perp = A^{-1}(\rho + \delta u_\perp). \tag{13}$$

It enables to obtain the inequality

$$\left\|u_h^{(k)} - \tilde{u}\right\| \le \left\|u_h^{(k)} - u_\perp\right\| = R_h \tag{14}$$

The centre $C_h = u_h^{(k)}$ and radius $R_h = \left\|u_h^{(k)} - u_\perp\right\|$ determine the numerical solution in the sense of the Prager&Synge.

Unfortunately, this algorithm (Eqs. 8–14) is extremely complicated from the algorithmic viewpoint (mainly due to the application of the adjoint solver) and unstable (due to the differentiation of a non regular function by the adjoint solver). The numerical tests demonstrated highly pessimistic estimations (too great estimates of the error norm), so, the above discussion is mainly of the heuristic value and illustrates the existence of auxiliary solution $u_\perp$ that determines the solution in sense of Prager&Synge.

4.2 Approximation Error Based Estimation of the Solution in Prager&Synge Sense

We may construct the auxiliary solution $u_\perp$ in another way using the superposition of N numerical solutions (obtained by independent algorithms on the same grid)

$$u_\perp = \sum_{i=1}^{N} w_i u_h^{(i)} = w_i u_h^{(i)}, \tag{15}$$

$$\sum_{i=1}^{N} w_i = 1. \tag{16}$$

First, we select some basic numerical solution $u_h^{(0)}$ (centre of the hypersphere) and estimate corresponding approximation error $\Delta u_h^{(0)} = A_h^{-1} \delta u_h^{(0)} \approx \Delta \tilde{u}_h^{(0)} = u_h^{(0)} - \tilde{u}_h$ using the defect correction approach. Second, we search for the auxiliary solution $u_\perp$ that is defined by the approximation error $\Delta u_\perp$ orthogonal to the error of the basic numerical solution

$$(\Delta u_h^{(0)}, \Delta u_\perp) = 0. \tag{17}$$

Under conditions (15) $\Delta u_\perp = w_i \Delta \tilde{u}_h^{(i)}$. We assume the existence of $\Delta u_\perp$ due to Eq. (11).

We search for weights $\{w_i\}$ that ensure the orthogonality condition in the variational statement

$$\{w_i\} = \arg \min(\varepsilon(\vec{w})), \tag{18}$$

$$\varepsilon(\vec{w}) = (\Delta u_h^{(0)}, w_i \Delta u_h^{(i)})^2 / 2. \tag{19}$$

The gradient of expression (19) has the appearance

$$\nabla \varepsilon_k = \frac{\partial \varepsilon}{\partial w_k} = (\Delta u_h^{(0)}, (w_i \Delta u_h^{(i)})) \cdot (\Delta u_h^{(0)}, \Delta u_h^{(k)}). \tag{20}$$

The steepest descent is used for the calculation of weights $\{w_i\}$.

$$w_k^{n+1} = w_k^n - \tau \nabla \varepsilon_k. \tag{21}$$

In order to avoid the shift of the exact solution the normalization is used past the termination of iterations

$$\tilde{w}_k^{n+1} = w_k^{n+1} / S^{n+1}, \tag{22}$$

$$S^{n+1} = \sum_{i=1}^{N} w_k^{n+1}. \tag{23}$$

The angle between $\Delta u_\perp$ and $\Delta u_h^{(0)}$

$$\phi = \arccos \left(\frac{(\Delta u_h^{(0)}, w_i \Delta u_h^{(i)})}{\left| \Delta u_h^{(0)} \right| \cdot \left| w_i \Delta u_h^{(i)} \right|} \right) \tag{24}$$

is used as the orthogonality criterion applied to check the quality of numerical results (convergence of iterations (Eq. 21).

The basis $\Delta u_h^{(i)} = A_h^{-1} \delta u_h^{(i)} = u_h^{(i),corr} - u_h^{(i)}$ contained from 2 to 5 vectors in numerical tests. The estimation of $\Delta u_h^{(i)}$ at first step of the algorithm is similar to the defect correction approach. However, the expression $u_\perp = w_i u_h^{(i)}$ contains the superposition of true approximation errors. We apply only angles between approximation errors $\Delta u_h^{(i)} =$

$A_h^{-1} \delta u_h^{(i)}$ obtained by the defect correction and we search not for the refined solution, but for the hypersphere, containing the projection of true solution on the selected grid. We use the ensemble of $N + 1$ numerical solutions. In order to enhance the reliability we consequently select a solution from the ensemble and define it as the basic solution (centre) $C_h = u_h^{(0)}$. After this we estimate the radius $R = \left\| u_h^{(0)} - u_\perp \right\|$. Finally, we select the maximum radius and the corresponding centre point over all tries as the solution in the Prager&Synge sense.

4.3 Nonintrusive Option for Estimation of Prager&Synge Solution

The inequalities similar to the Eqs. (5) and (14) are obtained in [13, 21, 22] (triangle inequality) and [14] (width of ensemble). These approaches may be treated as some approximations of the Prager&Synge method, since the distance between solutions is used as the majorant of error, which circumstance is valid for errors that are close to orthogonal. Unfortunately, the approximation errors $\Delta u_h^{(k)}$ are correlated near discontinuities (the errors involve waves with the positive and negative parts) and, by this reason, are not exactly orthogonal. In several cases the lack of the rigorous orthogonality may be compensated by some additional information. As such information we consider, herein, the maximum distance (ensemble width) between solutions over the ensemble of N independent solutions

$$d_{\max} = \max_{i,k} \left\| u^{(k)} - u^{(i)} \right\| i, k = 1, ..., N \tag{25}$$

and assume

$$\left\| \Delta u^{(k)} \right\| \le d_{\max}. \tag{26}$$

The numerical tests by [14] confirm that the inequality (26) becomes more reliable as the number of the ensemble elements increases. So, the estimation of the width of the ensemble (Eq. 25) and inequality (26) may be considered as some approximation of the Prager&Synge solution.

5 Test Problem and Numerical Algorithms

The above considered methods for the estimation of the Synge solution (*a posteriori* error estimation) are verified by the numerical tests. The numerical solutions for the Euler equations

$$\frac{\partial \rho}{\partial t} + \frac{\partial (\rho U_k)}{\partial x_k} = 0,$$

$$\frac{\partial (\rho U_i)}{\partial t} + \frac{\partial (\rho U_k U_i + P \delta_{ik})}{\partial x_k} = 0,$$

$$\frac{\partial (\rho E)}{\partial t} + \frac{\partial \left(\rho U^k h_0 \right)}{\partial x^k} = 0,$$

describing the flow of the inviscid compressible fluid, are used. The system is two-dimensional, $h_0 = (U_1^2 + U_2^2)/2 + h$, $h = \gamma e$, $E = (e + (U_1^2 + U_2^2)/2)$ are enthalpy and energies, $P = \rho RT$ is the state equation, $\gamma = C_p/C_v$ is the specific heats relation.

The Edney-I and Edney-VI flow structures [15] are analyzed due to the existence of analytical solutions used for the estimation of the "exact" discretization error $\Delta \tilde{u}^{(k)} = u^{(k)} - \tilde{u}_h$. This error is obtained by the subtraction of the numerical solution from the projection of the analytical one onto the computational grid. Figure 2 provides the spatial density distribution for Edney-I flow pattern ($M = 4$, flow deflection angles $\alpha_1 = 20°$ (upper) and $\alpha_2 = 15°$ (lower)). Figure 3 provides the density distribution for Edney-VI ($M = 3.5$, angles $\alpha_1 = 15°$ and $\alpha_2 = 25°$). Figures 2 and 3 illustrate two patterns for the interaction of two shock waves (inflow at left boundary). The pattern by Fig. 2 corresponds crossing shocks, the pattern by Fig. 3 corresponds two merging shocks.

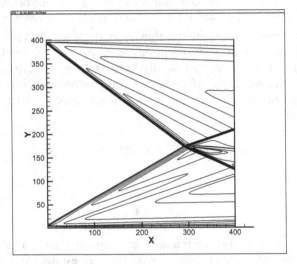

Fig. 2. Density isolines for Edney I flow structure

We analyze the set of numerical solutions obtained by the numerical methods covering the range of approximation order from one to four. The methods are used that follow. First order algorithm by Courant-Isaacson-Rees [23], second order MUSCL based algorithm that uses approximate Riemann solver by [24], second order relaxation based algorithm [25], third order algorithm based on the modification of Chakravarthy-Osher method [26], fourth order algorithm by [27], second order algorithm by [28].

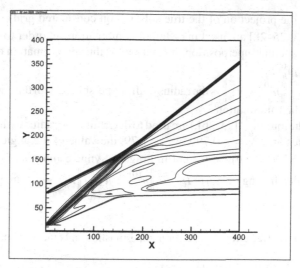

Fig. 3. Density isolines for Edney VI flow structure

6   The Results of Numerical Tests

The quality of *a posteriori* error estimation is described by the effectivity index

$$I_{eff} = \frac{\|\Delta u_h\|}{\|\Delta \tilde{u}_h\|}. \tag{27}$$

The condition for existence of the Prager&Synge solution is related with the efficiency index by the inequality $I_{eff} \geq 1$. The relation $3 > I_{eff} \geq 1$ is stated by [7] for the finite-element applications, which imply sufficiently smooth solutions.

We performed numerical experiments for several cases including different flow structures (Edney-I, Edney-VI), different flow parameters (Mach numbers in the range $3 \div 5$, shock angles in the range $10 \div 30°$) and uniform grids of 100×100, 200×200 and 400×400 nodes. Below several typical results are presented as the illustrations of different methods for calculation of the Prager&Singe solution.

The truncation error based estimation (Eqs. 8–14) provides nonrealistic pessimistic results for the efficiency index I_{eff} $10 \div 20$. In combination with the extremely algorithmic complexity these results make this option non competitive.

The approximation error based estimation (Eqs. 15–24) provides acceptable results.

The results provided by nonintrusive option (Eq. 25, 26) are close in the values of the efficiency index to results obtained by the approximation error based estimation (Eqs. 15–24).

Herein, we consider the numerical tests for the algorithm described in the Sect. 4.2 and based on the superposition of independent numerical solutions. The Edney-VI shock interaction pattern ($M = 3.5$, $\alpha_1 = 15°$, $\alpha_2 = 25°$) was used for the numerical tests on the grids containing 100×100 and 400×400 nodes.

At first step the solution obtained by the method [24] is used as the basic solution $u_h^{(0)}$, which is considered as the centre of the hypersphere containing the true solution (more correctly, the projection of the true solution on considered grid). The additional solutions $u_h^{(1)}$ [23, 25–28] are used in different combinations in order to generate some orthogonal error and the superposition of solutions as the approximation of the auxiliary solution $u_\perp = w_i u_h^{(i)}$.

The value $R = \left\| u_h^{(0)} - u_\perp \right\|$ is the radius of the hypersphere with the centre $C_h = u_h^{(0)}$, which contains the true solution

First, two solutions ($u_h^{(1)}$ [23, 26],) are used to determine the radius of the hypersphere. The weight values are $w_1 = -0.790$, $w_2 = 1.79$, the value of the angle between $\Delta u_\perp$ and $\Delta u_h^{(0)}$, $\phi = 87.6$. The true error norm, Prager&Synge estimates of error norm and the efficiency index having the form $I_{eff} = \left\| u_h^{(0)} - u_\perp \right\| / \left\| u_h^{(0)} - \tilde{u} \right\|$ are provided in the Table 1.

Table 1. True error norm, error norm estimation and efficiency index

$\left\| u_h^{(0)} - \tilde{u} \right\|$	$\left\| u_h^{(0)} - u_\perp \right\|$	I_{eff}
0.114	1.115	9.780

Second, three solutions ($u_h^{(1)}$ [23, 26, 27]) are used. The weight values are $w_1 = -0.579$, $w_2 = 0.654$, $w_3 = 0.925$, the angle $\phi = 90.0$ and $I_{eff} = 3.38$.

Third, four solutions ($u_h^{(1)}$ [23, 26–28]) are used. The corresponding weights are: $w_1 = -0.415$, $w_2 = 0.315$, $w_3 = -0.475$, $w_4 = 0.624$, the angle $\phi = 89.9°$ and $I_{eff} = 1.741$.

Fourth, five solutions ($u_h^{(1)}$ [23, 25–28]) are used. The corresponding weights are: $w_1 = -0.453$, $w_2 = 0.239$, $w_3 = 0.391$, $w_4 = 0.532$, $w_5 = 0.290$, angle $\phi = 89.8°$ and $I_{eff} = 1.438$.

At the next steps all other numerical solutions are consequently selected as the basic solution $u_h^{(0)}$ and the radius $R = \left\| u_h^{(0)} - u_\perp \right\|$ is computed.

The numerical tests on the grids containing 400×400 nodes are performed in order to study the influence of the mesh step on results.

The minimum value the effectivity index over all tests was greater the unit that confirms the successive estimation of the Prager&Synge solution.

The maximum value the effectivity index over all basic solutions is presented by the Table 2 in the dependence on the number of additional solutions for two different grids.

Table 2. The efficiency index in dependence on the number of solutions

N of solutions	2	3	4	5
I_{eff} (100×100)	10.810	3.680	1.931	1.583
I_{eff} (400×400)	9.081	1.754	1.437	1.329

One may see from the Table 2 that the magnitude of the effectivity index decreases as the number of solutions in use is enhanced. In general, two auxiliary solutions provide too pessimistic estimations of the approximation error. In the range of 3–5 solutions the values of the effectivity index are quite acceptable. Some saturation of the effectivity index is observable at increasing of the number of solutions. The grid resolution rather weakly affects the effectivity index.

In general, the numerical tests demonstrate the following situation.

The version of the Prager@Synge method based on the adjoint postprocessor and the defect correction (Eqs. 8–14) overestimates the error and provides the effectivity index in the range I_{eff} $10 \div 20$ that is too pessimistic.

The version of the Prager@Synge method based on the superposition of solutions (Eq. 15) provides the effectivity index in the range I_{eff} $1.3 \div 3.7$.

The effectivity index, based on the ensemble width (Eq. 26), is in the range I_{eff} $1.1 \div 2.5$ for the considered ensemble of numerical solutions engendered by six algorithms.

7 Conclusion

The original concept of the numerical solution is the main idea of the Prager&Synge method. This solution is not based on asymptotics at grid step diminishing and enables to avoid the modern "tyranny" of mesh refinement. Instead of the sequence of solutions, occurring at the grid size diminishing, the Prager&Synge solution deals with the hypersphere containing the projection of the true solution on a computational grid. The numerical solution is the centre of this hypersphere. The Prager&Synge solution provides a natural way to *a posteriori* error estimation and natural criterions for the mesh refinement termination related with the required tolerance of valuable functionals.

The limited domain of application is the main drawback of the Prager&Synge method in the original form. The universal versions of Prager&Synge method that may by applied to an arbitrary PDE system are discussed. They include intrusive and nonintrusive options for the calculation of the Prager&Synge numerical solution.

The intrusive version is based on the superposition of numerical solutions and is specified by the moderate complexity of the algorithm.

The nonintrusive version (based on the width of ensemble) is some approximate variant of the intrusive version with the minimum complexity of the algorithm.

These approaches to the computation of the numerical solution in the sense of the Prager&Synge are compared for the two dimensional compressible flows with the shock waves and demonstrated the acceptable value of efficiency index for the error norm estimation.

References

1. Guide for the Verification and Validation of Computational Fluid Dynamics Simulations, American Institute of Aeronautics and Astronautics, AIAA-G-077-1998, VA (1998)
2. Standard for Verification and Validation in Computational Fluid Dynamics and Heat Transfer, ASME V&V 20-2009 (2009)

3. Prager, W., Synge, J.L.: Approximation in elasticity based on the concept of function spaces. Quart. Appl. Math. **5**, 241–269 (1947)
4. Synge, J.L.: The hypercircle in mathematical physics. CUP, London (1957)
5. Synge, J.L.: The Hypercircle method, pp. 201–217. In Studies in Numerical Analysis, Academic Press, London (1974)
6. Babuska, I., Rheinboldt, W.: A posteriori error estimates for the finite element method. Int. J. Numer. Methods Eng. **12**, 1597–1615 (1978)
7. Repin, S.I.: A posteriori estimates for partial differential equations, vol. 4. Walter de Gruyter (2008)
8. Ainsworth, M., Oden, J.T.: A posteriori error estimation in finite element analysis. Wiley, NY (2000)
9. Richardson, L.F.: The approximate arithmetical solution by finite differences of physical problems involving differential equations with an application to the stresses in a masonry dam. Trans. Royal Soc. London, Ser. A **2**(10), 307–357 (1908)
10. Alexeev, A.K., Bondarev, A.E.: On some features of richardson extrapolation for compressible inviscid flows, pp. 42–54. Mathematica Montisnigri, XL (2017)
11. Roy, C.: Grid convergence error analysis for mixed-order numerical schemes. AIAA J. **41**(4), 595–604 (2003)
12. Carpenter, M.H., Casper, J.H.: Accuracy of shock capturing in two spatial dimensions. AIAA J. **37**(9), 1072–1079 (1999)
13. Alekseev, A.K., Bondarev, A.E., Navon, I.M.: On Triangle Inequality Based Approximation Error Estimation. arXiv:1708.04604 (2017)
14. Alekseev, A.K., Bondarev, A.E., Kuvshinnikov, A.E.: On uncertainty quantification via the ensemble of independent numerical solutions. J. Comput. Sci. **42**, 10114 (2020)
15. Edney, B.: Effects of shock impingement on the heat transfer around blunt bodies. AIAA J. **6**(1), 15–21 (1968)
16. Alekseev, A.K., Bondarev, A.K.: The numerical solution in the sense of Prager&Synge, arXiv: 2403.06273, (2024)
17. Alekseev, A.K., Bondarev, A.E.: On a posteriori error estimation using distances between numerical solutions and angles between truncation errors. Math. Comput. Simul. **190**, 892–904 (2021)
18. Shokin, Yu. I.: Method of differential approximation. Springer-Verlag (1983)
19. Alekseev, A.K., Navon, I.M.: A Posteriori error estimation by postprocessor independent of flowfield calculation method. Comput. Math. Appl. **51**, 397–404 (2006)
20. Linss, T., Kopteva, N.: A posteriori error estimation for a defect-correction method applied to convection-diffusion problems. Int. J. Numer. Anal. Model. **1**(1), 1–16 (2009)
21. Alekseev, A.K., Bondarev, A.E.: Estimation of the distance between true and numerical solutions. Comput. Math. Math. Phys. **59**(6), 857–863 (2019)
22. Alekseev, A.K., Bondarev, A.E.: On a posteriori estimation of the approximation error norm for an ensemble of independent solutions. Numer. Anal. Appl. **13**(3), 195–206 (2020)
23. Courant, R., Isaacson, E., Rees, M.: On the solution of nonlinear hyperbolic differential equations by finite differences. Comm. Pure Appl. Math. **5**, 243–255 (1952)
24. Sun, M., Katayama, K.: An artificially upstream flux vector splitting for the Euler equations. JCP **189**, 305–329 (2003)
25. Trac, H., Pen, U.-L.: A Primer on Eulerian Computational Fluid Dynamics for Astrophysics, arXiv:0210611v2 (2002)
26. Osher, S., Chakravarthy, S.: Very high order accurate TVD schemes. ICASE Report. N. **84–144**, 229–274 (1984)
27. Yamamoto, S., Daiguji, H.: Higher-order-accurate upwind schemes for solving the compressible Euler and Navier-Stokes equations. Comput. Fluids **22**, 259–270 (1993)
28. MacCormack, R.W.: The Effect of Viscosity in Hypervelocity Impact Cratering, AIAA Paper 69–354 (1969)

Enhancing Out-of-Distribution Detection Through Stochastic Embeddings in Self-supervised Learning

Denis Janiak[✉][ID], Jakub Binkowski[ID], Piotr Bielak[ID], and Tomasz Kajdanowicz[ID]

Wroclaw University of Science and Technology, Wrocaw, Poland
denis.janiak@pwr.edu.pl

Abstract. In recent years, self-supervised learning has played a pivotal role in advancing machine learning by allowing models to acquire meaningful representations from unlabeled data. An intriguing research avenue involves developing self-supervised models within an information-theoretic framework, e.g., feature decorrelation methods like Barlow Twins and VICReg, which can considered as particular implementations of the information bottleneck objective. However, many studies often deviate from the stochasticity assumptions inherent in the information-theoretic framework. Our research demonstrates that by adhering to these assumptions, specifically by employing stochastic embeddings in the form of a parametrized conditional density, we can not only achieve performance comparable to deterministic networks but also significantly improve the detection of out-of-distribution examples, surpassing even the performance of supervised detectors. With VICReg, specifically, we achieve an average AUROC of 0.858 for the stochastic unsupervised detector, compared to 0.796 for the supervised baseline. Remarkably, this improvement is achieved solely by leveraging information from the underlying embedding distribution.

Keywords: self-supervised learning · out-of-distribution detection · stochastic embeddings · uncertainty estimation · information theory

1 Introduction

Self-supervised learning (SSL) is an approach to learning representations of data without labels, often utilizing the data itself as a supervisory signal. In recent years, such methods have gained increasing popularity in computer vision and have demonstrated significant success in various downstream tasks [29]. The primary goal of SSL is to bring similar samples closer while pushing dissimilar samples further apart. This objective enhances the model's ability to discriminate between different data classes, contributing to its overall effectiveness.

L. Franco et al. (Eds.): ICCS 2024, LNCS 14838, pp. 337–351, 2024.
https://doi.org/10.1007/978-3-031-63783-4_25

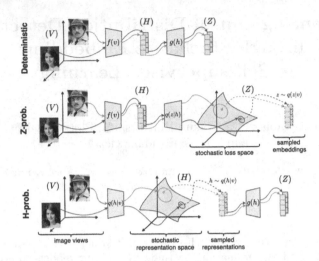

Fig. 1. This schematic illustrates the multi-view encoding process. It begins with image views (V), which are transformed via deterministic or stochastic pathways. The deterministic pathway processes the input through encoder function $f(\cdot)$, leading to deterministic representation space (H) and loss space (Z). The probabilistic pathways (H-prob. and Z-prob.) introduce stochasticity by sampling from associated spaces using stochastic encoders $q(h|v)$ and $q(z|v)$.

One effective strategy for learning meaningful representations involves maximizing the similarity between various views of augmented images, thereby ensuring invariance to these augmentations [7]. However, this approach risks encountering trivial solutions (i.e., where all embeddings collapse into a single point). Methods such as Barlow Twins [40] and VICReg [4] employ feature decorrelation mechanisms to address this issue. Interestingly, they are closely linked to an information-theoretic framework as their objective can be derived using the information bottleneck principle [2,33]. Nevertheless, the information-theoretic framework typically assumes a source of stochasticity (noise) within the model. The aforementioned methods do not fulfil this stochastic condition, as they are simplified using deterministic networks and rely on a proxy objective. By aligning with the assumption of the information-theoretic framework (source of noise within the model), we could effectively benefit from stochasticity for tasks such as uncertainty estimation and out-of-distribution detection (OOD) [39]. Recent advancements in machine learning underscore the importance of quantifying these uncertainties and identifying distributionally shifted (OOD) samples, particularly in safety-critical applications such as medical imaging, active learning, and autonomous driving [11].

Our study directly introduces stochasticity into the feature decorrelation-based methods by parameterizing the conditional density of a model, i.e., predicting the parameters of the embeddings' distribution. In essence, our goal is twofold: first, to adhere to the stochastic assumptions required by the information-theoretic framework, and second, to harness this stochasticity to enhance our ability to detect OOD examples and handle uncertainty. We demonstrate that leveraging stochastic embeddings enables us to outperform other supervised and deterministic methods for OOD detection while also achieving comparable results when evaluated across various downstream tasks (linear classification, semi-supervised learning, and transfer learning).

Our contributions are as follows:

1. We employed stochastic embeddings to feature decorrelation-based methods, adhering to the theoretical underpinnings of the information-theoretic framework.
2. We showcased the effectiveness of our approach in detecting OOD samples by leveraging the embedding distribution, outperforming traditional supervised detectors.
3. We explored novel strategies for exploiting stochastic embeddings to accurately identify OOD examples.
4. We conducted a comprehensive empirical evaluation to assess the impact of stochastic embeddings on downstream tasks.

The paper's remaining sections are organized as follows: Sect. 2 reviews related works in SSL, focusing on representation learning methods and addressing challenges like trivial solutions and lack of stochasticity. Section 3 details our approach for introducing stochasticity into feature decorrelation-based methods. In Sect. 4, we present experiments evaluating our approach's effectiveness in downstream tasks and OOD detection. Lastly, Sect. 5 offers conclusions and summarizes key findings.

2   Related Works

The primary objective of **self-supervised learning** is to optimize a specific loss function tailored to capture meaningful patterns or relationships within unlabeled data. One approach, proven to be very successful in vision tasks, is contrastive learning [7], which aims to maximize the agreement between positive (similar) pairs of data samples while minimizing it for negative (dissimilar) pairs. More recent avenues are non-contrastive methods that adopt various mechanisms to prevent representation collapse, eliminating the need for negative samples [34]. These mechanisms could include architectural constraints [13], clustering-based objectives [6], or feature decorrelation [4,40]. These non-contrastive methods indirectly optimize the uniformity property of the representation, aiming to prevent representation collapse [36]. In this context, we focus on feature decorrelation-based techniques, particularly the Barlow Twins [40] and VICReg [4]. The Barlow Twins method involves computing the cross-correlation matrix from embeddings of augmented images. The objective is to minimize the off-diagonal elements, encouraging feature decorrelation while promoting data invariance by aiming to set the diagonal elements to one. VICReg introduces an additional term into the loss function, which controls the variance of each dimension within the embeddings. Consequently, this facilitates straightforward computation of the covariance of the embeddings and eliminates the necessity for normalization.

OOD detection methods in machine learning are crucial for identifying instances during the testing phase that deviate semantically from the categories encountered in the training data, thereby preventing misclassification [11]. While supervised detectors [16,24] have demonstrated success, they rely on label information and a classifier to derive their scores. On the other hand, distance-based methods [23,31] utilize representations (features) for detecting OOD examples, yet many of them still necessitate computing class-conditional statistics from the training data. In fully unsupervised OOD detection, density-based and reconstruction-based methods leverage data density or reconstruction techniques, often incorporating generative models [39].

Our research focuses on **self-supervised** methods for **OOD detection**. Previous studies have explored various avenues, including combining self-supervised and discriminative objectives [16], employing hard data augmentations for sample separation [35], leveraging contrastively learned features [3,32], and integrating probabilistic modelling to estimate uncertainty [19,26]. For instance, [19] utilizes the von Mises Fischer distribution to model embeddings, with the concentration parameter serving as an uncertainty metric. Similarly, [26] investigates SimSiam [8] within the variational inference framework, employing a power spherical distribution to characterize the embedding distribution. Our study adopts a similar approach, leveraging embedding distribution characteristics to identify uncertain and OOD examples. However, we consider feature decorrelation-based methods within the information-theoretic framework, representing embeddings with a different distribution than the aforementioned studies and learning a different objective (Sect. 3.2).

The **information-theoretic perspective** offers crucial insights into the underlying mechanics of self-supervised learning. While models like InfoMax [17] prioritize capturing maximal data information, the Information Bottleneck (IB) principle emphasizes balancing informativeness and compression [2]. In this context, Barlow Twins exemplifies an IB aiming to maximize information between the image and representation while minimizing information about data augmentation, rendering the representation invariant to these augmentations. Another approach, the multi-view information bottleneck (MIB) framework [10], seeks to capture predictive information shared across different data views. This is achieved by maximizing mutual (shared) information between views in their embeddings while minimizing redundant information not shared between them. Shwartz-Ziv et al. [33] demonstrated that the VICReg objective can also be derived from an information-theoretic standpoint, exploiting a lower bound derived from the MIB framework.

3   Methodology

In this section, we begin by outlining the setup of the feature decorrelation-based SSL framework for deterministic (point estimate) embeddings. Following this, we extend this setup to incorporate stochasticity and introduce stochastic (probabilistic) embeddings with regularization. Lastly, we introduce methods to leverage the stochastic nature of the embeddings, offering stochastic OOD metrics. Figure 1 illustrates the workflow of deterministic and stochastic self-supervised learning variants.

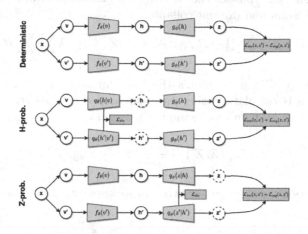

Fig. 2. Diagram of three approaches to SSL: (1) **Deterministic**, with deterministic mappings using f_θ and g_ϕ, (2) **H-prob**, introducing stochasticity with q_θ before deterministic projection g_ϕ, and (3) **Z-prob**, with deterministic mapping f_θ and stochastic projection q_ϕ. Losses $\mathcal{L}_{\text{inv}}(z, z')$ and $\mathcal{L}_{\text{reg}}(z, z')$ are computed using the embeddings. $\mathcal{L}_{\text{div}}$ loss is the stochastic regularization loss (see Sect. 3.3).

3.1 Deterministic Self-supervised Learning

We sample an image x from a dataset $\mathcal{D}$ and create two views v and v' by applying transforms t and t' sampled from a distribution $\mathcal{T}$. These views are then fed into an encoder f_θ, parameterized by θ, to create representations $h = f_\theta(v)$ and $h' = f_\theta(v')$. Next, the representation vectors are passed through the projector g_ϕ, parameterized by ϕ, to obtain embeddings $z = g_\phi(h)$ and $z' = g_\phi(h')$. In a batch represented by $Z = [z_1, \ldots, z_n]$ and $Z' = [z'_1, \ldots, z'_n]$, each containing n embedding vectors of dimension d, corresponding to embedded image views, we define $z_{(i)} = [z_{1(i)}, \ldots, z_{n(i)}]$ as the i-th variable across all samples in the batch. The loss function $\mathcal{L}(Z, Z')$ is then applied to these embeddings Z and Z'.

In **Barlow Twins**, the loss function is computed using the cross-correlation matrix R on embeddings, which are mean-centred along the batch dimension:

$$R_{ij} = \text{corr}(z_{(i)}, z'_{(j)}) = \frac{\text{cov}(z_{(i)}, z'_{(j)})}{\sigma_{z_{(i)}} \cdot \sigma_{z'_{(j)}}}, \tag{1}$$

where $z_{(i)}$ and $z'_{(j)}$ represent the i-th and j-th component of embedding vectors across all samples in the batch, while $\sigma_{z_{(i)}}$ and $\sigma_{z'_{(j)}}$ denote the standard deviations of $z_{(i)}$ and $z'_{(j)}$ respectively. From the cross-correlation matrix, we compute: (1) invariance term $\mathcal{L}_{\text{inv}}$ that optimizes the diagonal elements to be close to 1, aiming to enforce invariance to data augmentations; and (2) regularization term $\mathcal{L}_{\text{reg}}$ that pushes the off-diagonal elements towards 0 to promote feature decorrelation and prevent collapse:

$$\mathcal{L}_{\text{inv}}(Z, Z') = \sum_i (1 - R_{(i,i)})^2, \quad \mathcal{L}_{\text{reg}}(Z, Z') = \lambda \sum_i \sum_{j \neq i} R_{(i,j)}^2$$

In contrast, **VICReg** computes the loss function with three terms. The invariance term is calculated using mean-squared error loss scaled by α coefficient and divided by number of samples in batch n:

$$\mathcal{L}_{\text{inv}}(Z, Z') = \frac{\alpha}{n} \sum_{b=1}^{n} \|z_b - z'_b\|_2^2. \tag{2}$$

The regularization term comprises two components - covariance and variance:

$$\mathcal{L}_{\text{cov}}(Z) = \frac{1}{d} \sum_i \sum_{j \neq i} C_{(i,j)}, \quad \mathcal{L}_{\text{var}}(Z) = \frac{1}{d} \sum_{i=1}^{d} \max(0, \gamma - \sigma_{z_{(i)}+\epsilon}), \tag{3}$$

where $C_{(i,j)}$ is the element of the covariance matrix, i.e., $C_{(i,j)} = \text{cov}(z_{(i)}, z_{(j)})$. The covariance term involves summing the squared off-diagonal coefficients of the covariance matrix. Meanwhile, the variance term is a hinge function that operates on the standard deviation of the embeddings across the batch dimension. Both regularization terms are calculated separately for Z and Z' using τ and ν scalars as loss coefficient:

$$\mathcal{L}_{\text{reg}}(Z, Z') = \tau[\mathcal{L}_{\text{var}}(Z) + \mathcal{L}_{\text{var}}(Z')] + \nu[\mathcal{L}_{\text{cov}}(Z) + \mathcal{L}_{\text{cov}}(Z')], \tag{4}$$

The final loss function in both Barlow Twins and VICReg is then formulated as the sum of invariance and regularization terms: $\mathcal{L}_{\text{SSL}} = \mathcal{L}_{\text{inv}} + \mathcal{L}_{\text{reg}}$.

3.2 Stochastic Embeddings

Drawing inspiration from [10] and [33], we propose to reformulate our self-supervised objective as an information maximization problem and extend it to stochastic embeddings. We aim to maximize the mutual information between the views V and V' and their corresponding embeddings Z and Z', i.e., $I(Z; V')$ and $I(Z'; V)$, respectively. We utilize the lower bound from [33]:

$$\begin{aligned} I(Z; V') &= \mathcal{H}(Z) - \mathcal{H}(Z|V') \\ &\geq \mathcal{H}(Z) + \mathbb{E}_{v'}[\log q(z|v')] \\ &\geq \mathcal{H}(Z) + \mathbb{E}_{z|v}[\mathbb{E}_{z'|v'}[\log q(z|z')]] \end{aligned} \tag{5}$$

where $\mathcal{H}(Z)$, is implicitly optimized by the regularization term $\mathcal{L}_{reg}$, while the expectations $\mathbb{E}_{z|v}[\mathbb{E}_{z'|v'}[\log q(z|z')]]$ (square log loss) are optimized by the invariance term $\mathcal{L}_{inv}$. To compute the expected loss, we evaluate these expectations over empirical data distribution. Specifically, we backpropagate through K Monte Carlo (MC) samples using the reparametrization trick [18]:

$$\mathbb{E}_{z|v}[\mathbb{E}_{z'|v'}[\log q(z|z')]] \simeq \frac{1}{nK} \sum_{i=1}^{n} \sum_{k=1}^{K} \log q(z_{ik}|z'_{ik}). \tag{6}$$

We introduce two variations of the model, which differ in terms of choice of the stochastic space and, therefore, the variational conditional density $q(z|v)$. Figure 2 depicts the workflow of probabilistic (stochastic) variations of self-supervised learning.

Stochastic Loss Space (Z-prob.). In this model variant, we introduce the stochasticity into the loss space by parametrizing the projector conditional density $q_\phi(z|h)$, i.e., we make the projector stochastic (see Fig. 2 at the bottom). Specifically, we employ a two-step process for encoding image views. Initially, we utilize a deterministic encoder f_θ to transform the image view v into a representation h, i.e., $h = f_\theta(v)$. Subsequently, we employ a stochastic projector $q_\phi(z|h)$ to sample latent variables z based on h. Our conditional density is defined as $q_{\phi,\theta}(z|v) = q_\phi(z|f_\theta(v))$, and the sampling process is represented as $z \sim \mathcal{N}\left(z|\mu_\phi(h), \sigma_\phi^2(h)I\right)$, where $\mu_\phi(h)$ and $\sigma_\phi^2(h)$ denote the mean and variance functions determined by the stochastic projector, respectively. The same procedure is applied to the second image view v' to generate the representation h' and the corresponding embedding z', utilizing identical encoder and projector parameters denoted by θ and ϕ.

Stochastic Representation Space (H-prob.). In this model variant, we shift the stochasticity from the loss space Z to the representation space H (see Fig. 2 in the middle). We define the conditional density as $q_{\phi,\theta}(z|v) = g_\phi(q_\theta(h|v))$ and sampling process as $h \sim \mathcal{N}(h|\mu_\theta(v), \sigma_\theta^2(v)I)$. Then, we obtain the embedding z by mapping the representation h with a projector, g_ϕ. We apply the same procedure for the second image view v' to produce the representation h' and the embedding z', utilizing the same encoder and projector parameters θ and ϕ. In particular, to utilize the bound from Eq. 5, we must also account for the presence of h. We decompose the joint distribution as $q(v, h, z) = q(z|h)q(h|v)q(v)$, where z depends on h. The computation of $q(z|v)$ requires the marginalization of h, i.e. $q(z|v) = \int dh\, q(z, h|v) = \int dh\, q(z|h)q(h|v)$. Consequently, we can take expectations with respect to $q(h|v)$ and lower bound term from Eq. 5 using Jensen's inequality:

$$\mathbb{E}_{v'}[\log q(z|v')] \geq \mathbb{E}_{h'|v'}[\log q(z|h')]. \tag{7}$$

By taking the expectation over both Z and Z', we will obtain the final objective:

$$\mathbb{E}_{z|h,h|v}[\mathbb{E}_{z'|h',h'|v'}[\log q(z|z')]], \tag{8}$$

which involves an additional expectation step that we take using MC estimation.

3.3 Stochastic Regularization

Moving from point estimates to stochastic embeddings, we introduce an additional layer of uncertainty, which helps capture the inherent ambiguity and variability in the data. However, it also raises the challenge of regularizing this stochasticity to prevent trivial solutions and obtain reliable uncertainty estimates. To address this issue, we follow [2] and formulate an additional regularization term to the loss function in the form of a KL divergence between the stochastic embeddings $q(\cdot|v)$ and $q(\cdot|v')$, and a predefined prior $\hat{q}(\cdot)$, typically $\mathcal{N}(0,1)$:

$$\mathcal{L}_{\text{div}}(\cdot,\cdot) = \frac{\beta}{2}[\text{KL}(q(\cdot|v)\|\hat{q}(\cdot)) + \text{KL}(q(\cdot|v')\|\hat{q}(\cdot))], \qquad (9)$$

where $(\cdot,\cdot)$ is either (h,h') or (z,z'). This regularization, controlled by β parameter, acts as a bottleneck, constraining the capacity of our stochastic embeddings, which proved to be effective in previous work [1,2] in terms of improving robustness and disentanglement.

Consequently, the overall objective for our stochastic self-supervised learning framework is defined as $\mathcal{L}_{\text{Stochastic-SSL}} = \mathcal{L}_{\text{inv}} + \mathcal{L}_{\text{reg}} + \mathcal{L}_{\text{div}}$.

3.4 Stochastic OOD Detectors

We aim to utilize the stochastic nature of our embeddings to improve their ability to distinguish between in-distribution and OOD samples. To achieve this, we introduce **new stochastic scoring methods** that exploit the inherent variability in the embeddings. Notably, these methods do not require any training labels and do not depend on the information of OOD data. We propose the following scoring methods:

- **Log-prob prior (LogP)**: This score is derived by assessing the prior density $\hat{q}$ at the test input x^*, denoted as $\hat{q}(x^*)$. As we use KL regularization, the OOD examples may not be pushed towards the prior, and the log-prob for these examples could be higher.
- **Sigma mean (Sigma)**: This score computes the mean sigma value of the embedding, expressed as $\frac{1}{d}\sum_{i=1}^{d}\sigma(x^*)_{(i)}$, where i denotes the index of the embedding vector with dimension d, and σ is the variance predictor of the conditional density model. Higher sigma values indicate greater uncertainty in the embedding distribution, potentially signalling an unfamiliar example.
- **KLD-Kth-nearest (KLD-KN)**: In this approach, based on the test input x^*, we find its distance to the K-th nearest example x_k from the training samples $\mathcal{D}$, measured by the KL divergence, i.e., $\text{KL}(q(\cdot|x_k)\|q(\cdot|x^*))$. This method operates under the assumption that OOD examples, which the model has not encountered, may appear in a "hole" on a manifold. Thus, examining nearby examples can serve as a meaningful scoring metric.

- **Euclidean-Kth-nearest (Euclid-KN)**: This score serves as the deterministic counterpart to the KLD-KN score. We utilize the Euclidean distance instead of KL divergence, i.e., $||x_k - x^*||_2$. We implement this score to assess whether the KLD-KN score effectively exploits the embedding variance.

4 Experiments

We pretrain our model in a self-supervised manner (without labels), adopting the same image augmentations and closely adhering to the original works in determining the loss coefficients [4,40]. Due to computational constraints, we opt for the smaller ResNet-18 [14] architecture as our backbone encoder and a smaller non-linear projection head (3-layer MLP, each of 1024 dimensions). We train the model using the AdamW [25] optimizer with a batch size of 256. We make our code publicly available at https://github.com/graphml-lab-pwr/stochastic-embeddings-ssl.

4.1 Downstream Tasks Evaluation

Setup. In these tasks, the model is pretrained <u>once</u> for 100 epochs on the ImageNet dataset [30]. Our experiments and previous works showed that Barlow Twins and VICReg exhibit low sensitivity to model intialization [4]. Moreover, we utilize the $\mathcal{N}(0,1)$ prior $\hat{q}(\cdot)$ and 12 MC samples. Next, we employ three downstream tasks: linear classification, semi-supervised and transfer learning [12]. For linear classification (`Linear`), we train a linear classifier (single linear layer) on the frozen representations from our pretrained backbone encoder and corresponding image labels. A similar process is repeated for `Transfer learning` task, where we employ the SUN397 [37] and the Flowers-102 [28] datasets. In the `Semi-supervised` learning task, we fine-tune <u>both</u> the backbone encoder and the linear classifier. We utilize subsets of the ImageNet dataset corresponding to 1% and 10% of the labels [8].

Table 1. Comparison of top-1 accuracy (Acc@1) and expected calibration error (ECE) for ImageNet tasks. Both Barlow Twins and VICReg exhibit low variance, allowing for a single-run performance comparison.

		Linear		Semi-supervised				Transfer learning			
				1%		10%		SUN397		flowers-102	
		Acc@1(↑)	ECE(↓)	Acc@1(↑)	ECE(↓)	Acc@1(↑)	ECE(↓)	Acc@1(↑)	ECE(↓)	Acc@1(↑)	ECE(↓)
VICReg	Deterministic	0.490	0.011	0.315	0.044	0.509	0.055	0.477	0.112	0.649	0.445
	H-prob	0.451	0.013	0.313	0.220	0.498	0.114	0.460	0.038	0.622	0.315
	Z-prob	0.484	0.012	0.310	0.042	0.507	0.054	0.478	0.112	0.629	0.435
Barlow	Deterministic	0.495	0.008	0.316	0.040	0.518	0.138	0.482	0.115	0.645	0.428
	H-prob	0.451	0.010	0.309	0.217	0.495	0.111	0.457	0.023	0.637	0.302
	Z-prob.	0.489	0.010	0.313	0.039	0.506	0.054	0.481	0.111	0.645	0.447

Results. We present the results in Table 1. Our observations show that stochastic embeddings, particularly those in the loss space (Z-prob.), exhibit competitive performance compared to deterministic embeddings. For linear classification and transfer learning tasks, Z-prob. embeddings tend to outperform stochastic embeddings in the representation space (H-prob.) across both the Barlow Twins and VICReg methods in terms of accuracy. However, the difference is lower in the semi-supervised task, especially for 1% of available labels. Noticeably, in transfer learning, H-prob embeddings maintain lower ECE scores for both datasets, indicating better calibration, but they tend to exhibit higher ECE for semi-supervised. The performance variation across different datasets underscores the importance of dataset characteristics in model evaluation. We hypothesize that the superior performance of Z-prob. embeddings compared to H-prob. is attributable to the representation bottleneck created by the H-prob. model, which leads to an early representation compression, potentially removing data invariance at the representation level. As demonstrated in previous studies [5], such premature compression can negatively impact the performance of SSL models.

4.2 OOD Detection

Setup. We utilize the same setup for the backbone as in the ablation study and select the best-performing model (see Sect. 4.3). Next, we investigate the OOD capabilities of our methods, following a similar evaluation procedure to [38] and report the results with the commonly used AUROC metric. We consider the original test set of CIFAR-10 as IN data and assess its ability to distinguish between other Near (MNIST [22], SVHN [27], Places365 [41], Textures [9]) and Far (CIFAR-100 [20], TinyImageNet [TIN] [21]) OOD datasets. We evaluate our proposed stochastic detectors (LogP, Sigma, KLD-KN) against commonly used methods in OOD detection problems. Specifically, we compare them with classification-based methods such as MaxSoftmax probability (MSP) [15] and ODIN [24], as well as distance-based methods like Gram matrices [31] and Mahalanobis distance (MDS) [23]. Contrary to our detectors, these methods require label information from the training data: MaxSoftmax and ODIN rely on a trained classifier, while Gram and Mahalanobis necessitate the computation of class-conditional statistics. Moreover, we provide a comparison between stochastic KLD-KN and its deterministic counterpart Euclid-KN (we select K based on the hyperparameter search [38]). Finally, we evaluate our methods against SSD [32], a framework for unsupervised OOD detection in self-supervised learning, which leverages Mahalanobis distance on k-means detected clusters.

Results. Table 2 presents the results from our experiment on stochastic embeddings. As observed, leveraging the intrinsic properties of stochastic embeddings, such as their variance (Sigma) or latent space manifold (KLD-KN), can be highly effective as an OOD detector. For both VICReg and Barlow Twins, the KLD-KN detection score surpasses the performance of supervised detectors (MSP, ODIN) and significantly exceeds that of the Euclidean-KN, its deterministic counterpart. Furthermore, simple `Sigma` provides better scoring than the distances-based methods such as `Mahalanobis`, `Gram` and `SSD` for both VICReg and Barlow Twins while reaching the performance of classification-based detectors for VICReg, thereby demonstrating its ability to take into account the stochasticity of the embeddings.

Table 2. The AUROC performance of OOD detection methods. * denotes supervised detectors, while † denotes unsupervised detectors requiring fitting labels.

	Detector	Near OOD				Far OOD		
		MNIST	SVHN	Places365	Texture	CIFAR-100	TIN	Avg.
VICReg	MSP*	0.837	0.790	0.780	0.769	0.799	0.800	0.796
	ODIN*	0.862	0.668	0.806	0.778	**0.818**	**0.819**	0.792
	Gram†	0.946	0.927	0.599	0.727	0.545	0.583	0.721
	MDS†	0.935	0.928	0.577	0.851	0.506	0.546	0.724
	SSD	0.935	0.928	0.565	0.850	0.514	0.538	0.722
	LogP	0.926	0.742	0.651	0.550	0.658	0.598	0.688
	Sigma	0.890	0.858	0.815	0.645	0.730	0.769	0.784
	KLD-KN	**0.954**	**0.972**	**0.821**	**0.874**	0.742	0.783	**0.858**
	Euclid-KN	0.936	0.907	0.776	0.805	0.706	0.731	0.810
Barlow Twins	MSP*	0.888	0.812	0.768	0.783	0.787	0.790	0.805
	ODIN*	0.929	0.672	0.800	0.821	**0.809**	0.815	0.808
	Gram†	0.956	**0.960**	0.594	0.771	0.551	0.598	0.738
	MDS†	0.983	0.935	0.589	0.817	0.508	0.537	0.728
	SSD	0.978	0.946	0.544	**0.856**	0.504	0.530	0.726
	LogP	0.830	0.749	0.628	0.549	0.648	0.572	0.663
	Sigma	0.858	0.913	0.727	0.791	0.570	0.611	0.745
	KLD-KN	0.906	0.910	**0.840**	0.759	0.705	0.759	**0.813**
	Euclid-KN	**0.992**	0.954	0.721	0.825	0.668	0.688	0.808

Table 3 compares the performance of probabilistic and deterministic embeddings as an average over all datasets. We can see that classification-based detectors work best for deterministic embeddings, as their performance is often correlated with downstream performance. However, other distance- and feature-based detectors have higher AUROC for the stochastic embeddings, meaning we have a latent space more suited for detecting examples outside of IN distribution.

4.3   Ablation Study

Setup. The model is pretrained three times with different seeds for 200 epochs each time on the CIFAR-10 dataset [20] to evaluate different model hyperparameters. In particular, we assess the impact of various priors, β scales, and the number of MC samples. We compare the standard normal prior, $\mathcal{N}(0, 1)$,

Table 3. Comparison of AUROC performance (averaged over all datasets) for deterministic and stochastic embeddings in OOD detection.

		MSP	ODIN	Gram	MDS	SSD	LogP	Sigma	KLD-KN	Euclid-KN
VICReg	Deterministic	**0.806**	**0.805**	0.694	0.695	0.695	0.000	0.000	0.000	0.769
	H-prob	0.792	0.790	**0.721**	**0.723**	0.720	**0.688**	0.736	**0.858**	**0.810**
	Z-prob	0.796	0.792	0.704	**0.724**	**0.722**	0.683	**0.784**	0.785	0.796
Barlow	Deterministic	**0.808**	**0.812**	**0.739**	0.693	0.694	0.000	0.000	0.000	0.774
	H-prob	0.805	0.808	**0.738**	0.687	0.688	0.663	**0.745**	**0.813**	0.762
	Z-prob	0.803	0.802	0.719	**0.728**	**0.726**	0.659	0.687	0.774	**0.808**

with a Mixture of Gaussians (MoG),[1] aiming to assess the impact of employing a more expressive distribution for modelling stochastic embeddings. Additionally, we explore how the β scale influences the bottleneck and, consequently, the capacity of the embeddings.[2] Finally, we explore the advantages of utilizing multiple MC samples to estimate the expectation from Eq. 6. Like downstream task evaluations, we train a linear classifier on the fixed representation from our pretrained backbone encoder.

Table 4. Comparison of top-1 accuracy (Acc@1) and expected calibration error (ECE) for the ablation study.

	Embeddings	Beta	Prior (# of MC samples)							
			Normal(1)		Normal(12)		MoG(1)		MoG(12)	
			Acc@1($\uparrow$)	ECE($\downarrow$)	Acc@1($\uparrow$)	ECE($\downarrow$)	Acc@1($\uparrow$)	ECE($\downarrow$)	Acc@1($\uparrow$)	ECE($\downarrow$)
VICReg	Z-prob.	1e−3	0.824	0.022	0.826	0.021	0.793	0.021	0.793	0.022
		1e−4	0.834	0.020	0.831	0.021	0.830	0.019	0.832	0.020
		1e−5	0.834	0.018	0.831	0.022	0.836	0.021	0.830	0.021
	H-prob.	1e−3	0.804	0.018	0.817	0.025	0.802	0.010	0.810	0.012
		1e−4	0.823	0.010	0.828	0.011	0.826	0.009	0.825	0.010
		1e−5	0.826	0.009	0.829	0.011	0.824	0.011	0.824	0.011
Barlow Twins	Z-prob.	1e−1	0.821	0.022	0.819	0.026	0.748	0.021	0.746	0.025
		1e−2	0.827	0.031	0.827	0.033	0.817	0.020	0.821	0.019
		1e−3	0.823	0.031	0.826	0.025	0.823	0.020	0.824	0.019
	H-prob.	1e−2	0.790	0.014	0.805	0.015	0.788	0.011	0.801	0.010
		1e−3	0.799	0.010	0.809	0.011	0.782	0.031	0.804	0.012
		1e−4	0.801	0.011	0.803	0.009	0.796	0.013	0.800	0.010

---

[1] The MoG prior has the following form: $\frac{1}{M}\sum_{m=1}^{M} \mathcal{N}(\mu_m, \operatorname{diag}(\sigma_m^2))$, where M denotes the number of mixtures, while μ_m and σ_m denote trainable parameters of a specific Gaussian in the mixture model.

[2] The variability in the loss function's magnitude and method-specific sensitivities necessitated the selection of distinct beta (β) scale hyperparameters for each approach, as documented in Table 4.

Results. We report the results for the classification task in Table 4. Contrary to our initial expectations, the influence of MoG on performance appears insignificant, often leading to a deterioration in the model's efficacy. We have observed that smaller values of β tend to yield superior model performance, whereas higher values may degrade efficacy. However, excessively reducing β results in a corresponding reduction in the variance of the embeddings, rendering them more deterministic. In the case of H-prob. embeddings, increasing the number of MC samples enhances model performance, particularly with higher values of β. This suggests that employing more MC samples provides a more accurate and less biased estimation of expectations. Conversely, we found that the number of MC samples has a less significant effect on the performance of Z-prob. embeddings. While Z-prob. embeddings generally outperform H-prob. embeddings in terms of accuracy, the H-prob. embeddings offer better calibration, measured through the ECE.

5   Conclusions

In our study, we make significant strides in advancing the field by integrating stochastic assumptions directly into the information-theoretic-based self-supervised methods. Specifically, we introduce stochastic embeddings within feature decorrelation-based methods, demonstrating their potential to achieve performance competitive with the fully deterministic networks. Additionally, we delve into innovative strategies for effectively leveraging stochastic embeddings to identify OOD examples accurately. Our findings reveal that our methods exhibit robust OOD sample detection capabilities, surpassing traditional supervised detectors' performance. Moreover, we provide a comprehensive empirical evaluation, elucidating the impact of various hyperparameters on the training process. This showcases the potential of our approach and suggests avenues for future research in self-supervised learning optimization.

References

1. Achille, A., Soatto, S.: Emergence of invariance and disentanglement in deep representations. J. Mach. Learn. Res. **19**(1), 1947–1980 (2018)
2. Alemi, A.A., Fischer, I., Dillon, J.V., Murphy, K.: Deep variational information bottleneck. arXiv:1612.00410 [cs, math], October 2019
3. Ardeshir, S., Azizan, N.: Uncertainty in contrastive learning: on the predictability of downstream performance. arXiv:2207.09336 [cs, eess, stat], July 2022
4. Bardes, A., Ponce, J., LeCun, Y.: VICReg: variance-invariance-covariance regularization for self-supervised learning. arXiv:2105.04906 [cs], January 2022
5. Bordes, F., Balestriero, R., Garrido, Q., Bardes, A., Vincent, P.: Guillotine regularization: why removing layers is needed to improve generalization in self-supervised learning. arXiv preprint arXiv:2206.13378 (2022)
6. Caron, M., Misra, I., Mairal, J., Goyal, P., Bojanowski, P., Joulin, A.: Unsupervised learning of visual features by contrasting cluster assignments. Adv. Neural. Inf. Process. Syst. **33**, 9912–9924 (2020)

7. Chen, T., Kornblith, S., Norouzi, M., Hinton, G.: A simple framework for contrastive learning of visual representations. In: International Conference on Machine Learning, pp. 1597–1607. PMLR (2020)

8. Chen, X., He, K.: Exploring simple Siamese representation learning. In: Proceedings of the IEEE/CVF Conference on Computer Vision and Pattern Recognition, pp. 15750–15758 (2021)

9. Cimpoi, M., Maji, S., Kokkinos, I., Mohamed, S., Vedaldi, A.: Describing textures in the wild. In: Proceedings of the IEEE Conference on Computer Vision and Pattern Recognition, pp. 3606–3613 (2014)

10. Federici, M., Dutta, A., Forré, P., Kushman, N., Akata, Z.: Learning robust representations via multi-view information bottleneck. arXiv:2002.07017 [cs, stat], February 2020

11. Gawlikowski, J., et al.: A survey of uncertainty in deep neural networks. Artif. Intell. Rev. **56**, 1513–1589 (2021)

12. Goyal, P., Mahajan, D., Gupta, A., Misra, I.: Scaling and benchmarking self-supervised visual representation learning. In: Proceedings of the IEEE/CVF International Conference on Computer Vision, pp. 6391–6400 (2019)

13. Grill, J.B., et al.: Bootstrap your own latent-a new approach to self-supervised learning. Adv. Neural. Inf. Process. Syst. **33**, 21271–21284 (2020)

14. He, K., Zhang, X., Ren, S., Sun, J.: Deep residual learning for image recognition. In: Proceedings of the IEEE Conference on Computer Vision and Pattern Recognition, pp. 770–778 (2016)

15. Hendrycks, D., Gimpel, K.: A baseline for detecting misclassified and out-of-distribution examples in neural networks. arXiv:1610.02136 [cs], October 2018

16. Hendrycks, D., Mazeika, M., Kadavath, S., Song, D.: Using self-supervised learning can improve model robustness and uncertainty. In: Advances in Neural Information Processing Systems, vol. 32 (2019)

17. Hjelm, R.D., et al.: Learning deep representations by mutual information estimation and maximization. arXiv:1808.06670 [cs, stat], February 2019

18. Kingma, D.P., Welling, M.: Auto-encoding variational Bayes. CoRR abs/1312.6114 (2013)

19. Kirchhof, M., Kasneci, E., Oh, S.J.: Probabilistic contrastive learning recovers the correct aleatoric uncertainty of ambiguous inputs. In: Proceedings of the 40th International Conference on Machine Learning. JMLR.org (2023)

20. Krizhevsky, A.: Learning Multiple Layers of Features from Tiny Images (2009)

21. Le, Y., Yang, X.S.: Tiny ImageNet Visual Recognition Challenge (2015)

22. LeCun, Y., Cortes, C., Burges, C.: The MNIST database of handwritten digits (1998)

23. Lee, K., Lee, K., Lee, H., Shin, J.: A simple unified framework for detecting out-of-distribution samples and adversarial attacks. In: Advances in Neural Information Processing Systems, vol. 31 (2018)

24. Liang, S., Li, Y., Srikant, R.: Enhancing the reliability of out-of-distribution image detection in neural networks. arXiv:1706.02690 [cs, stat], August 2017

25. Loshchilov, I., Hutter, F.: Decoupled weight decay regularization. arXiv:1711.05101 [cs, math], January 2019

26. Nakamura, H., Okada, M., Taniguchi, T.: Representation uncertainty in self-supervised learning as variational inference. In: Proceedings of the IEEE/CVF International Conference on Computer Vision, pp. 16484–16493 (2023)

27. Netzer, Y., Wang, T., Coates, A., Bissacco, A., Wu, B., Ng, A.Y.: Reading digits in natural images with unsupervised feature learning. In: NIPS Workshop on Deep Learning and Unsupervised Feature Learning (2011)

28. Nilsback, M.E., Zisserman, A.: Automated flower classification over a large number of classes. In: 2008 Sixth Indian Conference on Computer Vision, Graphics & Image Processing, pp. 722–729, December 2008

29. Ozbulak, U., et al.: Know your self-supervised learning: a survey on image-based generative and discriminative training. arXiv:2305.13689 [cs], May 2023

30. Russakovsky, O., et al.: ImageNet large scale visual recognition challenge. arXiv:1409.0575 [cs], January 2015

31. Sastry, C.S., Oore, S.: Detecting out-of-distribution examples with gram matrices. In: Proceedings of the 37th International Conference on Machine Learning, pp. 8491–8501. PMLR, November 2020

32. Sehwag, V., Chiang, M., Mittal, P.: SSD: a unified framework for self-supervised outlier detection. arXiv preprint arXiv:2103.12051 (2021)

33. Shwartz-Ziv, R., Balestriero, R., Kawaguchi, K., Rudner, T.G.J., LeCun, Y.: An information-theoretic perspective on variance-invariance-covariance regularization. arXiv:2303.00633 [cs, math], March 2023

34. Shwartz-Ziv, R., LeCun, Y.: To compress or not to compress—self-supervised learning and information theory: a review. arXiv:2304.09355 [cs, math] (2023)

35. Tack, J., Mo, S., Jeong, J., Shin, J.: CSI: novelty detection via contrastive learning on distributionally shifted instances. Adv. Neural. Inf. Process. Syst. **33**, 11839–11852 (2020)

36. Wang, T., Isola, P.: Understanding contrastive representation learning through alignment and uniformity on the hypersphere. In: International Conference on Machine Learning, pp. 9929–9939. PMLR (2020)

37. Xiao, J., Hays, J., Ehinger, K.A., Oliva, A., Torralba, A.: SUN database: large-scale scene recognition from abbey to zoo. In: Computer Society Conference on Computer Vision and Pattern Recognition, pp. 3485–3492. IEEE, June 2010

38. Yang, J., et al.: OpenOOD: benchmarking generalized out-of-distribution detection. Adv. Neural. Inf. Process. Syst. **35**, 32598–32611 (2022)

39. Yang, J., Zhou, K., Li, Y., Liu, Z.: Generalized out-of-distribution detection: a survey. arXiv:2110.11334 [cs], January 2024

40. Zbontar, J., Jing, L., Misra, I., LeCun, Y., Deny, S.: Barlow Twins: self-supervised learning via redundancy reduction. In: International Conference on Machine Learning, pp. 12310–12320. PMLR (2021)

41. Zhou, B., Lapedriza, A., Khosla, A., Oliva, A., Torralba, A.: Places: a 10 million image database for scene recognition. IEEE Trans. Pattern Anal. Mach. Intell. **40**(6), 1452–1464 (2017)

Enhancing the Parallel UC2B Framework: Approach Validation and Scalability Study

Zineb Ziani[1,2]([✉]) [ID], Nahid Emad[1], Miwako Tsuji[3], and Mitsuhisa Sato[3]

[1] University of Paris Saclay/Li-PaRAD/MDLS, Paris, France
ziani.zineb.zz@gmail.com
[2] Numeryx, 17 Rue Jeanne Braconnier, 92360 Meudon, France
[3] RIKEN Center for Computational Science, Kobe, Hyogo 650-0047, Japan

Abstract. Anomaly detection is a critical aspect of uncovering unusual patterns in data analysis. This involves distinguishing between normal patterns and abnormal ones, which inherently involves uncertainty. This paper presents an enhanced version of the parallel UC2B framework for anomaly detection, previously introduced in a different context. In this work, we present an extension of the framework and present its large-scale evaluation on the Supercomputer Fugaku. The focus is on assessing its scalability by leveraging a great number of nodes to process large-scale datasets within the cybersecurity domain, using the UNSW-NB15 dataset. The ensemble learning techniques and inherent parallelizability of the Unite and Conquer approach are highlighted as key components, contributing to the framework's computational efficiency, scalability, and accuracy. This study expands upon the framework's capabilities and emphasizes its potential integration into an existing Security Orchestration, Automation, and Response (SOAR) system for enhancing cyber threat detection and response.

Keywords: Anomaly Detection · Linear Algebra · Unite and Conquer Approach · Machine Learning · High performance computing · Ensemble learning · Uncertainties · UC2B · UCEL · Cybersecurity

1 Introduction

Anomaly detection is a crucial element in data analysis and has gained widespread recognition for its ability to identify patterns or behaviors significantly deviating from normal or expected observations, with diverse applications across various domains [7].

In the field of finance, anomaly detection plays a pivotal role in detecting fraud and suspicious financial activities [4]. By identifying unusual transactions, atypical spending patterns, or fraudulent behaviors, anomaly detection contributes to fortifying the security and protection of financial assets. Likewise, in manufacturing, it is employed to monitor production processes, identifying failures or unexpected variations to enhance product quality, optimize operations,

L. Franco et al. (Eds.): ICCS 2024, LNCS 14838, pp. 352–366, 2024.
https://doi.org/10.1007/978-3-031-63783-4_26

and minimize downtime [24]. Additionally, in healthcare, anomaly detection aids in detecting unusual symptoms, identifying rare diseases, and analyzing medical images for precise diagnoses and timely interventions [29].

In cybersecurity, anomaly detection serves as a vital tool for safeguarding computer systems against malicious attacks [26]. It plays a crucial role in real-time threat detection and prevention by identifying abnormal behaviors on networks, data breaches, hacking activities, and intrusion attempts, thereby enhancing system security and minimizing the impact of cyber threats.

Despite advancements in anomaly detection driven by large-scale datasets and sophisticated machine learning algorithms, challenges persist due to the increasing complexity and size of modern datasets [12]. Efficient processing and analysis of data require substantial computational power, with deep learning models relying on high-performance GPUs or specialized hardware accelerators for training and fine-tuning. Real-time anomaly detection, on the other hand, demands rapid analysis of incoming data streams, necessitating the processing speed and scalability of modern hardware. Organizations must invest in powerful computational resources to fully harness the potential of these advanced techniques for effective anomaly detection in complex data.

To address these challenges, we have developed the parallel UC2B Framework in [30], an acronym for Unite and Conquer with Bagging and Boosting. Unite and Conquer is an iterative method that involves making several iterative methods collaborate by sharing their information. Bagging involves training multiple models on different data subsets and combining their predictions, and boosting improves a model's accuracy by emphasizing misclassified examples [25]. In previous experiments utilizing the parallel UC2B framework for anomaly detection, specifically on the smallest data set of UNSW-NB15, notable efficiency was demonstrated, achieving a detection rate ranging from 97% to 99% [30]. This framework leverages the Unite and Conquer methodology, integrating Bagging and Boosting techniques to primarily enhance prediction accuracy. However, these experiments have brought to light scalability concerns, particularly in managing the computational demands of extensive datasets. The number of nodes is confined by the number of co-methods, as each co-method necessitates a dedicated node for training. This interdependency introduces inefficiencies, affecting both memory usage and computational complexity. Addressing these challenges is crucial for the seamless implementation of the synchronous version of Parallel UC2B in real-world anomaly detection solutions.

In this paper, we introduce an enhanced version of the UC2B framework, a parallel anomaly detection system designed for cybersecurity threat detection. Conducting experiments on the Supercomputer Fugaku with up to 40 nodes, our versatile framework adeptly addresses diverse cybersecurity threats, with a specific emphasis on analyzing the biggest dataset of UNSW-NB15, which comprises over 2.5 million samples [18]. The Parallel UC2B extension integrates multi-level parallelism and double bagging, significantly enhancing processing efficiency. Our objectives encompass advancing the framework, fortifying defenses against emerging cyber threats, and facilitating its integration into existing SOAR systems. We validate its high accuracy, assess its effectiveness through confidence score calculations, and study its scalability under both weak and strong loads, including its behavior with larger databases. Key elements of our work include:

- Enhanced exploration of an optimized configuration incorporating multi-level parallelism, a fusion of double bagging and boosting, complemented by a restarting strategy inspired by the unite and conquer method for anomaly detection.
- Integration of a diverse array of components, encompassing various ML models, with inherent load balancing potential, distributed computation capabilities, and a fault-tolerant implementation strategy.
- Adoption of both model parallelism and data parallelism.
- Implementation of a parallel framework designed to harness the performance capabilities of high-performance computing architectures.
- Validation of the framework's efficacy through a series of experiments executed on the supercomputer Fugaku using 40 nodes.
- Specialized focus on the application of the framework within the cybersecurity domain, leveraging the UNSW-NB15 dataset for evaluation.
- Integrating a robust uncertainty metric deepens predictive insights, bolstering confidence in model performance and decision-making.

2   State of the Art

The state-of-the-art in anomaly detection within the field of cybersecurity has been advancing rapidly in recent years. Numerous studies and approaches have been proposed to address the challenge of detecting unusual and potentially harmful behavior in computer systems and networks. Some machine learning-based techniques applied to anomaly detection, including Bagging (which involves training multiple models on different data subsets and combining their predictions) and Boosting methods (that improve a model's accuracy by emphasizing misclassified examples [5]), run alongside spectral calculations [16] that involve analyzing eigenvalue and eigenvector values.

More recently, Diop et al. applied the Unite and Conquer approach [11] used in linear algebra to ensemble learning. The resulting technique, called UCEL, iteratively boosts a set of methods that work like bagging, and iterations of this boosting continue until the desired accuracy is achieved [8,9]. This extended method shows improved performance. Another combination of these techniques was presented in the article [30] to improve the results of UCEL in terms of detection rate.

Moreover, there have been significant efforts in evaluating these methods and comparing their performance on various data sets, including the widely recognized UNSW-NB15 data set [18]. The UNSW-NB15 data set, with its large number of simulated network traffic instances, is commonly used for evaluating the performance of anomaly detection algorithms in a realistic setting. It contains a wide range of attack types and is characterized by its high volume and high dimensionality, making it a challenging data set for anomaly detection algorithms.

In addition to the previously mentioned Bagging and Boosting methods and spectral calculations, other notable methods include Variational Autoencoders (VAE), which learn a probabilistic representation of normal data and identify anomalies based on the reconstruction probability [3]. Generative Adversarial Networks (GAN) have been applied to anomaly detection, where a generator reproduces normal data and a discriminator distinguishes between real and generated data [2]. Hidden Markov Models (HMM) have been employed for anomaly detection, extending the one-class support vector machine (SVM), by leveraging latent dependency structures [13]. The approach achieves superior anomaly detection performance compared to traditional one-class SVM, as demonstrated through empirical evaluations on diverse datasets in computational biology and computational sustainability domains. Recurrent Neural Networks (RNN), such as LSTM, have been effective in capturing sequential dependencies for anomaly detection in time series data [17]. These methods, along with preprocessing techniques for feature selection and data normalization, have contributed to the advancement of anomaly detection in cybersecurity.

As the application of anomaly detection techniques expands beyond the cybersecurity domain, researchers are actively exploring their adaptability to various specific application fields. This progression is exemplified by recent studies proposing innovative approaches to address real-time monitoring challenges in complex systems.

To solve the problem of real-time monitoring of the signals produced by the accelerators, a fault detection method is proposed in [14]. This method, based on data from the beam position monitoring system, can identify anomalies in SLAC's radio frequency (RF) stations and detect more events while reducing false positives compared to diagnostics of existing RF stations.

Moreover, the method CoAD proposed in [15], trains anomaly detection models on unlabeled data, based on the expectation that anomalous behavior in one sub-system will produce coincident anomalies in downstream sub-systems.

Furthermore, the lack of structured parallel implementation in anomaly detection poses a significant challenge for the field [12]. Anomaly detection algorithms often involve complex computations and deal with large datasets, making them computationally demanding. While parallel computing has the potential to accelerate these tasks by distributing the workload across multiple processing units, achieving efficient parallel implementations is not straightforward [6,23]. Many anomaly detection methods are not inherently parallelizable due to their sequential nature and data dependencies, requiring substantial modifications for parallel processing. Load imbalance among processing units, caused by the irregularity of anomaly occurrence in data, further complicates the parallelization process. Additionally, the absence of standardized parallel frameworks tailored explicitly for anomaly detection hinders progress [10]. To address these issues, focused research, collaboration between anomaly detection and parallel computing experts, and the development of specialized parallel frameworks are essential to unlock the benefits of parallel computing in advancing anomaly detection capabilities.

3   Software Architectures of Enhanced Parallel UC2B

In various scientific disciplines, the escalating data generation surpasses computational capacities, compelling the integration of modeling, analysis, and high-performance computing [21]. These challenges, spanning diverse fields, are rooted in applied mathematics, including linear algebra and statistics, alongside artificial intelligence, which encompasses machine learning methods and high-performance computing techniques. Within cybersecurity, the evolving subtlety of security breaches extends investigation times, demanding a discerning approach to distinguish authentic alerts from false alarms. Expertise and timely validation of 'false alerts' are crucial in a Security Operations Center (SOC) [20]. This undertaking seeks to contribute to the resolution of these challenges, exemplified through practical applications of data analysis in securing information systems within organizations, such as advanced technology enterprises.

As outlined in the state-of-the-art section, the application of the Unite and Conquer approach to Ensemble Learning methods, is another anomaly detection technique proposed by Diop et al. in [8,9], called UCEL. In this paper, we propose an enhanced version of UCEL which improves its performance. To distinguish this extension from UCEL, we call it Parallel UC2B for Unite and Conquer with Bagging-Boosting. The presence of several levels of boosting as well as that of multi-level intrinsic parallelism in UC2B partly explain its better performance relative to UCEL, in addition to a double bagging. Other characteristics such as the heterogeneity of its components, its fault tolerance as well as its potential for load balancing make UC2B a technique very well suited to recent parallel and/or distributed architectures.

3.1   Unite and Conquer Approach

"Unite and Conquer" is a problem-solving paradigm that orchestrates multiple iterative methods, or co-methods, to collectively address complex problems, particularly in linear algebra [11]. Applied in resolving expansive, sparsely populated linear systems and eigenvalue predicaments, this approach accelerates convergence by aggregating intermediate outcomes from each co-method. The strategic restarting approach plays a pivotal role in providing a better starting point for each new cycle, enhancing overall convergence. Co-methods exchange intermediate solutions to determine effective restarting conditions, resulting in swifter global convergence. With intrinsic advantages like multi-level parallelism, robust fault tolerance, adaptability to component heterogeneity, asynchronous communication capabilities, and inherent load balancing potential, the "Unite and Conquer" approach is well-suited for cutting-edge computational architectures. It optimally allocates computational resources, enhancing efficiency and parallel processing benefits, accelerating problem resolution, and maximizing resource utilization.

The Unite and Conquer algorithm can be expressed in a mathematical form as the following. Let P be the large linear algebra problem to be solved, $L_1, L_2, ..., L_l$ be a set of iterative methods that can solve P, I_i^k the the initial condition (with $k = 0$) and restarting condition (with $k > 0$) of L_i, and θ be the threshold value. Let f be a function defining the restarting strategy according to the intermediate results $(S_1^k, ..., S_\ell^k)$ with S_i^k the approximated solution obtained by L_i at the end of i-th iteration/cycle. An algorithm of Unite and Conquer can be defined as follows:

Algorithm 1. Unite and Conquer Algorithm

Initialize Choose a starting matrix $[I_1^0, \ldots, I_\ell^0]$, let $k = 0$.
For $i = 1$ to ℓ **do in parallel**
    Compute S_i^k by applying L_i to P with initial condition I_i^k.
    If S_i^k is sufficiently accurate, STOP all ℓ process and return S_i^k as the solution of P.
    Share S_i^k information with all other processes j ($j = 1, \ldots, \ell$ and $j \neq i$).
Update and Restart $[I_1^{k+1}, \ldots, I_l^{k+1}] = f(S_1^k, \ldots, S_l^k)$ and increment k.

Essentially, this approach boasts a simple yet versatile framework applicable to various iterative methods, as exemplified in this paper. We specifically explore integrating boosting techniques within bagging methodologies, introducing a second level of parallelism to enhance the model's adaptability and performance across diverse datasets and scenarios.

3.2   Parallel UC2B Insights

The Parallel UC2B framework aims to enhance anomaly detection accuracy and efficiency by integrating the Unite and Conquer problem-solving approach with Bagging, Boosting, and multi-level parallelism. The objective is to iteratively improve accuracy, ensure high confidence scores, and expedite anomaly identification. Collaboration among parallel co-methods refines their performance through multiple training cycles, culminating in a convergence state with substantial and stable improvements. Each co-method undergoes parallel training in inner bags of the dataset, emphasizing a multi-level parallelism approach.

In light of the constraint that LM models in scikit-learn cannot be trained on multiple nodes, we adopt a double bagging approach. This involves partitioning the database into outer bags through 'Node-based dataset partitioning' (cf. Fig. 1), ensuring the number of outer bags aligns with the total number of nodes divided by the number of co-methods (ML models). Subsequently, we thoroughly evaluate co-method performance using a validation set. Co-methods share their misclassified data (False Positives/Negatives), incorporating the boosting principle to adjust weights for misclassified samples during iterations based on co-method performance metrics. This iterative process heightens the likelihood of selecting crucial samples for constructing training data in subsequent cycles.

Resulting in inner bags of the original training data size from the boosted training dataset, this collaborative process allows each co-method to learn from its peers and gain insights into challenging data samples. The joint effort contributes to the gradual refinement and improvement of the models.

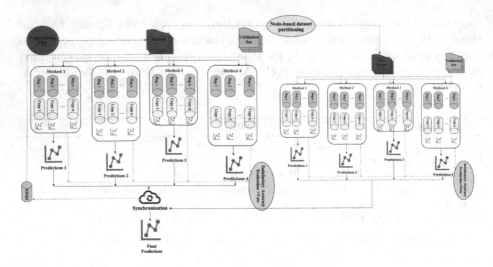

Fig. 1. Enhanced Parallel UC2B Architecture. (Color figure online)

The goal of Parallel UC2B is versatility, addressing a wide spectrum of attacks, whether internal or external, and anomalies, while maintaining reasonable execution time for practical deployment. In tackling the challenge of detecting sophisticated threats and anomalies, we seek to leverage insights from each co-method. Given that UC approach learns the underlying global structure of data, we provide the entire dataset to all methods in parallel. Each co-method then creates duplicates of itself (depicted by yellow cylinders in Fig. 1) and segments the dataset into multiple bags (illustrated by orange cylinders in Fig. 1), training each copy of the co-method on a bag. This approach ensures that each co-method learns from the entirety of the dataset and collaborates synchronously with the other co-methods by sharing their outputs, updating the weights of misclassified instances (FP/FN chunk), and checking if satisfactory accuracy has been achieved by testing on the validation dataset (purple arrows). In contrast, in UCEL [9], the dataset is divided into bags, with each bag exclusively assigned to a single co-method, limiting the number of bags. In Parallel UC2B, the number of bags is independent of the co-methods, providing flexibility with "n" bags for each method.

In Unite and Conquer, our focus is primarily on synchronous communications among co-methods, with asynchronous communications also accommodated. The collaborative mechanism integrates bagging and boosting techniques for diverse data treatment, effectively balancing bias mitigation and variance management. The training process incorporates feedback from all co-methods, addressing performance metrics and instances of FP/FN. Furthermore, inherent parallelism optimally utilizes computational resources, enhancing efficiency, especially in 'parallel UC2B,' where thread Parallelism and SIMD operations drive concurrent task handling, data processing, and collaborative sharing among co-methods. This streamlined approach, complemented by efficient data Input/Output, ensures timely information exchange, supports boosting mechanisms, and yields significant performance gains.

3.3   Algorithm of Enhanced Parallel UC2B

In the realm of machine learning, the choice of data analysis methods hinges on the nature of available information, whether it's labeled, unlabeled, or imbalanced. In corporate environments, routine activities prevail, leading to datasets predominantly skewed towards normal behavior. The prevalence of normal data introduces challenges for anomaly detection.

In this implementation, we employ Gaussian Naive Bayes (GNB), Isolation Forest (IForest), Decision Tree (DT), and Random Forest (RF) as co-methods in the Parallel UC2B framework. This collaborative approach addresses challenges faced by traditional supervised and unsupervised methods when dealing with limited abnormal examples. Unsupervised techniques, adept at handling imbalanced data, primarily focus on identifying deviations without delving into their underlying causes. In contrast, supervised methods excel in scenarios with balanced and labeled datasets, but achieving such balance is often impractical in real-world applications.

Algorithm 2. Enhanced Parallel UC2B

---

1 **Input:**
2 Data set D.
3 Number of bags I.
4 Number of all process iterations n.
5 Number of learners M.
6 Sample weights W initialized to ones.
7 **for** $i \leftarrow 1$ to n **do:**
8   **for** $j \leftarrow 1$ to M **do in parallel:**
9   **for** $k \leftarrow 1$ to I **do in parallel:**
10             $B_k \leftarrow$ Bags Bootstrap sample from D with replacement.
11             $y_k \leftarrow$ Vector label issued L_j training on the bags B_k.
12   Predictions[j] $\leftarrow$ Prediction using y_k.
13   Calculate misclassification rates using Predictions and true labels.
14             $\beta \leftarrow 1.1 \times$ misclassified $+ 0.9 \times$ classified
15   **Sync** and **Share** β and the results with all other processes.
16   **Check** for desired accuracy; if met, stop all processes and exit.
17 **Restart by Updating** the input data with adjusted weights for the next iteration:
18 $W = W \times \beta$
19 $D \leftarrow$ Updated D with adjusted sample weights W.
20 **Output:**
21 Obtain the boosted predictions after the desired iterations.

Our proposed approach begins with meticulous pre-processing of a dataset containing more than 2.5 million samples. This includes crucial steps such as data cleaning, feature selection, as well as scaling and normalization procedures. Subsequently, from this refined dataset, distinct sets for training, validation, and testing are carefully curated.

Following the node-based dataset partitioning, each model is trained in parallel to the others in inner parallel bags. Based on the predictions obtained, the coefficient β is calculated using the formula $\beta = 1.1 \times$ misclassified $+ 0.9 \times$ classified. This coefficient is then used to update the instances that were misclassified for a subsequent boosted iteration.

4   Experiments and Analysis

In this section, we present results from our experiments on the Supercomputer Fugaku, utilizing 40 nodes for assessment. We'll explore the Fugaku hardware specifics to align with our implementation settings, followed by validation in the first subsection and performance demonstration in the second.

Fugaku is a supercomputer that boasts a highly advanced hardware architecture, positioning it as the most powerful supercomputer in the world in 2020 and 2021 [27,28], and it is currently ranked as the number 2 supercomputer in 2023 [1]. It incorporates a state-of-the-art hardware design aimed at delivering exceptional performance and efficiency [19]. At its core, Fugaku utilizes the A64FX processor [22], which is based on the ARM architecture. Each A64FX chip comprises 48 computing cores, and each core is equipped with two 512-bit wide SIMD units. Powered by the A64FX chip, incorporates high-bandwidth memory (HBM2) modules, delivering substantial capacity and impressive bandwidth. To facilitate swift data transfer and node communication, Fugaku utilizes the custom-designed Tofu-D interconnect system. This network, based on a 6-dimensional mesh/torus topology, ensures efficient and low-latency interactions between nodes, enabling seamless data exchange and synchronization during parallel computations.

4.1   Validation of the Approach

The goal of this validation is to showcase that the Parallel UC2B approach achieves high accuracy with a robust confidence score.

We initiate our experiments by displaying the accuracy obtained on the training set during the UC iterations. As a reminder, the UC iterations involve the re-injection of False Positives (FP) and False Negatives (FN), updating misclassified instances through the β factor. Subsequently, we plot the curve obtained on the test dataset, which has never been seen by the framework.

Accuracy is chosen for performance evaluation in our context as it represents correct predictions relative to the total sample count. As depicted in Fig. 2, illustrating accuracies obtained in the training set, IForest emerges as the weakest among the co-methods. While GNB, Decision Tree, and Random Forest exhibit high accuracies, they lack stability. In contrast, Parallel UC2B demonstrates convergence over UC iterations, achieving the highest stable accuracy.

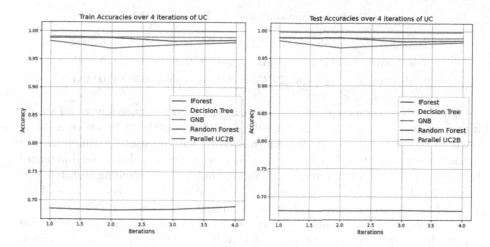

Fig. 2. Train & Test Accuracies: 4 Co-methods & Parallel UC2B (4 UC Iterations)

Transitioning to the test set graph, a similar pattern emerges. Other models display good accuracy but lack stability, while Parallel UC2B maintains its superiority in accuracy and stability even on data unseen by the framework.

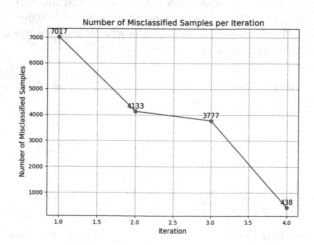

Fig. 3. The reduction of misclassified instances throughout the UC iterations.

This graph Fig. 3 illustrates the reduction of misclassified instances during the UC iterations on a node. As the data is divided based on nodes, this graph visually depicts the decrease in the number of misclassified instances, attributed to the β factor updating these instances in each iteration (Table 1).

To measure the model's effectiveness, we employed the formula:

$$Confidence\ Score = 1 - |Train\ Accuracy - Test\ Accuracy|$$

Table 1. Score confidence of Parallel UC2B

Train Accuracy of Parallel UC2B	0.9999	0.9999	0.9999	0.9999
Test Accuracy of Parallel UC2B	0.9979	0.9979	0.9979	0.9980
Confidence Score of Parallel UC2B	0.9980	0.9980	0.9980	0.9981

This score serves as a metric for assessing the model's consistency and stability across both training and test datasets. A higher Confidence Score, converging towards 1, indicates comparable performance on both datasets, highlighting the model's robustness. Conversely, a score closer to 0 suggests substantial differences between training and test data performances, potentially signaling instability or overfitting. In our specific case, the confidence scores for various iterations of the Parallel UC2B model consistently hover around 0.998. This steadfastness underscores a robust alignment between accuracy on training and test data, affirming the model's capability to generalize effectively to new data, which is a critical characteristic for model reliability.

4.2 Performance Demonstration of the Approach

After validating the accuracy of the approach, achieving a 99% accuracy on previously unseen test data and obtaining a very high confidence score, which improves upon the results obtained with the UCEL [8] and the previous version of Parallel UC2B [30], we will now focus on studying and evaluating the scalability of the framework, both strong and weak, along with its speedup and accuracy stability across different data sizes.

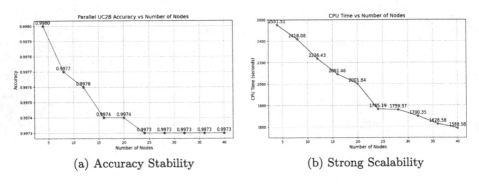

(a) Accuracy Stability   (b) Strong Scalability

Fig. 4. Accuracy Stability and Strong Scalability with Fixed Data-size

The graph Fig. 4a illustrates the accuracy achieved on a dataset comprising over 2.5 million samples as the number of nodes increases. Remarkably, the accuracy remains stable, showing no more than a 0.007% degradation from 4 to 24 nodes. Beyond 24 nodes, the accuracy plateaus at 0.9973, persisting even up to 40 nodes. This suggests that scaling the framework with additional nodes has negligible impact on accuracy.

As for graph Fig. 4b, it depicts the study of strong scalability, where we maintain a fixed problem size, increase the number of nodes, and evaluate speedup. In our case, the problem size exceeds 2.5 million, and even though the speed from 4 to 40 nodes doesn't quite double, it shows almost linear scalability up to 20 nodes. For node 24, we observe a speedup that deviates from linearity, followed by a near-linear recovery at node 28. The speed increase, while not doubling or more, can be attributed to synchronous communications between co-methods. Scaling from 4 to 40 nodes increases the number of communications tenfold. Additionally, the β factor, updating misclassified samples, is implemented in a way that each co-method needs to receive β for the entire 2.5 million data samples. This explains the suboptimal speedup obtained in this figure.

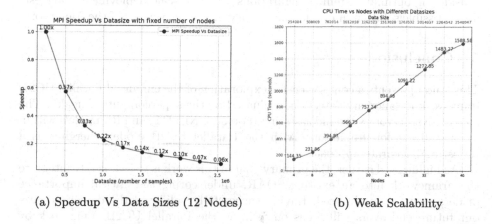

(a) Speedup Vs Data Sizes (12 Nodes)   (b) Weak Scalability

Fig. 5. Speedup with Fixed and Various Nodes and Various Data-sizes

The Fig. 5a illustrates the behavior of speedup using a fixed number of nodes, 12 in our case, while increasing the database size. We observe a degradation in speedup as the database size grows, which is expected given the constant number of nodes. However, starting from a database size of 1 million samples, we notice that the speedup does not degrade significantly. Interestingly, the execution time for 2.3 million samples is nearly the same as that for 2.6 million, indicating that the framework can effectively handle very large databases.

The graph in Fig. 5b illustrates weak scalability, which involves increasing the problem size in proportion to the addition of nodes. Ideally, for this experiment, the execution time would remain constant, as the increase in the database size is accompanied by a corresponding increase in the number of nodes. However, due to the same phenomenon explained in strong scalability, when the number of nodes increases, the number of synchronous communications also increases with the growth of the database size, leading to an increase in the size of β. Nevertheless, the obtained curve is almost linear, indicating that the execution time is proportional to the addition of nodes and datasize.

In summary of the interpreted results, the framework stands out for its robust detection capability, marked by a high confidence score. Regarding its scalability performance, the framework exhibits remarkable adaptability to large-scale databases, maintaining stable accuracy even with an increased number of nodes. While strong scalability shows a proportional trend in some sections, weak scalability displays an almost linear trajectory. However, it is important to note an impact on speed performance. As the database expands, synchronous communications and the size of β increase, contributing to a gradual rise in execution time. This observation underscores the delicate balance between expanding computational resources and addressing challenges related to increased inter-node communication. The synchronous implementation of the framework is identified as the source of these observations. These findings highlight the importance of considering optimal system configurations for large-scale deployments, suggesting a possible solution in developing the asynchronous version of the framework.

5 Conclusion

This study presents a comprehensive exploration of the enhanced Parallel UC2B framework for anomaly detection, evaluated on the supercomputer Fugaku. The core analysis revolves around assessing IForest, GNB, DT, and RF models within the parallelization framework, with the Parallel UC2B model emerging as a robust and accurate approach.

As highlighted earlier, the primary goal of this work is to seamlessly integrate this framework into an existing SOAR, underscoring the critical importance of detection rate and speed. Having successfully validated the detection rate, our future endeavors will focus on refining the Parallel UC2B framework by incorporating asynchronous communication capabilities. This enhancement aims to further reduce execution time while leveraging more nodes for superior strong and weak scalability. Additionally, we plan to conduct extensive testing across diverse applications such as healthcare and finance.

Acknowledgment. This research used computational resources of the supercomputer Fugaku provided by the RIKEN Center for Computational Science. We sincerely thank Research engineer Martial Mancip for his kind assistance, which greatly aided our study.

References

1. Supercomputer Fugaku - Supercomputer Fugaku, A64FX 48C 2.2 GHz, Tofu Interconnect D. https://www.top500.org/system/179807/
2. Akcay, S., Atapour-Abarghouei, A., Breckon, T.P.: GANomaly: semi-supervised anomaly detection via adversarial training. In: Jawahar, C., Li, H., Mori, G., Schindler, K. (eds.) Computer Vision–ACCV 2018: 14th Asian Conference on Computer Vision, Perth, Australia, 2–6 December 2018, Revised Selected Papers, Part III 14. LNCS, Vol. 11363, pp. 622–637. Springer, Cham (2019). https://doi.org/10.1007/978-3-030-20893-6_39

3. An, J., Cho, S.: Variational autoencoder based anomaly detection using reconstruction probability. Special Lect. IE **2**(1), 1–18 (2015)

4. Anandakrishnan, A., Kumar, S., Statnikov, A., Faruquie, T., Xu, D.: Anomaly detection in finance: editors' introduction. In: KDD 2017 Workshop on Anomaly Detection in Finance, pp. 1–7. PMLR (2018)

5. Bukhari, O., Agarwal, P., Koundal, D., Zafar, S.: Anomaly detection using ensemble techniques for boosting the security of intrusion detection system. Procedia Comput. Sci. **218**, 1003–1013 (2023). https://doi.org/10.1016/j.procs.2023.01.080

6. Cappello, F., Geist, A., Gropp, W., Kale, S., Kramer, B., Snir, M.: Toward exascale resilience: 2014 update. Supercomputing Front. Innov. **1**(1), 5–28 (2014). https://doi.org/10.14529/jsfi140101, https://superfri.org/index.php/superfri/article/view/14

7. Chandola, V., Banerjee, A., Kumar, V.: Anomaly detection: a survey. ACM Comput. Surv. **41**(3), 1–58 (2009). https://doi.org/10.1145/1541880.1541882

8. Diop, A., Emad, N., Winter, T.: A parallel and scalable framework for insider threat detection. In: 27th IEEE International Conference on High Performance Computing, Data, and Analytics, HiPC 2020, Pune, India, 16–19 December 2020, pp. 101–110. IEEE (2020)

9. Diop, A., Emad, N., Winter, T.: A unite and conquer based ensemble learning method for user behavior modeling. In: 39th IEEE International Performance Computing and Communications Conference, IPCCC 2020, Austin, TX, USA, 6–8 November 2020, pp. 1–8. IEEE (2020)

10. Du, Q., Tang, B., Xie, W., Li, W.: Parallel and distributed computing for anomaly detection from hyperspectral remote sensing imagery. Proc. IEEE **109**(8), 1306–1319 (2021). https://doi.org/10.1109/JPROC.2021.3076455

11. Emad, N., Petiton, S.G.: Unite and conquer approach for high scale numerical computing. J. Comput. Sci. **14**, 5–14 (2016)

12. Ghiasvand, S., Ciorba, F.M.: Anomaly detection in high performance computers: a vicinity perspective. In: 2019 18th International Symposium on Parallel and Distributed Computing (ISPDC), pp. 112–120 (2019). https://doi.org/10.1109/ISPDC.2019.00024

13. Görnitz, N., Braun, M., Kloft, M.: Hidden Markov anomaly detection. In: International Conference on Machine Learning, pp. 1833–1842. PMLR (2015)

14. Humble, R., et al.: Beam-based RF station fault identification at the SLAC Linac coherent light source. Phys. Rev. Accel. Beams **25**, 122804 (2022). https://doi.org/10.1103/PhysRevAccelBeams.25.122804

15. Humble, R., Zhang, Z., O'Shea, F., Darve, E., Ratner, D.: Coincident learning for unsupervised anomaly detection (2023)

16. Komolafe, T., Quevedo, A.V., Sengupta, S., Woodall, W.H.: Statistical evaluation of spectral methods for anomaly detection in static networks. Netw. Sci. **7**(2), 238–267 (2019)

17. Malhotra, P., Vig, L., Shroff, G., Agarwal, P., et al.: Long short term memory networks for anomaly detection in time series. In: ESANN, vol. 2015, p. 89 (2015)

18. Moustafa, N., Slay, J.: The evaluation of network anomaly detection systems: statistical analysis of the UNSW-NB15 data set and the comparison with the KDD99 data set, pp. 1–14, January 2016. https://doi.org/10.1080/19393555.2015.1125974

19. Nakao, M., Ueno, K., Fujisawa, K., Kodama, Y., Sato, M.: Performance of the supercomputer Fugaku for breadth-first search in Graph500 benchmark. In: Chamberlain, B.L., Varbanescu, AL., Ltaief, H., Luszczek, P. (eds.) High Performance Computing: 36th International Conference. ISC High Performance 2021, Virtual Event, 24 June–2 July 2021, Proceedings, pp. 372–390. Springer, Heidelberg (2021). https://doi.org/10.1007/978-3-030-78713-4_20

20. Nehinbe, J.O.: A simple method for improving intrusion detections in corporate networks. In: Chamberlain, B.L., Varbanescu, AL., Ltaief, H., Luszczek, P. (eds.) Information Security and Digital Forensics: First International Conference, ISDF 2009, London, United Kingdom, 7–9 September 2009, Revised Selected Papers 1, pp. 111–122. Springer, Cham (2010). https://doi.org/10.1007/978-3-642-11530-1_13

21. Reed, D.A., Dongarra, J.: Exascale computing and big data. Commun. ACM **58**(7), 56–68 (2015)

22. Sato, M., et al.: Co-design for A64FX manycore processor and "Fugaku". In: SC20: International Conference for High Performance Computing, Networking, Storage and Analysis, pp. 1–15 (2020). https://doi.org/10.1109/SC41405.2020.00051

23. Shanbhag, S., Wolf, T.: Accurate anomaly detection through parallelism. IEEE Network **23**(1), 22–28 (2009). https://doi.org/10.1109/MNET.2009.4804320

24. Stojanovic, L., Dinic, M., Stojanovic, N., Stojadinovic, A.: Big-data-driven anomaly detection in industry (4.0): an approach and a case study. In: 2016 IEEE International Conference on Big Data (Big Data), pp. 1647–1652. IEEE (2016)

25. Syarif, I., Zaluska, E., Prugel-Bennett, A., Wills, G.: Application of bagging, boosting and stacking to intrusion detection. In: Perner, P. (eds.) Machine Learning and Data Mining in Pattern Recognition: 8th International Conference, MLDM 2012, Berlin, Germany, 13–20 July 2012, Proceedings 8, pp. 593–602. Springer, Cham (2012). https://doi.org/10.1007/978-3-642-31537-4_46

26. Ten, C.W., Hong, J., Liu, C.C.: Anomaly detection for cybersecurity of the substations. IEEE Trans. Smart Grid **2**(4), 865–873 (2011). https://doi.org/10.1109/TSG.2011.2159406

27. TOP500: Top500 list - November 2020 (2020). https://www.top500.org/lists/top500/2020/11/. Accessed 26 July 2023

28. TOP500: Top500 list - June 2021 (2021). https://www.top500.org/lists/top500/2021/06/. Accessed 26 July 2023

29. Ukil, A., Bandyoapdhyay, S., Puri, C., Pal, A.: IoT healthcare analytics: the importance of anomaly detection. In: 2016 IEEE 30th International Conference on Advanced Information Networking and Applications (AINA), pp. 994–997. IEEE (2016)

30. Zineb, Z., Nahid, E., Ahmed, B.: A novel approach to parallel anomaly detection: application in cybersecurity. In: 2023 IEEE International Conference on Big Data (BigData), pp. 3574–3583. IEEE (2023)

Towards Modelling and Simulation of Organisational Routines

Phub Namgay and David Johnson[⊠]

Department of Informatics and Media, Uppsala University,
Box 513, 751 20 Uppsala, Sweden
{phub.namgay,david.johnson}@im.uu.se

Abstract. Organisational routines are repetitive, recognisable patterns of interdependent action by human and digital actors to accomplish tasks. Routine Dynamics is a theoretical base that informs discussion and analysis of such routines. We note that there is a knowledge gap in the literature on organisational routines to consolidate the constructs and ontologies of routines into an abstract data model. In this paper, we design and implement a data model of routines using the Unified Modelling Language. We present a demonstration to illustrate our data model's use, and how one can then use instantiations of the model to analyse and simulate organisational routines based on real-world data. This example examines recurrent patterns of action inferred from data in the GitHub issue tracking system about the open-source software project, *scikit-learn*. Our study extends the theoretical/empirical understanding and knowledge base of Routine Dynamics by laying the groundwork towards examining organisational routines from a model-driven perspective that gives rise to simulating the dynamics of routines.

Keywords: organisational routines · Routine Dynamics · modelling and simulation

1 Introduction

Organisational routines are defined as '... repetitive, recognizable patterns of interdependent actions, carried out by multiple actors.' [1]. Routine Dynamics [1,2] is an increasingly applied theoretical perspective of organisational routines that account for recurrent patterns of interdependent action by sociotechnical actors in digital ecosystems [3]. It is a foundational theory for studying routines in workflows and work practices, particularly in the fields of organisational science and information systems research. Numerous studies on routines thus far, such as the effect of emergent routines in open-source software (OSS) development [4], have fostered the development of Routine Dynamics as a robust tool to investigate processes in sociotechnical phenomena. However, there has yet to be a consolidation and formalisation of Routine Dynamics' constructs and ontologies into a standard model of routines, and thus there lie uncertainties when

© The Author(s), under exclusive license to Springer Nature Switzerland AG 2024
L. Franco et al. (Eds.): ICCS 2024, LNCS 14838, pp. 367–379, 2024.
https://doi.org/10.1007/978-3-031-63783-4_27

dealing with related data. The application of Routine Dynamics to research and practice is typically done in an ad-hoc fashion, where different individual studies re-interpret and apply the theoretical concepts heterogeneously. As a consequence, comparisons of studies relating to organisational routines becomes cumbersome, and there are no agreed upon approaches to studying routines and related data.

The paper is structured as follows: In Sect. 2, we discuss background literature relating to organisational routines, and describe our motivations for the work reported in this paper. In Sect. 3, we provide an overview of our data model of Routine Dynamics. In Sect. 4, we demonstrate how we can instantiate the data model using a popular OSS project as the subject of study to perform analysis and simulations of routines. In Sect. 5, we discuss the opportunities and challenges of modelling routines, and finally conclude the paper in Sect. 6.

2   Background and Motivation

Routines are essential to organised work [1]. They are observed behaviours following recognised action patterns through multiple interdependent actors [30]. The theory underscores action over actors and does not distinguish between human and material agency. Feldman and Pentland [1] note, 'A pattern of action that occurs only once is not a routine.' Sociotechnical actors acquire knowledge [29] in enacting routines to accomplish tasks through learning from experience. Thus far, the literature on routines has primarily focused on theoretical and empirical studies. Hayes, Lee and Dourish [5] explored the enactment of processes and routines via a narrative network perspective [6] with narrative fragments as nodes bridged by an action [7]. The processual dynamics of routines have been studied to account for digitised processes through a combination of humans and material agents [8]. Studies on routines investigating an invoice processing system as 'patterns of action' suggest that the action patterns differ across organisations despite using the same systems [9]. Gaskin et al. [10] conducted a study examining and modelling the lexicons of sociomaterial routines using the sequence analysis. Table 1 summarises the key concepts from Routine Dynamics theory that capture routines by enabling or constraining the enactment of processes [8].

In our review of the literature, we do not find any evidence of research that has explored routines from the perspective of standardising the approaches to expressing data about routines for the purposes of modelling and simulation. Therefore, we set out to build a first model of Routine Dynamics.

3   Proposed Data Model for Routine Dynamics

Our proposed data model of Routine Dynamics draws on the concepts extracted from the routines literature summarised in Table 1, where we use an object-oriented design formalisation expressed in UML. Figure 1 shows our data model as a UML class diagram based on the concepts from Table 1 and how the concepts are interrelated.

Table 1. Concepts and definitions of organisational routines.

Concept	Definition
Routines	Recurrent, recognisable patterns of action enacted by multiple interdependent actors [1], such as routines to resolve issues on the GitHub issue tracking system
Actor	Human or digital agents that perform a task, which implies the agency of agents [1], such as different sociotechnical actors involved in the GitHub issue tracking system [11]
Action	Steps taken to accomplish a task and things actors do, such as different phases and iterations for resolving issues on GitHub
Action pattern	Steps to accomplish a task that repeats over time, such as repetitive action patterns to flag, report, and resolve issues on the GitHub issue tracking system
Activity	Basic unit of carrying out a single function by actors and produces an outcome in routines [10], and activities in routines vary in structure and dynamics
Ostensive	Abstract codified repetitive, recognisable patterns of action [9], which embody the structure of routines [1,4] and are implicit [7], such as codified standard operating procedure, manual or taken-for-granted norms that underpin actual performances of routine work (see [5])
Performative	Actual performance of routines by interdependent actors, which bring routines to life across spatiotemporal dimensions [1], such as human actors resolving issues on the GitHub issue tracking system
Generativity	Create or produce something new or adapt existing routines through the agency of sociotechnical actors in routines, and thus 'generative capacity' [7,12,13]
Intermediary	An actor involved in maintaining connections or relaying data or information with minimal performative power while being engaged in a task [13], such as an application programming interface (API) that is passively involved in facilitating connection and communication on GitHub
Mediator	An actor that transforms, translates, and modifies meaning in action [13] in a network of processes to accomplish tasks, such as a contributor on GitHub who is actively involved in analysing and resolving issues
Live routine	Repetitive, recognisable patterns of action enacted by humans [5,14] through an individual agency to improvise and learn from experience, such as manual processes to resolve issues on GitHub issue tracking system
Dead routine	Recurrent patterns of action emerging from artefacts that are characterised as 'rigid, mundane, mindless, and codified' [15], such as action patterns of algorithmic agents for resolving issues through event logs of user communities and contributors on GitHub issue tracking system

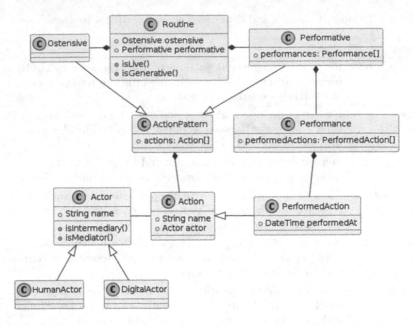

Fig. 1. UML class diagram of the routines data model, where concepts are drawn from literature as described in Table 1.

Routines embody a duality of structure and agency. An ostensive aspect embodies codified structure [5]. The `Routine` class is composed of both `Ostensive` and `Performative` classes. Routines can be guided by one or many ostensive facets and performative facets. For example, standard operating procedures, norms, and rules are ostensive features of routines as they evoke a simplified view of tasks in practice. The actual performance of routines by interdependent sociotechnical actors represents agency [1]. Ostensive and performative share a mutual recursive association [1,7], where the action patterns that compose of each influence each other in practice. Consequently, we model the `Ostensive` and `Performative` as extensions of an `ActionPattern`, where `ActionPatterns` are a composition of a sequence of `Actions`. `Actions` are carried out by `Actors`, which may be `HumanActors` (representing real people performing an action) or `DigitalActors` (representing a non-human agents performing an action, often a computer program). The `Performative` aspect concerns actual `Performances`, which are composed of a sequence of `PerformedActions`. One can distinguish an `Action` and `ActionPattern` from a `PerformedAction` and `Performance` by thinking of the former as prototypical patterns, while the latter are indicative of actual instances of a routine having been in action. Hence, `PerformedActions` have a timestamp while `Actions` do not. Performativity closely connects with generativity. Human and digital actors have an endogenous capacity to generate innovative outcomes, retain novel action patterns, or adapt action patterns [4,13]. The `Routine` class therefore contains the `isGenerative()`

interface to enable implementations of generativity assertion. However, it is not a compulsion for routines to manifest generativity.

In routines, an actor can be a human or digital artefact, as indicated by the 'is a' inheritance relationship. The `Actor` class holds some common properties shared by its child classes. A task is taken up by actors, and subsequent routines to perform the task can be 'live' routines or 'dead' routines [14,15] or an interplay of the two. The `Routine` class therefore contains an interface to enable the implementation of an `isLive()` method to allow implementations of assertions on the type of routine (live or dead). For example, GitHub collaborators can enact 'live' routines for resolving issues on repositories, projects, or codes, where each and every `Action` is carried out by a `HumanActor`. In contrast, the GitHub platform's software actively facilitates services via autonomous bots, such as continuous integration, where a `Routine` is performed entirely by `DigitalActor`. Depending on their generative potential, an actor can act as an intermediary or mediator in task routines [9,13], and thus `Actors` also contain corresponding assertion interfaces in `isIntermediary()` and `isMediator()`. In itself, digital artefacts can act as a mediator. However, when its service is used in another system, it might serve as a passive digital agent, thus an intermediary role in routines [13]. How these interfaces are implemented is intentionally left undefined in the data model in order to allow flexibility of interpretation on the concepts that are harder to describe as data tangibly.

4   Demonstration of Routines Data Model In-Use

In order to demonstrate our routines data model, we replicated the approach used by Deng et al. [4] for analysing routines inferred from issues posted in the GitHub issue tracking system. In particular, we use GitHub issues as trace data [16] that record specific action patterns, and thus we can discover recurrent action patterns (i.e. performative routines) that are used to resolve issues in projects on GitHub. When able to identify performative routines in trace data, we can then utilise extant quantifiable data relating to the performances for further analysis or modelling. In this section, we present how we map GitHub's representation of issues and associated events into the routines data model and an example of how one can then use the instantiated data model for simulation purposes.

4.1   *Scikit-learn* Development as a Data Source

Scikit-learn is a robust Python library for machine learning and statistical modelling [11]. As an OSS project hosted on GitHub, *scikit-learn* has an active development community with code contributions going back as far as 2010 and over 33,000 issues and pull requests tracked. The issue resolution process on GitHub [17] can be considered as a transparent and accountable assemblage of humans, integrated development environments, algorithmic processes, APIs, data, norms, and policies in a constant interplay as 'live' and 'dead' routines [14].

Resolving issues through discussion ensures that all the activities are legitimate from authorised users, thereby ensuring technical integrity in open environments [18]. The issue resolution routines on the *scikit-learn* repository present an ideal real-world case to map the dynamics of sociotechnical actors' patterns of action for undertaking collaborative tasks to our routines data model. In addition, such a study also presents an opportunity to examine the ostensive, performativity, and generativity of human and digital actors in practice.

Data related to the routines for resolving issues in an OSS project on GitHub can be viewed from two perspectives through the lens of Routine Dynamics theory. Ostensive routines can be thought of as the idealised normative examples of recurrent action patterns, often determined through the qualitative analysis of authoritative sources of data, such as participatory or direct observations, documentation and expert interviews. With respect to OSS contributions and issue tracking, guidelines provided by GitHub themselves or project-specific documentation, as illustrated in Table 2, could be qualitatively analysed to determine the ostensive aspect of issue resolution routines.

Table 2. Documents that provide qualitative data to infer the ostensive aspect of *scikit-learn* repository routines (see [1,5]).

Policy document	Description
Scikit-learn governance and decision-making [19]	The document formalises the governance mechanism for managing the *scikit-learn* project and clarifies decision-making and interaction among various elements in the *scikit-learn* community. The document establishes a decision-making structure that considers feedback from all community members
Scikit-learn community code of conduct [20]	The document adopts a guideline to define community standards to foster a welcoming and inclusive *scikit-learn* project. It also facilitates a healthy and constructive community and social atmosphere for projects to develop and achieve their desired goals
GitHub Docs - Issues [21]	Prescriptive guidelines for reporting issues for ideas, feedback, tasks, or bug reporting on GitHub
GitHub Docs - Collaborating with pull requests [22]	Prescriptive guidelines for proposing and reviewing changes in pull requests on GitHub

On the other hand, the performative aspect of routines looks at how routines are actually enacted or performed, the patterns of which are ascertained by observing so-called performances. Ostensive routines theoretically guide how a routine should be performed, but this does not mean the corresponding performances conform to the norms in practice. These two aspects together (i.e. the ostensive and the performative) allow us to analyse routines and their dynamics deeply. Routine performances may be evidenced through, again, participatory or direct observation, logging (i.e. diaries/journals or digital logs), and increasingly by computationally processing digital trace data. In the context of this exemplar, as in [4], we use GitHub issues themselves as the trace data from which to infer routine performances.

4.2   Research Replication Using Our Data Model

Our demonstration analysis is inspired by the work reported in [4], where Deng et al. show the impact of routine change in OSS development. They use a methodological approach that leverages digital trace data from GitHub to focus on how changes in emergent development routines influence OSS project popularity. The authors selected a stratified sample of 271 OSS projects based on their popularity, measured via forks and stars, to ensure a diverse representation of projects, ultimately building a dataset of over 20 million events and inferring over 3 million performances. They then use sequence analysis [23], a technique similar to gene sequence analysis in bioinformatics, but where categorical data are organised chronologically, and hidden Markov models (HMM) [24] to identify and quantify emergent routines and their changes within these projects. Routine diversity is then measured by the variety of routines, while routine change is assessed both in terms of its magnitude and frequency. What their study demonstrates is that large-scale computational analysis of trace data can be used to generate knowledge about routines. The authors use the GitHub API's data model of events as the units of observation by which to construct routine performances and thus infer 'contextualised routines' that describe specific patterns of action, where they then quantitatively analyse these patterns.

In Deng's paper, the authors present an example of 'contextualised routines' for a single OSS project, *Angular.js*. To demonstrate the application of our data model, we replicate this analysis presented in that paper but for the routines found by mining *scikit-learn*'s issues.

4.3   Modelling Routines in the *scikit-learn* OSS project

We use the GitHub REST API to extract issues relating to software development from the *scikit-learn* OSS project repository. This was done using the PyGitHub v2.2.0 library [25], which provides access to the REST API via typed interactions in Python programming code. We collected 25,799 GitHub issues and pull requests marked as closed (as of 19 April 2024). The GitHub REST API provides access to download each issue, indicated with an `IssueEvent` object. Each `IssueEvent` has a related timeline of constituent events that is accessed using PyGitHub with a `.get_timeline()` method. This method gives us a chronological sequence of events, the types of which are listed in Table 3, and their metadata such as related users, timestamps, state, etc.

By iteratively retrieving issues and their timelines using the GitHub API, and retrieving their constituent event objects, we construct a dataset of 617,440 events, spanning from 31 August 2010 to 17 April 2024, that we map into

Table 3. GitHub event types relating to development teamwork, as per [4], and related GitHub data entities, and the corresponding mapping into our routines data model classes.

GitHub API model	Inferred action	Routines model
IssueEvent	Issue is created by a user (human or digital)	PerformedAction
PushEvent	Code is committed and pushed to a repository	PerformedAction
CloseEvent	Issue or pull requests is closed	PerformedAction
ReopenEvent	Issue or pull request that had been closed is reopened	PerformedAction
PullRequestEvent	User request new code to be pushed to a repository	PerformedAction
CommentEvent	Comment is created on issue or pull request	PerformedAction
CommitCommentEvent	Comment is created on a commit	PerformedAction
PullRequestReviewEvent	Review comment is created on a pull request	PerformedAction
MergeEvent	Pull request is accepted and code is merged to a repository	PerformedAction
SubscribeEvent	Issue or pull request is subscribed to by a user	PerformedAction
UnsubscribeEvent	Issue or pull request is unsubscribed by a user	PerformedAction
ReferenceEvent	Issue or pull request is referenced	PerformedAction
MentionEvent	User is mentioned	PerformedAction
AssignEvent	Issue is assigned to a user	PerformedAction
Timeline	Collection of events ordered temporally	Performance
User	A GitHub user such as a human contributor	HumanActor
User	A GitHub user that is a bot	DigitalActor
	Recurrent action pattern derived from performances	Performative

25,799 `Performance`s made up of 617,440 `PerformedAction`s as expressed in our routines data model. From these 25,799 `Performance`ss, we identify 8,474 recurrent `ActionPattern`s as performative routines (similar to [4]'s presented contextualised routines). It was done by identifying and reducing the repeated `PerformedAction` types to describe patterns of `Action` that occur in a particular chronological sequence. For each `PerformedAction`, we also identify and link an `Actor`, either human (`HumanActor`) or digital artefact (`DigitalActor`), as found in the GitHub event's metadata referencing a GitHub user. Table 4 summarises the top 10 most commonly identified performative routines from *scikit-learn*'s issue data.

While the routines can be expressed as patterns in the style shown in Table 4, we can further look at the performative routines as stochastic models. Given their

Table 4. Summary statistics about the top 10 occurring recurrent action patterns (performative routines) from *scikit-learn* issues. **C = commented, CL = closed, O = opened, L = labelled, RV = reviewed, M = mentioned.**

Recurrent action pattern	No. events	No. comments	No. unique actors	Avg. duration
O/C+/CL	26007	11764	1735	105 days 12:33
O/CL	14000	2308	1166	24 days 10:57
O/C/CL	13439	3442	1113	55 days 04:41
O/L/C/CL	3636	676	391	78 days 17:36
O/RV+/CL	5084	189	260	4 days 07:45
O/RV/CL	3246	203	295	3 days 02:17
O/L+/RV/CL	3739	114	184	4 days 14:45
O/L+/RV+/CL	4544	72	149	5 days 13:57
O/C+/M/C+/CL	6462	3060	408	165 days 13:57
O/L/C+/CL	3203	1214	405	76 days 09:19

nature, the representation of routines as a sequence of actions means that we can straightforwardly model routines from the perspective of states and transitions, as simple Markov chains. Given that we have observations of routine performances represented in our data model, we can calculate transition probabilities from action-to-action and populate a transition matrix accordingly (see Table 5 and corresponding visualisation in Fig. 2). Consequently, we can now perform simulations of routines by running the Markov chain.

Table 5. Transition matrix showing probabilities of the state transitions between different routine actions in *scikit-learn* development issues, based on the observed trace data from GitHub.

	commented	opened	labeled	reviewed	assigned	unlabeled	closed	mentioned
commented	0.5444	0.0000	0.0177	0.0505	0.0018	0.0066	0.0864	0.2926
opened	0.4210	0.0000	0.3018	0.0909	0.0019	0.0051	0.0770	0.1023
labeled	0.3755	0.0000	0.3159	0.2006	0.0024	0.0390	0.0563	0.0102
reviewed	0.3437	0.0000	0.0461	0.3887	0.0026	0.0069	0.1880	0.0240
assigned	0.4435	0.0000	0.0601	0.1555	0.0124	0.1590	0.1661	0.0035
unlabeled	0.3375	0.0000	0.4014	0.0422	0.0049	0.0804	0.1300	0.0036
closed	0.8691	0.0000	0.0195	0.0124	0.0000	0.0080	0.0089	0.0821
mentioned	0.5835	0.0000	0.0317	0.0894	0.0013	0.0082	0.0486	0.2374

Modelling routines as Markov chains provides further insights into how routines are performed based on real-world observations, where we can use the simulations to better understand the effects of different sequences of actions to fulfil a given routine. Taking this further, we may decide to model subsets of routines for comparative study, and can also integrate other data into the modelling

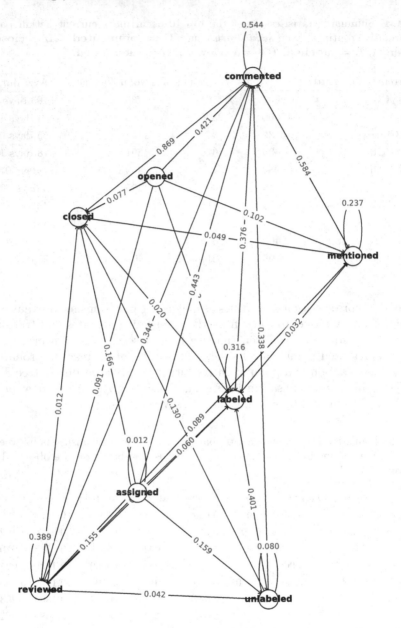

Fig. 2. Visualisation of the Markov chain model of issue resolution routines in *scikit-learn*, based on the transition probabilities shown in Table 4.

and simulation, such as conformance to the ostensive routines (i.e. comparing distances between action patterns of the ostensive and performative), sequence variability, effects of different actor types and roles, performance factors and so on, or other underlying processes that are not directly observable perhaps using HMMs. Uncertainty modelling becomes vital when considering the performative aspect of routines.

5 Discussion

5.1 Opportunities for Modelling Routines

The uptake of Routines Dynamics in the mainstream benefited in expanding the constructs and ontologies of routines. However, using an established modelling framework to model routines has not received much scholarly attention. There remains much uncertainty in the interpretation of the constructs in Routine Dynamics [1,13] since they are abstract and complex, especially to novice researchers. As such, we argue that the formalised representation of routines as a data model may ease the comprehension of routines' ontologies, structure, and behaviour. Moreover, modelling enables one to have a deeper insight into the underlying intricate dynamics and assumptions of real-world routines. The normative approach of discourse on Routine Dynamics is found inadequate, especially adopting as a theoretical framework for researchers or analytical tools for practitioners. Therefore, model-driven approaches to studying routines aid in grasping the inner workings and multifaceted dynamics of routines, as illustrated in the analyses and modelling in Tables 4 and 5, and Fig. 2. Indeed, such a conceptual representation and physical simulation allows one to visualise the duality of routines-structure and agency of sociotechnical actors. Due to its abstractness, some core constructs in Table 1 of routines are demanding to think of in agentic terms [1]. Thus, the UML [26,27] perspective to examine and model the structural and behavioural aspects of routines unfolding in the real world adds a novelty to the knowledge base of routines.

5.2 Challenges of Modelling Routines

The routines for accomplishing tasks will likely change in a dynamic environment with multi-agent systems [3]. Given the autonomous, decentralised, and heterogeneous nature of actors in digital ecosystems, capturing routines enacted by humans or encoded in digital artefacts and corresponding micro-level knowledge embedded in patterned action is difficult. Winter [28] notes, 'knowledge advances cumulatively'. Hence, abstract properties of sociotechnical actors' routines such as expertise [1], knowledge [29], repertoire [30], and memory [4] in digitised processes are challenging to grasp and represent in a model. Additionally, one cannot take action patterns observed in a phenomenon at face value and treat them as routines without understanding context and questioning assumptions. The routines data model presented in this study is not static and can evolve as new constructs of routines are added to the knowledge base and refined over time.

6   Conclusion

This study unpacked routines' intricate structure and dynamics for resolving GitHub *scikit-learn* repository issues to develop a routines model using UML. Routines in an organisational environment entail complex agency, action, and interplay of interdependent sociotechnical actors. The UML class diagram captures routines' structural and behavioural facets. The knowledge base of routines lacks scholarly discussion and articulation of modelling routines that consolidate the constructs and ontologies into a conceptual, logical, and physical model. This study demonstrated that modelling frameworks such as UML could provide such an account of routines to further our understanding of and augment the knowledge base on routines. We suggest researchers and practitioners employ a modelling and simulation perspective to investigate the constructs, structure, and dynamics of routines. The source code and data for this paper is found in: https://github.com/UppsalaIM/Towards_Modelling_Organisational_Routines/.

Acknowledgements. Phub Namgay is supported in part by the Swedish Research School of Management and IT (Forskarskolan management och IT).

Disclosure of Interests. The authors have no competing interests to declare that are relevant to the content of this article.

References

1. Feldman, M.S., Pentland, B.T.: Reconceptualizing organizational routines as a source of flexibility and change. Adm. Sci. Q. **48**(1), 94–118 (2003)
2. Feldman, M.S., Pentland, B.T., D'Adderio, L., Dittrich, K., Rerup, C., Seidl, D.: What is routine dynamics? In: Feldman, M.S., Pentland, B.T., D'Adderio, L., Dittrich, K., Rerup, C., Seidl, D. (eds.) Cambridge Handbook of Routine Dynamics. Cambridge University Press (2021)
3. Li, W., Badr, Y., Biennier, F.: Digital ecosystems: challenges and prospects. In: MEDES 2012: Proceedings of the International Conference on Management of Emergent Digital EcoSystems. ACM, New York (2012)
4. Deng, T., Robinson, W.N.: Changes in emergent software development routines: the moderation effects of routine diversity. Int. J. Inf. Manage. **58**, 102306 (2021)
5. Hayes, G.R., Lee, C.P., Dourish, P.: Organizational routines, innovation, and flexibility: the application of narrative networks to dynamic workflow. Int. J. Med. Informatics **80**(8), 161–177 (2011)
6. Pentland, B.T., Feldman, M.S.: Narrative networks: patterns of technology and organization. Organ. Sci. **18**(5), 781–795 (2007)
7. Howard-Grenville, J., Rerup, C.: A process perspective on organizational routines. In: Langley, A, Haridimos T. (eds.) The SAGE Handbook of Process Organization Studies. SAGE (2016)
8. Pentland, B.T., Liu, P., Kremser, W., Hærem, T.: The dynamics of drift in digitized processes. MIS Q. **44**(1), 19–47 (2020)
9. Pentland, B.T., Hærem, T., Hillison, D.: Comparing organizational routines as recurrent patterns of action. Organ. Stud. **31**(7), 917–940 (2010)

10. Gaskin, J., Berente, N., Lyytinen, K., Yoo, Y.: Toward generalizable sociomaterial inquiry. MIS Q. **38**(3), 849–A812 (2014)
11. Pedregosa, F., et al.: Scikit-learn: machine learning in Python. J. Mach. Learn. Res. **12**, 2825–2830 (2011)
12. Thomas, L.D., Tee, R.: Generativity: a systematic review and conceptual framework. Int. J. Manag. Rev. **24**(2), 255–278 (2022)
13. Sele, K., Grand, S.: Unpacking the dynamics of ecologies of routines: mediators and their generative effects in routine interactions. Organ. Sci. **27**(3), 722–738 (2016)
14. Sammon, D., Nagle, T., McAvoy, J.: Analysing ISD performance using narrative networks, routines and mindfulness. Inf. Softw. Technol. **56**(5), 465–476 (2014)
15. Cohen, M.D.: Reading Dewey: reflections on the study of routine. Organ. Stud. **28**(5), 773–786 (2007)
16. Pentland, B.T., Recker, J., Wolf, J.R., Wyner, G.: Bringing context inside process research with digital trace data. J. Assoc. Inf. Syst. **21**(5), 5 (2020)
17. Bissyandé, T.F., Lo, D., Jiang, L., Réveillere, L., Klein, J., Le Traon, Y.: Got issues? Who cares about it? A large scale investigation of issue trackers from GitHub. In: IEEE 24th International Symposium on Software Reliability Engineering. IEEE (2013)
18. Tsay, J., Dabbish, L., Herbsleb, J.: Let's talk about it: evaluating contributions through discussion in GitHub. In: Proceedings of the 22nd ACM SIGSOFT International Symposium on Foundations of Software Engineering. ACM, New York (2014)
19. Scikit-learn governance and decision-making on scikit-learn.org. https://scikit-learn.org/stable/governance.html#governance. Accessed 26 Apr 2024
20. Scikit-learn community code of conduct on scikit-learn.org. https://github.com/scikit-learn/scikit-learn/blob/main/CODE_OF_CONDUCT.md. Accessed 26 Apr 2024
21. GitHub Docs - Issues. https://docs.github.com/en/issues/tracking-your-work-with-issues/about-issues. Accessed 26 Apr 2024
22. GitHub Docs - Collaborating with pull requests. https://docs.github.com/en/pull-requests/collaborating-with-pull-requests. Accessed 26 Apr 2024
23. Abbott, A., Tsay, A.: Sequence analysis and optimal matching methods in sociology: review and prospect. Sociol. Methods Res. **29**(1), 3–33 (2000)
24. Rabiner, L.R.: A tutorial on hidden Markov models and selected applications in speech recognition. Proc. IEEE **77**(2), 257–286 (1989)
25. PyGithub v2.2.0 release on GitHub. https://github.com/PyGithub/PyGithub/. Accessed 14 Mar 2024
26. Vasilakis, C., Lecznarowicz, D., Lee, C.: Developing model requirements for patient flow simulation studies using the Unified Modelling Language (UML). J. Simul. **3**(3), 141–149 (2009)
27. Dobing, B., Parsons, J.: How UML is used. Commun. ACM **49**(5), 109–113 (2006)
28. Winter, S.G.: Understanding dynamic capabilities. Strateg. Manag. J. **24**(10), 991–995 (2003)
29. Cohen, M.D., Bacdayan, P.: Organizational routines are stored as procedural memory: evidence from a laboratory study. Organ. Sci. **5**(4), 554–568 (1994)
30. Hansson, M., Hærem, T., Pentland, B.T.: The effect of repertoire, routinization and enacted complexity: explaining task performance through patterns of action. Organ. Stud. **44**(3), 473–496 (2021)

Teaching Computational Science

Enhancing Computational Science Education Through Practical Applications: Leveraging Predictive Analytics in Box Meal Services

Ilona Jacyna-Golda[1] , Pawel Gepner[1]([envelope]) , Jerzy Krawiec[1] ,
Kamil Halbiniak[3] , Andrzej Jankowski[2], and Martyna Wybraniak-Kujawa[1]

[1] Faculty of Mechanical and Industrial Engineering,
Warsaw University of Technology, Warsaw, Poland
{ilona.golda,pawel.gepner,jerzy.krawiec,martyna.kujawa}@pw.edu.pl
[2] Intel Corporation, Santa Clara, USA
andrzej.jankowski@intel.com
[3] Faculty of Mechanical Engineering and Computer Science,
Czestochowa University of Technology, Czestochowa, Poland
khalbiniak@icis.pcz.pl

Abstract. This paper presents a student project carried out in collaboration with a major industry partner, demonstrating the simultaneous novel application of predictive analytics, in particular machine learning (ML), in the domain of boxed meal services, and explores its implications for IT education. Drawing from a validated ML model trained on data collected from box meal companies, this study showcases how predictive analytics can accurately predict customer sociodemographic characteristics, thereby facilitating targeted marketing strategies and personalized service offerings. By elucidating the methodology and results of the ML model, this article demonstrates the practical utility of computational techniques in real-world electronic services. Moreover, it discusses the pedagogical implications of incorporating such case studies into computational science education, highlighting the opportunities for experiential learning, interdisciplinary collaboration, and industry relevance. Through this exploration, the article contributes to the discourse on innovative teaching methodologies in computational science, emphasizing the importance of bridging theory with practical applications to prepare students for diverse career pathways in the digital era.

Keywords: Computational Science Education · Predictive Analytics · Machine Learning · Box Meal Services · Experiential Learning · Interdisciplinary Collaboration · Industry Relevance

© The Author(s), under exclusive license to Springer Nature Switzerland AG 2024
L. Franco et al. (Eds.): ICCS 2024, LNCS 14838, pp. 383–397, 2024.
https://doi.org/10.1007/978-3-031-63783-4_28

1   Introduction

The aim of this project was to create an IT system overseen by a staff member from an eminent IT company. Its purpose was to showcase genuine challenges faced by users in practical scenarios, while also illustrating the intricacies and techniques involved in such operations. The project was meticulously chosen and managed to address real-world issues without compromising sensitive information or jeopardizing business affiliations.

Understanding the nutritional preferences of users, along with the freshness of the food offered and logistic efficiency, is a crucial aspect of the box catering service. Unlike customers ordering food in bars and restaurants, where the products can be seen and tasted, box catering customers rely solely on descriptions and images. Moreover, box catering typically involves ordering meals for multiple days, necessitating a deep understanding of customer preferences to ensure satisfaction over time [21]. The challenge lies in aligning the offered meal boxes with customer expectations to reduce service cancellations after the initial trial period.

To address this challenge, this study leveraged machine learning (ML) methods to infer customer sociodemographic characteristics based on their box selections and relevance to various service factors. The dataset utilized for this analysis, sourced from a survey conducted during the Covid-19 pandemic by a Polish meal box catering company and available on Kaggle, provided insights into customer behavior and preferences amidst changing circumstances. By employing ML techniques, the study aimed to classify and estimate customer profiles to enhance service personalization and customer satisfaction.

While previous research has explored customer profiling using artificial neural network techniques, few studies have comprehensively examined the impact of the Covid-19 pandemic on customer behavior in the context of box meal catering. Existing literature demonstrates the efficacy of ML models in profiling customers across various industries, including food tourism and online delivery services [22]. However, the specific application of ML in meal box customer profiling remains underexplored.

This study contributes to filling this gap by analyzing the factors influencing customers' choices in meal box services. By gaining insights into customer profiles and behaviors, businesses can tailor their services to better meet customer needs and preferences. Additionally, the findings shed light on the pedagogical implications of incorporating such case studies into computational science education, emphasizing the importance of bridging theory with practical applications to prepare students for diverse career pathways in the digital era. Through this exploration, the study aims to contribute to the discourse on innovative teaching methodologies in computational science and promote interdisciplinary collaboration between academia and industry.

2   Related Work

This section provides an overview of key research fields pertinent to the study, including meal box catering, customer profiling, and machine learning applications.

2.1   Meal Box Catering and Online Food Delivery

The food delivery market has experienced substantial growth, with online orders comprising a significant portion of all food deliveries [9]. Third-party delivery platforms such as GrubHub, UberEats, and DoorDash have played pivotal roles in expanding the scope of online food delivery services [17]. These platforms offer a wide range of meal options, including breakfasts, lunches, dinners, and all-day packages, catering to diverse customer preferences [1,17]. The COVID-19 pandemic further accelerated the adoption of online food delivery services, particularly meal box subscriptions, as people sought convenient and safe meal solutions [21].

The proliferation of meal box catering services during the pandemic highlighted the importance of understanding customer preferences and tailoring services accordingly. This shift in consumer behaviour underscores the need for robust customer profiling strategies to enhance service personalization and satisfaction.

2.2   Customer Profiling

Customer profiling is a fundamental aspect of customer relationship management (CRM), enabling businesses to understand customer needs and behaviours [10]. By leveraging data mining techniques, companies can extract valuable insights from customer data to improve service offerings and enhance customer retention [2]. Effective customer profiling facilitates targeted marketing efforts and fosters customer engagement and loyalty [20].

2.3   Machine Learning and Artificial Neural Networks

Machine learning (ML) and artificial neural networks (ANNs) have emerged as powerful tools for analyzing complex datasets and making data-driven predictions [5]. While machine learning is fueling technology that can help workers or open new possibilities for businesses, there are several things business leaders should know about machine learning and its limits. One is explainability, meaning what the machine learning models do and how they make decisions [7]. ANNs, inspired by the human nervous system, are particularly effective in solving pattern recognition and classification tasks [4].

The rapid advancement of ML techniques has led to their widespread adoption across various domains, including business, healthcare, and automotive industries [19,24]. With the increasing complexity of ML algorithms and the availability of computational resources, these techniques are finding novel applications and fueling innovation across industries [18,23].

3   Materials and Methods

This section describes the dataset's characteristics and how the raw data has been collected and utilized for this experiment. We built and tested various machine learning models to learn and understand the feature importance and correlation between them. We have also experimented with these models to predict customer classification profiles using their characteristics and motivations in decision-making.

3.1   Dataset

The study is based on the publicly available dataset posted on Kaggle, and it contains available data reported by one of the meal box providers in Poland. The data was collected at the beginning of the 2020 COVID-19 pandemic. The available data set contains 842 records with 35 columns. Each column precisely describes the characteristics of the question it belongs to. Initially, the data was collected in Polish, but the dataset on Kaggle was translated into English. We can find 3 categories of the information collected. The first section provides social-demographics information like Age, Gender, Marital Status, Occupation, Monthly Income, Educational Qualifications, Place of Residence, Family size, Height, and Weight. The second section is focused on the box itself and type of delivery, payment preferences, special needs or requirements and physical activity profile. The third section is focused on the motivation and various factors of importance to choosing the meal box. The data collected in this third section is the most interesting because we know very little about this aspect and the role that fear of catching COVID-19 plays in the decision-making process, which factors influence this and what needs to be considered to properly understand consumer behavior.

3.2   Machine Learning Algorithms

We implemented an ML learning algorithm for customer classification in a meal box catering service. A training set (training data + answers) provided information about each client and his/her box for correct classification. The analyzed data set provides information about each client, including gender, age, education, income, profession, BMI-Body Mass Index, the type of food offered, and the importance they assign to various motivating factors that determine this choice. Responses to the following set of questions/features motivating the dietary decision allowed us to connect the sociodemographic characteristics of users to the type of answers given:

- Diet Type,
- For which period the order was placed,
- Motivation to choose a diet,
- Physical activity,
- Special needs,

- Do you find it cheaper than cooking at home?
- Do you find it easier and more convenient than preparing meals at home?
- Do you find it a time-saving alternative?
- What type of payment do you select?
- Do you value more diet offers and flexibility in changing them, and are you ready to pay for them?
- Is the food of the same or better quality as in the restaurant?
- Do you value a box diet from a health concern perspective?
- Do you see a problem with offered delivery times, and are you ready to pay for this?
- Do you value diversity in the menu, and are you ready to pay for this?
- Do you consider the quality and aesthetics of the packaging and the method of delivery important, and are you ready to pay for it?
- Do you value the possibility of replacing the meals, and are you ready to pay for this?
- How do you rate the safety of such a diet in terms of Covid 19 - no physical contact, no need to go shopping?
- How did you hear about this kind of service?
- Do you combine this diet with a special training program?
- Do you have a personalised diet created by a dietitian?
- How long are you going to use a diet,
- Are you going to extend it?
- Would you recommend this type of food to others?

The ML algorithm learns with the training set. When a new customer arrives at the E-Commerce portal and begins an order by answering questions, the algorithm can classify them based on the knowledge about sociological classification it has acquired. During our research phase, we built our machine learning models based on theoretical foundations for multiclass and multi-output algorithms. This classification joining method is known as multi-output and is also called multitask classification. We have experimented with and verified a few models for which we have used TensorFlow and ScikitLearn. Fundamentally, most classification algorithms aim to capture dependencies between input variables x_j and the output variables y_j. The prediction $y' = f(x)$ of a scoring classifier f is often regarded as an approximation of the conditional probability $P_r(y = y'/x)$, i.e., the probability that y' is the true label for the given instance x. In multi-output classification models, dependencies may exist between x and each target y_i and between the labels $y_1, ..., y_n$ themselves. Traditionally, a supervised multi-output learning paradigm simultaneously predicts multiple outcomes with one entry, which means much more complex decision problems. Compared to traditional single output learning, multi-output learning is multi-dimensional, and the results can be complex interactions that can only be handled by structured inference. Multi-output machine learning algorithms can be used in many ways in various fields [26] e.g.:

- binary output values - related to multi-label classification problems,
- nominal output values - related to multidimensional classification problems,

- ordinal output values - studied in label ranking problems,
- real-valued outputs - considered in multi-target regression problems.

Many studies have been conducted in the field of multi-output classification, and the same problem of funnel model and theory for understanding consumer decision-making and behavior has been discussed by many authors. Predicting consumer behavior becomes a prerequisite for marketing decision-making, and it is considered very important, especially in online shopping. In the work of Jungwon Lee [15], we can deeply analyze the suitable machine learning model for predicting online consumer behavior. This work analyzed 8 models comparing suitable machine learning algorithms and sampling methods in the context of online consumer behavior. Those models are:

- Classification tree (TREE);
- Artificial neural network (NNET);
- K-Nearest-Neighbor (KNN);
- Logistic Regression (LOGIT);
- Support vector machine with the linear kernel (SVML);
- Random forest (RF);
- Gradient Boosting Algorithm (GBM);
- eXtreme Gradient Boosting (XGB).

The authors validated some of them and proposed the best model adapted to the tested data. In our analysis, we looked at their solutions and employed the proposed algorithms in data analysis and model building in a customized fashion for our study. As mentioned earlier, in our study, we built a model capable of combining multi-output classification with multi-label classification. Ultimately, in our scenario, each multi-output class also utilizes multiple labels. In other words, our model classifies an object into different class categories simultaneously, giving them a specific label from the pool of available labels for a given class category. We have been building our multi-output neural network model utilizing Tensor-Flow framework, which has been used to classify our social-demographics items using 8 separate forks in the architecture:

- one fork is responsible for classifying the age of a given input;
- the second fork is responsible for classifying the gender;
- next fork classifies marital status;
- next fork occupation accuracy;
- next fork monthly income;
- next fork educational qualifications;
- next fork place of residence;
- the last fork is responsible for classifying family size.

Finally, we have trained the network to classify the training dataset and obtain the multi-output classification results.

4 Results

4.1 Environment Setup

Using Jupyter Notebook, a seed has been initialized to ensure reproducibility in all experiments.

4.2 Data Collection and Structure

The data was collected from the Warsaw Concept company, which used catering services from miodmalina.eu, the collection period lasted from March 2020 to November 2020, and the data was published as a dataset on Kaggle.com. The dataset was cleaned before publication, and there are complete data and answered questions. From the data type perspective, we only have one data category in floating point format. This value was calculated as BMI, obtained from height and weight values, which are integers. Box calories are also in integer format. The remaining data represents the type of the data type object. All other data are not in numerical format but can be easily converted to a numeric scale assigned to a given response category. We have answered the questions and grouped them into sets 1 to 5 or 1 to 6 or 1 to 7, or 1 to 9 categories. Psychometric questions include "Do you value the possibility of replacing the meals and are you ready to pay for this?" These are typical Likert-type questions, ranging from 1 to 5 or 1 to 7, where 1 means "strongly disagree" and 5 "strongly agree". An example of the data categorization method is shown in Fig. 1, which shows the age distribution of the studied dataset grouped into 6 categories. The count represents the number of users of the meal box services assigned to each age category group. After reading and cleaning the data, categorical variables have been converted to one-hot encoding representation for all models (neural networks, trees, GBMs) except for Logistic Regression (the targets for regression have been encoded as integers). Next, we selected the target data columns. We have extracted the columns from the data set that can be used to do the sociodemographic characteristics of the clients of box catering services. From the entire

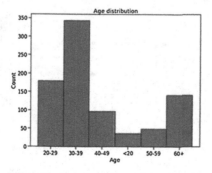

Fig. 1. Age distribution of the meal box catering users.

dataset, we have selected the following data types that we use to categorize and predict users:

- Age;
- Gender;
- Marital Status;
- Occupation;
- Monthly Income;
- Educational Qualifications;
- Place of residence;
- Family size.

The next step was to divide the dataset into training and evaluation parts. The data has been split into training (631 records) and testing (211 records) sets.

4.3   Model Performance Evaluation

This section evaluates the performance of various machine learning models in predicting the sociodemographic characteristics of customers using box catering services based on their box selection behaviour. We first discuss the performance of a multi-label Multi-Layer Perceptron (MLP) model and then compare it with other classic machine learning algorithms.

Multi-Label Multi-Layer Perceptron (MLP) Model We initially constructed a multi-label MLP model to predict specific sociodemographic properties. The model was trained using a training dataset and validated using test data. Figure 2 illustrates the confusion matrix for the "Occupation" category, demonstrating the model's accuracy in predicting different occupations.

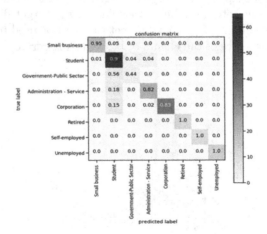

Fig. 2. Confusion matrix for the "Occupation" category

The MLP model exhibited commendable accuracy, particularly in predicting categories such as "Student" or "Small business," achieving over 90% accuracy for these categories.

Hyperparameter tuning played a crucial role in optimizing the model's performance. We manually tuned the hyperparameters, considering the dataset's size and model complexity, resulting in satisfactory accuracy for real-world applications.

The topology of the MLP model, depicted in Fig. 3, comprised hidden layers with ReLU activation and an output layer with softmax activation. The model achieved an accuracy of 0.8934 with a loss of 0.3123.

```
Model: "sequential_5"

Layer (type)              Output Shape             Param #
=================================================================
dense_63 (Dense)          (None, 128)              12672

dense_64 (Dense)          (None, 32)               4128

dropout_5 (Dropout)       (None, 32)               0

dense_65 (Dense)          (None, 8)                264
=================================================================
Total params: 17,064
Trainable params: 17,064
Non-trainable params: 0
```

Fig. 3. Single output multi-label MLP model summary

Multi-Output Neural Network Model to obtain a comprehensive understanding and analyze correlations between multiple output scenarios, we developed a multi-output neural network model using TensorFlow with Keras API. The model comprised eight branches, each dedicated to predicting a specific sociodemographic feature.

The model architecture consisted of dense layers with ReLU activation feeding into each branch. We monitored training using validation data and compiled the model using categorical cross-entropy loss and the Adam optimizer.

The achieved accuracy for individual categories and the mean absolute loss functions are summarized in Table 1. The model exhibited an impressive coefficient of accuracy across all features, with an average accuracy of 0.8933.

Comparison with Classic Machine Learning Algorithms We compared the performance of our neural network models with several classic machine learning algorithms, including Decision Trees, Random Forest, eXtreme Gradient Boosting (XGB), Support Vector Machines (SVM), and Logistic Regression.

The results, presented in Table 2, highlight the superior accuracy achieved by our models compared to those discussed in prior studies [15]. Notably, our multi-output MLP model outperformed other algorithms, achieving an accuracy of 0.8934.

Our experiments demonstrate the effectiveness of neural network models, particularly multi-output MLP, in predicting the sociodemographic characteristics of box catering service customers. These models outperformed classic machine learning algorithms and provided valuable insights for targeted marketing and customer profiling in the food service industry.

Table 1. Accuracy and loss for multi-output classification

Feature	Accuracy	Loss
Age	0.9289	0.1939
Gender	0.9242	0.1929
Marital Status	0.9005	0.2447
Occupation	0.8768	0.5240
Monthly Income	0.8341	0.3685
Educational Qualifications	0.8910	0.2904
Place of Residence	0.9100	0.2929
Family Size	0.8815	0.3899

Table 2. Comparison of accuracy with classic ML algorithms

Model	Accuracy (Our Experiment)	Accuracy (Prior Study)
MLP	0.8934	0.8367
Decision Trees	0.8524	0.8133
Random Forest	0.8750	0.8544
XGB	0.8728	0.8643
SVM	0.8181	0.8318
Logistic Regression	0.8833	0.7386

5 Discussion

In this section, we will explore how this student's project is making an impact on computational science education, highlighting experiential learning opportunities, interdisciplinary collaboration and relevance to industry.

5.1 Experiential Learning Opportunities

Applying Kolb's Experiential Learning Cycle to educational practices in computational science, such as machine learning, offers a structured framework for understanding how students engage with and internalize complex concepts [13]. Kolb's model delineates four stages: concrete experiences, reflective observation, abstract conceptualization, and active experimentation. In this discussion, we explore how each stage aligns with the hands-on activities involved in teaching machine learning and computational science, drawing from existing literature to support our argument.

Concrete Experiences: Engaging students in hands-on activities, such as building and evaluating machine learning models, serves as the foundation for concrete experiences in Kolb's model. This aligns with Kim et al., who advocate for experiential learning in computational science education [6]. By actively participating in tasks like data preprocessing, model selection, hyperparameter tuning, and performance evaluation, students directly interact with real-world datasets and algorithms. This hands-on approach not only enhances understanding but also allows students to grasp the practical implications of theoretical concepts.

Following concrete experiences, students engage in reflective observation, a stage characterized by introspection and analysis of one's experiences. Larson and Keiper emphasize the importance of reflection in experiential learning, highlighting how it enables learners to identify patterns and connections within their experiences [14]. In the context of machine learning education, students reflect on their data preprocessing strategies, model selection criteria, and the outcomes of their experiments. This reflective process encourages deeper comprehension and facilitates the transition from concrete experiences to abstract conceptualization.

Abstract conceptualization involves synthesizing observations and experiences to develop theoretical frameworks or mental models. Through reflection, students internalize concepts and principles underlying machine learning algorithms and methodologies. They begin to generalize their experiences, recognizing recurring patterns and principles that govern the behavior of computational systems. This aligns with Kolb's notion of abstract conceptualization and underscores the transformative potential of experiential learning in computational science education.

Active experimentation represents the culmination of the learning cycle, wherein students apply their newfound understanding to solve real-world problems or explore novel scenarios. By deploying machine learning models on authentic datasets or designing computational experiments, students demonstrate mastery of concepts and techniques learned throughout the learning process. This iterative approach to learning reinforces understanding and fosters innovation, as students refine their skills through repeated cycles of experimentation and reflection.

Applying Kolb's Experiential Learning Cycle to machine learning education facilitates a comprehensive and iterative approach to skill development and conceptual understanding. By providing students with opportunities for hands-on experimentation, reflection, conceptualization, and application, educators, especially those with deep knowledge and experience in industry, can foster deep learning experiences that prepare students to tackle the complexities of computational science effectively. This pedagogical approach not only enhances students' technical proficiency but also cultivates critical thinking abilities essential for success in data-driven fields.

5.2   Interdisciplinary Collaboration

Leveraging Vygotsky's Social Development Theory posits that learning is inherently social and occurs through interactions with more knowledgeable peers or mentors [3]. In the context of computational science education, interdisciplinary collaboration with the top expert from an international IT company mirrors this social learning process, enabling students to benefit from diverse perspectives and deep expertise knowledge [11].

By working collaboratively on projects involving data scientists, domain experts, and industry practitioners, students engage in peer-to-peer learning that enhances their cognitive development and problem-solving abilities [11]. Through shared experiences and collective problem-solving, students co-construct knowledge and develop critical thinking skills essential for success in computational science and beyond [16].

5.3   Industry Relevance

Integrating Bloom's taxonomy of learning domains provides a structured framework for categorizing educational objectives into cognitive, affective, and psychomotor domains [8]. This study primarily focuses on cognitive objectives within the context of computational science education, aligning with revision of Bloom's Taxonomy.

By infusing industry-relevant case studies, datasets, and projects into the curriculum, educators elevate learning experiences to higher levels of cognitive complexity. This approach enables students not only to acquire knowledge but also to comprehend, apply, analyze, synthesize, and evaluate information within the context of real-world problems. McKendree et al. emphasize the importance of authentic tasks in promoting deep learning, as they challenge students to engage in higher-order cognitive processes [25].

Through active engagement with industry-relevant challenges, students develop critical thinking skills essential for success in data-driven fields. Koedinger and many others advocate for project-based learning approaches that immerse students in authentic problem-solving contexts, allowing them to develop domain expertise while honing their analytical and evaluative abilities [12].

In the context of machine learning projects, students are tasked with analyzing datasets, selecting appropriate algorithms, optimizing model performance, and interpreting results. By navigating these complex tasks, students engage in cognitive processes spanning the spectrum of Bloom's Taxonomy. They not only acquire theoretical knowledge but also apply it to practical scenarios, analyze data to extract meaningful insights, synthesize findings to inform decision-making, and evaluate the effectiveness of their solutions.

Moreover, exposure to industry-relevant projects equips students with transferable skills that are highly valued in data science, machine learning, and related fields. By tackling authentic challenges, students develop resilience, adaptability, and problem-solving abilities that are crucial for success in dynamic professional environments.

Integrating Bloom's Taxonomy into machine learning education enhances the cognitive rigor of learning experiences, fostering the development of critical thinking skills and domain expertise. By immersing students in authentic problem-solving contexts and challenging them to engage with real-world datasets and scenarios, educators prepare students for the demands of careers in data science and computational science. This approach not only enriches learning but also empowers students to make meaningful contributions to the field of machine learning and beyond.

6   Conclusion

In conclusion, this study not only contributes to advancements in machine learning and business analytics but also underscores the transformative potential of computational science education in fostering innovative teaching and learning practices. By leveraging machine learning algorithms to analyze sociodemographic characteristics and motivations of customers in the catering box services industry, we have demonstrated the practical utility of computational methodologies in informing business strategies and catering service offerings.

However, beyond the technical contributions of our study, it is essential to emphasize the educational innovations and pedagogical insights gained throughout the research process. Our project served as a compelling educational case study, providing students with hands-on experience in data analysis, model development, and interdisciplinary collaboration. The collaboration between industry instructors and students was instrumental in bridging the gap between academic theory and real-world application, providing students with valuable insights into industry practices and challenges.

By integrating real-world projects like the one presented in this study into curricula, we have created experiential learning opportunities that not only bridge the gap between theory and practice but also foster collaboration and knowledge exchange between academia and industry. This cooperative approach not only enhances the relevance and applicability of educational experiences but also prepares students for seamless transitions into the workforce.

Moreover, our study highlights the importance of feedback loops between research findings, industry practices, and the teaching and learning process. Through reflection on our experiences and insights gained from the project, we can identify areas for curriculum enhancement, pedagogical innovation, and student engagement. By embracing educational feedback loops and fostering a culture of collaboration, we can continuously refine our educational approach and adapt to the evolving needs of students and industries.

In terms of our contribution to educational innovation, it is essential to highlight the novel approach employed in integrating machine learning projects into computational science education. By emphasizing hands-on, experiential learning opportunities and fostering collaboration between industry instructors and students, we have created a dynamic learning environment that equips students with the skills and knowledge needed to thrive in an increasingly data-driven world.

Moving forward, future research directions should focus on further refining our educational approach, incorporating larger and more diverse datasets, and exploring the broader implications of machine learning in computational science education. By continuing to prioritize educational innovation and fostering collaboration between academia and industry, we can ensure that computational science education remains at the forefront of preparing students for the demands of the modern workforce.

In summary, while our study contributes to advancements in machine learning and business analytics, it also underscores the transformative potential of computational science education in fostering educational innovation and empowering students. By embracing feedback loops, prioritizing experiential learning opportunities, and fostering collaboration between academia and industry, we can cultivate a dynamic learning environment that prepares students to excel in the digital age.

Acknowledgements. This paper and the research behind it would not have been possible without the exceptional support of Graphcore Customer Engineering and Software Engineering team. We would like to express our very great appreciation to Hubert Chrzaniuk, Krzysztof Góreczny and Grzegorz Andrejczuk for their valuable and constructive suggestions connected to testing our algorithms and developing this research work. This research was partly supported by PLGrid Infrastructure at ACK Cyfronet AGH, Krakow, Poland.

References

1. Annaraud, K., Berezina, K.: Predicting satisfaction and intentions to use online food delivery: what really makes a difference? J. Foodserv. Bus. Res. **23**(4), 305–323 (2020)
2. Anshari, M., Almunawar, M.N., Lim, S.A., Al-Mudimigh, A.: Customer relationship management and big data enabled: personalization & customization of services. Appl. Comput. Inform. **15**(2), 94–101 (2019)
3. Berger, M.: Vygotsky's theory of concept formation and mathematics education. Int. Group Psychol. Math. Educ. **2**, 153–160 (2005)
4. Beucler, T., Pritchard, M., Rasp, S., Ott, J., Baldi, P., Gentine, P.: Enforcing analytic constraints in neural networks emulating physical systems. Phys. Rev. Lett. **126**(9), 098302 (2021)
5. Boutaba, R., et al.: A comprehensive survey on machine learning for networking: evolution, applications and research opportunities. J. Internet Serv. App. **9**(1), 1–99 (2018)
6. Braad, E., Degens, N., Ijsselsteijn, W.: Designing for metacognition in game-based learning: a qualitative review. Transl. Issues Psychol. Sci. **6**, 53–69 (2020). https://doi.org/10.1037/tps0000217
7. Brown, S.: Machine Learning, Explained. MIT Sloan School of Management, Cambridge (2021), https://mitsloan.mit.edu/ideas-made-to-matter/machine-learning-explained
8. Gogus, A.: Bloom's Taxonomy of Learning Objectives, pp. 469–473. Springer US, Boston, MA (2012). https://doi.org/10.1007/978-1-4419-1428-6_141
9. Gunden, N., Morosan, C., DeFranco, A.: Consumers' intentions to use online food delivery systems in the USA. Int. J. Contempor. Hosp. Manag. **32**, 1325–1345 (2020)
10. Hassan, M., Tabasum, M.: Customer profiling and segmentation in retail banks using data mining techniques. Int. J. Adv. Res. Comput. Sci. **9**(4), 24–29 (2018)
11. Johnson, D., Johnson, R.: Making cooperative learning work. Theory Pract. **38**, 67–73 (1999). https://doi.org/10.1080/00405849909543834

12. Koedinger, K.R., D'Mello, S., McLaughlin, E.A., Pardos, Z.A., Rosé, C.P.: Data mining and education. WIREs Cognit. Sci. **6**(4), 333–353 (2015). https://doi.org/10.1002/wcs.1350
13. Konak, A., Clark, T.K., Nasereddin, M.: Using kolb's experiential learning cycle to improve student learning in virtual computer laboratories. Comput. Educ. **72**, 11–22 (2014). https://doi.org/10.1016/j.compedu.2013.10.013
14. Larson, B., Keiper, T.: Instructional strategies for middle and secondary social studies: methods, assessment, and classroom management. In: Instructional Strategies for Middle and Secondary Social Studies: Methods, Assessment, and Classroom Management, pp. 1–290 (2011). https://doi.org/10.4324/9780203829899
15. Lee, J., Jung, O., Lee, Y., Kim, O., Park, C.: A comparison and interpretation of machine learning algorithm for the prediction of online purchase conversion. J. Theor. Appl. Electron. Commer. Res. **16**(5), 1472–1491 (2021)
16. Leslie, D.: The ethics of computational social science. In: Bertoni, E., Fontana, M., Gabrielli, L., Signorelli, S., Vespe, M. (eds.) Handbook of Computational Social Science for Policy, pp. 57–104. Springer, Cham (2023). https://doi.org/10.1007/978-3-031-16624-2_4
17. Li, C., Mirosa, M., Bremer, P.: Review of online food delivery platforms and their impacts on sustainability. Sustainability **12**(14), 5528 (2020)
18. Lu, J., Wang, G., Tao, X., Wang, J., Törngren, M.: A domain-specific modeling approach supporting tool-chain development with Bayesian network models. Integr. Comput. Aided Eng. **27**(2), 153–171 (2020)
19. Ma, L., Sun, B.: Machine learning and AI in marketing-connecting computing power to human insights. Int. J. Res. Mark. **37**(3), 481–504 (2020)
20. Mach-Król, M., Hadasik, B.: On a certain research gap in big data mining for customer insights. Appl. Sci. **11**(15), 6993 (2021)
21. Mazurkiewicz, P.: Rewolucja w gastronomii, pude?ka warte miliard z?otych. https://www.rp.pl/biznes/art342111-rewolucja-w-gastronomii-pudelka-warte-miliard-zlotych/ (2021). Accessed 20 Sept 2021
22. Moral-Cuadra, S., Solano-Sánchez, M.Á., López-Guzmán, T., Menor-Campos, A.: Peer-to-peer tourism: tourists' profile estimation through artificial neural networks. J. Theor. Appl. Electron. Commer. Res. **16**(4), 1120–1135 (2021)
23. Saxena, A., et al.: A review of clustering techniques and developments. Neurocomputing **267**, 664–681 (2017)
24. Štrumbelj, E., Kononenko, I.: Explaining prediction models and individual predictions with feature contributions. Knowl. Inf. Syst. **41**(3), 647–665 (2014)
25. Team, I.E.: 10 must-have tutoring skills (2023). https://www.indeed.com/career-advice/career-development/tutoring-skills
26. Xu, D., Shi, Y., Tsang, I.W., Ong, Y.S., Gong, C., Shen, X.: Survey on multi-output learning. IEEE Trans. Neural Netw. Learn. Syst. **31**(7), 2409–2429 (2019)

Teaching High–performance Computing Systems – A Case Study with Parallel Programming APIs: MPI, OpenMP and CUDA

Pawel Czarnul[(✉)], Mariusz Matuszek, and Adam Krzywaniak

Faculty of Electronics, Telecommunications and Informatics, Gdansk University of Technology, Narutowicza 11/12, Gdansk 80-233, Poland
`pczarnul@eti.pg.edu.pl`

Abstract. High performance computing (HPC) education has become essential in recent years, especially that parallel computing on high performance computing systems enables modern machine learning models to grow in scale. This significant increase in the computational power of modern supercomputers relies on a large number of cores in modern CPUs and GPUs. As a consequence, parallel program development based on parallel thinking has become a necessity to fully utilize modern HPC systems' computational power. Therefore, teaching HPC has become essential in developing skills required by the industry. In this paper we share our experience of conducting a dedicated HPC course, provide a brief description of the course content, and propose a way to conduct HPC laboratory classes, in which a single task is implemented using several APIs, i.e., MPI, OpenMP, CUDA, hybrid MPI+Pthreads, and MPI+OpenMP. Based on the actual task of verifying Goldbach's conjecture for a given range of numbers, we present and analyze the performance evaluation of students' solutions and code speed-ups for MPI and OpenMP. Additionally, we evaluate students' subjective assessment of ease of use of particular APIs along with the lengths of codes, and students' performance over recent years.

Keywords: teaching HPC · parallel computing · HPC education · MPI · CUDA · OpenMP

1 Introduction

In recent years, a considerable increase in computational power has become possible primarily due to the incorporation of an increasingly larger number of cores into both CPUs and accelerators, such as GPUs, and improvements in memory sizes and bandwidth [11]. This applies to all the segments of computing devices, i.e., the data center, server, workstation, desktop, and mobile CPUs and GPUs. As a consequence, the development of parallel programs has become a necessity in order to make the most of the computational power of the computing devices

L. Franco et al. (Eds.): ICCS 2024, LNCS 14838, pp. 398–412, 2024.
https://doi.org/10.1007/978-3-031-63783-4_29

within each computer/node as well as in clusters of machines. Teaching HPC has then become the issue of the utmost importance. A number of Application Programming Interfaces (APIs) exist for general purpose programming in HPC, i.e., OpenMP, Pthreads for shared memory systems and multithreaded programming for multi-core CPUs, as well as offloading to accelerators, NVIDIA CUDA and OpenCL for GPUs, OpenCL for programming shared memory CPU+GPU systems, and Message Passing Interface (MPI) for distributed memory systems [11]. Use of frameworks allows for easier programming at a higher level of abstraction but it comes at the cost of performance overheads. For example, Kernel-Hive allows parallelization of computations among the nodes of a cluster using OpenCL as computational kernels and Java based management layer(s) supporting workflow model execution. The authors of [28] demonstrated overheads up to around 11% compared to the highly tuned MPI+OpenCL solution. Access to HPC resources can be granted either via ssh, remote shell [10], or through various middlewares, including Web interfaces [12], which are, however, often adapted over time [34]. Courses in HPC can also be delivered using container virtualization and open-source cloud technologies [2].

In this paper, we provide subjective assessment of the programming difficulty using a selected set of APIs that a population of programmers face. We also analyze the execution times and scalability of parallel codes written by them for various input data sizes and for the codes programmed with selected APIs. For this purpose, we propose a methodology that uses the same programming task, i.e., the implementation of Goldbach's conjecture for a range of input numbers, which is the material for a separate student laboratory, and focuses on various programming APIs such as: OpenMP, MPI, CUDA, and selected hybrid combinations. Our research results allowed us to conclude how the performance of codes, written by various programmers to solve a particular task using a given technology, differs. Additionally, we provide information on the length of codes for the analyzed APIs as well as track our students' performance over recent years.

The outline of the paper is as follows. In Sect. 2 we discuss the related works, while in Sect. 3 we present the motivation for our work. Section 4 contains a detailed description of the High Performance Computing Systems course from which we gathered the data analyzed in this paper. In Sect. 5 we provide details of our evaluation of the programmers and codes, including the proposed methodology, students' subjective evaluation data, practical evaluation of the students' codes, using objective metrics, and add subsequent discussion. Finally, Sect. 6 contains the study conclusions and the scope identified for further research.

2   Related Work

Importance of HPC Education. HPC education has become essential in recent years, especially that parallel computing on HPC systems enables modern machine learning models to grow in scale [9]. Accelerating deep learning is the key for future large models development [4]. However, as the authors of [27] claim, computing educators are only beginning to recognize the need for HPC education. The industry demand for talented sofware developers with experience in the HPC area has raised a concern if HPC skills should not be introduced earlier in the university education, i.e., for undergraduates [17], or even for secondary school students [6]. In this context, teaching HPC has become essential in developing relevant skills required by the industry, and already over a decade ago HPC was recommended to be taught as part of Computer Science Engineering education [29] at all the university degree levels.

Teaching HPC Academic Standards. The most common technologies used in teaching HPC are C and C++, with significantly less frequent practice of JAVA, Fortran or Python [7]. The most commonly used software APIs for HPC education is OpenMP, MPI, and CUDA. As confirmed by the most recent review [32], C++ along with MPI have become primary tools in teaching computational science, specifically when demonstrating parallel programming. In this context, we want to emphasize that our experience and conclusions from dealing with the method of teaching HPC presented in this paper are based on the technologies that are the industry and academic standard, i.e., C++, MPI, OpenMP, and CUDA.

As stated in [19], based on the review of 94 papers related to teaching HPC, the predominant method is still traditional learning, whereas projects (problem based learning), gamification, and collaborative learning are employed in far fewer cases. Also, the teaching objective is almost always programming, while the architecture or parallel thinking focus are rare. In this paper, we present the experience from conducting a course which combines project based teaching with more focus on architecture and parallel thinking, specifically emphasizing the differences between three parallel APIs and CPU vs GPU architecture.

Evolution of HPC Courses. In the past two decades, teaching HPC has evolved from Parallel Programming courses, where MPI and OpenMP, among others, were used as exemplary APIs for writing multithread and multi–process programs [25]. Also, Parallel Distributed Computing has gained more attention and various combinations of parallel APIs have been used to demonstrate the capabilities of diverse platforms and architectures, as reported by [30], i.e., using not only MPI and OpenMP for HPC on CPU clusters, but also employing GPGPU and APIs, such as CUDA, or presenting distributed processing of large data sets with Hadoop. Although the majority of courses use C and C++ as primary programming languages, natively supported by OpenMP, CUDA or MPI, some of them present a different approach with OpenMP-like directives used with Java and Pyjama compiler [20].

Early courses in HPC encountered different barriers such as lack of access to typical HPC clusters. The problems were addressed with using small educational clusters such as Mozart, which unblocked teaching HPC [5]. In developing countries, e.g. in Mexico, as reported by [33], even nowadays one of the barriers, besides deficiencies in HPC educational infrastructure, is still lack of interest from students in learning HPC.

On the other hand, many HPC courses with the well established scope tend to explore innovative approaches to HPC education, such as focus on gaining cluster configuration skills instead of just introducing HPC APIs, proposed by [22], or emphasizing parallel distributed computing patterns, proposed by [1]. The authors of [14] explore how using a higher level of abstraction, i.e., the Thrust framework for parallel GPU programing, enhances programmer productivity. Our HPC course proposed for postgraduates [13] introduces such concepts as: blocking and non-blocking communication, overlapping of compute and communication, and dynamic load balancing. The method described in this paper is the extension of the previous studies, focusing on the practical incorporation of various parallel APIs (OpenMP, MPI, CUDA, hybrid MPI+OpenMP), into the solution for a well defined task and running its various implementations in several system types (multicore CPU, cluster of CPUs, and GPU).

Methods of Teaching HPC. Following the challenges arising with the evolution of HPC education, novel methods enhancing HPC courses are being proposed. At the early stages of education, teaching Computational Thinking was proposed by [21] for elementary and middle school students. For undergraduates, the authors of [24] propose teaching inexperienced programmers with a set of analogies helpful in understanding the basic concept of parallelism. On the other hand, the authors of [23] propose a challenge–based approach inspired by Parallel Programming Marathons in Brazil.

The authors of [8] identified three major issues in HPC education and addressed them by creating their own HPC curriculum system by empowering students' innovation capability with a project–based Parallel Programming course, involving real–life examples from the HPC field, such as Computational Fluid Dynamics, or 3-D simulations of human ventricular tissues. Lack of suitable HPC environments dedicated to teaching HPC skills is often addressed by proposing small–scale clusters. For instance, [26] proposed a 25-node cluster based on Raspberry Pi nodes, while [3] presented their small–scale cluster based on Odroid boards.

3 Motivations

Considering the evolution of HPC courses in the past two decades and the shift of focus in HPC education towards more parallel thinking and project–based classes, we were motivated to propose a new method for the laboratory part of our original HPC course [13]. It assumed the implementation of the same programming task using several representative parallel programming APIs. While we did not aim at performing an inter-API or inter-software stack performance comparison, the method allowed us to provide meaningful assessment of the programmers' relative performance within each API. Additionally, the obtained results provided information on the most important general purpose parallel programming APIs because these cover the most important types of computer systems and compute devices, i.e., shared-memory single node systems with multi-core CPUs (programmed with OpenMP) and GPUs (programmed with CUDA), and distributed memory multi-node systems, programmed with: MPI, MPI+OpenMP, and MPI+Pthreads. For this purpose, we incorporated both theoretical knowledge on parallelization concepts, paradigms and APIs as well as practical tasks using particular APIs into our High Performance Computing Systems (HPCS) course, offered to all the computer science first semester MSc level students. For the practical part, the same programming task – verification of Goldbach's conjecture – was given to students for implementation using the aforementioned APIs.

4   High Performance Computing Systems Course

4.1   Aims, Structure and Scope

The goal of the course is to acquire the neccessary knowledge and skills that allow for designing, implementation, deployment as well as testing parallel implementations, and solving a given computational problem. The primary objective for the student is to be able to develop scalable parallel programs that would obtain high speed-ups, taking advantage of modern parallel environments. The latter include both shared and distributed memory systems (clusters) with multi-core CPUs and GPUs. Assessment of students' work involves taking and passing a theoretical exam (with a score of at least 50%) as well as obtaining at least 50% of the points on lab exercises. Both the theoretical and the practical parts contribute to the final grade equally.

The teaching approach involves getting to know parallel processing paradigms representative of computationally demanding applications and, subsequently, discussing specific techniques allowing for their efficient parallelization. The following paradigms are considered [11]:

1. Dynamic master-slave – the master process/thread initially partitions the input data into chunks which are distributed among the slaves dynamically.
2. Geometric Single Program (Stream) Multiple Data (Stream) – the computations are performed in iterations over a domain which is partitioned geometrically among the processes/threads for parallel updates of their subdomains, followed by the synchronization/communication (boundary data exchange). This paradigm typically refers to simulations of physical phenomena.
3. Pipelining – data chunks of the input stream are passed through a pipeline and parallel processing of particular chunks is performed.
4. Divide and conquer – the original problem is partitioned into subproblems recursively down to the leaves of either a balanced or an imbalanced tree. Subtrees are processed in parallel by processes/threads with synchronization corresponding to data merging and subsequent passing up towards the root of the tree.

After the students have learnt these concepts and techniques, mapping relevant solutions onto specific APIs and environments is discussed during this course.

The mapping is done mainly with distributed memory systems in mind. For this purpose, MPI is used. Additionally, the HPCS course features the basics of shared memory programming, including multithreaded programming for multi-core CPUs - using OpenMP, and for GPUs - using NVIDIA CUDA.

4.2 Lecture

The lecture includes the following components:

1. Introduction to computing in parallel environments: model of a parallel application: basic parameters of an application and a parallel system, topologies; assignment criteria; latency and bandwidth; specifications of real systems.
2. Parallel architectures: Shared Memory, Distributed Shared Memory, Distributed Memory.
3. Parallel processing paradigms: master-slave, SPMD, pipelining, divide-and-conquer; examples.
4. MPI: model of an application; execution, various implementations, queueing systems; send communication modes; non-blocking communication; blocking communication; collective vs. point-to-point communication, creation of data types and packing; communicators; dynamic load balancing: repartitioning, "ghost nodes"; examples of applications in MPI – performance on real clusters; spawning processes dynamically, one-way communication; checkpointing of parallel aplications; MPI+multi-threading; Parallel I/O in MPI; mapping paradigms to MPI.
5. OpenMP – API: directives and functions; synchronization constructs; access to shared resources; minimization of synchronization and overheads; using OpenMP for accelerators; examples; mapping paradigms to OpenMP.
6. CUDA – model of an application, grid parameters, performance vs. usage of resources; optimized memory access patterns; using GPU shared memory; examples; mapping paradigms to CUDA.

4.3 Laboratory Tasks

Typically, there are 6 laboratory classes (90-minute each) during which students are given tasks that usually require modifications of and building on the provided sample code and the corresponding manual. Each student is expected to complete the task either during a given lab session or during the following meetings (penalties apply in case of delay). We offer the following manuals along with the corresponding exemplary codes:

1. **MPI Embarrassingly parallel computations** with parallel reduction of results by the designated process. *Example:* parallel computations of a series such as for computing the pi number. Even though partitioning of the series among the processes for load balancing of computations is straightfoward, it could still be done in various ways, which might potentially affect the solution accuracy, depending on what elements are added in the process (for instance, adding numbers with similar absolute values in each process vs. adding both large and small numbers).

2. **MPI dynamic master-slave example**. The master partitions the input data into chunks which are then distributed among slave processes. The latter return results to the master, waiting for subsequent data chunks. This scheme, assuming that the ratio of time spent on computations to communication is high, and the number of data chunks is considerably larger than the number of slave processes, allows for effective load balancing and high speed-ups. *Example:* parallel numerical integration of a function on a given range.

3. **CUDA basics** of parallel programming based on programming operations on vectors and matrices such as weighted addition of vectors, multiplication of a matrix and a vector, multiplication of a sparse matrix by a vector, dot product. Gains from using CUDA shared memory are demonstrated.

4. **OpenMP dynamic master-slave** code in which threads fetch the available packet identifiers in a critical section, fetch the packet afterwards, and then process these in parallel. The supplied manual further discusses and demonstrates that the critical section may become a bottleneck when the number of threads is very large, and proposes to use named critical sections to improve the performance for smaller thread groups. *Example:* parallel numerical integration - the initial range is partitioned into a number of data packets at least a few times (4 to 5+) larger than the number of threads.

5. **MPI dynamic master-slave** example with overlapping communication and computations using non-blocking MPI communication routines. This code extends the previous example with initial sending of a data packet per slave, followed by non-blocking sends of another data packet per slave. This facilitates overlapping computations in a slave by sending a new data packet from the master, and receiving the previous result packet from the slave. For applications using larger data packets this shall lead to the minimization of the total application execution time [11]. MPI_Isend, MPI_Irecv and MPI_Wait(all) routines are used.

6. **MPI+Pthreads**. Extension of Task 1 with multithreading. Multithreading modes in MPI are discussed and Pthreads are spawned in each process for multithreaded processing on a multi-core CPU or CPUs. Inter-thread communication using MPI_THREAD_MULTIPLE is discussed vs. optimized reduction within the processes and using MPI_Reduce. *Example:* parallel computations of a series, e.g., computing the pi number.

5   Evaluation

5.1   Methodology

Within this work, we aim at the following targets:

1. Individual assessment of technologies by students (difficulty, usefulness).
2. Assessment of performance in a programmers' population sample by measuring the differences in times as well as speed-ups. The differences can stem from either a better selection of the algorithm and/or its parallel design and/or implementation.

3. Measuring the lengths of codes using respective APIs.

For these purposes, we collected both subjective opinions of our students (survey) and objective measures, i.e., execution times of the codes for various technologies and for various input data sizes. We shall note that for the latter we used an isolated environment, the same that was used by the students to develop and test the codes. What is important, the same application was implemented by the students using various parallel programming APIs.

5.2   Computational Task

Goldbach's conjecture states that any positive even integer number can be written as a sum of two prime numbers [15]. Goldbach's conjecture can benefit significantly from the initial generation of the sieve of Eratosthenes, which is used in the parallel verification of the conjecture. The best strategy for verification of the conjecture for an $[a, b]$ range might depend on the a and b values, as well as $b - a$. Several strategies and parallel implementations were investigated by the authors of [31], including the ones using OpenMP, MPI hybrid MPI-OpenMP parallelism, as well as OpenACC GPU accelerated computing. The evaluation of CUDA's Unified Memory (UM) and dynamic parallelism (DP) for Goldbach's conjecture was verified in [18], assuming that the verification is performed for a vector of input numbers. The second mechanism, tested for numbers in the order of 10^{10}, resulted in performance loss compared to the standard, non-DP version. On the other hand, verification of Goldbach's conjecture for 10,000 numbers from 10^8 to 10^{12} using Unified Memory resulted in practically the same performance as in the standard non-UM version. In this case, this is a positive result since UM-based implementation is typically more straightforward and shorter.

In this paper, we use Goldbach's conjecture to evaluate students' approaches to parallel implementation and assess the group's relative performance. We asked students to carry out 5 laboratory tasks, each meant to implement the verification of Goldbach's conjecture, using the following 3 technologies: MPI, OpenMP, CUDA, and their 2 combinations: MPI+OpenMP, MPI+Pthreads. Each program was expected to verify the conjecture for all the numbers in the given input $[a, b]$ range.

5.3   Students' Subjective Evaluation of Parallel Programming APIs

Figure 1 presents students' subjective evaluation of how easy it was to program with a given parallel programming API (higher grade means easier). It can be seen that OpenMP, allowing for a relatively straightfoward extension of sequential codes with directives and library calls, is clearly perceived as the easiest to use, followed by CUDA and MPI with practically the same median values.

5.4   Practical Evaluation of Students' Codes Using Objective Metrics

Given the limited laboratory time, students devised various implementations that consequently resulted in various execution times, as presented next. Fig-

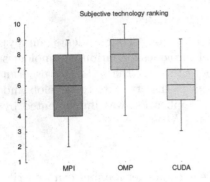

Fig. 1. Students' subjective evaluation of programming easiness using a particular programming API

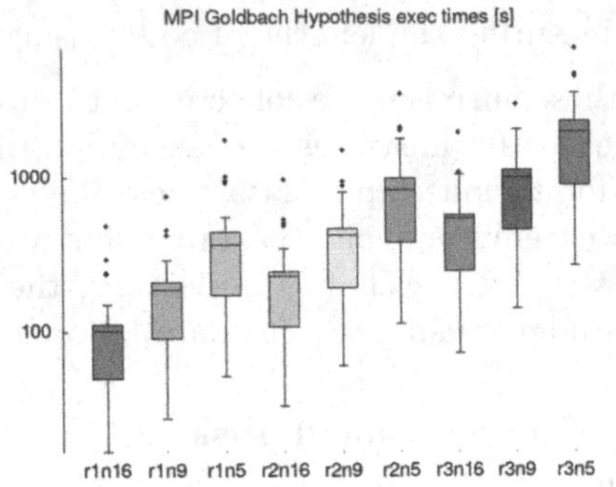

Fig. 2. Execution times for MPI codes

ures 2, 3 and 4 present boxplot type charts for execution times for the codes that were not based on the Eratosthenes sieve. This approach was adopted by the majority of students for the codes implemented using a single technology: MPI, CUDA, and OpenMP respectively.

The testbed environment consisted of 16 workstations, each equipped with an Intel Core i7-7700 CPU with four physical cores and Hyperthreading, 16GB of RAM and interconnected by a 1GBit Ethernet network. In the following charts, rx, i.e., r1, r2, and r3, denotes three different input data ranges of numbers to be verified against the Goldbach conjecture, respectively. They are: r1=[50.000.000:100.000.000], r2=[50.000.000:150.000.000], and r3=[50.000.000: 250.000.000].

Additionally, nx in the charts indicates x number of nodes used in the MPI environment (with a single MPI process slot on each node) while tx indicates x number of threads used in the OpenMP version of the code.

After running the students' codes in our environment, we present the execution times for the analyzed parallel programming APIs and the ranges in Fig. 2 (MPI), Figs. 3 (CUDA), and 4 (OpenMP). Figures 5 and 6 present the results for hybrid technologies, i.e., MPI+Pthreads and MPI+OpenMP respectively, for configurations employing 8 threads per CPU and 16 MPI processes, across all the input data ranges.

It should be noted that the objective of these tasks was to develop parallelized and scalable (versus the number of processes or/and threads) code using a given technology, rather than performing a direct comparison of the technologies performance.

Apart from the aforementioned results without the sieve, Table 1 summarizes the results that utilized the sieve-based approach. It includes the proportions of students utilizing a given technology who developed a sieve-based solution, and the best times for the sieve-based approaches. Efficient sieve-based solutions were proposed by the students for two technologies: OpenMP and MPI.

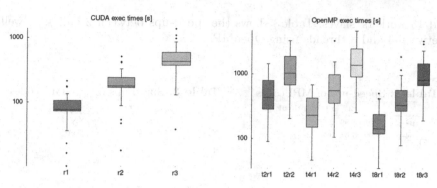

Fig. 3. Execution times for CUDA codes

Fig. 4. Execution times for OpenMP codes

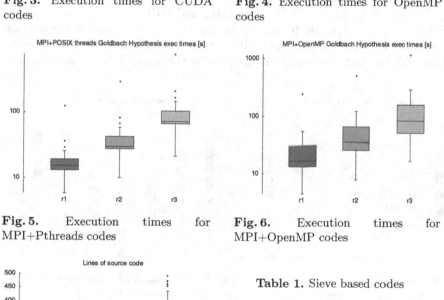

Fig. 5. Execution times for MPI+Pthreads codes

Fig. 6. Execution times for MPI+OpenMP codes

Fig. 7. Numbers of lines for the analyzed codes

Table 1. Sieve based codes

Version	OpenMP	MPI
Percentage of sieve-based codes [%]	9.6%	6.9%
r1 best [s]	1.44	3.98
r2 best [s]	2.32	5.18
r3 best [s]	4.13	6.74

Additionally, we assessed the speed-ups of the students' codes. Table 2 presents the average speed-ups of MPI codes between 4 and 8, as well as between

4 and 15 workers/slaves. Table 3 shows the speed-ups between 2 and 4, as well as between 4 and 8 threads using OpenMP.

Table 2. Speed-ups of MPI codes

	r1	r2	r3	optimal
p4-p8	1.95	1.99	2.00	2.00
p4-p15	3.47	3.61	3.72	3.75

Table 3. Speed-ups of OpenMP codes

	r1	r2	r3
t2-t4	1.90	1.88	NA
t4-t8	1.41	1.43	1.43

Figure 7 depicts boxplot graphs of lines of codes for the implementations using 3 major technologies: CUDA, OpenMP, and MPI.

Furthermore, Figs. 8 and 9 present the grading results from the theoretical and laboratory course components of those students who passed the course, over the period of the last 7 years (0.5 is the passing threshold, 1 is the maximum).

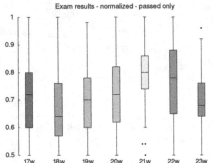

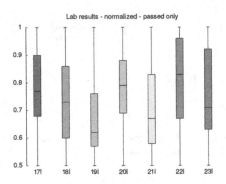

Fig. 8. Evaluation of theoretical component in successive years

Fig. 9. Evaluation of laboratory component in successive years

5.5 Discussion

We observed a systematic decrease in the population sample size during the experiment progress, i.e., at the beginning of the course all the students attempted to solve the assignments (MPI and OpenMP). As soon as they secured the minimum required to pass the course, we observed a significant drop in the submitted assignments (CUDA, MPI+OpenMP, MPI+Pthreads). We shall note that during the initial verification of solutions handed in by the students during the laboratory classes, the test ranges were smaller than these studied in the paper. Consequently, some of the solutions resulted in timeouts in this analysis.

In terms of execution times for a particular API and the range size, we see observe that the performance of students' codes can differ by an order of magnitude practically for all the APIs.

From Table 1 we can conclude that for MPI and OpenMP there is a small number of much faster codes that benefit from implementations using the Eratosthenes sieve and other optimizations. We shall note that time could play an important role in the submission of assignments, as students are generally expected to deliver their solutions by the end of a 90-minute laboratory session, on which the assignment was given. Later submissions resulted in penalties. Consequently, students apparently leaned to simpler solutions first, in order to secure passing a given assignment, and attempted the optimized versions when time permitted. We conclude that the proportions listed in Table 1 reflect the work of the most ambitious students. Additionally, we see that in the case of CUDA and hybrid codes students were not able to submit faster sieve-based codes which might indicate that these technologies appeared more challenging to develop optimized solutions, given the time limit.

In terms of speed-ups shown in Table 2, we can conclude that, since the presented speed-ups are the average values that are so close to the optimal ones (assuming the ideal scaling), using a provided well scaling template (for the master-slave MPI exercise) results in very good speed-ups, especially since the Goldbach conjecture code features a large compute/communication time ratio. Still, we can observe the anticipated growth in speed-ups with the increasing problem size. In the case of speed-ups obtained for the OpenMP codes shown in Table 3, we should realize that they were run on an Intel i7-7700 CPU with 4 physical cores and HyperThreading (HT) for a total of 8 logical processors. Thus, we can see that between 2 and 4 threads the speed-ups are very good – 1.88 and 1.9 (for r3 and 2 threads we did not run the code due to its long execution times). Using the HT technology resulted in approx. 41–43% gain over a physical core performance, which is in line with other reference results [11].

Boxplots presenting the line counts of programmers' codes using OpenMP, CUDA, and MPI, shown in Fig. 7, reveal that, OpenMP codes were generally the shortest, followed by those in CUDA, and then in MPI, the latter being considerably longer. For the MPI codes, many students used the provided master-slave template that could be filled in with master and slave content specific to Goldbach's conjecture implementation.

From Figs. 8 and 9 we can see that both the exam and laboratory results were relatively stable over the period of 7 years. However, some differences might be indicated. We can observe that for the 2021 COVID edition, the exam median was noticeably higher and the interquartile range was narrower. Additionally, only for that year the outliers were close to the 50% passing threshold, which denotes that the vast majority of the results was higher than 60%. We attribute this observation to the remote/online mode of exam taking at that time. Interestingly, students achieved relatively lower scores for the laboratory assignments then.

Furthermore, for each technology, we measured the correlation between students' subjective view of ease of use of a particular technology, and their code execution times for that technology. We found out that there is no meaningful correlation between the two, as the correlation coefficients obtained for MPI, OpenMP, and CUDA were: -0.055, 0.152, and 0.066 respectively.

6   Conclusion and Future Work

In the paper, we presented details of the High Performance Computing Systems course that has been conducted at Gdańsk University of Technology for several years already. Based on a specific laboratory exercise – verification of Goldbach's conjecture for a range of numbers, we analyzed the performance of students' parallel codes solving that problem, implemented using several parallel programming APIs, including: MPI, OpenMP, CUDA, MPI+Pthreads, and MPI+OpenMP. We not only analyzed the execution times of a group of programmers with medians, boxplots and outliers giving relative spread of students' performance, but also studied speed-ups for MPI and OpenMP for three input range sizes. We distinguished sieve and non-sieve based solutions. We found out that the students subjectively evaluate the ease of programming as greater for OpenMP, compared to MPI and CUDA. We also provided medians and boxplots of the codes' line counts, and of students' performance in the theoretical and laboratory components of the course over the past 7 years.

Based on our findings, we concluded that using the proposed course format with a single task problem allowed students to focus on the differences between APIs and technologies used, rather than spend a considerable amount of time getting to know the details of specific problems. In terms of evaluation, we confirmed that using the same algorithmic approach, the performance of students' codes can differ by an order of magnitude, which is consistent with other findings on programmers' performance [16].

In the future, we plan to repeat the experiments with other exercises as well as look into the possible optimizations such as overlapping communication and computations for codes with considerable communication and synchronization times.

References

1. Adams, J., Brown, R., Shoop, E.: Patterns and exemplars: compelling strategies for teaching parallel and distributed computing to CS undergraduates. In: 2013 IEEE International Symposium on Parallel & Distributed Processing, Workshops and Ph.d. Forum, pp. 1244–1251 (2013). https://doi.org/10.1109/IPDPSW.2013.275
2. Al-Jody, T., Aagela, H., Holmes, V.: Inspiring the next generation of HPC engineers with reconfigurable, multi-tenant resources for teaching and research. Sustainability 13(21), 11782 (2021). https://doi.org/10.3390/su132111782
3. Alvarez, L., Ayguade, E., Mantovani, F.: Teaching HPC systems and parallel programming with small-scale clusters. In: 2018 IEEE/ACM Workshop on Education for High-Performance Computing (EduHPC), pp. 1–10 (2018).https://doi.org/10.1109/EduHPC.2018.00004

4. Ben-Nun, T., Hoefler, T.: Demystifying parallel and distributed deep learning: an in-depth concurrency analysis. ACM Comput. Surv. **52**(4), 3320060 (2019). https://doi.org/10.1145/3320060

5. Bernreuther, M., Brenk, M., Bungartz, H.J., Mundani, R.P., Muntean, I.L.: Teaching high-performance computing on a high-performance cluster. In: Sunderam, V.S., van Albada, G.D., Sloot, P.M.A., Dongarra, J.J. (eds.) Computational Science - ICCS 2005, pp. 1–9. Springer, Heidelberg (2005). https://doi.org/10.1007/11428848_1

6. Brođanac, P., Novak, J., Boljat, I.: Has the time come to teach parallel programming to secondary school students? Heliyon **8**(1), e08662 (2022). https://doi.org/10.1016/j.heliyon.2021.e08662

7. Carneiro Neto, J.A., Alves Neto, A.J., Moreno, E.D.: A systematic review on teaching parallel programming. In: Proceedings of the 11th Euro American Conference on Telematics and Information Systems. EATIS 2022, Association for Computing Machinery, New York, NY, USA (2022). https://doi.org/10.1145/3544538.3544659

8. Chen, J., Impagliazzo, J., Shen, L.: High-performance computing and engineering educational development and practice. In: 2020 IEEE Frontiers in Education Conference (FIE), pp. 1–8 (2020). https://doi.org/10.1109/FIE44824.2020.9274100

9. Coates, A., Huval, B., Wang, T., Wu, D.J., Catanzaro, B.C., Ng, A.Y.: Deep learning with COTS HPC systems. In: International Conference on Machine Learning (ICML) (2013)

10. Czarnul, P.: Integration of compute-intensive tasks into scientific workflows in BeesyCluster. In: Alexandrov, V.N., van Albada, G.D., Sloot, P.M.A., Dongarra, J. (eds.) Computational Science - ICCS 2006, pp. 944–947. Springer, Heidelberg (2006). https://doi.org/10.1007/11758532_127

11. Czarnul, P.: Parallel Programming for Modern High Performance Computing Systems. CRC Press, Taylor & Francis (2018). ISBN 9781138305953

12. Czarnul, P.: Teaching high performance computing using BeesyCluster and relevant usage statistics*. Procedia Comput. Sci. **29**, 1458–1467 (2014). https://doi.org/10.1016/j.procs.2014.05.132

13. Czarnul, P., Matuszek, M.: Use of ICT infrastructure for teaching HPC. In: 2019 IEEE 14th International Conference on Computer Sciences and Information Technologies (CSIT), vol. 1, pp. xvii–xxi (2019).https://doi.org/10.1109/STC-CSIT.2019.8929841

14. Daleiden, P., Stefik, A., Uesbeck, P.M.: GPU programming productivity in different abstraction paradigms: a randomized controlled trial comparing CUDA and thrust. ACM Trans. Comput. Educ. **20**(4), 3418301 (2020). https://doi.org/10.1145/3418301

15. Guy, R.: Unsolved Problems in Number Theory. Springer Science & Business Media, New York (2013). https://doi.org/10.1007/978-0-387-26677-0

16. Guzdial, M.: Is there a 10x gap between best and average programmers? And how did it get there? (November 2014), Communications of the ACM. https://cacm.acm.org/blogs/blog-cacm/180512-is-there-a-10x-gap-between-best-and-average-programmers-and-how-did-it-get-there/fulltext

17. Holmes, V., Kureshi, I.: Developing high performance computing resources for teaching cluster and grid computing courses. Procedia Comput. Sci. **51**, 1714–1723 (2015). https://doi.org/10.1016/j.procs.2015.05.310

18. Jarzabek, L., Czarnul, P.: Performance evaluation of unified memory and dynamic parallelism for selected parallel CUDA applications. J. Supercomput. **73**(12), 5378–5401 (2017). https://doi.org/10.1007/s11227-017-2091-x

19. de Jesus Oliveira Duraes, T., Sergio Lopes de Souza, P., Martins, G., Jose Conte, D., Garcia Bachiega, N., Mazzini Bruschi, S.: Research on parallel computing teaching: state of the art and future directions. In: 2020 IEEE Frontiers in Education Conference (FIE), pp. 1–9 (2020).https://doi.org/10.1109/FIE44824.2020.9273914

20. Kurniawati, R.: Teaching parallel programming with java and Pyjama. In: Proceedings of the 53rd ACM Technical Symposium on Computer Science Education, vol. 2, p. 1109. SIGCSE 2022, Association for Computing Machinery, New York, NY, USA (2022). https://doi.org/10.1145/3478432.3499115

21. Lamprou, A., Repenning, A.: Teaching how to teach computational thinking. In: ITiCSE. Association for Computing Machinery, New York, NY, USA (2018).https://doi.org/10.1145/3197091.3197120

22. López, P., Baydal, E.: Teaching high-performance service in a cluster computing course. J. Parallel Distrib. Comput. 117, 138–147 (2018). https://doi.org/10.1016/j.jpdc.2018.02.027

23. Marzulo, L., Bianchini, C., Santiago, L., Ferreira, V., Goldstein, B., França, F.: Teaching high performance computing through parallel programming marathons. In: IPDPSW, pp. 296–303 (2019) https://doi.org/10.1109/IPDPSW.2019.00058

24. Neeman, H., Lee, L., Mullen, J., Newman, G.: Analogies for teaching parallel computing to inexperienced programmers. SIGCSE Bull. 38(4), 64–67 (2006). https://doi.org/10.1145/1189136.1189172

25. Pan, Y.: Teaching parallel programming using both high-level and low-level languages (2001)

26. Pfalzgraf, A.M., Driscoll, J.A.: A low-cost computer cluster for high-performance computing education. In: IEEE International Conference on Electro/Information Technology, pp. 362–366 (2014). https://doi.org/10.1109/EIT.2014.6871791

27. Raj, R.K., et al.: High performance computing education: current challenges and future directions. In: ITiCSE-WGR 2020. Association for Computing Machinery, New York, NY, USA (2020). https://doi.org/10.1145/3437800.3439203

28. Rościszewski, P., Czarnul, P., Lewandowski, R., Schally-Kacprzak, M.: Kernelhive: a new workflow-based framework for multilevel high performance computing using clusters and workstations with cpus and gpus. Concurr. Comput. Pract. Exp. 28(9), 2586–2607 (2016). https://doi.org/10.1002/cpe.3719

29. Rüde, U., Willcox, K., McInnes, L.C., Sterck, H.D.: Research and education in computational science and engineering. SIAM Review 60(3), 707–754 (2018). https://doi.org/10.1137/16M1096840

30. Shamsi, J.A., Durrani, N.M., Kafi, N.: Novelties in teaching high performance computing. In: 2015 IEEE International Parallel and Distributed Processing Symposium Workshop, pp. 772–778 (2015) https://doi.org/10.1109/IPDPSW.2015.88

31. Shaw, A., Varon, D.: Numerical verification of GoldBach's conjecture with parallel computing (2018). CS205: Extreme Scale Data and Computational Science, Harvard. https://github.com/djvaron/Goldbach

32. Sitsylitsyn, Y.: A systematic review of the literature on methods and technologies for teaching parallel and distributed computing in universities. Ukrainian J. Educ. Stud. Inf. Technol. 11(2), 111–121 (2023). https://doi.org/10.32919/uesit.2023.02.04

33. Trejo-Sánchez, J.A., et al.: Teaching high-performance computing in developing countries: a case study in Mexican universities. In: IPDPSW, pp. 338–345 (2022). https://doi.org/10.1109/IPDPSW55747.2022.00066

34. Xu, Z., Chi, X., Xiao, N.: High-performance computing environment: a review of twenty years of experiments in China. Natl. Sci. Rev. 3(1), 36–48 (2016). https://doi.org/10.1093/nsr/nww001

Evaluating Teacher's Classroom Performance

Raul Ramirez-Velarde[(⊠)] ⓘ, Laura Hervert-Escobar ⓘ, and Neil Hernandez-Gress ⓘ

Tecnologico de Monterrey, Eugenio Garza Sada 2501 Sur, Col. Tecnológico,
64849 Monterrey, N. L., Mexico
`rramirez@tec.mx`

Abstract. Teacher evaluation in the classroom is a multifaceted challenge, with no one-size-fits-all solution. While student evaluations provide valuable feedback, they are limited by students' inability to accurately assess their own learning. This reliance on student evaluations alone can lead to biased assessments, grade inflation, and misconceptions about learning. To address these issues, a multi-dimensional approach to teacher evaluation is proposed. This approach incorporates various assessment methods, including on-site evaluations, standardized tests, portfolio reviews, and student surveys, to provide a comprehensive view of a teacher's performance.

Keywords: Teacher evaluation · Classroom assessment · Student feedback

1 Introduction

As in any professional environment, the evaluation of professors at any university is an exercise that not only aids the institute in improving its educational processes but also assists professors in discovering their strengths and areas for improvement. This paper will exclusively discuss the evaluation of professors in the classroom, without considering their roles in administrative positions, scientific production, or participation in projects.

The evaluation of professors in universities plays a crucial role in the educational process as it aims to ensure the quality and effectiveness of teaching. This mechanism allows academic institutions to gather valuable feedback from students regarding the performance of teachers, thereby opening the door to continuous improvement. Evaluation has become an essential tool for measuring aspects such as clarity in communication, the ability to motivate students, and currency in content knowledge, among others.

However, its implementation poses challenges, such as the inherent subjectivity in student opinions, mapping the personal and professional characteristics of the professor to a score in the evaluation, intellectual production, and the need to balance quantitative and qualitative evaluation. In this context, the evaluation of professors emerges as a dynamic process aimed at raising educational standards and promoting academic excellence.

This paper aims to conduct a scientific evaluation of performance evaluation of professors in the classroom and the role that the Students Evaluation of Teaching (SET)

L. Franco et al. (Eds.): ICCS 2024, LNCS 14838, pp. 413–426, 2024.
https://doi.org/10.1007/978-3-031-63783-4_30

plays in it by reviewing recent available literature. Based on the research findings, a reference model for the classroom performance evaluation of professors is proposed.

The assertion that the outcome of the SET significantly influences professor evaluation, salary increments, and future career trajectory is widely acknowledged. SETs serve as the primary means to measure a professor's performance in their interaction with students. There is a prevailing assumption that a higher evaluation score on the SET correlates with better student learning outcomes. However, is this true?

In this paper, we will address the following points:

1. Students' perception of their own learning.
2. Evaluation of professors and actual learning outcomes.
3. The illusion of learning: a definition of teaching fluency.
4. Factors biasing the evaluation of professors.
5. The persistence of professor evaluation: correlations over time and space.
6. The relationship between professor evaluation and their level of rigor.
7. The relationship between learning, professor evaluation, and grades.
8. Does professor evaluation incentivize poor teaching practices?
9. The role of social networks.
10. Alternatives for evaluating professors' performance in the classroom: A Tec Model.

To address these topics, a literature review has been conducted, summarized as follows.

2 Students' Perception of Their Own Learning

How proficient are students in assessing their own learning? It is reasonable to assume that students are adept at estimating how much they have effectively learned, especially considering that they have spent decades learning and dedicating most of their time to education. Initially, in the mid-20th century, experiments seeking to measure students' perception of their own learning found substantial evidence that students were adept at gauging their level of learning [1]. However, more careful measurements conducted later revealed that these results were **influenced by three biases**: a) publication bias, b) small sample bias, and c) disciplinary bias. **Publication bias** occurs because **results with a positive correlation are more likely to be published** in a journal than those with a null or negative correlation. **Small sample bias** implies that **correlations found in small or non-carefully selected samples must be very large to be statistically significant**. That is, if the sample size is small and the correlation is not close to 1, the result is not valid because the confidence interval in small samples is very wide. The most significant bias is **disciplinary bias**. The field **of Pedagogy consistently**, suffering from the two biases mentioned above, **finds a positive correlation between perception of learning and actual learning, while disciplinary areas such as business, science, and engineering have found results that are precisely the opposite**. See Clayson's meta-analysis (2009) [1] and Uttl et al.'s meta-analysis of meta-analyses (2017) [2]. Perhaps initially, students could accurately evaluate their learning, and then this changed as the teaching system evolved. Thus, experiment after experiment has shown that students are poor at estimating their own learning. Under normal conditions, students consistently overestimate their learning, resulting in generally negative reactions when their actual performance is communicated to them. Not only that. Students may believe they have learned more than they have if the instructor speaks fluently and enthusiastically. And they may believe they have learned less than they have if the instructor speaks less fluently and lacks enthusiasm. The literature on this matter is vast. In addition to Clayson and Uttl et al., we have Papamitsiou and Economides (2014) [3], Deslauriers et al. (2019) [4], Clauss and Geedey (2010) [5], Stroebe [6], and many more references included within these articles. Uttl et al. is a meta-analysis of previous meta-analyses.

A very interesting result is found in Claus et al., where the difference between perceived learning and assessed learning reached a value of up to 36 on the Kruskal-Wallis test, when the limit is 8 [5]. The problem with this study is the small sample size. The sample contains about 80 evaluations when the minimum sample size is 384. More robust and interesting results are found in Papamitsiou et al., where, using pre- and post-tests, the difference between perceived learning both pre and post, compared against the measurement of actual learning, is up to 6.56 standard deviations, whereas the limit for statistical significance is approximately 2 standard deviations [3]. **This means that students are not bad at assessing their own learning.**

What does this mean for the evaluation of professors? We will discuss this in the following section.

3 Professor Evaluation and Actual Learning

The first article on student evaluations of their professors (SET) was published over 80 years ago, in 1927. In the 1960s and 1970s, the use of student evaluations expanded significantly in the United States, and currently, nearly all North America employs SET to assess the effectiveness of professors. Such evaluations are used for hiring, salary increases, promotions, selection for special projects, classification, department and school evaluations, among other purposes.

There are several reasons for utilizing SET. Among them, a) it is a quick, relatively inexpensive, and numerical way to evaluate teaching, b) it allows students to have a voice in the evaluation of their professors and academic departments, and c) students are the primary witnesses to how professors conduct teaching [1, 2].

Although there is abundant literature defending SET as a good indicator of learning, this literature is subject to **publication bias**, in the sense that it mainly comes from the field of pedagogy and typically has methodological issues such as lack of statistical rigor in data treatment and small sample sizes [1, 2]. A vast array of studies, including a couple of meta-analyses that reassess the conclusions of several previous studies [1–4, 7, 8], demonstrate that **the correlation between professor evaluation and actual learning is at best very low, commonly null, and even occasionally negative**. Uttl et al. found that **SET can only explain 1% of measured learning and that if the student's prior learning is considered, the relationship between SET and learning is not statistically different from zero** [2]. The positive correlation, typically low, occurs on rare occasions and is highly dependent on disciplinary area, professor, course level, and multiple other factors [1]. And in general, **the more objectively learning is assessed, for example, in analytical classes using standardized exams, the relationship between learning**

and SET disappears [1]. To clarify, a negative correlation means that a professor with certain speaking characteristics gives a false sense of learning but teaches less than a professor who is rated worse. This will be discussed in the following section.

4   The Illusion of Learning: A Definition of Teaching Fluency

If students are poor at estimating their learning within a class, what are we evaluating? Actually, **SETs measure student satisfaction with the course**, which, while a genuinely useful measurement, is not a measurement of learning. SET is based on the student's perception of their professor's characteristics, such as their level of subject knowledge, clarity of expression, organization, enthusiasm, likability, fairness, availability, accessibility, sense of humor, personal treatment, etc. [2]. But there are many other factors, all outside the control of the professor, that influence student satisfaction with the subject matter and therefore influence SET. For example, whether the student has received the grades they believe they deserve, the student's interest in the topics of the subject, whether the subject is elective or the student is forced to take it, whether the course is analytical or quantitative or not, the level of the class in the student's career profile, the class time and duration, whether the student has had any disputes with the professor, whether the student finds the professor's accent strange, whether the professor is male or female, and even the well-known effects of height and attractiveness on perceived competence [2, 8].

As we have already mentioned, students are poor at estimating their own learning. Derived from this, multiple studies have shown that students feel they have learned more when the lesson is taught by a professor who achieves fluency in presentation than when the lesson is taught by a professor who disrupts fluency for active learning. Even when it is the same professor. Carpenter et al. call this the "**Illusion of Learning**".

There are a couple of experiments by Carpenter et al. [7] and Deslauriers et al. [4]. In Carpenter's experiment, the same professor taught two classes. In the first group, the professor gave the class with great fluency, and in the second group with little fluency. In both experiments, a fluid class is defined as one in which the presenter stands upright, maintains eye contact with the students, does not consult their notes, and is expressive in their gestures. Additionally, the professor must speak clearly, and the presentation must be well-organized with a defined structure. In the non-fluid class, the professor had a slouched posture, somewhat hiding behind furniture, did not maintain eye contact with the students, consulted their notes constantly, spoke in a monotone, and was not expressive in their gestures. The students gave a much better evaluation to the fluent version of the professor and expressed that they felt they learned a lot with this professor, while the students of the non-fluid version rated the professor much lower and expressed having learned little. Surprisingly, the **exams showed that the learning of both groups was the same.**

In Deslauriers' experiment, two classes were set up again. One was a fluid class, without interruptions for activities, taught by an excellently well-rated star professor, and the other class was taught using active learning considering the best practices of the discipline. This learning included difficult activities that students should perform within the classroom. Again, the **students of the fluid class rated the professor up to 20%**

higher. However, when tests were conducted to assess learning, it was found that the students in the active class achieved higher grades, as expected. In both experiments, the results passed the test of statistical significance.

There is another experiment conducted in 1973 [8], called the "Dr. Fox" experiment, and its results are called the "Dr. Fox Effect." In this experiment, Naftulin, Ware, and Donnelly offered a course on the applications of game theory to human behavior that would be taught by the expert Dr. Myron L. Fox. Dr. Fox was a Hollywood actor who knew nothing about the topic. Additionally, the actor was asked to make the presentation less precise, including vague material, contradictory statements, redundancies, and some humor unrelated to the topic. Dr. Fox delivered the presentation in an enthusiastic and passionate manner. How was the presentation evaluated? 90% of the feedback stated that the presentation was well-organized, interesting, and contained interesting examples.

It is evident that students prefer a professor with fluent speech over one who engages students with activities. It is also evident that most students do not learn much in class and prefer to be entertained. There is a statement that, although an exaggeration, summarizes what we have seen. In 1965, journalist Donald R. P. Marquis said, "If you make people think they're thinking, they'll love you. If you actually make them think, they'll hate you" (quoted by Morley & Everett, 1965, p. 237). In short, the **best teacher is not the one who produces the best learning**, indicating that the problem of teaching and learning is truly complicated.

5   Factors Biasing Professor Evaluation. Honesty in Evaluation.

Apart from fluency and enthusiasm, there are numerous factors that influence professor evaluation that are unrelated to learning. Some of these factors may be controlled by the professor, while others are not.

We have already mentioned some factors. For example, whether the student has received the grades they believe they deserve, the student's interest in the subject topics, whether the subject is elective or mandatory, whether the course is analytical or quantitative, the level of the class in their career profile, class time, and duration, whether the student has had any disputes with the professor, whether the student finds the professor's accent strange, whether the professor is male or female, and even the well-known effects of height and attractiveness on perceived competence [1, 2, 8, 9].

In Carpenter et al. [8], studies on biases in SET are mentioned containing many bibliographic references that we will not mention here. The following biases have been found:

1. One of the most documented biases is **gender**. Most of the time, men are better evaluated than women, although occasionally it is the other way around. It all depends on what is being asked. If the question is exclusively about effectiveness, men are better evaluated. However, in interactions, women are usually better evaluated.
2. **Leniency and severity** in grading also affect SET. We will address this topic in the following section.
3. The **age** of the professor is negatively correlated with SET. That is, younger professors are usually better evaluated than older ones.

4. The **appearance** of the professor affects SET. More attractive professors receive better evaluations than those perceived as less attractive.
5. **Race and accent**. Multiple studies in the United States show that Caucasian professors or those who speak without an accent are better evaluated than non-Caucasians or those who speak English with the accent of a non-native English speaker.
6. And although Carpenter et al. do not document it we also have **height**, it is known that taller people in general are perceived with more authority and competence.
7. Even **chocolate**. In an experiment conducted by Youmans and Jee in 2007, where students taking two statistics courses and one research methods course were enrolled in either of two discussion groups on course topics. Halfway through the course, it is evaluated using the discussion groups. Just before evaluating the course, one discussion group received chocolate and the other did not. **The students in the discussion group that received chocolate rated the professors better than those that did not receive chocolate**. There was no statistically significant difference in grades between discussion groups in each course [8].

SET is partly a memory exercise. Students must remember relevant aspects that have happened in the class over a period of several months. And it is a well-known fact that feelings influence memory [10]. Facts with emotional connotations are remembered more easily. Therefore, students will mainly remember facts that emotionally affected them positively or negatively. If there was no emotional connotation, it is likely that students will not remember enough of the class to conduct an objective SET.

In another experiment mentioned in [8], conducted by Uijtdehaage and O'Neal in 2015, in a medical course normally taught by several professors, **fictitious professor names were included**. And although students could select "**Not Applicable**," only **34% of students selected this option**. The rest, **66%, provided creative evaluations of the fictitious professors**. In an additional experiment conducted in 2008 by Brown, it was found that some students used SET to **retaliate** against professors for not receiving the grade they expected, that **36.5% of students personally reported knowing other students who made evaluations with false comments** about the professor because they didn't like them, and it is estimated in the same study **that 30% of SETs contain deliberately false information introduced by students**.

6   Persistence of Professor Evaluation: Correlations Over Time and Across Groups

SET exhibits persistence within the same group and across groups, as well as over time. This means that professors are often evaluated similarly at the beginning and end of the course. If they teach other courses, they are also likely to be evaluated similarly. It is rare for a professor to receive a low evaluation in one course and a high evaluation in another. The latter can only occur if there are significant differences in the courses. For example, students who take a course voluntarily tend to rate their professors better than those who are compelled by the academic program. Students in advanced semesters tend to rate their more demanding professors better, etc. As mentioned, SET depends on many factors that cannot be controlled by the professor. However, this persistence in evaluation indicates that **SET is based on relatively stable aspects and therefore difficult for the professor** to change. Evidence indicates that **each professor has a baseline evaluation related to their appearance, background, and personality,** which has nothing to do with student learning. Therefore, some professors may need to make significant changes to alter their baseline evaluation, or they may need to exert more effort than other professors to compensate for their inherent disadvantages. Ewing found that in SET, under the same conditions, simply changing the professor altered the intercept of regression models, i.e., the baseline evaluation. This is known in statistics as a **fixed-effect parameter,** indicating that the instructor has a **predetermined average evaluation** [9].

7   The Relationship Between Professor Evaluation and Their Level of Rigor

One of the oldest questions that has been raised is the relationship between a professor's rigor and Student Evaluation of Teaching (SET). Is it true that more demanding professors are evaluated worse than more lenient ones?

The quick answer to this question is yes. The **correlation between a professor's rigor and SET is negative.** This can be observed in [1, 8, 9], and [11], along with many references to experiments contained in these articles. Generally, when a professor is perceived as demanding in grading and in designing activities (creating more challenging activities and grading them rigorously), they will receive a lower SET than a less demanding professor in the same circumstances (and with the same students).

Furthermore, the effects of rigor and lack of fluency in teaching can combine to create the perfect storm. Under certain circumstances, for example, in advanced elective courses where students voluntarily enroll, students may accept rigor as fair and concede that it enhances learning. But if students do not perceive significant learning, as would be the case with a professor who teaches with little fluency, they will consider high grading rigor very unfair and evaluate this professor poorly.

Now we know that rigor affects SET, but to what extent? This is difficult to estimate due to the challenge of measuring rigor in evaluation while keeping all other factors constant. However, it has been established that the negative correlation between rigor and SET is clear and has been demonstrated through various experiments. However, a relationship that is much more documented is the relationship between SET and grades. Since grades are easily available, there are many studies exploring this relationship.

8   The Relationship Between Learning, Professor Evaluation, and Grades

Multiple studies show a positive relationship between SET and student grades. Not only that, but it has also been proven that there is a relationship between SET and **the grade that students expect to receive in the course.** And this expected grade serves as a proxy for rigor. In other words, students expect lower grades with demanding professors, and they tend to give them a lower evaluation.

This relationship between the grade and the expected grade and SET is discussed in [1, 8, 9, 12, 13], and other studies mentioned within these articles show a **positive**

relationship, ranging from low to moderate, between grades and SET. However, Ewing and Clayson and others have found **a strong relationship** between the grade expected by the student and SET [1, 9], as well as Lindsay [12] and Eiszler [13].

This association between grades and SET, which has been widely documented, is like the chicken or the egg problem. The relationship has been extensively discussed, and two hypotheses have been proposed. A) **Reciprocity**. Better-evaluated professors are better teachers, and therefore, students achieve better grades. Students feel they learn more and reward the professor with a better evaluation. B) **Leniency**. Students reward professors who grade leniently because the grade they expect will be higher, closer to the grade they believe they deserve. Both effects, reciprocity and leniency, blend and are difficult to separate. However, there is a moment in time when the effects do not blend, and that is the future. Both Clayson [1], Uttl et al. [2], and Carpenter and Witherby [8] found that students who took introductory courses with professors with high SET, getting high grades on average, significantly lowered their grades in subsequent courses. So, there is evidence of leniency, although it cannot be said that all professors with high SET are lenient. There is also evidence, at least to some extent, that the grade students expect influences SET positively. And there is compelling evidence that rigor negatively affects SET. The same articles mentioned also found evidence that **students who participated in classes where the assessment was rigorous in introductory courses performed better in subsequent courses** on the same topic, even if SET was not as high.

9 Does Teacher Evaluation Encourage Poor Teaching Practices? Grade Inflation and Student Work Deflation

The use of SET has its positive aspects. As previously mentioned, it is a numerical instrument from which information is easily obtainable directly from the main participants of the process. The cost of obtaining this information is moderate and allows students to express and reflect on their experiences in the teaching-learning process. It is a quality control instrument intended to improve teachers' classroom performance.

However, despite the advantages of SET, the imperfections of the instrument discussed earlier **lead to a series of problems in the teaching-learning process**. These are:

1. **Promoting enthusiastic and entertaining lessons that have little influence on students' true learning, reducing student engagement during class, and being generous in evaluation** [4, 8]. Students favor teachers who give fluid, entertaining, and enthusiastic classes. This way of teaching promotes passivity in students, a fact that has been proven in multiple studies. Remember that passive learning, even if interesting and entertaining, leads to inferior learning compared to learning obtained in a class where students have periods of active learning. Naturally, giving a fluid class is not bad and can be good, if this oratory is accompanied by effective learning activities in which students actively apply themselves to their learning.

2. **Promoting learning at the lower levels of Bloom's taxonomy at the expense of higher levels** [5]. Clauss et al. observed that students were more widely mistaken in their perception of their level of learning at the intermediate (and perhaps higher) levels of Bloom's taxonomy. This means that as teachers raise the cognitive level,

and therefore the effort required to learn, students may feel that they are being asked too much and that they are not being graded properly [5]. This could lower teachers' evaluations and incentivize them to reduce cognitive demand to the lower levels of Bloom's taxonomy. Uttl et al. (pp. 40) report that **teachers who teach courses with more quantitative content receive lower evaluations on average than teachers who teach courses without quantitative content** [2]. While Butcher et al. report that departments teaching humanities and social sciences courses generally receive higher SETs than departments teaching economics and sciences [14].

3. **Poor student decisions regarding study habits.** Carpenter et al. report that students who took classes with fluent-speaking professors believe they have learned more, therefore, they dedicate less time to study [7]. While Babcock reports that the expected grade significantly affects study time. Students who took classes where the average grade was A, dedicated 50% less time to studying than students who took classes in courses where the average grade was C [15].

4. **Lower performance in advanced courses.** Uttl et al.'s meta-study reports that the limited follow-up given to students' performance in advanced courses after taking basic courses indicates that students who took classes with fluent-speaking professors continuously had lower performance than students who took courses with less fluent professors that included more activities in the classroom [1, 2, 8].

5. **Deliberate selection of courses where a higher grade can be obtained with less effort.** Multiple studies indicate that students are much more likely to select courses where the expected grade is higher [1, 9, 14]. This is a bad study habit that can have consequences for student performance in more advanced courses.

6. **Grade inflation.** Grade inflation is a phenomenon that occurs in almost all higher education institutions in North America primarily but has also been observed in Europe. It is a widely studied phenomenon, which is why it is proven and documented. Grade inflation has multiple causes, and the relationship between SET and grades as a probable cause has been studied since the first SETs began. There is strong evidence that SET is, at least in part, a cause of grade inflation [1, 2, 8, 9, 12, 16, 17]. The degree of this relationship is unclear. However, although the relationship between SET and grades can be debated, there is no doubt that both **teachers and students believe that there is a positive correlational relationship** between them [1]. This belief alone can be one of the most powerful causes of grade inflation. Undoubtedly, grade inflation has multiple causes beyond SETs, and consequences, which will be studied in a separate document. But we can advance consequences, for example, a) loss of confidence that grades reflect students' preparation, b) reduction of time invested by students in their studies, c) promotion of bad study habits, d) false sense of achievement, e) unhealthy competition between departments and institutions, etc.

10   The Role of Social Media

An aspect not addressed in the literature is the influence of social media on teacher evaluation. As mentioned earlier, SET is persistent across courses and over time. Regarding social media and SET, the **first effect** they have is to **serve as a long-term social memory**. It is entirely possible for students to enter with an already made idea of what to expect from the teacher's teaching style and the grade they can expect based on the multiple comments that can be read on social media. This will make the persistence over time of teacher evaluation even greater. The teacher will have to work for a long period to ensure that a negative collective memory about their teaching style is gradually erased, replaced by a positive memory.

It is known that social media has negative effects on young people. These effects have been acknowledged by the companies that created these networks [18][1]. Social media is present in SET through real-time discussions among active students of the teacher. In this discussion held on social media, all aspects of the teaching-learning process will be discussed, but the teacher's opinion will not be necessary. In general, the social network consists only of students. Why is this important? This is the **second effect** of social media on education. We are talking about the **"influencer" effect** on SET. We define "influencer" as a person with the ability to communicate and influence decisions on specific topics of other people using social media. In Ding et al., and in many other easily accessible references, it is reported that an "influencer" can radicalize the opinions of others to one side or the other [19]. Ding et al.'s article presents research focused on politics. But it is easy to see how their conclusions extend to the educational field. Multiple studies show that social media can affect people's behavior, beliefs, and engagement in various areas of society. There is no evidence that education is different.

Another important aspect related to the **possibility of changing opinions or their perseverance**, which would be the **third effect**. This can be read in Lee et al. [20] and in Swale et al. [21]. In these studies, it is shown that when people are novices in a certain topic, that is, without a fixed opinion, they seek information about the topic and naturally consult the social media at their disposal; for example, a social network explicitly formed to exchange information about a class or a teacher. In this situation, their opinion can be changed by "influencers". However, after a certain time, the opinion will solidify and will not change because the person will only accept information that reinforces their opinions. Both effects can be good or bad for teacher evaluation.

Let's talk about the negative effect of the "influencer" on teacher evaluation. In the case of opinions that can be changed because students are not sure of their opinion, for example, when the teacher is new or at the beginning of their career, only a negative opinion from an "influencer" within the classroom is required to change the opinion of multiple students in the same direction. Hence, removing evaluations with extreme numerical values (i.e., eliminating the lowest and highest evaluations) in an SET is useless. The "influencer" moves multiple opinions, possibly many, not just the lowest or highest. And in the case of a well-known teacher, and who has had previous experiences with the students of his current class or who is somehow known (for example, in the memory left in social networks), if the students' opinion regarding the teacher's teaching style is negative, this opinion will hardly change even if the teacher does a good job. Students with solidified opinions will selectively choose facts during the class that reinforce their opinion of the teacher. Hence the false or exaggerated comments that teachers often read when receiving their evaluations.

Naturally, if the influencer's opinion is positive, it will sway multiple opinions of their followers in the same direction. And if a person holds a positive opinion of a teacher

---

[1] Social media also has positive effects on young adults and adolescents. See the advisory on social media from the Surgeon General USA and the American Psychological Association.

that is already established, the teacher will have to have significant flaws for that opinion to change. This is a well-known effect of popularity, called the "Halo effect."

A **fourth aspect** is the teacher's involvement in social media. Students prefer their teacher to have the characteristics of an influencer: to be an active presence on social media, to have positive opinions on these platforms, to receive many likes, to be followed by many people, and to be seen interacting with other entities, individuals, and recognized companies. It is entirely possible that, in the future, young people will obtain both their learning and their news through social media, bypassing entirely traditional educational environments and media sources. It is not a bad idea for teachers to increase their presence on social media platforms.

11 Alternatives for Assessing Teacher Performance in the Classroom: A Tec Model

The SET is a survey, not a teacher evaluation. If it is intended to make SET a teacher evaluation, several aspects of its design must be changed.

1. Currently, the **SET cannot answer the following questions: a)** Why would a teacher with 20 years of experience and several periods of teaching the class receive a score of 0% or 10%? **b)** What behavior or lack of skills corresponds to a rating lower than 50%? **c)** Why can a teacher who teaches the same subject to different groups, with the same syllabus, materials, and teaching techniques, receive statistically significant differences in evaluations? As long as these questions remain unanswered, the SET is too limited a tool to function as a classroom performance evaluation.

2. An even more intriguing question. It has already been mentioned that specific questions within the SET are not related to the final evaluation. **Why can a teacher receive high ratings in knowledge, fairness, and even teaching method, but receive a global low rating?** This point, along with the previous one, suggests that the evaluation may be based on emotional rather than strictly subjective factors.

3. **Reverse mapping.** Assume a teacher receives a low rating in the teaching method or any other variable in the SET. How can she correct this? What actions can be taken? The answer to this question is crucial to help teachers with poor evaluations.

4. Above all, if the SET is an evaluation, there must be **accountability**. When a student receives a rating of 67%, the teacher must account for each point missing from 100%. **Why can a student give a rating of 0% or 10% to a teacher without justifying anything?** If the SET is an evaluation, then it should be confidential, not anonymous. If a student makes a claim about the teacher, the student should be able to say when and where the incident occurred, and it should be possible to verify it with recordings or witnesses. **How is it possible for a student to lie in SET with impunity, affecting the reputation and career of the teacher without consequences?**

5. **Comparing teacher evaluation with the evaluation of other professionals.** Who evaluates a doctor? Other doctors. Certainly, the opinion of a patient is considered; however, the performance evaluation of a doctor is done by tracking the progress of patients over time and considering the opinion of other doctors. How are engineers evaluated? Similarly, by reviewing the quality of their products over time and considering the opinion of other engineers.

However, just like in other professions, it is **universally observed that the primary article in the evaluation of professionals is the outcomes**. And, above all, **long-term outcomes**. Therefore, outcome measurement should be as objective as possible, although in education it is truly difficult to achieve due to the role that feelings and ideologies play in each of the instruments that can be used.

Below are several instruments proposed for measuring teacher performance in the classroom in addition to the SET. No evaluation instrument is free of biases. What is proposed here is a proposal that hopes that by using multiple instruments to evaluate the teacher, the set of instruments will decrease the uncertainty caused by the biases that each instrument has.

Other Alternatives for Teacher Evaluation:

1. **On-site evaluation or through recordings carried out by entities responsible for pedagogy**. **Bias**: Pedagogy specialists typically are not experts in the topic of the class and therefore find it difficult to assess the effectiveness of the method used by the teacher.
2. **On-site evaluation or through recordings carried out by other teachers in the same area of specialization**. **Bias**: Each teacher has their teaching method and generally believes that their method is the best. A teacher may disagree with another's methods. To determine if one method is superior to another, a statistical test with strict control of variables must be conducted.
3. Since grades are not a solid instrument for evaluating student learning, **using standardized tests**. **Bias**: In the US, standardized tests have been used for 10 years at the high school level, and although the vast majority of teachers are evaluated very well, students have not learned more [22]. Why? Because teachers know the content of standardized tests and prioritize that content over the rest of the material.
4. **Evaluate multiple times in the academic period**. It has been observed that, although the SET is highly correlated over time, there are variations depending on when the evaluation is conducted. At 20% of the school period, the student already knows if the teacher presents information clearly and fluently. This estimation is free of teacher-student conflicts. At 80% of the school period, the student now has an estimate of what their final grade will be and may have had conflicts with their teacher. And at 100% or more, the student can map the efficiency of the teacher's method with their result. Three evaluations are proposed at the mentioned time periods.
5. **Reviewing the teacher's portfolio**. This is probably the method with the least bias. It involves reviewing notes, presentations, examples, exercises, exams, etc., generated by the teacher. **Bias**: It's necessary to see how teachers apply these resources.
6. **Student surveys**. The student surveys always provide valuable information that can show serious errors and give an idea of students' course satisfaction.

12   Conclusions

Assessing a teacher's performance in the classroom is complex, and there is no single solution. Students are very poor at evaluating their own learning, so teacher evaluation by students is far from being a reliable tool for assessing a teacher's performance in the classroom. This evaluation, although it is a numerical instrument that provides valuable information for improving teaching practice, is affected by multiple biases that are sometimes difficult to overcome. Furthermore, it tends to favor fluent classes that are not necessarily the most effective for teaching complex concepts. Evaluating teachers' performance in the classroom using only student evaluation can incentivize bad teaching practices, grade inflation, and a mistaken idea in students about their own learning. A method is proposed to improve teacher evaluation for use in the current environment, and a multidimensional and comprehensive model of teacher evaluation is proposed.

References

1. Clayson, D.E.: Student evaluations of teaching: Are they related to what students learn? A meta-analysis and review of the literature. J. Mark. Educ. **31**(1), 16–30 (2009)
2. Uttl, B., White, C., Gonzalez, D.W.: Meta-analysis of faculty's teaching effectiveness: student evaluation of teaching ratings and student learning are not related. Stud. Educ. Eval. **54**, 22–42 (2017)
3. Papamitsiou, Z., Economides, A.: students' perception of performance vs. Actual performance during computer-based testing: a temporal approach. In: INTED 2014 Proceedings, Valencia (Spain) (2014)
4. Deslauriers, L., McCarty, L., Miller, K., Kestin, G.: Measuring actual learning versus feeling of learning in response to being actively engaged in the classroom. Proc. Natl. Acad. Sci. **39**(116), 19251–19257 (2019)
5. Clauss, J.M., Geedey, C.K.: Knowledge surveys: students ability to self-assess. J. Scholarship Teach. Learn. **10**(2), 14–24 (2010)
6. Stroebe, W.: Student evaluations of teaching encourages poor teaching and contributes to grade inflation: a theoretical and empirical analysis. Basic Appl. Soc. Psychol. **42**(4), 276–294 (2020)
7. Carpenter, S.K., Wilford, M.M., Kornell, N., Mullaney, K.M.: Appearances can be deceiving: instructor fluency increases perceptions of learning without increasing actual learning. Psychon. Bull. Rev. **20**, 1350–1356 (2013)
8. Carpenter, S.K., Witherby, A.E., Tauber, S.K.: On students' (mis) judgments of learning and teaching effectiveness. J. Appl. Res. Mem. Cogn. **2**(9), 137–215 (2020)
9. Ewing, M.: Estimating the impact of relative expected grade on student evaluations of teachers. Econ. Educ. Rev. **31**(1), 141–154 (2012)
10. Zull, J.E., Zull, J.E.: The art of changing the brain: enriching the practice of teaching by exploring the biology of learning, p. 2020. Stylus Publishing, LLC (2020)
11. Carpenter, S.K., Witherby, A.E., Tauber, S.K.: On students' (mis) judgments of learning and teaching effectiveness: where we stand and how to move forward. J. Appl. Res. Mem. Cogn. **9**(2), 181–185 (2020)
12. T. K. Lindsay, Combating the "other" inflation: Arresting the cancer of college grade inflation, Texas Public Policy Foundation, 2014
13. Eiszler, C.F.: College students' evaluations of teaching and grade inflation. Res. High. Educ. **43**, 483–501 (2002)
14. Butcher, K.F., McEwan, P.J., Weerapana, A.: The effects of an anti-grade inflation policy at Wellesley College. J. Econ. Perspect. **28**(3), 189–204 (2014)
15. Babcock, P.: Real costs of nominal grade inflation? New evidence from student course evaluations. Econ. Inq. **48**(4), 983–996 (2010)
16. Bar, T., Kadiyali, V., Zussman, A.: Grade information and grade inflation: the Cornell experiment. J. Econ. Perspect. **23**(3), 93–108 (2009)

17. Park, B., Cho, J.: How does grade inflation affect student evaluation of teaching? Assess. Eval. High. Educ. **48**(5), 723–735 (2023)
18. "Social Media's Concerning Effect on Teen Mental Health," The Annie E. Casey Foundation (2023). https://www.aecf.org/blog/social-medias-concerning-effect-on-teen-mental-health. Accessed 19 Jan 2024
19. Ding, C., Jabr, W., Guo, H.: Electoral competition in the age of social media: the role of social media influencers. MIS Quar. **47**(4), 2023 (2023)
20. Lee, W., Yang, S.G., Kim, B.J.: The effect of media on opinion formation. Phys. A **595**, 127075 (2022)
21. Swalé, F.I.: Inception: Social Media's Influence On Your Opinion. Forbes (2023). https://www.forbes.com/sites/forbeseq/2023/02/27/inception-social-medias-influence-on-your-opinion/?sh=77aa7db9cf13. Accessed 19 Jan 2024

Analyzing Grade Inflation in Engineering Education

Raul Ramirez-Velarde(⊠) ⓘ, Laura Hervert-Escobar ⓘ, and Neil Hernandez-Gress ⓘ

Tecnologico de Monterrey, Eugenio Garza Sada 2501 Sur, Col. Tecnológico, 64849 Monterrey, N. L., Mexico
rramirez@tec.mx

Abstract. Grade inflation is a complex phenomenon observed predominantly in institutions like the Tecnológico de Monterrey and English-speaking universities, with sporadic instances in certain European regions. Its causes are multifaceted, ranging from competition among universities for prestigious opportunities to shifts in teaching and evaluation methods. Efforts to curb grade inflation, such as limiting grade distribution and publishing course averages, have had mixed success. We carry out both ChatGPT led and traditional bibliographical research to understand its causes, identify successful strategies, and adopt a collaborative, systemic approach that can effectively control grade inflation without compromising teaching standards or exacerbating student stress.

Keywords: Grade inflation · Education Evaluation · Teaching standards

1 Introduction

Grade inflation is defined as the tendency for average grades to increase over time, resulting in a higher proportion of students receiving high grades compared to previous periods. Grade inflation is present at all levels of higher education all around the world. Is a global phenomenon that has many negative implications on the quality of engineering education. As we will see, many prestigious universities have tried to control it with limited success.

In this paper, we first explore grade inflation on a global level, and then make recommendations for controlling grade inflation within Tecnológico de Monterrey. This is conducted in two parts. **First**, a **conversation with ChatGPT** is undertaken to quickly summarize the problem and attempt to obtain bibliographic references. In the **second** part, a bibliographical review is conducted without ChatGPT assistance using **traditional bibliographic sources**. Finally, in the third part, we produce recommendations for grade inflation control policies that can be adopted by higher education institutions.

2 First Part: A Conversation with ChatGPT

Researcher: At university level, what is understood as grade inflation?

ChatGPT: Grade inflation at the university level refers to the phenomenon where there is a **consistent increase** in average grades over time without a **corresponding increase in the quality of student performance** or academic standards. In simpler terms, it means that students are receiving higher grades for the same level of achievement compared to previous years. This trend can occur across various disciplines and institutions and may have several underlying causes, including changes in grading practices, academic policies, student expectations, and institutional pressures. Grade inflation can have implications for academic integrity, credibility of qualifications, and perceptions of student achievement.

Researcher: What are the characteristics of grade inflation at university level?

ChatGPT: Grade inflation at the university level typically exhibits several characteristics:

1. **Consistent Increase in Average Grades**: One of the primary characteristics is a steady rise in the average grades awarded to students over time. This increase often occurs across multiple courses, departments, or even entire institutions.
2. **Increasing Disproportion of high Grades**: There is a disproportionate increase in the number of A grades compared to lower grades. A higher percentage of students are receiving A grades, contributing to an overall upward shift in the grade distribution.
3. **Narrowing Grade Distribution**: Grade inflation often leads to a compression of the grading scale, with a higher proportion of students receiving top grades (A's or equivalent) and fewer students receiving lower grades (C's or below). This narrowing of the grade distribution makes it challenging to differentiate between levels of student achievement. This is called Grade Compression.
4. **Decrease in Academic Rigor**: Grade inflation may coincide with a perceived decrease in academic rigor, where it becomes easier for students to earn higher grades without demonstrating a commensurate level of knowledge or skill mastery.
5. **Inconsistent Performance**: Despite receiving higher grades, students may not necessarily demonstrate improved academic performance or mastery of course material. This discrepancy between grades and actual performance is a hallmark of grade inflation.
6. **Normalization of High Grades**: There may be a societal expectation or norm that students should achieve high grades, leading to pressure on instructors to assign inflated grades to avoid student dissatisfaction or complaints.
7. **It's Pervasive**: Grade inflation may not be limited to specific departments or disciplines but can be observed across various fields of study and institutions, albeit to varying degrees.

Researcher: What are the negative effects of grade inflation in university level engineering education?

ChatGPT: Grade inflation in university-level engineering education can have several negative effects:

1. **Diminished Academic Rigor**: Grade inflation may lead to a decrease in academic rigor within engineering programs, as students may perceive that they can achieve

high grades with minimal effort. This can compromise the quality of education and the development of essential engineering skills.

2. **Impact on Motivation**: Students may have less motivation to strive for excellence if they perceive high grades as easily attainable. This can affect overall motivation to engage deeply with the material and pursue academic challenges.

3. **Undermining Student Effort**: In an environment of grade inflation, students may perceive that they can achieve success with minimal effort. This undermines the value of hard work and may contribute to a culture of academic entitlement.

4. **Misleading Transcripts and Mismatch between Grades and Competence**: Inflated grades may no longer accurately reflect students' actual competence and mastery of engineering concepts and skills. Employers and graduate programs rely on grades as indicators of proficiency, and grade inflation can lead to a misrepresentation of students' abilities.

5. **Erosion of Credibility**: Grade inflation undermines the credibility of academic credentials earned in engineering programs. If grades are inflated, employers, licensing bodies, and professional organizations may question the rigor and quality of engineering education, potentially leading to a loss of trust in graduates' abilities.

6. **Reduced Motivation for Excellence**: When students perceive that high grades are easily attainable, they may become less motivated to strive for excellence and may prioritize achieving high grades over genuine learning and skill development. This can hinder their long-term success in engineering careers where competence and innovation are essential.

7. **Decreased Differentiation among Students**: Inflated grades make it challenging for employers and graduate programs to differentiate among engineering graduates based on academic performance. This can result in difficulties in identifying the most qualified candidates for employment or further academic pursuits.

8. **Decreased Accountability from Faculty**: Faculty members may face reduced accountability for maintaining high academic standards if there is pressure to inflate grades. This can impact the overall quality of the education provided.Impact on Accreditation and Ranking: Grade inflation may affect the accreditation status and rankings of engineering programs. Accreditation bodies assess the quality and rigor of engineering education, and grade inflation may raise concerns about the program's ability to adequately prepare students for professional practice.

9. **Mismatch in Student Expectations**: Grade inflation can create a mismatch between students' expectations of their academic performance and the actual demands of the engineering profession. Students may enter the workforce with inflated perceptions of their abilities, leading to challenges in meeting employer expectations and performing job responsibilities effectively.

10. **Long-Term Implications for Engineering Profession**: If grade inflation persists unchecked, it can have long-term implications for the engineering profession as a whole. A workforce with inflated academic credentials may struggle to meet the evolving demands of the industry, potentially affecting innovation, safety, and public trust in engineering solutions.

Addressing grade inflation in engineering education is crucial to maintaining the quality and integrity of engineering programs and ensuring that graduates are adequately prepared for the challenges of professional practice.

Researcher: According to published bibliography, what are the main causes of grade inflation?

ChatGPT: According to published literature, several main causes of grade inflation have been identified. These include:

1. **Changing Grading Standards**: Some studies suggest that there has been a shift in grading standards over time, with instructors becoming more lenient in assigning grades. This leniency may be influenced by a variety of factors, including a desire to keep students satisfied or institutional pressures.

2. **Student Expectations and Consumer Mentality**: The perception that higher education is a service that students pay for has led to a consumer-oriented mentality. Instructors may feel pressure to meet students' expectations for high grades to ensure positive evaluations and maintain enrollment.

3. **Grade Compression**: Grade compression occurs when the range of grades awarded becomes narrower, making it more difficult to distinguish between high and low performers. Some argue that this compression is a result of grade inflation, as higher grades become more prevalent.

4. **External Pressures and Accountability**: Universities face external pressures for accountability and may be influenced by factors such as rankings, accreditation, and funding. In response, institutions may be inclined to maintain or increase average grades to present a positive image.

5. **Student Evaluations of Teaching (SETs)**: Research has shown a correlation between lenient grading and positive student evaluations. Instructors may be motivated to give higher grades to receive positive feedback from students, which can impact their teaching evaluations and, in turn, their professional standing within the institution.

6. **Competitive Environment**: Concerns about student retention and attracting new students, particularly in a competitive higher education market, may drive institutions to maintain high average grades to appeal to prospective students and their families.

7. **Pressure on Faculty**: Faculty members may face pressure to give higher grades to avoid negative evaluations, especially in institutions where student feedback significantly influences faculty evaluations and promotions.

8. **Lack of Standardization**: Differences in grading practices and standards among courses and departments within an institution may contribute to grade inflation. The lack of standardization makes it difficult to maintain consistent grading practices.

9. **Merit-Based Financial Aid**: The relationship between high grades and financial aid or scholarships can create incentives for students to seek higher grades. This, in turn, can lead to increased pressure on faculty to inflate grades.

10. **Faculty-Student Relationship**: Positive relationships between faculty and students may lead to reluctance to assign lower grades. This is sometimes referred to as the "pleasing the professor" phenomenon.

11. **Generational Changes**: Changes in the characteristics and expectations of incoming student cohorts can influence grading practices. Some argue that newer generations of students may expect higher grades for their efforts.

These causes are often interconnected and may vary in significance depending on the specific context of each institution. Additionally, new research may continue to shed light on additional factors contributing to grade inflation.

Researcher: What are the main published bibliographical references for grade inflation?

ChatGPT: ChatGPT's answer was 6 bibliographical references that were verified using Google Scholar and turned out to be non-existent.

Researcher: What policy and methodological changes have been successfully implemented to reduce grade inflation?

ChatGPT: Addressing grade inflation often requires a multifaceted approach involving both policy changes and methodological adjustments. While there isn't a one-size-fits-all solution, several strategies have been suggested or implemented to mitigate grade inflation. It's important to note that the effectiveness of these measures can vary across institutions. Here are some policy and methodological changes that have been considered or implemented:

Several policy and methodological changes have been implemented to successfully reduce grade inflation. These include:

1. **Standardized Grading Guidelines**: Implementing clear, institution-wide grading standards to ensure consistency in evaluation criteria. Providing faculty with guidelines on how to assign grades based on specific learning outcomes helps maintain transparency and fairness in grading practices.
2. **Regular Grade Audits**: Conducting regular audits of grading patterns to identify departments or courses with unusually high grade distributions. Reviewing and discussing grading practices in departments helps ensure alignment with institutional standards and promotes accountability among faculty members.
3. **Grade Distributions Transparency**: Making grade distributions public to increase transparency and accountability. Sharing information on grade distributions with students, faculty, and administrators fosters a culture of openness and encourages discussions on grading practices.
4. **Faculty Development**: Providing training and workshops for faculty on effective grading practices. Faculty members can benefit from learning about research-based strategies for fair and consistent grading, as well as techniques for providing constructive feedback to students.
5. **Revision of Grading Policies**: Reconsidering and revising grading policies, including the use of curve grading systems. Implementing policies that discourage grade inflation and emphasize the importance of maintaining academic rigor can help address systemic issues contributing to inflated grades.
6. **Institutional Culture Shift**: Promoting a culture that values high academic standards and emphasizes learning outcomes over grades. Recognizing and rewarding

faculty for maintaining rigor in their courses can help reinforce a commitment to excellence in teaching and assessment.

7. **Use of External Examiners**: Introducing external examiners or peer review processes to evaluate the fairness and appropriateness of grading practices. Seeking external input on grading standards helps ensure objectivity and provides valuable insights for improving assessment practices.

8. **Incentives for Honest Grading**: Recognizing and rewarding faculty who maintain high standards in their grading. Ensuring that promotion and tenure decisions consider a commitment to maintaining academic rigor encourages faculty members to prioritize fairness and integrity in their grading practices.

9. **Implementation of Plus-Minus Grading Systems**: Transitioning to a plus-minus grading system, which provides more granularity in assessing student performance and can reduce grade inflation by distinguishing between marginal differences in achievement.

10. **Assessment of Learning Outcomes**: Shifting the focus from grades to a more comprehensive assessment of learning outcomes. Emphasizing the importance of formative assessment and feedback throughout the course helps ensure that grades accurately reflect students' mastery of course content and skills.

11. **Grade Normalization**: Implementing policies to normalize grades across courses or departments. Adjusting grades based on historical performance data to account for variations in grading practices.

12. **Incentivizing Rigor**: Creating incentives for faculty to maintain high academic standards. Recognizing and rewarding faculty who uphold rigorous grading practices, possibly linking evaluations and promotions to the maintenance of academic rigor.

It's essential to recognize that these measures may have different impacts depending on the specific context of each institution. Successful strategies often involve a combination of policies and practices that are tailored to the unique characteristics of the institution and its academic programs. Regular assessment and adaptation of these strategies are crucial to ensure their effectiveness over time.

End of Part 1, ChatGPT assisted research.

3 Second Part. Traditional Bibliographical Search

Although ChatGPT **couldn't find bibliographic references on grade inflation**, only on **effective assessment techniques** (not shown); conducting a traditional bibliographical research did yield a significant number of references. Grade inflation has been under discussion since the 1970s, with the topic resurfacing each decade. The bibliographic references presented span from 1991 to 2021, encompassing 30 years of discourse on grade inflation. The discussion is structured into the same sections as with ChatGPT:

1. The causes of grade inflation
2. The effects of grade inflation
3. Policies and methodologies aimed at controlling grade inflation.

We will find that the results seen in the bibliography are largely contained within the responses provided by ChatGPT. To continue this investigation, we define grade inflation as the increase in the average final grade across various courses required for obtaining a university degree and the rise in the number of maximum grades awarded.

3.1 Causes of Grade Inflation

In the early 1990s and at the beginning of the millennium, when discussions about grade inflation began, some research indicated that grade inflation did not exist. For instance, Kohn in 2002 [1].

There are also studies suggesting that the primary cause of grade inflation is simply that students are getting better, such as Jephcote et al. [2] and Bar et al. [3]. However, much of the literature mentions that rigorous statistical studies do not show that students are improving; rather, there is evidence indicating that students are spending less time studying and that other factors are at play.

Regarding other causes, two important references are Lindsay, and O'Halloran and Gordon. Based on these two references, we will compile a list of causes and indicate authors who agree. According to Lindsay, the main causes are a list of beliefs that rigorous statistical studies **have shown to be mistaken** [4]:

1. Grades do not influence faculty evaluation.
2. Student evaluations of their professors are a reliable measurement of institutional effectiveness.
3. Abundance of high grades in a course indicates high student achievement.
4. The selection of programs and courses is not affected by the grade expected by students.
5. In loosely regulated environments, the meaning of a numerical grade is not consistently and objectively the same across different classes, departments, majors, and institutions.

In reality, research has shown the following [4]:

1. Differences in grading practices by professor's **bias students' evaluations of the professor**. There is a relationship between the grade students expect and the expected grade point (EGP), Eiszler C. F. [5].
2. Student evaluations of professors are **not a reliable indicator of a professor's effectiveness in the classroom** and can only explain a small portion of the variance in learning.
3. High grades in students **cannot be associated with high levels of achievement** and learning.
4. Differences in grading leniency have a substantial impact on student enrollment and **cause fewer students to select fields that are more demanding in grading**.
5. There are systematic **differences in grading leniency among disciplines, departments, and instructors**, and these differences cause serious inequities in establishing student competency levels.

According to O'Halloran and Gordon [6], the causes of grade inflation can be classified based on their origin into **endogenous** and **exogenous** factors and based on the related actors into **social, institutional, departmental, and individual factors**:

Exogenous Factors

Exogenous factors are generally social factors, which are due to the environment in which higher education institutions operate.

1. **Regulatory environment**. Accreditation processes establish minimum standards in terminal efficiency, time to first employment, acceptance percentage in graduate school.
2. **Competitive environment**. Institutions compete with institutions at their same level, trying to demonstrate that they improve learning outcomes. A rigorous assessment of an institution may disadvantage its graduates when compared to graduates from another institution with higher grades for employment and graduate school access. Nordin et al. [7], Finefter-Rosenbluh and Levinson [8] agree on this.

The rest of the factors are endogenous, or internal to the institutions.

Institutional Factors

3. Allowing students to **drop courses** until very late in the semester. Typically, the courses that are dropped are those in which the student has obtained low grades.
4. **Poor course distribution**, which allows students to **take fewer courses in mathematics, sciences, foreign languages**, etc., which typically give lower grades.
5. The use of **student evaluations of faculty** as the sole source of performance evaluation in the classroom.

Departmental Factors

6. The difference in the average grades obtained in different departments causes students **not to select subjects** in which their average may decrease. Also, Lindsay [4].

Individual Factors

7. The need for students **to maintain their GPA to retain scholarships** drives them to select majors and courses where they can achieve higher grades with less risk.
8. **Student evaluations of professors**. Professors who are not full-time or those who are just starting tend to give higher grades to avoid confrontations with students and not jeopardize their professional careers by receiving poor evaluations. Additionally, professors, departments, and institutions are penalized when grades are low, Finefter-Rosenbluh and Levinson [8], Germain and Scandura [9], Stroebe [10].
9. **Grade distribution is not merit-based** and responds to rather vague notions.
10. The **use of curves** to improve exam outcomes.
11. Allowing students **to select the best grades to calculate the final average**. For example, if a student submits 4 assignments, allowing them to receive a low grade on one and only selecting 3 for the final average.
12. Providing **multiple opportunities** to submit assignments and take exams.

13. **Avoiding conflicts with students and parents**. Students have an idea of how much they have learned and exerted themselves and may initiate conflicts with professors to raise their grades to what they believe they deserve or need. Sometimes parents support their children. Also, Donaldson et al. [11].

14. **Reducing academic requirements** to meet outcome expectations. Professors simply lower academic requirements so that high averages and outstanding grades align with expectations.

To these factors, according to Donaldson et al. [11], the following should be added:

15. **Lack of training** for evaluators to use concrete merit-based criteria.

16. **Evaluator-student relationship**. As evaluators get to know the students better, they are reluctant to give low grades or do not wish to harm their future prospects. Paskausky et al. [12], Finefter-Rosenbluh and Levinson [8] agree on this.

17. **Design of assessment tools**. Sometimes, assessment tools lend themselves to giving high evaluations, such as peer evaluations or rubrics with very generic indications in the delivery of practical work.

18. Many students failing the course **reflects poorly** on the professor's image.

The Effects of Grade Inflation

Grade inflation has a plethora of negative effects that can be challenging to identify because they are not always evident. Let's begin with Lindsay [4]:

1. **Dilutes the value of schools, departments, and faculty with high academic standards**. According to Paskausky et al., it casts doubt on the competencies of graduates [12]. And according to Finefter-Rosenbluh and Levinson, a) universities cannot use grades to identify outstanding students, and employers cannot distinguish them, and b) it reduces academic standards and undermines faculty integrity [8]. According to Stroebe, it incentivizes poor teaching practices, and grades lose their value as a measure of teaching effectiveness [10].

2. **Discourages effort in students**. And according to Bar et al., students know which professors give higher grades and which professors are better evaluated, and they prefer them [13]. According to Finefter-Rosenbluh and Levinson, a) Poor effort receives average grades, b) Students have an inflated idea of their actual capabilities. Stroebe agrees [10], as do Rojstaczer and Healy [14].

3. **Makes it difficult for employers to distinguish student performance levels**. Tyner also agrees [15]. Consequently, according to Paskausky et al., it casts doubt on the competencies of graduates [12].

4. **Discourages the recruitment of students in engineering and sciences**. Sabot and Wakeman-Linn agree [16], as do Stroebe [10], and Rojstaczer and Healy [14].

Additionally, according to Sabot and Wakeman-Linn [16]:

5. **Creates differentiation in the average grades at the departmental level**. Departments of humanities, political science, philosophy, and art have higher averages than departments of economics, chemistry, and mathematics.

6. Students whose majors belong to departments with high grading averages **respond less to incentives to achieve good grades**.

7. Grades of students in majors in departments with high averages are **less predictive of future success** and do not allow differentiation in student competencies such as skill, knowledge, organization, motivation, etc. On this, Stroebe and multiple other references indicate that student effectiveness in subsequent courses is reduced [10].

Policies and Methodologies Attempting to Control Grade Inflation

The literature indicates that only a handful of grade inflation control practices have been implemented and have worked, albeit at a significant cost. Some, like the control of the maximum number of grades that can be awarded, have even been abandoned after a while. Let's first review the proposals. Let's begin with Lindsay [4]:

1. **Dialogue between institutions and areas of knowledge**. That is, a comprehensive reform in the evaluation process that has winners and losers. Sabot and Wakeman-Linn mention [16], related to this, a) policies must be developed to equalize the averages of different departments, and b) differences in averages between departments should only be due to a design that responds to specific needs.
2. **Training for teachers and establishment of rules and best grading and evaluation practices**. In other words, evaluation workshops. Donaldson et al. agree [11], suggesting that workshops train teachers to use reasoned evaluation, providing evidence for grading decisions and specific criteria. Even O'Halloran and Gordon suggest measuring academic rigor [6]. There is a zone of rigor that is the middle ground. If students feel that it is necessary to attend class to learn, the teacher will be better evaluated. For example, if techniques are used explicitly designed to prepare students to achieve the best result in their exams, students will be more convinced that they are being treated fairly. Additionally, establishing evaluation rules clearly from the beginning also convinces the student that the teacher is treating them fairly.
3. **Limiting grade distribution**: a) The average must fall within an interval (e.g., 85 ± 5) or b) the accumulation of grades must be limited. Butcher et al. agree [17], as do O'Halloran and Gordon [6].
4. **Reporting adjusted grades on transcripts or alternatively the student's position in the group**. Stroebe also agrees [10]. Adjustment sometimes involves reducing the grade by a certain percentage, something that students naturally see as very unfair. Bar et al. add [3] that the average assessment of each professor should be published. This measure was implemented and abandoned because in the Cornell University system, students have a lot of freedom to choose their courses. Students preferred courses where the reported average was higher, which further fueled grade inflation. Finefter-Rosenbluh and Levinson suggest doing more [8]: a) include on the report card not only the grade but also the average, median, mode, grade distribution, and the student's position in the class (or at least whether their grade is above or below average), and b) pay more attention to the differences between the maximum and minimum grades than to the group average.
5. **Control the maximum number of high grades** that can be awarded in a group (e.g., 35% of grades between 90 and 100). This eliminates cooperation among students and increases stress because it creates a competitive environment.
6. **Decrease the role of SETs in evaluating teachers' performance in promotions and salaries to eliminate the incentive to give better grades in exchange for better**

evaluations. O'Halloran and Gordon add [6], a) do not base teacher evaluation solely on student opinion but use a multi-source approach that takes into account not only student opinion (which is a valuable tool) but also the learning resources portfolio and visits from specialists and other teachers in the classroom. Additionally, they suggest b) not using the same evaluation method for teachers at all levels. Teachers at the beginning of their careers are more vulnerable than teachers with an established career.

And according to Finefter-Rosenbluh and Levinson [8]:

7. **Use standardized exams**. Gershenson also suggests [18].
8. **Increase rigor**. That is, be stricter in the criteria for achieving certain grades.
9. **Combine the grade awarded by the teacher with the grade achieved on a standardized exam using adjustable percentages**. However, it is mentioned that these percentages can be sources of bias if not properly adjusted.

The traditional bibliographical research results just presented agree with the results given by ChatGPT. There is no contradiction and most of ChatGPT results have been validated with reviewed bibliographical references.

4   Third Part: Recommendations for Controlling Grade Inflation

One of the main issues affecting institutions that have implemented controls on grade inflation is that, indeed, the grade average of graduates decreases and puts them at a disadvantage against graduates from other universities of the same level. This indicates that the problem is systemic. The solutions implemented must involve the largest possible number of participants. In other words, the problem should preferably be addressed by initiating a dialogue between similar institutions with the aim of disadvantaging those institutions that do manage to control inflation. The same applies to schools, departments, and naturally individuals. An institution that effectively controls grade inflation may experience a decrease in enrollments, as demonstrated by multiple references mentioned above. A department that implements control measures may find that faculty evaluations decrease, that some courses have fewer students, and that student averages also decrease. There are multiple references that prove that a strict professor will face tougher student evaluations unless they control several circumstances that can mitigate the impact.

If measures were to be implemented to control grade inflation, focus should be on the most mentioned causes: **Evaluation instruments and student evaluations of professors. Institutional policies** would come next.

Evaluation Instruments

1. **Assessment workshops**. To standardize evaluation criteria across subjects, departments, and schools, it would be convenient to develop workshops where professors practice grading different assessment instruments, such as challenges, engineering exams, humanities and administrative sciences exams, and final projects. It is important that rubric designs contain 100% clear and realistic criteria. It is absurd for 80% of the group to receive an outstanding evaluation.

2. **Standardized final exams, especially in the early academic periods.** On every occasion where student grades did not correspond to their actual learning, it was because they were compared against standardized exams, such as the SATs in the U.S. Therefore, developing standardized exams for a larger number of subjects can help control grade inflation. This would probably be more effective in the initial semesters to ensure high academic levels in basic subjects.

3. **Exams developed by other professors.** If it is not possible to develop standardized exams for all subjects, at least the final exam presented should not be developed or graded by the instructor, but by another professor.

4. **Audit or reflection on evaluations.** It would be convenient to routinely conduct an exercise to review and reflect on the final grade awarded by the professor to a group of students, listening to the professor's reasoning and contrasting it against the opinions of other professors. This should not necessarily mediate any incidents or dissatisfaction but rather serve as a calibration reflection.

5. **External evaluators.** The caveat is that often, no training is provided to the training partner on evaluating the deliverable. Additionally, it is possible to bring in external evaluators for all types of teaching methods, even for traditional classes with assignments and exams. Naturally, the cost of bringing in external evaluators is high, so careful selection must be made of those subjects that require an external evaluator with a high level of excellence and academic rigor.

6. **In the early courses, perhaps the first two years of the degree, standardize the obtained grade and report the group's average.** Grade standardization involves subtracting the mean and dividing by the standard deviation. Suppose for a subject α, Professor A assigns the following grades: 73, 76, 77, 78, 81, and Professor B assigns the following grades: 86, 86, 87, 88, 88. The average of group A is 77, while the average of group B is 87. The dispersion of grades is also greater in group A than in group B. In group A, the difference between the highest and lowest grades is 8, while in group B, it is 2. The worst graded student in group B, student 6, has a grade of 86, which is higher than the best grade in group A, which is 81 from student 5. Under these circumstances, it seems that student 6 is a better student than student 5. However, if we subtract the mean from each grade and divide it by the standard deviation, standardization returns the grade in standard deviations from the mean. Thus, group A has -1.37, -0.34, 0, 0.34, and 1.37, while group B has -1, -1, 0, 1, and 1. In this way, student 5 in group A has a grade that is 1.37 standard deviations **above the mean**, while student 6 in group B has a grade that is -1, that is, one standard deviation **below the mean**. In group A, the minimum passing grade of 70 has a standardization of -2.4 (standard deviations from the mean), so in a group of 30 students, the probability of failing is 0.012. In group B, the standardization of 70 is $6.38 \times 10{-17}$, which makes the probability of failing zero.

It is important to clarify *that we cannot know if the difference in the group average is because Professor A is stricter in grading or because Professor B is a teacher who inspires students to excel.* The only way to know would be to a) compare against standardized exams, b) review by professors who teach the same course, or c) track student grades in subsequent courses. However, we no longer have the image that student 6 is surely better than student 5. Perhaps, it is quite the opposite.

7. **In advanced courses, perhaps in the last two years of the degree, control distribution by limiting the number of maximum grades that can be awarded**. In courses with many projects, limiting the number of maximum grades can create an environment of healthy competition that can showcase those teams that truly make an exceptional effort.

Teacher Evaluation

8. **Retain student evaluations of their professors as a valuable feedback tool**. This evaluation provides very valuable information that must be considered to improve groups. However, it must be understood that this evaluation measures student satisfaction in the course, not their learning or the teacher's performance in the classroom. This evaluation gives students a voice that must be heard and considered, as long as it is truthful and well-intentioned.

9. **Use a portfolio of learning resources and visits from pedagogy specialists and colleagues from the field of knowledge in the classroom (or record the session)**. The literature clearly indicates that using student evaluations of professors as the sole instrument to measure a professor's performance in the classroom, in such a way that it is a very important factor in hirings, promotions, and salary increases, is one of the main causes of grade inflation. It is important, therefore, to develop a multidimensional system for evaluating a professor's performance in the classroom. This should include portfolio review, student results in standardized exams, student performance in subsequent courses, evaluation by both pedagogues and professors from the same area, etc.

10. **Celebrate rigor and high academic standards**. There are many types of awards and distinctions given to professors within and outside the institute. Some individuals are rightly distinguished for their excellent results in student evaluations, others for being leaders in their area of development, others for their ability to attract students, etc. But do we celebrate rigor and high academic standards? How could we measure this? After all, a professor's job, above all, is to achieve in the student a deep understanding of complex phenomena. How exactly does a university measure the level of rigor and demand, and its effects over time, and how are professors who have largely been responsible for the institution's great prestige celebrated? This is an open question.

Institutional Policies

11. **Do not allow students to drop a course in which they are failing too late in the academic period**. Naturally, eliminating courses with poor performance will increase the average grades. It has the same effect as allowing a student to choose their best graded assignments to calculate their grade in the subject.

12. **Do not allow grade curves on exams**. Curves favor students with better performance. And if a grading correction system is developed that helps students with

poor performance in the same way (for example, adding X points to all students' grades), it will also raise the institution's average grades.

13. **Limit flexibility in selecting elective courses to prevent students from hunting for grades to improve their average**. Make a proper combination of disciplines when choosing subjects to not allow students to enroll subjects to improve grades.

5 Conclusions

Grade inflation is a phenomenon primarily observed all around the world. The causes of grade inflation are diverse, including competition among universities for positions in prestigious graduate programs and employment opportunities, changes in evaluation methods resulting from shifts in teaching methods—from predominantly final exam-based assessments to a combination of continuous assessment tools, projects, and team-work—and student evaluations of their professors, among other factors. Setting limits on grade inflation is not straightforward, and ideas such as controlling grade distribution by capping high grades, establishing a range for the group's average grade, or publishing course averages, historical group medians, or students' positions within the group alongside their grades have either failed, had limited success, or fostered an undesirable development environment. However, by understanding the causes of grade inflation, and identifying successful cases, it is possible to control the phenomenon without increasing student stress or creating a hostile environment.

References

1. Kohn, A.: The dangerous myth of grade inflation. Chron. High. Educ. **49**(11), B7 (2002)
2. Jephcote, C., Medland E., Lygo-Baker, S.: Grade inflation versus grade improvement: are our students getting more intelligent? Assess. Eval. High. Educ. **46**(4), 547–571 (2021)
3. Bar, T., Kadiyali, V., Zussman, A.: Online Posting of Teaching Evaluations and Grade Inflation. University of Connecticut, Department of Economics (2014)
4. Lindsay, T.K.: Combating the "other" inflation: Arresting the cancer of college grade inflation, Texas Public Policy Foundation (2014)
5. Eiszler, C.F.: College students' evaluations of teaching and grade inflation. Res. High. Educ. **43**, 483–501 (2002)
6. O'Halloran, K.C., Gordon, M.E.: A synergistic approach to turning the tide of grade inflation. High. Educ. **68**, 1005–1023 (2014)
7. Nordin, M., Heckley, G., Gerdtham, U.: The impact of grade inflation on higher education enrolment and earnings. Econ. Educ. Rev. **73**, 101936 (2019)
8. Finefter-Rosenbluh, I., Levinson, M.: What is wrong with grade inflation (if anything)? Philosoph. Inquiry Educ. **23**(1), 3–21 (2015)
9. Germain, M.L., Scandura, T.A.: Grade inflation and student individual differences as systematic bias in faculty evaluations. J. Instruct. Psychol. **32**(1) (2005)
10. Stroebe, W.: Student evaluations of teaching encourages poor teaching and contributes to grade inflation: a theoretical and empirical analysis. Basic Appl. Soc. Psychol. **42**(4), 276–294 (2020)
11. Donaldson, J.H., Gray, M.: Systematic review of grading practice: is there evidence of grade inflation? Nurse Educ. Pract. **12**(2), 101–114 (2012)

12. Paskausky, A.L., Simonelli, M.C.: Measuring grade inflation: a clinical grade discrepancy score. Nurse Educ. Pract. **14**(4), 374–379 (2014)
13. Bar, T., Kadiyali, V., Zussman, A.: Grade information and grade inflation: the Cornell experiment. J. Econ. Perspect. **23**(3), 93–108 (2009)
14. Rojstaczer, S., Healy, C.: Grading in American colleges and universities. Teach. College Rec. **4**, 1–6 (2010)
15. Tyner, A., Gershenson, S.: Conceptualizing grade inflation. Econ. Educ. Rev. **78**, 102037 (2020)
16. Sabot, R., Wakeman-Linn, J.: Grade inflation and course choice. J. Econ. Perspect. **5**(1), 159–170 (1991)
17. Butcher, K., McEwan, P.J., Weerapana, A.: The effects of an anti-grade inflation policy at Wellesley College. J. Econ. Perspect. **28**(3), 189–20 (2014)
18. Gershenson, S.: Grade Inflation in High Schools (2005–2016). Thomas B. Fordham Institute (2018)

Author Index

L. Franco et al. (Eds.): ICCS 2024, LNCS 14838, pp. 443–444, 2024.
https://doi.org/10.1007/978-3-031-63783-4

Printed in the United States
by Baker & Taylor Publisher Services